Computational Nanophotonics
Modeling and Applications

Computational
Nanophotonics
Modeling and
Applications

edited by
Sarhan M. Musa

CRC Press
Taylor & Francis Group
Boca Raton London New York

CRC Press is an imprint of the
Taylor & Francis Group, an **informa** business

CRC Press
Taylor & Francis Group
6000 Broken Sound Parkway NW, Suite 300
Boca Raton, FL 33487-2742

First issued in paperback 2017

© 2014 by Taylor & Francis Group, LLC
CRC Press is an imprint of Taylor & Francis Group, an Informa business

No claim to original U.S. Government works

Version Date: 20130620

ISBN 13: 978-1-4665-5876-2 (hbk)
ISBN 13: 978-1-138-07344-9 (pbk)

Library of Congress Cataloging-in-Publication Data

Computational nanophotonics : modeling and applications / editor, Sarhan M. Musa.
    pages cm
    Includes bibliographical references and index.
    ISBN 978-1-4665-5876-2 (hardcover : alk. paper) 1. Nanophotonics--Data processing. I. Musa, Sarhan M., editor of compilation.

TA1530.C66 2013
621.36′5--dc23   2013019890

Visit the Taylor & Francis Web site at
http://www.taylorandfrancis.com

and the CRC Press Web site at
http://www.crcpress.com

To my father Mahmoud, my mother Fatmeh, and my wife Lama

Contents

Preface

Nanophotonics is a new, rapidly growing field that provides opportunities in understanding the interaction between light and matter on a nanoscale level smaller than the wavelength of radiation. It is the science and engineering of photonics at the nanometer scale. In other words, it is a combination of photonics and nanotechnology. Nanophotonics involves the interaction of light with structures smaller than about 100 nm. Indeed, it offers challenging opportunities for studying the fundamental processes of interaction between the radiation field and matter on a scale much smaller than the wavelength of radiation. The evolution and development of today's nanophotonics technology require new computational nanophotonics techniques. Nanophotonics represents the convergence of many fields and therefore demands a new computational nanophotonics—modeling and applications that can be used for new development and future interdisciplinary research by engineers, scientists, and business managers.

The book provides an introduction to key concepts in a manner that is simple and easily digestible to a beginner in the field and to a broader audience. This book can also serve as a single source of reference to the veteran in the field. It provides a complete understanding of computational nanophotonics in multidisciplinary fields. It is intended for a broad audience working in the fields of physics, chemistry, biology, engineering, and medicine. It is also meant for professionals, researchers, and students who would like to discover the recent challenges and the great opportunities concerning the development of the next generation of nanoscale computational nanophotonics—modeling and applications. In addition, it emphasizes the importance of computational nanophotonics in the development of efficient nanoscale systems. Indeed, experimental studies of nanophotonics systems greatly depend on theoretical guidance for the interpretation of measurements and the design of experiment. Therefore, computational techniques of nanophotonics become essential for solving nanophotonics problems. Undoubtedly, the design and modeling of nanophotonics systems require in-depth knowledge of the underlying physical and engineering principles, together with efficient computational efforts. Therefore, Computational Nanophotonics: Modeling and Applications, aims to provide a detailed account of major advances that have taken place in the field with an emphasis on the most recent trends.

This book comprises 13 chapters and 3 appendices. It also describes a wide variety of recent applications of microscale and nanoscale modeling systems such as COMSOL multiphysics, CAMFR, CST MW 2010, MIT Meep, MIT MPB, and MATLAB®.

Chapter 1 presents modeling and simulations of light coupling in waveguides through the optical microprism coupling method using the finite element method (FEM) with COMSOL multiphysics and provides an overview of optical nanoprism computation.

Chapter 2 presents a review of computation modeling from micro- to nanophotonics in the design, optimization, and operation of optical networks and a future prospective of nanophotonics components technology.

Chapter 3 provides the practical photonic applications of nanomaterials and nanostructures, focusing mainly on semiconductor nanowires. It investigates the enhanced light absorption and tunability of resonant modes in silicon (Si) nanowires by analytical solutions and numerical simulations. It also examines the modal characteristics and nonlinear properties of nanowires using full-field electromagnetic simulations. In addition, it investigates the plasmonic resonant modes supported by subwavelength-scale nanocavities. It demonstrates numerical simulations and analytical solutions and the ability to accurately predict the unique optical characteristics of these nanomaterials and nanostructures. The authors conclude that miniaturization of photonic structures could enable enhanced device performance. A variety of computational methods can be used to design optimized nanostructures for practical applications. They also believe that nanomaterials and nanostructures are a versatile platform for the next generation of highly efficient photonic devices with extremely small thermal overhead.

Chapter 4 presents photonic nanowires and their fabrication techniques. The authors show that with air–silica nanowires they can generate a few optical cycles that are a few millimeters in length. They found that less than a single optical cycle can be generated in an 800 nm air–silica nanowire at low-input pulse energy and for a short interaction distance.

Chapter 5 provides the modeling optical properties of nanofibers/nanowires, as well as the applications of these 1D structures for photonic components or devices.

Chapter 6 presents cavity quantum electrodynamics with application to quantum state transfer through a nanophotonics wave guidance system using an optical fiber as the photon waveguide.

Chapter 7 addresses the nanopatterned photonic elements integrated on a single scanning probe for efficient light manipulations in the near field. Specifically, the authors propose a combination of an embedded metallic grating coupler and photonic crystal nanoresonator for near-field light confinement by using numerical modeling methods.

Chapter 8 presents a coupled mode theory based on complex modes with examples of short-/long-period gratings. Further, the authors formulated the complex coupled mode theory based on local mode and applied it to study the transition loss of the tapered waveguide structure. They also derived an analytical solution of radiation loss for long-period grating. The effectiveness as well as the accuracy has been validated through the study of transmissive waveguide grating with different surrounding materials.

Chapter 9 summarizes some important aspects of coupling of plasmon modes with waveguide modes arising from different coupled plasmonic nanostructures, along with a brief comparison with conventional structure wherever necessary. Some simple computational techniques described in the chapter provide information to carry out the simulation to compare different structures and optimize different layer thicknesses in order to meet certain application-oriented performance criteria. A type of simulation study such as this is necessary before the actual implementation as a pre-design procedure.

Chapter 10 discusses advanced issues in medical computational nanophotonics. It also introduces medical nanoplasmonics, which is an important specific part of medical nanophotonics.

Chapter 11 presents a brief background on the current applications of finite element analysis in nanomaterials and systems used in medicine and dentistry. The authors examine the processes used for the production of nanocoatings and further analysis related to the factors that affect the nanoindented biomaterials by finite element analysis.

Chapter 12 presents a comprehensive study that identifies viable nanophotonics applications, their relationship to military systems, and associated vulnerabilities, risks, and impacts on critical defense capabilities. In addition, it summarizes emerging nanotechnology trends to assure device operation at reduced power, as well as highly integrated information and enhanced spatial, contrast, temporal, and dynamic resolution for imaging and patterning, a new sensor of increased sensitivity and specificity, and miniaturized devices with enhanced figures of merit.

Chapter 13 presents principles of engineering and design of operation of several nanophotonics components, devices, systems, and techniques and relates the technical challenges to computational modeling. The chapter introduces key paradigms of nanophotonics and their operational design principles within different technological domains. Also, the authors discuss novel methodologies, systems, techniques and applications in the area of bionanophotonics for medical diagnostics, treatment and drug delivery aimed at highlighting the significance of computer modeling and simulations of complex nanoscale systems. Indeed, bionanophotonics is expected to open tremendous opportunities toward the design of novel and efficient diagnostic/therapeutic techniques and devices within the biological, biochemical and medical device fields, tending to become a potentially powerful tool for medical diagnostics, treatment and drug delivery.

Finally, the appendices conclude the book. Appendix A provides common material and physical constants, with the consideration that the material's constants values vary from one published source to another due to many varieties of most materials, and conductivity is sensitive to temperature, impurities, moisture content, as well as dependence of relative permittivity and permeability on temperature and humidity and the like. Appendix B provides equations for photon energy, frequency, and wavelength and electromagnetic spectra, including approximation of common optical wavelengths' ranges of light. In addition, it provides wavelengths of commercially available lasers. Appendix C provides common symbols and useful mathematical formulas.

COMSOL and COMSOL Multiphysics are registered trademarks of COMSOL AB. For details, visit www.comsol.com.

MATLAB® is a registered trademark of The Mathworks, Inc. For product information, please contact:

The MathWorks, Inc.
3 Apple hill Drive
Natick, MA, 01760-2098 USA
Tel: 508-647-7000
Fax: 508-647-7001
Email: inf@mathworks.com
Web: www.mathworks.com

Sarhan M. Musa

Acknowledgments

My sincere appreciation and gratitude to all the book's contributors. Thanks to Brain Gaskin, James Gaskin, Jim Stevenson, and Anna Barrera for their wonderful hearts and for being great American neighbors. It is my pleasure to acknowledge the outstanding help and support of the team at Taylor & Francis Group/CRC Press in preparing this book, especially from Nora Konopka, Michele Smith, Kari Budyk, Scott Shamblin, and Rachael Panthier. Thanks also to Arunkumar Aranganathan from SPi Global for his outstanding suggestions.

I thank Dr. John Burghduff and Professor Mary Jane Ferguson for their support and understanding and for being great friends. Thanks also to Dr. Fadi Alameddine for taking good care of my mother's health during the course of this project.

I thank Dr. Kendall Harris, my college dean, for his constant support. Finally, this book would never have seen the light of day if not for the constant support, love, and patience of my family.

Editor

Sarhan M. Musa, PhD, currently serves as an associate professor in the Department of Engineering Technology at Prairie View A&M University, Texas. He has been director of Prairie View Networking Academy, Texas, since 2004. Dr. Musa has published more than 100 papers in peer-reviewed journals and conference proceedings and is the editor of *Computational Nanotechnology Modeling and Applications with MATLAB®* and *Computational Finite Element Methods in Nanotechnology*. Dr. Musa is a senior member of the Institute of Electrical and Electronics Engineers (IEEE) and is also a LTD Sprint and a Boeing Welliver Fellow.

Contributors

S.M. Ashraf (retired)
Materials Research Laboratory
Department of Chemistry
Jamia Millia Islamia
New Delhi, India

Mahua Bera
Department of Applied Optics
    and Photonics
University of Calcutta
Kolkata, India

Rim Cherif
GreS'Com Laboratory
Engineering School of Communication of
    Tunis (Sup'Com)
University of Carthage
Ariana, Tunisia

P.K. Choudhury
Institute of Microengineering
    and Nanoelectronics
Selangor, Malaysia

Orion Ciftja
Department of Physics
Prairie View A&M University
Prairie View, Texas

Aditi Deshpande
Department of Biomedical
    Engineering
The University of Akron
Akron, Ohio

Tannaz Farrahi
Department of Electrical and Computer
    Engineering
The University of Akron
Akron, Ohio

George C. Giakos
Department of Electrical and Computer
    Engineering
and
Department of Biomedical
    Engineering
The University of Akron
Akron, Ohio

Chintha Chamalie Handapangoda
Department of Electrical and Computer
    Systems Engineering
Monash University
Melbourne, Victoria, Australia

Thomas J. Kempa
Department of Chemistry and Chemical
    Biology
Harvard University
Cambridge, Massachusetts

Sun-Kyung Kim
Department of Physics
Korea University
Seoul, South Korea

and

Department of Chemistry and Chemical
    Biology
Harvard University
Cambridge, Massachusetts

Ryan Koglin
Department of Biomedical
    Engineering
The University of Akron
Akron, Ohio

Mohit Kumar
Department of Electrical and Computer
    Engineering
The University of Akron
Akron, Ohio

Yinan Li
Department of Electrical and Computer
Engineering
The University of Akron
Akron, Ohio

Charles M. Lieber
Department of Chemistry and Chemical
Biology
and
School of Engineering and Applied
Sciences
Harvard University
Cambridge, Massachusetts

Anandi Mahadevan
Department of Biomedical Engineering
The University of Akron
Akron, Ohio

Chris Mela
Department of Biomedical
Engineering
The University of Akron
Akron, Ohio

Jianwei Mu
Massachusetts Institute of Technology
Cambridge, Massachusetts

Sarhan M. Musa
Department of Engineering Technology
Prairie View A&M University
Prairie View, Texas

Chaya Narayan
Department of Electrical and Computer
Engineering
The University of Akron
Akron, Ohio

Hong-Gyu Park
Department of Physics
Korea University
Seoul, South Korea

Malin Premaratne
Department of Electrical and Computer
Systems Engineering
Monash University
Melbourne, Victoria, Australia

Md. Mijanur Rahman
School of Computer and Communication
Engineering
Perlis, Malaysia

Mina Ray
Department of Applied Optics
and Photonics
University of Calcutta
Kolkata, India

Ufana Riaz
Materials Research Laboratory
Department of Chemistry
Jamia Millia Islamia
New Delhi, India

Amine Ben Salem
GreS'Com Laboratory
Engineering School of Communication
of Tunis (Sup'Com)
University of Carthage
Ariana, Tunisia

Suman Shrestha
Department of Electrical and Computer
Engineering
The University of Akron
Akron, Ohio

Jennifer Syms
Department of Physics
The University of Akron
Akron, Ohio

Limin Tong
State Key Laboratory of Modern Optical
Instrumentation
Department of Optical Engineering
Zhejiang University
Hangzhou, Zhejiang, People's Republic
of China

Lingyun Wang
Department of Electrical and Computer
  Engineering
The University of Texas at Texas
Austin, Texas

Yipei Wang
State Key Laboratory of Modern Optical
  Instrumentation
Department of Optical Engineering
Zhejiang University
Hangzhou, Zhejiang, People's Republic
  of China

Viroj Wiwanitkit
Hainan Medical University
Haikou, Hainan, People's Republic
  of China

and

Joseph Ayobabalola University
Ikeji-Arakeji, Nigeria

and

Wiwanitkit House
Bangkok, Thailand

Yasha Yi
Department of Electrical and Computer
  Engineering
University of Michigan
Ann Arbor, MI

and

MIcroPhotonics Center
Massachusetts Institute of Technology
Cambridge, MA

Mourad Zghal
Engineering School of Communication
  of Tunis (Sup'Com)
University of Carthage
Ariana, Tunisia

Lin Zhang
Department of Chemistry
Cleveland State University
Cleveland, Ohio

Xiaojing Zhang
Department of Biomedical
  Engineering
The University of Texas at Texas
Austin, Texas

1

Computational of Optical Micro-/Nanoprism

Sarhan M. Musa and Orion Ciftja

CONTENTS

1.1 Introduction

Many nanoscale materials have found wide applications in various scientific and engineering fields. Optical devices have also benefited from this trend. During the last decade, extensive research has been devoted to various metal nanoparticles. These structures manifest very interesting optical properties, which are strongly dependent on their shape and size. By artificially controlling their shapes and sizes, one can effectively tune the overall optical properties, a much desirable feature when building optical devices.

Though nanoparticles can come with various shapes and, generally, it is not easy to fully control their final shape, microprisms and nanoprisms represent a very important key element that has already been synthesized. Micro-/nanoblocks of optical material with flat polished faces arranged at precisely controlled angles are called micro-/nanoprisms. Light passing through micro-/nanoprisms is governed by the laws of light. In fact, the study of light passing through micro-/nanoprisms becomes very useful to nanophotonics and has a great significant use in building optical devices and formatting of images.

When light of an appropriate wavelength goes through such a structure, it can cause oscillation of conduction electrons in metal nanoparticles, such as those of silver and gold, inducing a special optical phenomenon. Differently from bulk optical phenomena, coupling of light and matter in this case is size dependent and scale dependent. The exploitation of the optical properties associated with these structures is based on a small number of discrete differences between features of the microprisms or nanoprisms and also quantum effects.

Today, light coupling and losses occurring as coupling through optical waveguides become an essential issue for optical communications. Optical prism is an optical device that refracts light. Light coupling through prism was used in [1–3]. Indeed, the optical prism coupling method was first established in [4]. Optical prism couplers allow for coupling light in and out of an optical waveguide without exposing the cross section of the waveguide [5]. In fact, the phase-match angles θ_0 (the angle of the incident light normal from the waveguide surface) [1] can be written as

$$\theta_0 = \cos^{-1}\left(\frac{\beta}{n_p k_0}\right) \tag{1.1}$$

where
  β is the mode propagation constant of the waveguide
  n_p is refractive index of the prism
  k_0 is the vacuum wave number

The vacuum wave number k_0 is given by

$$k_0 = \frac{2\pi}{\lambda_0} \tag{1.2}$$

where
  λ_0 is the wavelength of the light

The coupling of light into narrow optical waveguides is only possible if high coupling losses due to diffraction effects in the coupling region are accepted [6].

This chapter will focus on the results of modeling and simulations of light coupling in waveguides through microprism coupling method using finite-element method (FEM) with COMSOL Multiphysics to study and analyze the results.

This chapter provides a general perspective of key aspects of light–matter interaction in systems of optical microprisms. Though we focus our efforts mostly on an optical microprism model, we point out that such a model serves as a semiclassical precursor to studies of nanoprisms with smaller size.

1.2 Computation of Optical Microprism between Two Optical Waveguides

Microprism has been used in wideband optoelectronic interconnects. A microprism is used for reducing radiative losses in photonic waveguide bends. A low-loss optical waveguide bend with an effective photonic crystal microprism, consisting of the sub-wavelength periodic lattice, was designed using the 2D finite-difference time-domain (FDTD) in [7]. In the low-loss bending waveguide using the phase compensation rule, the bending radius of optical waveguide (R_c) can be presented as [8]

$$R_c = \left(\frac{n_w}{n_s - n_p} - 0.5\right) W_f \tag{1.3}$$

where
  n_w is the refractive index of the waveguide eigenmode
  n_s is the refractive index of waveguide substrate
  n_p is the refractive index of the photonic crystal microprism
  W_f is the wavefront width of waveguide fundamental mode

Also, most of optical power has been directed into the bend optical waveguide with the aid of the optical microprism except that the power outside the bend corner becomes radiation after bending. Indeed, as most of the optical power is confined within the effective waveguide width, the power outside the bend corner should be directed into the bent waveguide with the aid of the extended microprism [8]. Moreover, the relationship between the width of microprism and the effective width of optical waveguide can be written as

$$W_e = W_p + \frac{1}{\alpha_{1x}} + \frac{1}{\alpha_{2x}} \tag{1.4}$$

where
  W_e is the effective width of the waveguide
  W_p is the width of the microprism
  α_{1x} and α_{2x} are the attenuation coefficients of the symmetric structure and can be
    defined as

$$\alpha_{1x} = \alpha_{2x} = \sqrt{k_0 \left(n_e^2 - n_s^2 \right)} \tag{1.5}$$

where
  k_0 is the wave number in a vacuum
  n_e is effective index of the waveguide eigenmode

The transmitted power can be maximized by a certain value of α using Snell's law as given by [9]

$$\tan \alpha = \frac{n_1 - n_p \cos(\theta/2)}{n_p \sin(\theta/2)} \tag{1.6}$$

where
  n_1 is the refractive index for waveguide 1
  n_p is the refractive index for microprism

A structure for low-index diamond-like microprism on a single-mode symmetric Y-junction waveguide was proposed by [10].
  We illustrate here that by placing a microprism between two waveguides forming a sharp bend, light will be guided between the waveguides through the prism as shown in Figure 1.1 using FEM with COMSOL Multiphysics.

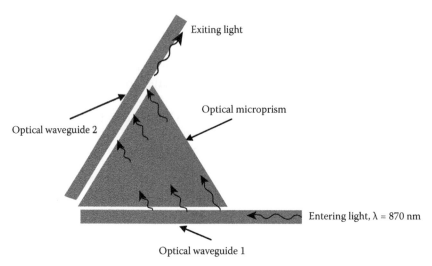

Exiting light

Optical microprism

Optical waveguide 2

Entering light, λ = 870 nm

Optical waveguide 1

FIGURE 1.1
Photonic microprism between two optical waveguides.

For a certain refractive index of the prism, the light propagating through the prism couples to the respective mode under just the appropriate resonance angle [11]. The optical microprism to the guide is the inverse to the light transfer from the guide to the microprism, when the initial field distribution does not diffract while propagating through the prism. Thus, the efficiency of the process is very high. The interface between the optical waveguide and the optical microprism must be sufficiently long to allow almost all the power to exit from the optical waveguide into the optical microprism and vice versa. In the meantime, to avoid diffraction, the size of the microprism should be kept as small as possible. In our modeling, we use perfectly matched layers (PMLs). PMLs are not boundary conditions but are domains with artificial absorption of the incident radiation that reduce nonphysical reflections. A PML provides good performance for a wide range of incidence angles and is not particularly sensitive to the shape of the wavefronts. The PML formulation can be deduced from Maxwell equations by introducing a complex-valued coordinate transformation under the additional requirement that the wave impedance should remain unaffected [12].

1.2.1 Modeling and Simulation of the Optical Microprism

The model is built using the 2D in-plane transverse electric (TE) wave application mode. The modeling takes place in the x–y plane. The in-plane wave application mode covers a situation where there is no variation in the z direction, and the electromagnetic field propagates in the modeling x–y plane.

As the field propagates in the modeling x-y plane, a TE wave has only one electric field component in the z direction, and the magnetic field lies in the modeling plane. Thus, the transient and time-harmonic fields can be written as

$$\mathbf{E}(x,y,z) = E_z(x,y,z) = E_z(x,y,z)\mathbf{e}_z e^{j\omega t} \tag{1.7}$$

$$\mathbf{H}(x,y,t) = H_x(x,y,t)\mathbf{e}_x + H_y(x,y,t)\mathbf{e}_y = \left[H_x(x,y)\mathbf{e}_x + H_y(x,y)\mathbf{e}_y \right]e^{j\omega t} \tag{1.8}$$

where
  E is the electric field
  H is the magnetic field
  ω is angular frequency

To be able to write the fields in this form, it is also required that the relative permittivity (ε_r), the conductivity (σ), and the relative permeability (μ_r) are nondiagonal only in the x–y plane. μ_r denotes a 2-by-2 tensor, and ε_{rzz} and σ_{zz} are the relative permittivity and conductivity in the z direction.

Given the earlier constraints,

$$\nabla \times \left(\mu_r^{-1} \nabla \times \mathbf{E} \right) - k_0^2 \varepsilon_{rc} \mathbf{E} = 0 \tag{1.9}$$

where

$$\varepsilon_{rc} = \varepsilon_r - j \frac{\sigma}{\omega \varepsilon_0} \tag{1.10}$$

where
  ε_{rc} is the complex relative permittivity

Substituting Equation 1.4 in Equation 1.3 gives

$$\nabla \times \left(\mu_r^{-1} \nabla \times E_z \right) - \left(\varepsilon_r - \frac{j\sigma}{\omega \varepsilon_0} \right) k_0^2 E_z = 0, \tag{1.11}$$

where
  μ_r denotes the relative permeability
  ω the angular frequency
  σ the conductivity
  ε_0 the permittivity of vacuum
  ε_r the relative permittivity
  k_0 the wave number

Equation 1.9 can be simplified to a scalar equation for E_z:

$$-\nabla \cdot \left(\tilde{\mu}_r \nabla E_z \right) - \varepsilon_{rczz} k_0^2 E_z = 0 \tag{1.12}$$

where

$$\tilde{\mu}_r = \frac{\mu_r^T}{\det(\mu_r)} \tag{1.13}$$

where
  $\tilde{\mu}_r$ is the relative permeability
  μ_r^T is the transpose relative permeability
  $\det(\mu_r)$ is the determinant of relative permeability

By using $\varepsilon_r = n^2$, where n is the refractive index, with assumption that $\mu_r = 1$ and $\sigma = 0$, we get

$$-\nabla \cdot \nabla E_z - n_{zz}^2 k_0^2 E_z = 0 \tag{1.14}$$

$$k_0 = \omega\sqrt{\varepsilon_0\mu_0} = \frac{\omega}{c_0} \tag{1.15}$$

where
  k_0 is the wave number in vacuum
  c_0 is the speed of light, $c_0 \approx 3 \times 10^m \, \mathrm{m/s}$
  ε_0 is permittivity of vacuum, $\varepsilon_0 = \dfrac{1}{c_0^2\mu_0} = \dfrac{1}{36\pi} \times 10^{-9}\,\mathrm{F/m} \approx 8.854 \times 10^{-12}\,\mathrm{F/m}$
  μ_0 is permeability of vacuum, $\mu_0 = 4\pi \times 10^{-7}\,\mathrm{H/m} \approx 12.6 \times 10^{-7}\,\mathrm{H/m}$

The equation in time domain can be presented as

$$\mu_0\sigma\frac{\partial \mathbf{A}}{\partial t} + \mu_0\varepsilon_0\frac{\partial}{\partial t}\left(\varepsilon_r\frac{\partial \mathbf{A}}{\partial t} - \mathbf{D}_r\right) + \nabla \times \left[\mu_r^{-1}\left(\nabla \times \mathbf{A} - \mathbf{B}_r\right)\right] = 0 \tag{1.16}$$

where
  A is the magnetic potential
  D is the electric displacement (or electric flux density)
  B is the magnetic flux density

Equation 1.10 can be simplified in a scalar equation for A_z as

$$\mu_0\sigma\frac{\partial A_z}{\partial t} + \mu_0\varepsilon_0\frac{\partial}{\partial t}\left(\varepsilon_r\frac{\partial A_z}{\partial t} - D_{rz}\right) + \nabla \times \left[\mu_r^{-1}\left(\nabla \times A_z - \mathbf{B}_r\right)\right] = 0 \tag{1.17}$$

We used here the following constitutive relations:

$$\mathbf{B} = \mu_0\mu_r\mathbf{H} + \mathbf{B}_r \tag{1.18}$$

and

$$\mathbf{D} = \varepsilon_0\varepsilon_r\mathbf{E} + \mathbf{D}_r \tag{1.19}$$

where **H** is the magnetic field intensity.
  By using $\varepsilon_r = n^2$, where n is the refractive index, with assumption that $\mu_r = 1$ and $\sigma = 0$ and only the constitutive relations for linear materials can be used, we get

$$\mu_0\varepsilon_0\frac{\partial}{\partial t}\left(n^2\frac{\partial A_z}{\partial t}\right) + \nabla \cdot \left(\nabla A_z - \mathbf{B}_r\right) = 0 \tag{1.20}$$

In photonics and optics applications, the refractive index is often used instead of the permittivity. In materials where $\mu_r = 1$, the relation between the complex refractive index, n_c,

$$n_c = n - j\kappa \tag{1.21}$$

and the complex relative permittivity is

$$\varepsilon_{rc} = n_c^2 \tag{1.22}$$

where κ is the damping of the electromagnetic wave.

The refractive index for the waveguides is 1.5 and for the microprism is 2.5. For the physical setting, we set the frequency (f) to $f = \dfrac{c}{\lambda} = \dfrac{3 \times 10^8 \, \text{m/s}}{870 \, \text{nm}}$.

In the domain, the dependent variable in this application mode is the z component of the electric field **E**. It obeys Equation 1.11. Different refractive indices are used for the prism and the guides. The wave is dampened by PMLs where the wave enters and exits the setup of the model. The whole geometry is surrounded by another PML, which decreases reflections from the nonphysical exterior boundary. The solution is calculated for IR light with a wavelength in vacuum of 870 nm.

For the boundary conditions, the exterior boundaries in this model, we use a scattering boundary condition to terminate the PML. Inside the geometry, continuity is applied everywhere except at the boundary where the wave is entering the structure. This boundary is excited with a cosine function fitted to match the width of the waveguide.

1.2.2 Results and Analysis for the Modeling and Simulation of the Optical Microprism

Based on our modeling and simulation, Figure 1.2 shows the 2D surface plot for the electric energy density of the model, while Figure 1.3 shows the 3D surface of it. These figures show the geometry and the solution of the model. The wave enters the horizontal guide from the right and exits at the top of the vertical guide. The circles surrounding the entry and the exit are PMLs.

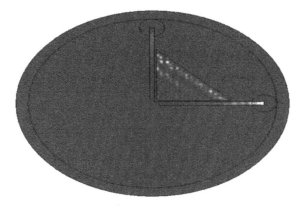

FIGURE 1.2
The 2D surface plot for the electric energy density.

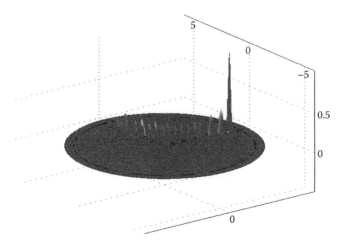

FIGURE 1.3
The 3D surface plot for the electric energy density.

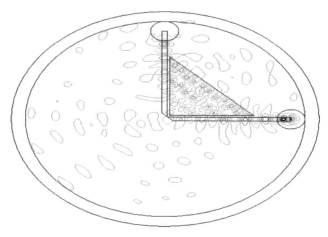

FIGURE 1.4
Contour plot for the electric energy density.

Several variables govern the transmission of the wave through the bend, for example, the relation between the refractive indices of the guide and the prism, the size of the prism, and the gap between the guides and the prism. The optimal values of the parameters also depend on the angle of the bend. Figure 1.4 presents the contour plot for the electric energy density of the model.

Figure 1.5 shows a mesh for the electric energy density of the model and Table 1.1 summarizes the mesh statistics. Also, the line plot for the mesh size is presented in Figure 1.6.

Figure 1.7 shows the 2D surface plot for the z component of the electric field. Figure 1.8 shows the line plot for the z component of the electric field at the entering light in optical waveguide 1, at the optical microprism, and the existing light at the optical waveguide 2.

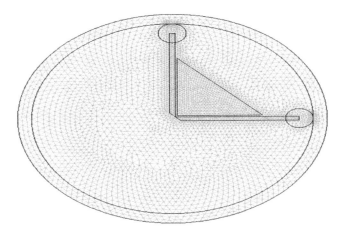

FIGURE 1.5
Mesh for the electric energy density.

TABLE 1.1

Mesh Statistics

Items	Value
Number of degrees of freedom	15,986
Total number of mesh points	4,033
Total number of elements	7,921
Triangular elements	7,921
Quadrilateral	0
Boundary elements	636
Vertex elements	28

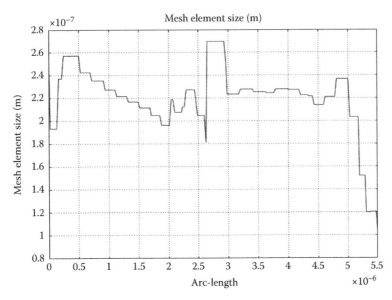

FIGURE 1.6
Line plot for the mesh size.

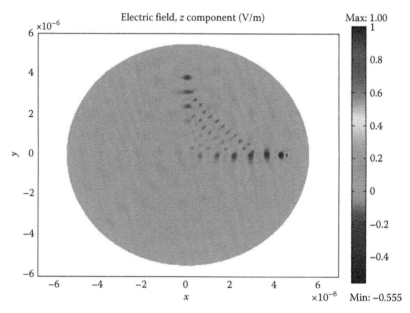

FIGURE 1.7
(See color insert.) The 2D surface plot for the z component of the electric field.

Figures 1.9 through 1.11 show the line plot for the power flow, time average for x component and y component, and norm, respectively.

Figures 1.12 and 1.13 show the line plot for the electric field and z component and norm of the model, respectively

Figures 1.14 and 1.15 show the line plot for the electric displacement and z component and norm of the model, respectively.

Figure 1.16 shows the line plot for the scattered electric displacement and z component, while Figure 1.17 shows the line plot for the scattered electric field and z component.

Figure 1.18 shows the line plot for the total energy density and time average of the model, while Figure 1.19 shows the line plot for the resistive heating and time average of the model.

1.3 Computational of Optical Nanoprism

Recent progress on synthesis methods has made optical nanoprisms readily available in a wide variety of sizes and compositions [13,14]. While the predetermination of the final shape is not always easy, there are now shape-selective synthesis procedures that yield any desirable shape. For instance, a common method based on photoinduction allows one to convert arbitrarily shaped silver nanoparticles into triangular silver nanocrystals, namely, a nanoprism [15]. Since a prism shape is intrinsically anisotropic, one expects very interesting optical properties when coupling to light occurs. Ensembles of nanoprisms have also been synthesized and studied at a wide range of lengths [16].

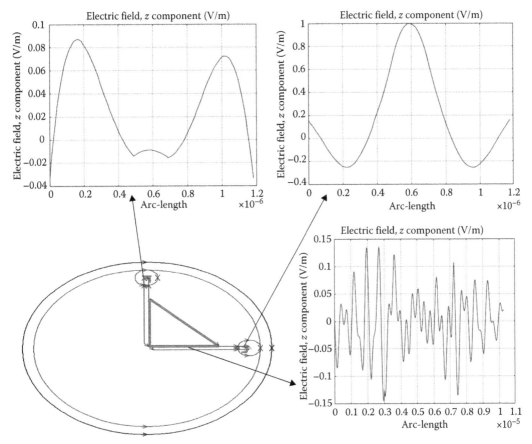

FIGURE 1.8

The line plot for the z component of the electric field.

Gold and silver triangular nanoprims are reviewed by describing a variety of solution-based methods for synthesizing gold and silver triangular prismatic structures in [17]. The review showed the UV–Vis–NIR spectra and corresponding solution of silver nanoprisms with varying edge length and orientation-averaged extinction efficiency for triangular nanoprism based on a 100 nm edge dimension with snips of 0, 10, and 20 nm. In addition, photoinduced conversion of silver nanospheres to nanoprisms is reported in [18]. Also, in optical near-field mapping of plasmonic nanoprisms studied in [19], the authors used the interformatic homodyne tip-scattering near-field microscopy for plasmonic near-field imaging of crystalline triangular silver nanoprism.

Coupled triangular nanoprisms have been investigated in [20,21]. Resonant excitation of the quadrupolar surface plasmon mode of the nanoprisms increases Raman scattering intensity from the substrate as the distance between the nanoparticle pairs decreases, as reported in [22]. Indeed, the authors used finite-element modeling, and plasmon coupling theory indicates that symmetry is reduced as the nanoparticles approach, resulting in increased dipole–quadrupole coupling. The plasmonic properties of single silver triangular nanoprisms are investigated using dark-field optical microscopy and spectroscopy, and they observed two distinct localized surface plasmon resonances (LSPR) in [23,24].

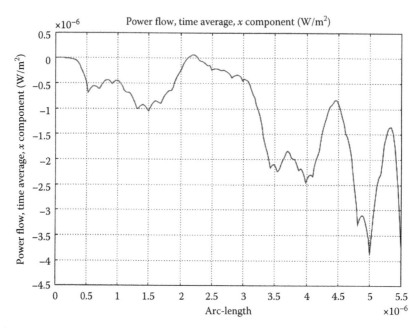

FIGURE 1.9
The line plot for the power flow, time average, and x component.

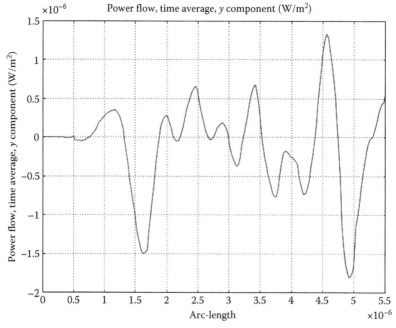

FIGURE 1.10
The line plot for the power flow, time average, and y component.

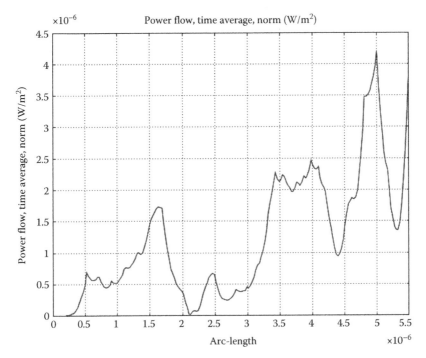

FIGURE 1.11
The line plot for the power flow, time average, and norm.

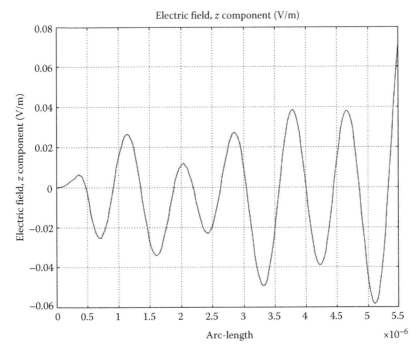

FIGURE 1.12
The line plot for the electric field and z component.

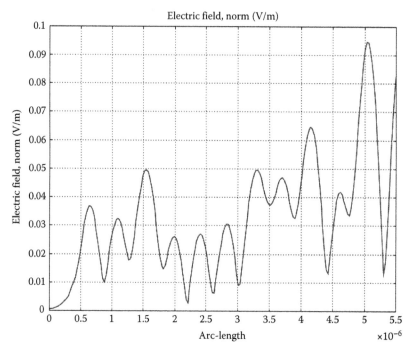

FIGURE 1.13
The line plot for the electric field and norm.

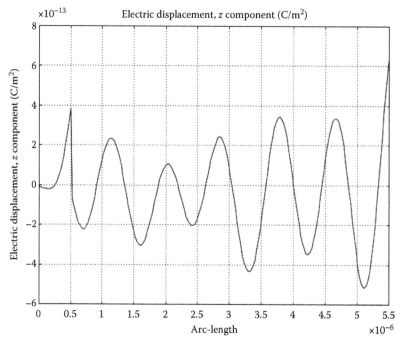

FIGURE 1.14
The line plot for the electric displacement and z component.

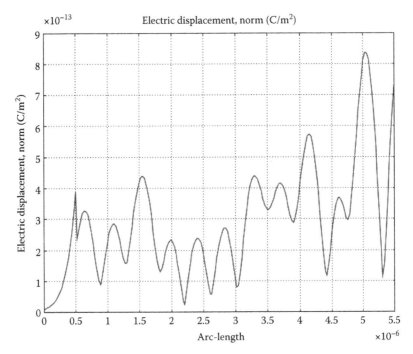

FIGURE 1.15
The line plot for the electric displacement and norm.

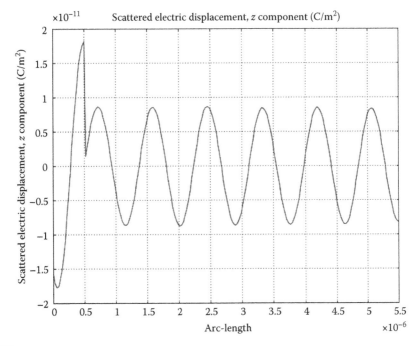

FIGURE 1.16
The line plot for the scattered electric displacement and z component.

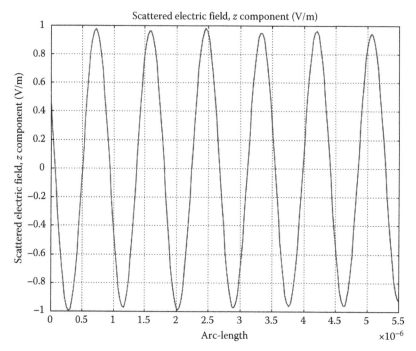

FIGURE 1.17
The line plot for the scattered electric field and z component.

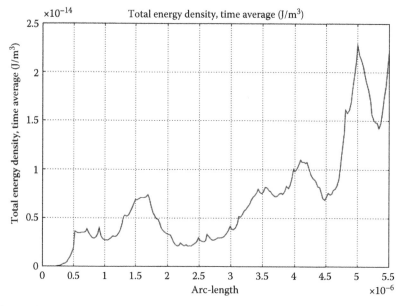

FIGURE 1.18
The line plot for the total energy density and time average.

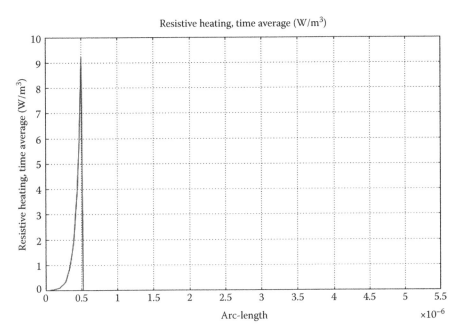

FIGURE 1.19
The line plot for the resistive heating and time average.

Size-dependent microscopic nonlinear optical properties of silver triangular nanoprisms in colloidal solution were reported in [25].

The fabrication process involves exposure of the nanoparticles to selected beams of light and seems to be driven by surface plasmons. By experimentally choosing a given illumination light wavelength, one can obtain as a final product a nanoprism with sharp geometric features. For a different choice of light, no stable nanoprism structure is obtained. Overall, the photoinduction process can be steadily controlled by well-known optical methods, thus allowing the observation of various intermediate semistable structures. The interesting outcome is that this light-controlled process leads to a structure with specific optical properties directly related to the nanoprism shape of the particles.

Theoretical models coupled with computational calculations [26] agree with experimental observations regarding the important role played by the nanoprism plasmon excitation bands. The prism shape, thus the expected anisotropy, has also a profound effect on the overall optical behavior of a nanoprism as seen in various Rayleigh scattering experiments. This can be useful in developing sensitive multicolor light detector devices on the basis of shape.

1.4 Conclusions

Optical properties of materials crucially depend on many factors such as size, structure, and composition, to mention a few. With modern tools, one can typically engineer or modify such properties by changing the relative influence of various components with respect to the characteristic size or dimension of the macroscopic bulk system. This approach is so straightforward and successful that many materials today are composites with micro- to nanodomain sizes. The optical properties of materials in this transition regime depend on the physical or chemical character of each domain. In our case, we focused our attention on microprisms and nanoprisms. For these length scales, there is a complex relation between the prism's size, its composition, and coupling to light. Overall, the manifested optical properties emerge from an intricate interplay of many factors. As a result, the final optical properties may depend very sensitively on the bulk and surface properties of each ingredient. In addition, unexpected optical properties may well display at the interface.

A detailed research process of the optical properties of microprisms and nanoprisms provides an important stepping-stone for studying other nanomaterials with different shapes. Thus, in principle, one can foresee building much larger artificial structures with amazing optical properties by controlling shape, size, and interactions during the fabrication process at different wavelength ranges.

We successfully obtained the results of modeling and simulations of light coupling in waveguides through microprism coupling method using FEM with COMSOL Multiphysics in this chapter. Also, we gave an overview of computational of optical nanoprism.

References

1. R. Hunsperger, *Integrated Optics*, Springer, Berlin, Germany, 1995.
2. C. R. Doerr, *Planar Lightwave Devices for WDM, Optical Fiber Telecommunications*, Ed. I. Kaminow and T. Li, Academic Press, London, U.K., 2002.
3. A. Yariv and P. Yeh, *Photonics: Optical Electronics in Modern Communications*, Chapter 3, Oxford University Press, Oxford, U.K., 2007.
4. P. K. Tien, R. Ulrich, and R. J. Martin, Modes of propagating light wave in thin deposited semiconductor films, *Applied Physics Letters*, 14, 291, 1969.
5. J. T. Andrews, R. R. Kumar, and P. K. Sen, Prism coupling of light in optical waveguide, *Proceedings of the COMSOL Conference 2010*, Bangalore, India, 2010.
6. W. A. Pasmoou, P. A. Mandersloot, and M. K. Smit, Prism-coupling of light into narrow planar optical waveguide, *Journal of Lightwave Technology*, 7(1), 175–180, 1989.
7. M. L. Wu, H. C. Lan, C. M. Wang, and J. Y. Chang, Design of wide-angle low-loss waveguide bends using phase compensated effective microprism, *Japanese Journal of Applied Physics*, 46(8B), 5426–5430, 2007.
8. H. B. Lin, J. Y. Su, P. K. Wei, and W. S. Wang, Design and application of very low-loss abrupt bends in optical waveguides, *IEEE Journal of Quantum Electronics*, 30(12), 2827–2835, 1994.
9. K. Hirayams and M. Koshiba, A new low-loss structure of abrupt bends in dielectric waveguide, *Journal of Lightwave Technology*, 10(5), 563–569, 1992.
10. W. C. Chang and H. B. Lin, A novel low-loss wide-angle Y-branch with a diamond-like microprism, *IEEE Photonics Technology Letters*, 11(6), 683–685, 1999.
11. U. Peschel, L. Leine, F. Lederer, and C. Wachter, CLEO 2001, *Conference on Lasers and Electro-Optics*, OSA Technical Digest (Optical Society of America, Washington, DC, 2001), pp. 129–130.
12. J. Jin, *The Finite Element Method in Electromagnetic*, 2nd edn., Wiley-IEEE Press, New York, May 2002.

13. J. E. Millstone, S. Park, K. L. Shuford, L. Qin, G. C. Schatz, and C. A. Mirkin, Observation of a quadrupole plasmon mode for a colloidal solution of gold nanoprisms, *Journal of the American Chemical Society*, 127, 5312–5313, 2005.
14. J. Zhao, L. J. Sherry, G. C. Schatz, and R. P. Van Duyne, Molecular plasmonics: Chromophore–plasmon coupling and single-particle nanosensors, *IEEE Journal of Selected Topics in Quantum Electronics*, 14(6), 1418–1429, 2008.
15. M. G. Blaber, A.-I. Henry, J. M. Bingham, G. C. Schatz, and R. P. Van Duyne, LSPR imaging of silver triangular nanoprisms: Correlating scattering with structure using electrodynamics for plasmon lifetime analysis, *Journal of the Physical Chemistry C*, 116, 393–403, 2012.
16. A. Farhang and O. J. F. Martin, Plasmon delocalization onset in finite sized nanostructures, *Optics Express*, 19(12), 11387–11396, 2011.
17. J. E. Millstone et al., Colloidal gold and silver triangular nanoprisms, *Small*, 5(6), 646–664, 2009.
18. R. Jin et al., Photoinduced conversion of silver nanospheres to nanoprisms, *Science*, 294(5548), 1901–1903, 2001.
19. M. Rang et al., Optical near-field mapping of plasmonic nanoprisms, *Nano Letters*, 8(10), 3357–3363, 2008.
20. K. L. Kelly et al., The optical properties of metal nanoparticles: The influence of size, shape and dielectric environment, *Journal of Physical Chemistry B*, 107, 668–677, 2003.
21. C. L. Haynes et al., Nanoparticle optics: The importance of radiative dipole coupling in two-dimensional nanoparticle arrays, *Journal of Physical Chemistry B*, 107(30), 7337–7342, 2003.
22. E. C. Dreaden et al., Multimodal plasmon coupling in low symmetry gold nanoparticle pairs detected in surface-enhanced Raman scattering, *Applied Physics Letters*, 98, 183115-3, 2011.
23. L. J. Sherry et al., Localized surface plasmon resonance spectroscopy of single silver triangular nanoprisms, *Nano Letters*, 6(9), 2060–2065, 2006.
24. J. N. Anker, W. P. Hall, O. Lyandres, N. C. Shah, J. Zhao, and R. P. Van Duyne, Biosensing with plasmonic nanosensors, *Nature Materials*, 7, 442–453, 2008.
25. A. K. Singh et al., Nonlinear optical properties of triangular silver nanomaterial, *Chemical Physics Letters*, 481, 94–98, 2009.
26. R. Marty, G. Baffou, A. Arbouet, C. Girard, and R. Quidant, Charge distribution induced inside complex plasmonic nanoparticles, *Optics Express*, 18(3), 3035–3044, 2010.

2

Role of Computational Intelligence in Nanophotonics Technology

Ufana Riaz and S.M. Ashraf

CONTENTS

Summary

Nanophotonics describes optical science and technology at the nanoscale that offers challenging opportunities for studying fundamental processes of interaction between the radiation field and the matter on a scale much smaller than the wavelength of radiation. The use of such confined interaction to spatially localized processes offers exciting opportunities to chemists. Nanophotonics is considered to be of immense technological significance. The evolution and development of the nanophotonics technology with the increased need for an internetworking broadband communication had lead to the development of more sophisticated optical networks. Those networks are now interconnected and cooperated with other network technologies such as wireless and satellite and have also expanded from the transport layer to the access layer as fiber to the home (FTTH) applications.

The scope of this chapter is to review the computation modeling of nanophotonics in the design, optimization, and operation of optical networks and the future prospective of nanophotonic components technology.

2.1 Introduction

Nanophotonics is defined as optical science and technology at the nanoscale that explores the fundamental processes of interaction between the radiation field and the matter on a much smaller scale than the wavelength of radiation. The investigation of optical wave phenomena at the nanoscale requires the application of rigorous numerical electrodynamics modeling. Advanced nanophotonic structures therefore require optical simulations to get a deeper understanding of light propagation and scattering. Advances in computing technology have resulted in greater use of data analysis algorithms in optical metrology. Computational optical measurement reveals how the computing technology plays a significant role in developing smarter techniques and systems. Phase measurement techniques, such as phase shifting and Fourier transform, are now synonymous with interferometry, whereas digital holography highlights image reconstruction with increasing versatility and widening application fields through computation. The simulation of good optoelectronic device requires an understanding of numerous physical aspects. The device to be analyzed can be considered as an inhomogeneous system of carriers, which interact with themselves, the lattice and the optical field. In certain conditions, the quantum properties of the carriers play the crucial role, whereas in other conditions, they can be accurately described by classical physics. One of the commonly used approaches is called the drift–diffusion approach. The main attraction of these simulation models is the speed and the simplicity of calculations by incorporating all the physical effects. As the dimensions of semiconductor and optoelectronic (photonic) devices scale down to nanoscale dimensions (few hundreds of angstroms), kinetic and quantum effects in the carrier transport become crucial for the device's design and operation. In such cases, carriers are treated as quantum many-body system with all possible interactions. Optoelectronic devices combine the electrical transport of carriers as well as the optical field interacting with the carriers. In such devices, spontaneous and stimulated emissions play an important role. To consistently describe spontaneous emission, the light field must also be quantized for which quantum-well lasers and quantum-cascade lasers are utilized for designing nanoscale optical devices.

2.2 Nanometer Confinement and Fabrication of Matter

Nanoscale confinement of matter is used to control the bandgap and the optical resonance and the excitation dynamics. Nanoscale confinement is used to produce periodic optical domains that lead to photon localization. Nanoscale confinement of domains to produce quantum dots, quantum wells, and quantum wires in inorganic semiconductors is well known and extensively studied [1]. Nanosize control of the local structure also helps in manipulating excitation dynamics by controlling the local phonon density of states. It is also used to influence energy transfer by controlling intermolecular interactions. Nanosize manipulation of molecular design and morphology is one of the most powerful tools to control the overall electronic and optical properties and the processibility of materials. The design and processing of nanostructured materials has recently emerged as one of the frontier areas of fundamental materials research. The electronic and photonic properties materials are strongly dependent on their bandgaps. Bandgap dependence has been well documented in the case of inorganic semiconductors, where nanostructure quantum dots of different sizes have been used to control the electronic, luminescence, and nonlinear optical properties [2–8]. The interest in photo processes is driven by the future needs of industrial applications, including integrated circuit designs, optoelectronic coupling, surface treatment, data storage, and lithography. Such applications require a major improvement in photosensitivity with a higher degree of precision and resolution. The spatial resolution in a conventional processing device depends on the spot size of the light source, which is diffraction limited. Thus, the transverse size of the produced structure is close to the submicron level. The near-field optical technique overcomes the diffraction-limited spot size and offers several advantages over the conventional processing technique in precision and fine resolution. It extends the capability of processing at nanometer scale. Although the applications of near-field scanning optical microscopy (NSOM) are still limited, immense progress has been made in the area of photonic processing and fabrication [9–14].

2.3 Application of Computational Intelligence in Photonics Technology

Despite optical fibers' enormous physical bandwidth, the development of optical networks for today's advanced and reliable services requires efficient management of the bandwidth together with an orthological use of optical components. These requirements can be fulfilled by the development of sophisticated photonic possessing increased functionality. Different design and optimization techniques have been applied to date, but as the complexity increases, the use of computational intelligence (CI) in these problems is becoming a unique tool of imperative value. The applicability of different CI classes (genetic algorithms (GAs) and evolution strategies, fuzzy systems, and artificial neural networks (ANNs)) in optical wavelength-division multiplexing (WDM) networks has to be identified and evaluated. The main challenge for the optical technology is the development of all-optical networks in order to fully exploit the available optical fibers' bandwidth. The current approach adopted for the development of all-optical networking is the WDM technology [15], where hundreds of wavelength channels can be established in a single-fiber link. Transmission capacities of Tb/s have been well demonstrated in optical test beds, and hence, WDM technology is expected to dominate the next-generation backbone transport networks.

CI is an area of fundamental and applied research [16,17] involving numerical information processing in contrast to the symbolic information processing techniques of artificial intelligence (AI). The main theoretical characteristics of a CI system are mainly the following [18]. It deals only with numerical (low level) data, has a pattern recognition component, and exhibits computational adaptability, computational fault tolerance, and speed approaching human-like turnaround. CI's major advantages like broad applicability, robustness to dynamic environments, and self-optimization capability established its role as an optimization and design tool in modern telecommunication networks [16–18]. As complexity of technology and networks' services increase, new challenging multi-combinatorial problems are emerging and consequently the CI applications are expected to be further enhanced.

Photonic components are based on fundamental physical mechanisms of light propagation or manipulation, but today, they have been evolved in quite complex systems that are usually multifunctional devices. Photonics itself is quite extended. Their fabrication platform can be also quite diverse—in contrast, for example, to microelectronics VLSI circuits where almost they exclusively use silicon technology—and can be bulk optics, fiber optics, integrated optics, crystal photonic, or nanophotonics, with an also diverse range of target functions such as active devices (lasers, amplifiers, wavelength converters) and passive devices (filters, multiplexers, optical cross-connects (OXCs), isolators, variable attenuators, etc.). The increased functionality of today's optical components together with the diversity of their physical characteristics and limitations can be proved as a challenging problem for their design. In this section, the extended range of CI applications in components design, potential new combined multifunctional components with respective multi-objective optimization challenges, and new areas of research components for all-optical networks is demonstrated.

Light polarization is an important parameter in optical communication as the given information can be coded in different polarization states and then can be multiplexed travel within a fiber. The major components are polarizing beam splitters that can be in a bulk optics form or in thin-film compact form. GAs are successfully applied for the design of multilayer polarizing beam splitters [19]. Another major component for optical systems is the isolator, a nonreciprocal device that protects expensive equipment in a link by strong signal back reflections. GAs have been employed for the design of an isolator based on a multilayer structure of layers of different magneto-optic, nonlinear, linear properties that function also as a switch [20].

WDM systems are moving toward integrated planar form in order to accommodate higher channel count for a given volume of the device. Volume is a major issue for optical components, and fiber-based devices cannot be compact enough. A planar structure introduces more degrees of freedom in the design process making; thus, the optimization problem is more demanding and challenging for CI. Although low scale of integration (LSI) has been used for the optimization of an optical power switch 1 · 2 coupler by using evolution strategies [21], variational optimization technique has been found to be more fast and efficient for a relatively simple problem of coupler optimization [22]. The 1 · 2 coupler optimization has now been extended to a more generic problem of 1 · 4 optical power splitter based on a multimode interference (MMI) coupler [23]. The GA used here optimizes the geometrical characteristics of the device in such a way so as to achieve the desired spatial interference of four simultaneous electric fields (modes) at the end face of the MMI, thereby producing equally splitting ratios. High-channel-count splitters, equipped with tenability functions, are being used for the construction of OXCs, key devices for switching in WDM networks. The development of multimedia applications such as multimedia conferencing, distance education, video distribution, and distributed games is largely driving the need for point-to-multipoint and multipoint-to-multipoint communication, referred to as multicasting. WDM optical networks efficiently support multicasting as splitting of light is inherently easier than copying data into an electronic buffer. The main operation in this technology is passive splitting of light in light trees in order to reach the final subscribers. Another important class of optical components is active devices such as optical amplifiers (OA). The main representative of all-OA is the erbium-doped fiber amplifiers (EDFA), and Raman amplifiers that exhibit quite broad amplification range have also been used. The desirable operational characteristics of amplifiers are high signal amplification, high saturation gain, large amplification band-width, good gain uniformity of the operational bandwidth, and low noise. The design and optimization process can be quite complex and generally deals with pure physical properties and parameters of the amplifying systems, such as fiber's composition, fiber's refractive-index contrast, core radius, fiber length, pumping power, and wavelength. GAs have been applied for the optimization of an active erbium-doped fiber and the EDFA [24]. In GA optimization, a multistage EDFA with complex structures is employed to obtain flattening-based filters [25]. The combination of the gain-flattening issues in the design process can produce a quite complex multi-objective optimization problem. Optimization of Raman amplifiers operation is focused essentially in the optimization of the pump-ing scheme that injects optical power into the fiber. A Raman amplifier is optimized for broadband operation and flattened operation. The Raman amplifier is based on a nonlin-ear photonic crystal (PC) fiber with inherent gain-flattening characteristics, together with the additional functionality of negative dispersion coefficient for dispersion compensation applications. Such complex systems provide a challenging in-field of CI optimization. The techniques have been widely applied in the optimization process of fiber-optic parametric amplifier (FOPA) [26,27].

WDM technology is a huge leap toward the use of optical bandwidth and is currently applied extensively in point-to-point and ring networks, reaching quite mature levels. This massive increase in networks bandwidth due to WDM drives the need for faster and more efficient switching. Point-to-point and ring WDM networks using electronic switching are evolving toward all-optical switching, based on the concept of optical-circuit switching (OCS) that is currently under way in transport networks. A network based on OCS consists of OXCs arranged in an arbitrary topology and provides interconnection to a number of subnetworks. An OXC can switch the incoming optical signal carried in a wavelength, on an output fiber link. The output can be in the same wavelength or in a different wavelength if OXC is equipped with a wavelength-converter device as the functionality of the latter case is much greater. The communication channel established in an OXCs network is called "light path" that is extended over a number of fiber links (multi-hop). If OXCs have no wavelength conversion (WC) capability, then the light path is associated with a single wavelength on each hop. Wavelength-routed networks (WRN) based on OXCS are significant in all-optical networks. The major drawback is that optical channels are reserved regardless data are being transmitted or not. To make efficient use of a light path capacity, there is a need for traffic grooming (TG) that is the main optimization issue. Because of the coarse granularity of the WRN, they inefficiently take up the re-configurability issue, in the absence of TG. Optical packet switching (OPS) [28] is inevitably a platform providing arbitrarily fine transmission and switching granularity, toward a flexible and efficient bandwidth. The implementation of OPS is challenging because of the significant problems in the required underlying photonics technology. The realization of OPS requires practical and cost-effective implementations for optical buffering and all-optical photonic switches that can perform all-optical packet address recognition [29]. Optical packet header recognition relies on specially designed Bragg gratings [30,31]. The current developing solution that is essentially an intermediate step between wavelength routing and OPS is optical burst switching (OBS) [28,32], where the basic switching entity is a variably sized burst. A burst is an aggregate flow of data packets with a single burst header. This is usually a group of IP packets heading for the same destination with the same quality of service (QoS) requirements. Grouping packets into bursts with a single burst header reduces header inspection duration and buffering at intermediate nodes. The three main switching techniques, that is, WRN, OBS, and OPS, have different application domains in the future optical networks [101]. The freedom to select the appropriate technology for specific applications is expected to make their implementation much more cost efficient.

2.3.1 Wavelength-Division Multiplexing

The efficient use of WDM leads to an increase in transmitted information and consequently to the need for flexible switching technologies (WRN, OBS, OPS) in order to accommodate the increased traffic and support advanced guaranteed services (QoS). The coexistence of all switching technologies cooperates in a generic control plane. Considering WRN problems, generally classical optimization problems are faced such as physical and logical topology design and routing and wavelength assignment (RWA). These problems are intractable, and therefore, CI and GAs specifically are employed in order to find a near-optimal solution that converges in polynomial time of search. The main CI issues in terms of the designing are

- Physical and logical topology
- Virtual topology reconfiguration
- RWA
- Multicast RWA
- Optimal placement of components
- TG
- Survivability
- Protection and restoration on optical layer
- Control and management
- QoS routing
- Services optimization

The design of WRN based on OXCs is a key issue, attracted intense research, in the implementation of optical networks. In order to reduce the complexity of the whole problem and make it more intuitive, the problem typically is divided in two subproblems [27]: network design and RWA. The network design part can be divided in two distinct problems, the physical topology design and the configuration design. For the WRN/OXC networks, the physical topology problems determine the count of OXCs and their interconnectivity, while the network configuration is related to the size of OXCs, the number of fibers, and the set of light paths. The network design also includes the placement of the key components such as amplifiers, wavelength converters, and power slitters, as well as the consideration of additional constraints like protection schemes due to OXC failures, geographical issues, etc. Establishing a set of light paths creates a logical topology over the physical topology, while the physical topology represents the physical interconnection of WDM nodes by actual fiber links in the WDM optical network. The links in the logical topology represent all-optical connections or light paths established between pairs of nodes. The purpose of RWA [33,34] issue is considered jointly. Routing is the establishment of a light path between two edge nodes by mapping such a path on the physical topology. The allocation of a wavelength for the connection of those two nodes via the established light path is called wavelength assignment. The RWA problem is more complicated than in electronic networks, because of the following constraints: wavelength continuity constraint, where light path must use the same wavelength along a multi-hop fiber link from the start to the destination node, and the distinct wavelength constraint, where all light paths on the same fiber link must use distinct allocated wavelengths. The RWA problem can be classified in two main classes, the static RWA where the traffic requirements are well known in advance and they remain unmodified and the dynamic RWA where light path requests from nodes follow a dynamic random pattern. The static RWA problems are usually referred as virtual topology design problem as they are essentially equivalent. The objective of RWA is often to minimize the number of wavelengths used or to maximize the number of light paths given the constraint of a limited number of wavelengths. A second objective can be the minimization of physical lengths of the corresponding physical paths, to minimize the impact of physical impairments such as fiber attenuation, fiber's dispersion, nonlinear effects, or accumulated amplifier noise. These physical impairments are a distinct characteristic of optical networks, and their consideration can produce more robust and realistic solutions, increasing, however, design's complexity. Photonics technology has demonstrated components based on semiconductor OA (SOA) or periodically poled lithium niobate (PPLN) capable of implementing a wavelength conversion operation. Wavelength converter (WC) works as a two-port device where information is carried in one wavelength and can be identically transformed in another wavelength at the output port. A WRN based on OXCs equipped with WCs network becomes more flexible in terms of routing choices and more efficient in terms of wavelength and bandwidth use. However, due to the increased cost of WCs, it is necessary to make careful use of WCs only in carefully selected OXCs or other wavelength convertible nodes of the network. A design strategy known as sparse wavelength conversion introduces another important design and optimization problem, the WC's optimal placement. The problem of optimal placement is more generic and can involve amplifiers placement, as well as splitters placement, in multicasting WDM networks. Optimization can target minimization of components count or their optimal placement. New design trends suggest that some of the aforementioned, individually presented problems, should be considered jointly, rather as independent problems. Xin et al. [27] considered in a unified approach the problem of physical and logical topology design in a large-scale OXC-based optical network. The RWA problem here was combined with the physical topology. Chu et al. [35] studied the wavelength-converter placement problem under different RWA algorithms and concluded that finally the two problems have to be considered jointly.

2.3.2 Wavelength Assignment

RWA [34] is essentially the core problem involving lots of variations in the design of a WDM network [33,36]. This problem has been the field of application. Ali et al. [37] tried to incorporate in the RWA problem, a GA in order to include realistic power consideration issues, arising from signal degradation by optical components. It was concluded that global search meta-heuristic approaches provide better solutions than previously used heuristic techniques. Generally, simple heuristics have limited applicability in RWA problems because they exhibit high time complexity and low scalability. To overcome this limitation, Talay and Oktug [38] employed a combined hybrid GA/heuristic approach to address the RWA problem in WDM networks, demonstrating the advantage of this technique compared to simple heuristics. Talay [39] presented a parallel implementation of the hybrid GA/heuristic technique, with improved results. Parallel GA techniques are in general a powerful tool, and recent advances in this area are expected to find solution to the complex problems.

2.3.3 Dynamic Routing and Wavelength Assignment

Dynamic RWA (DRWA) is becoming a major and challenging design issue, with an increased complexity when other issues like wavelength conversion are taken jointly into account [40]. Le et al. [41] proposed a GA for the DWRA problem with the wavelength continuity constraint by adopting an improved fitness function that took into account both the path length and the number of free wavelengths of the route. The GA exhibited better solutions than the initial GA. A novel and very interesting concept for the solution of DRWA problem in WDM networks with sparse wavelength conversion was proposed by Le et al. [42]. The available GA-based algorithms for DRWA are usually time consuming for the production of the first population of routes for a request, and this delay may result in the slow establishment of a path. This is a critical issue for dynamic RWA, and in order to overcome this problem, they introduced the cooperation of mobile agents into the network. By keeping a suitable number of mobile agents to cooperatively explore the network states and continuously update the routing tables, this combined algorithm can much faster determine the first population of routes for a new request, helping exactly the weak point of GA-based DRWA algorithm. This new hybrid GA/mobile agent algorithm performed better than the typical fixed-alternate routing algorithm, in terms of load balance and blocking probability. This combined technique suggested a novel proposal with possible new applications in problems like protection and restoration.

Hwang et al. [43] introduced a much different approach for the dynamic RWA problem in an IP/GMPLS over WDM network. They proposed a dynamic RWA scheme based on fuzzy logic control. The algorithm dynamically allocated network resources and reserved partial bandwidth based on current network information, like request bandwidth and average utilization of each channel. Durand [44] considered the impact of polarization mode dispersion—the effect due to different group velocities of an optical signal's two orthogonal polarization parts in a network with hybrid multiplexing technique, WDM, and optical code division multiplexing (OCDM). In this technology, virtual paths were established in both the wavelength domain (virtual wavelength path (VWP)) and the code domain (virtual code path (VCP)). For the establishment of VCP/VWP, a routing channel assignment method based on GA was used. The almost arbitrary technological complexity that can be formed in an optical network suggested the potential applicability of flexible CI optimization in such problems. The continuous expansion of multimedia applications, multimedia conferencing, distance education, video distribution, etc., over the optical backbone has led to the development of multicasting WDM systems.

Many of these multimedia applications require packets of information to be sent with a certain QoS [45], and thus, multicast (MC)-RWA can have additional constraints such as bounded end-to-end delay. The MC-RWA or even the multimedia multicasting routing in non-optical networks is an NP-complete problem [45,46]. Din [47] considered the problem on optimal multiple multicast (OMMP) in WDM ring networks without wavelength conversion capabilities. Being an NP-hard problem, Din proposed a number of different GAs and proved their applicability and robustness by experimental results. The QoS requirement introduces additional constraints in the multicast routing problem, such as the end-to-end delay. Employment of a GA in the problem of multicast routing under delay constraint proves to obtain a near-optimal solution [48]. Siregar [49] considered the multicast problem in large-scale WDM networks. For a given number of multicast requests and available number of wavelengths, the authors minimized the number of split-capable nodes. A genetic algorithm was employed in order to search for alternative shortest paths and minimize consequently the number of splitting capable nodes. The GA is applied in realistic network cases and succeeded 10% reduction of splitting nodes. A concept relevant to multicast in WDM networks is the Anycast routing with the associated Anycast RWA (ARWA) problem. Anycast implies the transmission of data from a source node to any edge node that serves a designated recipient in the network. Din [50] proposed the use of genetic algorithm to solve the NP-hard problem of ARWA, in a network under the wavelength continuity constraint. The goal was to minimize the required number of wavelengths. The performance of genetic algorithm was compared to that of two simplified heuristics based on graph-coloring method. The result suggested that the solution obtained by GA was superior to that obtained by the heuristics.

2.3.4 Protection and Restoration

Very critical issues for the implementation of reliable networks that serve real-world applications are survivability [51,52] or recovery after a link failure, in a reasonable amount of time without loss of information for the subscribers. Optical networks operate at 10 GB/s or at 40 GB/s at a single wavelength channel or light path considering that each fiber can carry several such multiplexed wavelengths. It is obvious that the impact of a single-fiber cut will affect the applications. This is also what creates the problem of IP survivability of IP over WDM given that a single WDM failure can affect massively the IP layer [53]. It is of critical importance to design WDM networks taking into consideration their survivability by a single-link failure. Survivability is implemented by two main ways—protection and restoration. In protection, there is a reserved backup route—light path—in order to redirect the traffic of a failed link. If this route is dedicated to a specific light path, the protection is called dedicated ("1 + 1"), where when a backup route is shared between many (N) primary routes is called shared ("1:N"). Restoration [54] on the other hand is the dynamic technique where after the failure, it uses spare capacity and recourses in order to accommodate the traffic by the failed link and restore the service. The objective in survivability design is to provide adequate protection at the minimum possible cost. Restoration is much slower procedure than protection. In WDM networks, there is the possibility that both the primary and the backup routes, as light paths, share the same fiber link and thus, in case of even a single cable failure, they will simultaneously fail. Modiano and Narula-Tam [55] considered this problem by introducing the concept of survivable RWA. The problem is NP-complete, and they considered it by an approximation of its integer linear program formulation. Depending on the recommended survivability solution, the design process considers either alternative physical routes providing a spare capacity or suitable logical topologies or a dynamic RWA solution. Following a different approach, the same researchers considered in [56] the survivability problem with a more robust solution by designing suitable physical topologies. The ILP problem was approached for the simplified topology of ring WDM networks. Xin and Rouskas [57] studied the problem of path protection in large-scale (up to 1000 nodes) optical networks based on OXCs. To design the path protection scheme, they considered both the physical topology design based on their previous work [27] and the survivable RWA problem. A genetic algorithm was used in order to search the space of two connected physical topologies that are formed with OXCs. Some authors [58] used genetic algorithms in order to design traffic restoration technique in a WDM network, by allocating spare wavelengths for the establishments of backup light paths. A generalized form of the previous work was considered in where authors optimized the restoration solution, by finding the minimum required spare capacity for a survivable WDM mesh network [59]. The consideration of protection problem is expected to dynamically emerge in multicast WDM networks [62]. Other authors considered this variant of protection problem for the first time, and a mathematic formulation was derived [112]. The inherent similarity of protection and routing inspired the application of neural networks to the restoration of optical networks. Gokisik et al. [60] employed a previously trained feed forward neural network, distributed over the optical communications network, which is used for the optimal RWA of the primary and backup capacity. In case of a network failure, the RWA is employed to redirect the traffic and restore the network. Goncalves et al. [61] applied the concept of ANN for fault prediction in an IP/DWDM network, by describing an environment (RENATA2) where ANN was able to generate proactive intelligent agents, which predicted failure trends in the network. By the early prediction of failure, the ANN could then approximate a variety of protection schemes, from dedicated $(1+1)$ to shared (1:N). The concept of survivability can have a slightly different meaning implying the robustness of the networks to future traffic and service demands, a very important issue for telecommunications industry and operators. For a network to be cost-effective and upgradeable, special care should be taken in the design process so to be able to expand its capacity when the demand is in a mature state and can use it effectively. Pickavet and Demesteer [89] considered a multi-period planning of a WDM network. They employed a genetic algorithm that allowed each time period of a given demand, to generate alternative network topologies, and selected from them (based on the shortest-path technique) the one that gave the least-cost network expansion plan for multi-time periods.

2.3.5 Optimization of Quality of Service Routing

QoS routing is of crucial importance in guaranteed type of service applications over optical networks [63]. Being an NP-hard problem, it introduces new optimization challenges for CI and heuristic methods. The objective of QoS routing is to identify network paths that meet the QoS requirements and to select the path that leads to high overall recourse efficiency for the network.

Recently, considerable research work has emerged in QoS routing in WDM networks. The problem is essentially the optimal selection of routing paths and wavelength assignment in order to meet specific QoS requirements. Jukan and Franzl [64] investigated quite complex realistic cases where also physical impairments, traffic, reliability, and policy issues were considered. Wang and colleagues in [65] developed an optimization strategy of multi-constrained QoS routing in optical networks, using genetic algorithms.

They employed an improved version of a genetic algorithm that used the optimal maintain operator (OMO) and demonstrated its efficiency and fast convergence compared to other GAs. Jia et al. [66] investigated the emerging problem of QoS multicast, in WDM networks. Given a set of QoS multicast, the target was to find a set of cost-optimal set of routing trees and assign wavelengths to them, with the objective to minimize the number of required wavelengths. Ravikumar [67] employed a genetic algorithm to consider the problem of routing in multicast WDM networks with the specific constraint of delay-bounded multicasting. The suitability of the genetic algorithm to the problem was validated by benchmarking case studies where different types of mesh physical topologies were considered, while the operation of the networks was simulated by broadcasting video of different information content and from various nodes of the mesh network. Another QoS multicast approach based on genetic algorithm was proposed for computer-supported cooperative work (CSCW) applications in IP over DWDM optical Internet [68]. The constraint faced in this investigation was a time-bound delay that was required by the CSCW group users. The GA solution was able to efficiently locate multicast routing trees and assignment wavelengths for the establishment of QoS multicast light trees. In multiservice networks, complex design problems arise due to multiple, conflicting QoS requirements. A genetic algorithm–based strategy was developed in order to optimize multiservice convergence in a metro WDM network [69]. The optimization strategy provides the physical topology of the network and the operation solution by the optimal parameterization of the medium access control (MAC) protocol. However, optical networks are continuously emerging new issues. When optical technology reaches a mature state and components like optical buffers or all-optical packet header recognition chips become cost-effective solutions, then OPS will become a reality. New design and optimization issues are required for packet-switched networks, employing heuristics and genetic algorithms [18,70]. The complexity of large communication networks like the Internet requires highly efficient and adaptive routing algorithms. Recent studies are examining novel methods for the consideration of adaptive routing, such as evolutionary game theory (EGT) [71]. Optical networks are also evolved toward the direction of all-optical networks and consequently to the realization of all-optical Internet. This could help in improving our understanding about future global all-optical networks.

2.4 Nanowire Photonics

Nanowire photonic circuitry is generally built from 1D building blocks. It offers opportunities for the development of next-generation optical information processors. The synthesis of 1D, single-crystalline semiconductor nanowire materials has a rich history, dating back to the work of Wagner and Ellis [72] at Bell Labs in the early 1960s with the vapor–liquid–solid (VLS) growth mechanism. Improvements in scanning and transmission electron microscopy (TEM) [73,74] have played a crucial role in the characterization of these materials particularly in guiding the rational growth of nanowires in this direction of materials research. Advances in organometallic vapor deposition [75] and other chemical [76,77] techniques have allowed the development of a vast array of inorganic nanowire compositions, including groups IV8, II–VI, and III–V compound and alloy crystal structures [78]. Laboratory-scale reactions typically take place in a horizontal or vertical tube

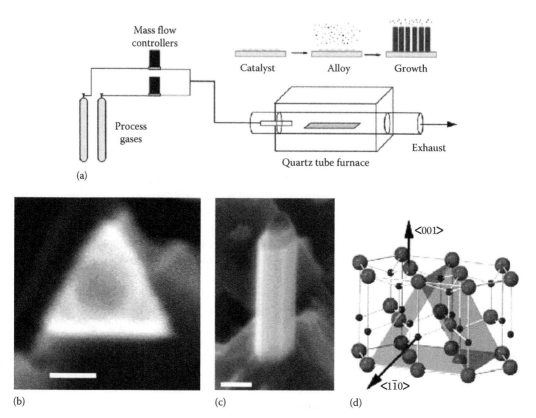

FIGURE 2.1

(a) Schematic of a horizontal hot-wall flow reactor used in the synthesis of various nanowire materials. Metallic clusters melt in the furnace, become saturated with process gases, and continuously precipitate single-crystalline nanowires. (b) Top-view scanning electron microscope (SEM) image of a GaN nanowire with triangular cross section growing in the [110] direction. The circular structure in the middle of the triangle is a Au catalyst droplet. Scale bar = 50 nm. (c) Side-view SEM image of a GaN nanowire growing in the [001] direction. Scale bar = 100 nm. (d) Schematic of GaN and ZnO's hexagonal wurtzite crystal structure. Arrows indicate observed growth directions. (Reprinted with Permission from Elsevier, 2006.)

furnace (Figure 2.1). Process gases are generally introduced and regulated by way of mass flow controllers, while metals such as Ga may be introduced either by organometallic precursors or by placing a metal pellet within the reactor.

The introduction of various process gases causes the saturation of the molten metal droplet, leading to the formation of continuous precipitation of single-crystalline nanowires (Figure 2.1a). The diameter of the nanowire is generally determined by the size of the alloy droplet, which is determined by the original size of the metallic cluster. By using monodispersed metal nanocrystals, a narrow diameter distribution of nanowires can be achieved [79]. The application of conventional epitaxial crystal growth techniques to the VLS process is possible to gain precise orientation control during nanowire growth. VLS epitaxy (VLSE) [80] is particularly useful for designing controlled high-quality nanowire arrays and single-wire devices [81].

The direct bandgap II–VI and III–V systems are of particular interest because of their high optoelectronic efficiencies relative to indirect bandgap group IV crystals. ZnO is a typical II–VI nanolaser material [82,83] possessing a wide bandgap of 3.37 eV and binding energy of 60 meV [82]. It is possible to fabricate a self-contained resonant nanowire cavity that achieves gain and ultraviolet (UV) lasing through an exciton–exciton collision mechanism at room temperature [83]. Simple chemical techniques such as thermal reduction [84] are also used to produce wide-coverage substrate of nanowires from gas-phase deposition of Zn and O precursors. The development of solution-phase chemistry has made it possible to coat large surface areas of arbitrary material with ZnO nanowires for light-emission applications [85]. Wurtzite crystal symmetry of ZnO makes it an interesting material for nonlinear second harmonic generation and wave mixing in nanoscale cavities [86]. An enormous body of literature has been produced for GaN because of its high Young's modulus, thermal conductivity, electron mobility, high melting point, and low chemical reactivity [87,88]. There are a number of different synthetic methods available for the growth of GaN nanowires, including pulsed laser ablation [78], metal-organic chemical vapor deposition [88,89] hydride vapor-phase epitaxy, molecular beam epitaxy, and conventional chemical vapor-phase transport [90]. The group III nitrides offer an additional advantage in that the bandgap is tunable [91] from the near infrared (InN) to the near UV. This suggests an opportunity for solid-state, white light-emitting diodes with low power and high efficiency [92], robust on-chip UV photo detectors, and variable-wavelength solid-state lasers for remote national security or biologically related sensing. One can image nanowires with visible-light microscopy because of their large dielectric constants. Even though the two dimensions of the wire's cross section are often well below the diffraction limit of light, Rayleigh scattering from the wires is significant, scaling as 1/"4, where" is the free-space wavelength of light. Dark-field microscopy is also an essential tool for initial assessments of wire growth, as well as in the physical manipulation of individual nanowires (Figure 2.2). Glancing angle excitation is used to excite photoluminescence, which is typically collected through a microscope objective.

The highly integrated light-based devices are used to assemble photonic circuits from a collection of nanowire elements that occupy different functions, such as light creation, routing, and detection. Si has been highlighted in recent reports of optical parametric gain [93], electro-optic (EO) modulators [94,95], and Raman lasing [96,97]. The infrared

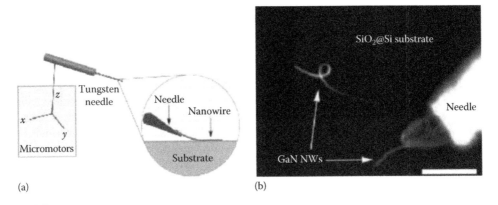

(a)   (b)

FIGURE 2.2

(a) Side-view schematic of nanowire manipulation instrument. A finely etched tungsten needle is used to pick up and transport single nanowires through mechanical and surface-adhesion forces. (b) Top-view dark-field optical microscope image of nanowires on a thermal SiO_2 surface. Adhesion to the surface allows the wire to maintain its bent conformation. Scale bar = 100 µm. (Reprinted with Permission from Elsevier, 2006.)

bandgap of Si makes it difficult to produce efficient photonic components in the visible and UV spectral regions. Chemically synthesized nanowires offer advantageous features that make them versatile in photonic building blocks exhibiting various optical and electrical properties, good size and composition control, low surface roughness, and, in principle, ability to operate both above and below the diffraction limit. The nanowire device elements include various types of transistors [98–100], light-emitting diodes [101], and lasers and photo detectors [102]. The diversity of freestanding crystalline nanowires to build prototype multiwire designs has been used for the manipulation and detection of light. Nanowires of binary oxides are usually employed because of their extreme mechanical flexibility and chemical stability. SnO_2 has recently been shown to act as an excellent sub-wavelength waveguide of both its own visible photoluminescence and that from other nanowires and fluorophores [103]. The waveguiding in these nanowires mimics conventional SiO_2 optical fiber.

Nonresonant waveguiding that is sub-bandgap light in these structures is achieved by focusing laser diodes on the end facet of the nanowire. The wires possess fairly uniform (±10%) rectangular cross sections with side dimensions as large as 2 μm by 1 μm and as small as 15 nm by 5 nm [104]. Most wires (>80%) have dimensions between 100 and 400 nm, an optimal size range to efficiently guide visible and UV wavelengths because of the high index of refraction of SnO_2 (n > 2). To synthesize these wires in milligram quantities with lengths greater than 3 mm, having precise control over cross-sectional size is still a great challenge. Optical linkages between active nanowires (GaN and ZnO) and passive nanowires (SnO_2) can be formed via tangential evanescent coupling. A staggered side-by-side configuration, in which the active and passive elements interact over a few microns, outperforms bridged or direct end-to-end coupling. Weaker coupling is achieved by staggering structures with a thin air gap (several hundred nanometers) between them, allowing communication via tunneling of evanescent waves [105]. It is possible to create more branched optical hubs and Mach–Zehnder interferometers (optical modulators) that use the EO effect for phase shifting using integration. To create effectively transducing and routing packets of optical information within an optical computer or communication device, the integration of high-frequency electrically driven lasers is required.

Many challenges remain to be overcome in order to create devices made from various materials. This synthetic limitation is especially true for nanowires because of issues with impurities, liquid-catalyst compatibility, and thermal decomposition. Langmuir–Blodgett assembly has been proved to be one of the powerful methods to organize nanowires over substrates with arbitrary composition [106–109]. But this technique is still unable to take into account single nanostructure and then position them with arbitrary precision. One possible way to overcome this limitation especially in case of nonmetallic materials is to manipulate nanowires with highly focused laser beams known as "optical traps." Optical traps act in situ in closed aqueous chambers and have potential applicability to a broad range of dielectric materials, spatial positioning accuracy (<1 nm); control of degree of intensity, wavelength, and polarization via tunable lasers; acousto-optic modulators; and holographic optical elements [110–112]. Optical confinement of metal nanocrystals in two and three dimensions was demonstrated in the mid-1990s, and single-beam optical traps have been used for almost two decades to manipulate and interrogate micro- and nanometer-sized objects [113–115]. Lately, birefringent crystals have been rotated in an optical trap by angular momentum transfer, and CuO nanorods have been manipulated in two dimensions with optical trap [116,117].

The work in infrared laser has shown that it is possible to trap semiconductor nanowires optically at room temperature, at both physiological pH and ionic strength [118]. Infrared wavelengths were selected in order to minimize heating and radiation damage to biomolecules and cells, and optically trapped nanowires can be positioned with respect to many other structures, such as living cells. For example, HeLa tissue culture cells were grown on lysine-coated quartz coverslips, and chambers were assembled with nanowire solutions at physiological ionic strength and pH and were trapped in GaN nanowire across the cell membrane, by placing one end of the nanowire against the cell membrane and maintaining the wire's position for different durations. The small cross section and very high aspect ratio of nanowires is used to deliver extremely localized chemical, mechanical, electrical, or optical stimuli to cells, based on the construction of integrated assemblies. In addition to the hetero structures that can be constructed from nanowires, optical trapping facilitates novel experiments for the in situ characterization of biological materials.

Single-crystalline, 1D structures are intriguing materials for photonic applications. Chemically synthesized nanowires and other geometric shapes offer a platform for producing photonic elements, including lasers, detectors, and passive waveguides. These are then integrated with existing photonic and sensing technologies to realize their full potential in optoelectronic devices. Since the range of material includes nonlinear, semiconducting inorganic crystals, as well as a rich variety of polymers, there exists a unique capability of designing photonic circuits from the bottom up. In conjunction with high-density integration and large refractive indexes, inorganic semiconductor nanomaterials also provide properties that can enhance the study of biological and pharmaceutical interactions within native fluid environments. Moreover, chemically synthesized nanowires and nano-ribbons are freestanding, mechanically flexible entities that can be integrated with micro-electromechanical systems to enable the investigation of physical and biological sciences.

2.5 Higher-Order Time-Domain Methods for the Analysis of Nanophotonic Systems

Time-domain computations play a very prominent role in the design of micro- and nanophotonic structures. Strong multiple scattering and/or near-field effects allow a far-reaching control over the propagation characteristics of light and its interaction with matter. Important systems that are currently being investigated include periodic nano-structures and plasmonic elements [119–121]. Among the various methods for time-domain computations of Maxwell equations, the finite-difference time-domain (FDTD) method [122] stands out for its efficiency and versatility. In fact, the basic FDTD method represents a conditionally stable algorithm with second-order accuracy in space and time. Within this approach, the Maxwell curl equations are discretized such that electric and magnetic fields are, respectively, evaluated on a uniform and staggered spatial grid, the so-called Yee grid. The time stepping is explained via a leap frog scheme. The precise treatment of material interfaces that are not grid aligned is a serious concern for linear optics; advanced methods have been developed very recently [123–126]. However, these methods either reduce the efficiency of the basic algorithm [123,124] or cannot be extended three dimensionally [125,126]. For the linear optics of curved interfaces between metals and dielectrics, notable problems, associated with spurious field enhancements, are still a major constraint. The accurate representation of arbitrary interfaces and the high-order spatial discretization are the trademarks of finite-element methods [127]. Despite these apparent advantages, there exists one essential drawback that has prevented traditional finite-element methods from becoming mainstream in large-scale time-domain computations: The resulting time-stepping algorithms are implicit such that for every time step, typically large system of equations needs to be solved. Dramatic progress has been made after the application of discontinuous Galerkin (DG) finite-element techniques to Maxwell equations [128]. The approach combines the attractive features of finite elements with explicit time-stepping capabilities. The treatment of arbitrary interfaces in combination with high-order discretizations in both space and time leads to performance characteristics that are ideally suited for applications in nanophotonic systems. The essentials of accurate treatment of point sources within DG time-domain (DGTD) approach have been described [129]. The DGTD implementation to the important problem of extraordinary transmission through sub-wavelength apertures in nanostructured metallic films has also been described.

A DG method in the nodal formulation of Hesthaven and Warburton is applied [128]. Maxwell equations can be expressed as

$$\delta_t(Q(\vec{r})q(\vec{r},t)) + \nabla \cdot F(\vec{q}) = 0 \qquad (2.1)$$

where the matrix $Q(\vec{r})$, the filed vector q, and the flux vector $F(\vec{q}) = (F1(q) + F2(q) + F3(q))^T$ are defined as

$$Q(\vec{r}) = \begin{pmatrix} \varepsilon(\vec{r}) & 0 \\ 0 & \mu(\vec{r}) \end{pmatrix}, \quad q(\vec{r},t) = \begin{pmatrix} E(\vec{r},t) \\ H(\vec{r},t) \end{pmatrix}, \quad \text{and}$$

$$F_i(q) = \begin{pmatrix} -\hat{e}i & x & H(\vec{r},t) \\ -\hat{e}i & x & E(\vec{r},t) \end{pmatrix}$$

$\hat{e}i$, $i = 1, 2, 3$ denote the Cartesian unit vectors. In order to solve this system, the computational domain is tessellated into K-conforming elements Ω^k. Typically, these elements are triangles in two dimensions and tetrahedra in three dimensions. On each element, the fields are then expanded in terms of interpolating Lagrange polynomials $Li(\vec{r})$:

$$q^k(\vec{r},t) \approx \sum_{i-1}^{N} q^k(\vec{r},t)Li(\vec{r}) = \sum_{i-1}^{N} q_i^{-k}(t)Li(\vec{r}) \qquad (2.2)$$

where N denotes the number of coefficients that have been utilized. For triangles, this number is connected to the polynomial expansion order n via $N_{\text{Tri}} = (n+1)(n+2)/2$, while for tetrahedra, the corresponding dependence is $N_{\text{Tet}} = (n+1)(n+2)(n+3)/6$. The vector $q_i^{-k}(t)$ contains the unknown field values that have to be solved for. A suitable set of interpolation nodes ri can be obtained via a proposed method [130]. The standard Galerkin approach consists in multiplying Equation 2.1 with $Li(\vec{r})$ and integrating over an element Ω^k, which yields

$$\int_{\Omega^k} \left(Q^k dt q^{-k} + \nabla \cdot \vec{F}\left(q^{-k}\right) \right) Li(\vec{r}) d\vec{r} = 0$$

To facilitate the coupling with neighboring cells, the next step is to employ integration by parts and to substitute in the resulting contour integral the physical flux $\vec{F}\left(q^{-k}(t)\right)$ with a so-called numerical flux $\vec{F}^*\left(q^{-k}(t)\right)$. A second integration by parts then results in the strong formulation

$$\int_{\Omega^k} \left(Q^k dt q^{-k} + \nabla \cdot \vec{F}\left(q^{-k}\right) \right) Li(\vec{r}) d\vec{r} = \int_{d\Omega^k} \hat{n} \left(\vec{F}\left(q^{-k}\right) - \vec{F}^*\left(q^{-k}\right) \right) Li(\vec{r}) d\vec{r} \tag{2.3}$$

where $\hat{n}$ is the outward-pointing normal vector of the contour. It has been proved that this procedure results in a stable and convergent scheme if the numerical flux $\vec{F}^*\left(q^{-k}(t)\right)$ is chosen properly. One suitable choice is the well-established up-winding flux [128] that leads to the expression

$$\hat{n}\left(\vec{F}\left(q^{-k}\right) - \left(\vec{F}^*\left(q^{-k}\right)\right)\right) = \begin{pmatrix} \dfrac{\Delta\vec{E} - \hat{n}\left(\hat{n}\Delta\vec{E}\right) + Z^+ \hat{n}\times\Delta\vec{H}}{Z} \\[2ex] \dfrac{\Delta\vec{H} - \hat{n}\left(\hat{n}\Delta\vec{H}\right) - Y^+ \hat{n}\times\Delta\vec{E}}{Y^+} \end{pmatrix} \tag{2.4}$$

Here, $\Delta\vec{E} = \Delta\vec{E}^+ - \Delta\vec{E}^-$ and $\Delta\vec{H} = \Delta\vec{H}^+ - \Delta\vec{H}^-$ denote the difference of the fields across the cell interface, and the superscript "+" denotes the neighboring element while the superscript "−" denotes local cell. The cell impedances $Z^\pm = \sqrt{\mu^\pm/\varepsilon^\pm}$ and conductances $Y^\pm = 1/Z^\pm$ together with corresponding summed values $Y = Y^+ + Y^-$ and $Z = Z^+ + Z^-$. Several types of relevant boundary conditions can be applied by setting the material parameters at the outer cell to $Z^+ = Z^-$. Explicit expressions for the fields are obtained by inserting the expansions (5.2) together with the numerical fluxes from Equations 2.4 to 2.3. If we assume constant material parameters ε^k, μ^k within each element, we find after certain algebraic transformations

$$\varepsilon^k \frac{d\vec{E}^k}{dt} = \vec{D}^k \times \vec{H}^k + \left(M^k\right)^{-1} F^k \left(\frac{\Delta\vec{E} - \hat{n}\cdot\left(\hat{n}\cdot\Delta\vec{E}\right) + Z^+ \hat{n}\times\Delta\vec{H}}{Z} \right) \tag{2.5}$$

$$\mu^k \frac{d\vec{H}^k}{dt} = -\vec{D}^k \times \vec{E}^k + \left(M^k\right)^{-1} F^k \left(\frac{\Delta\vec{H} - \hat{n}\cdot\left(\hat{n}\cdot\Delta\vec{H}\right) + Y^+ \hat{n}\times\Delta\vec{E}}{Y} \right) \tag{2.6}$$

The vectors $\vec{E}^k$ and $\vec{H}^k$ are introduced, where each component is an N vector of the respective field values in element k. A vector of differentiation matrices $\vec{D}^k = \left(\vec{D}_x^k, \vec{D}_y^k, \vec{D}_z^k\right)$, the mass matrix M^k, and the face matrix F^k according to

$$\left(D_m^k\right)_{ij} = d_m L_{j(\vec{r}_i)}$$

with

$$m \in \{x, y, z\}$$

$$\left(M^k\right)_{ij} = \int_{\Omega^k} L_i(\vec{r}) L_j(\vec{r})\, d\vec{r},$$

$$\left(F^k\right)_{ij} = \int_{d\Omega^k} L_i(\vec{r}) L_j(\vec{r})\, d\vec{r}$$

where

$$j \in \left\{ j \,\middle|\, \vec{r}_j \in d\Omega^k \right\}$$

The remaining step of integrating the semi-discrete system (2.5) and (2.6) in time is executed through a fourth-order low-storage Runge–Kutta scheme [131]. A Courant–Friedrichs–Levy (CFL) criterion has to be fulfilled in order to guarantee stability. In practice, we limit the time step according to

$$\Delta t \leq c_{\mathrm{CFL}} d_{\min} \frac{\min}{k}\left(r_{in}^k\right) \tag{2.7}$$

where $d_{\min}$ is the minimal distance between two interpolation nodes $\vec{r}_i$, r_{in}^k denotes the radius of the in circle or in sphere of element k, and c_{CFL} is a number of the order one. Its critical value can only be obtained empirically and depends on the dimensionality of the system as well as on the expansion order. However, the value $c_{\mathrm{CFL}} = 1$ leads to stable results. The reduced performance due to the CFL criterion is improved upon by using more advanced integration methods [132–134].

2.6 Implementation of Perfectly Matched Layers (PMLs), Point Sources, and Dispersive Media

For modeling of nanophotonic systems, extensions to the bare numerics are done. The first is of absorbing boundaries by applying Silver–Muller boundary conditions. But they are insufficient for the accurate modeling of open systems and are complemented by PMLs [122,135]). Lu et al. [135] demonstrated the implementation of uniaxial PMLs in two dimensions. To extend to fully 3D systems, general uniaxial formulation is applied [122,136], where a PML region is described as a dispersive, anisotropic material with susceptibility tensor $\underline{\underline{\epsilon}} = \epsilon \underline{\underline{\wedge}}$ and permeability tensor $\underline{\underline{\mu}} = \mu \underline{\underline{\wedge}}$. Here, the diagonal tensor $\underline{\underline{\triangle}}$ is explicitly given by

$$\underline{\underline{\triangle}} = \begin{pmatrix} \dfrac{s_y s_z}{s_x} & 0 & 0 \\ 0 & \dfrac{s_x s_z}{s_y} & 0 \\ 0 & 0 & \dfrac{s_x s_y}{s_z} \end{pmatrix} \text{ with}$$

(2.8)

$$S_m = 1 + \frac{\sigma_m}{i\omega}, \quad m \in \{x, y, z\}$$

Transforming Maxwell equations to the frequency domain and inserting the anisotropic material parameters, we get

$$i\omega\varepsilon \underline{\underline{\triangle}} \overset{\cup}{\vec{E}} = \nabla \times \overset{\cup}{\vec{H}}$$

(2.9)

$$i\omega\mu \underline{\underline{\triangle}} \overset{\cup}{\vec{H}} = \nabla \times \overset{\cup}{\vec{E}}$$

(2.10)

where "$\cup$" denotes the variable in frequency domain. We limit the discussion to the first component of $\vec{E}$:

$$i\omega\varepsilon \overset{\cup}{E}_x = -\frac{d}{dy}\overset{\cup}{H}_z + \frac{d}{dz}\overset{\cup}{H}_y - \underbrace{i\omega\varepsilon\left(\frac{s_y s_z}{s_x} - 1\right)\overset{\cup}{E}_x}_{x}$$

(2.11)

$$=: \overset{\cup}{J}$$

Polarization current $\overset{\cup}{j}_x$ can be simplified to

$$\overset{\cup}{J}_x = \frac{i\omega\varepsilon}{i\omega + \sigma_x}\left(\sigma_y + \sigma_z - \sigma_x + \frac{\sigma_y \sigma_z}{i\omega}\right)\overset{\cup}{E}_x$$

(2.12)

By introducing the new variable $\overset{\cup}{P}_x = \overset{\cup}{J}_x - \varepsilon(\sigma_y + \sigma_z - \sigma_x)\overset{\cup}{E}_x$, we get

$$i\omega \overset{\cup}{P}_x = -\sigma_x \overset{\cup}{P}_x + \varepsilon(\sigma_x^2 + \sigma_y \sigma_z - \sigma_y \sigma_x - \sigma_z \sigma_x)\overset{\cup}{E}_x$$

(2.13)

that can be transformed to the time domain. The resulting equations are obtained as

$$\frac{d}{dt}E_x = -\frac{d}{dy}H_z + \frac{d}{dz}H_y - \varepsilon(\sigma_y + \sigma_z - \sigma_x)E_x - P_x$$

(2.14)

$$\frac{d}{dt}P_x = -\sigma_x P_x + \varepsilon(\sigma_x^2 + \sigma_y \sigma_z - \sigma_y \sigma_x - \sigma_z \sigma_x)E_x$$

(2.15)

The auxiliary differential equation (ADE) for the other field components is analogous. The widths of the layers and their strengths σ_m are free parameters that are optimized for the best performance. There are two different ways to inject radiation into the system. The more obvious path is to add current density terms to Equations 2.5 and 2.6. An alternative is given by the so-called total-field/scattered-field (TF/SF) approach [122]. Typically, the current density is more suitable to introduce localized sources, while the TF/SF method can be employed to inject plane waves. One can easily implement the TF/SF by modifying the field differences in the numerical fluxes. The addition of current terms also presents no fundamental problem. The spatial profile is expanded into Lagrange polynomials. For delta-like point source, this expansion becomes rather intricate, and a refined mesh around the source is required to model it accurately. For the injection, a small contour around the desired source location is embedded into the mesh, and the outer area is defined as the total-field (TF) region. The required fields on the contour can be obtained by means of Green's functions [137]. For the 2D case in transverse magnetic (TM)-polarization using polar coordinates, the fields generated by a point source at the origin can be expressed as

$$E_z(\rho, \phi, t) = \frac{1}{2\pi} \int_{\rho}^{t} d\tau \, \frac{j'(t-\tau)}{\sqrt{\tau^2 - \rho^2}}$$

$$H_x(\rho, \phi, t) = \frac{\cos(\phi)}{2\pi\eta\rho} \int_{\rho}^{t} d\tau \, \frac{\tau j'(t-\tau)}{\sqrt{\tau^2 - \rho^2}} \tag{2.16}$$

$$H_y(\rho, \phi, t) = \frac{\sin(\phi)}{2\pi\eta\rho} \int_{\rho}^{t} d\tau \, \frac{\tau j'(t-\tau)}{\sqrt{\tau^2 - \rho^2}}$$

where $j(t)$ is the time dependence of the source and $j'(t)$ denotes its derivative. The aforementioned integrals exhibit a singularity at the left limit $\tau = \rho$. However, upon application of quadrature rule [138], we can get values at arbitrary locations with sufficiently high accuracy. For the modeling of silver, which in the visible frequency range as described by the Drude model [139], the frequency-dependent permittivity is given by

$$\varepsilon(\omega) = 1 - \frac{\omega_{pl}}{\omega(\omega + i\omega_{col})} \tag{2.17}$$

where ω_{pl} and ω_{col} denote the plasma and collision frequencies, respectively. The derivation of the ADEs for the Drude model leads to additional equations, one per spatial dimension that complements Maxwell equations:

$$\frac{d}{dt} \vec{j}_{\text{Drude}} = \omega_{pl}^2 \vec{E} - \omega_{col} \vec{j}_{\text{Drude}} \tag{2.18}$$

As long as the material parameters are constant for each element Ω^k, the inclusion of the corresponding auxiliary equations is straightforward.

2.7 Comparison of PML with FDTD

PMLs contain a free parameter s that needs to be determined numerically for the best performance. In the case of FDTD, extensive numerical studies [122] have shown that optimal performance is obtained when making s-position dependent. A typical choice is the polynomial grading

$$\sigma(x) = \left(\frac{x}{d}\right)^m \sigma_{max} \tag{2.19}$$

where d is the thickness of the PML layer while m and σ_{max} are free parameters. We can compare PMLs with FDTD, using the same polynomial grading with free parameters m and R. A point source is placed in the center of the system and is injected via a TF/SF contour. Adding a second layer improves the performance by roughly one order of magnitude and makes the system less sensitive to the parameters m, and restudying a multitude of different systems using different numbers of elements as well as different polynomial orders, we conclude that a minimum of $m=0$ is a universal feature. A flat absorption profile yields similar or better results than the polynomial grading usually employed in FDTD. Further, for most applications, it is sufficient to only use a single cell of PMLs around the system, and for optimal performance, one should choose R = 10. Most of these conclusions also hold for 3D systems.

Galerkin method can be employed to study the optical response of metallic nanostructures. Through the development of important add-ons such as optimized PMLs and point sources, the method has the same versatility as standard methods such as FDTD. Most importantly, through its ability to work on conforming meshes and higher-order spatial accuracy, the method exhibits distinct advantages over FDTD.

2.8 Modeling of Photonic Bandgap Structures

1D PCs are materials with a periodic modulation of the dielectric constant with a lattice constant comparable to the wavelength of light [140,141]. 1D PCs have attracted much attention from both fundamental point and practical point of views, and novel concepts such as photonic bandgap have been predicted and various new and interesting applications of 1D PCs have been proposed [142–144]. Such materials show a forbidden spectrum of photons where the electromagnetic waves cannot propagate for both polarizations through the structure. These structures allow manipulating the flow of light on the scale of wavelength of light. The most important feature of optics at the photonic band edges that makes these frequency regimes particularly interesting is rapid transition from low transmission to high transmission zone (i.e., the steep slope of the transmission curve at the band edges). The application possibilities of the rapid transition near the band edge is decreasing the group velocity of the light near the band edge to less than many orders of magnitude of the velocity of light in vacuum. This property is applicable in short-pulse lasers compression, all-optical buffering, switching components, and dispersion compensation for high bit rate (40 Gb/s) to prevent reshaping of the optical signals in all-optical communication systems [145–149]. By using the localization of light, we can increase the interaction of light with matter that causes many desirable phenomena such as nonlinear optics [150,151] or magneto-optics (Kerr and Faraday effects) phenomena [152]. By adjusting the bandgap width, the slow light can be generated in any desired magnitude and range of wavelength (flexible optical buffering), or omnidirectional reflection (ODR) components [153–155] can be designed in any angular range (as a flexible demultiplexer component) per requirements. All of the potential applications of 1D PCs can be controlled and adjusted via controlling the optical contrast ratio (the ratio of high refractive index to low refractive index of the medium), and consequently controllable PBG width can be achieved. But the choice of materials with refractive indexes intermediate to those of the highest and the lowest values is impossible in any spectral region. Also, selection of materials is further limited by a number of other practical considerations.

2.8.1 Band Structure Properties of 1D PCs

In Figure 2.3, n_H and n_L correspond to the high refractive index and the low refractive index of alternative materials, the physical thicknesses of each layers are denoted by dH and dL, and the optical thickness by $n_H d_H$ and $n_L d_L$, respectively. The period of the 1D PCs is shown by $\wedge$. The refractive-index distribution satisfies the following relationship:

$$n(z)=n(z+\Lambda) \tag{2.20}$$

The plane light wave propagation through the dielectric multilayer of 1D PCs along z-axis can be described by using the admittance matrix method. The propagation through a single low-index or high-index thin film can be expressed by a 2×2 matrix of M_L or M_H, respectively:

$$M = (M_L M_H)^q = \begin{pmatrix} a_{11} & a_{12} \\ a_{21} & a_{22} \end{pmatrix}^q \tag{2.21}$$

where q is the repetition period number. M_L and M_H are

$$
M_L = \begin{pmatrix} \cos\delta_L & i\sin\dfrac{\delta_L}{n_L} \\ in_L \sin\delta_L & \cos\delta_L \end{pmatrix}
$$

$$
M_H = \begin{pmatrix} \cos\delta_H & i\sin\dfrac{\delta_H}{n_H} \\ in_H \sin\delta_H & \cos\delta_H \end{pmatrix}
\tag{2.22}
$$

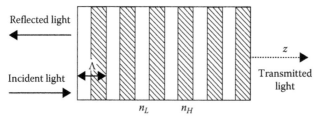

FIGURE 2.3
1D PC. (Reprinted with permission from Elsevier, 2008.)

where n_H and n_L are refractive indexes of high-refractive-index and low-refractive-index materials. δ_H and δ_L are phase thicknesses of high and low layers with $\delta_L = 2\pi n_L d_L / \lambda_0$ and $\delta_H = 2\pi n_H d_H / \lambda_0$, and λ_0 is the wavelength of incoming light. Introducing $X = (a_{11} + a_{22})/2$ into Equation 2.21, the characteristic matrix of the 1D PC can be expressed as

$$(M_L M_H)^q = S_{q-1}(X)(M_L M_H) - S_{q-2}(X) I \tag{2.23}$$

where I is the unit matrix and the Chebyshev polynomials $Sq(X)$ with order q are defined by

$$S_q(X) = \frac{\sin\left[(q+1)\cos^{-1} X\right]}{\sqrt{1 - X^2}} \tag{2.24}$$

1D PCs consist of high- and low-refractive-index layers with the same optical thickness $\lambda_0/4$, X as follows:

$$X = \cos^2 \beta - \frac{(n_H + n_L)^2}{2 n_H n_L} \sin^2 \beta \tag{2.25}$$

where $\beta = \dfrac{\pi}{2} \cdot \dfrac{\omega}{\omega_0}$ and ω_0 is the angular frequency corresponding to λ_0 in vacuum. Considering the Bloch theorem, wave propagating through periodic medium may exist only if it keeps the same amplitude but with a constant phase shift ϕ across each periodic unit, so the coefficient $e^{\pm i\phi}$ should be the eigenvalues of matrix $M_a M_b$; we get

$$X = \cos\phi = \cos(K\Lambda) \tag{2.26}$$

where K is the Bloch wave vector of the 1D PC along z direction, which is real for normal light while image for evanescent light. As a result, for $|X| > 1$, the reflectance increases steadily as q increases and tends to unity when q tends to infinity. This is the so-called PBG zone that light waves whose frequencies locating in it are forbidden to propagate in the crystal. The limit of the PBG zone is given by $|X| = 1$; the expression for the relative bandwidth $\Delta\omega/\omega_0$ of the PBG is shown in Figure 2.3 Schematic of a 1D PC can be derived as

$$\frac{\Delta\omega}{\omega_0} = \frac{4}{\pi} \sin^{-1} \frac{(x-1)}{(x+1)} \tag{2.27}$$

where $x = n_H/n_L$ is the optical contrast ratio. It is evident that when the optical contrast ratio is fixed, the width of the PBG zone remains a constant value.

Figure 2.4 shows the relative bandwidth of the PBG as a function of x. According to Equation 2.27, it is evident that by varying the optical contrast ratio x continuously, the relative width of PBG can be changed to any desired value less than its maximum value. It has been seen that a basic three-layer (or multilayer) symmetrical assembly can be represented mathematically by a single equivalent layer, which is also associated to an equivalent characteristic matrix $M_e(n_e, \delta_e)$ with a given equivalent phase thickness de and refractive index n_e as shown in Figure 2.5 [156].

The combination of the three symmetrical layers can be expressed as the form of (aL) (bH) (aL).

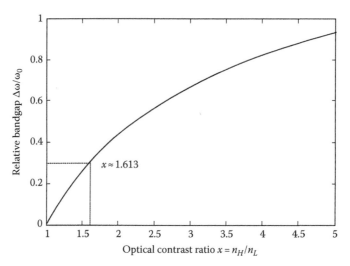

FIGURE 2.4
Relative PBG width in the 1D-PCs versus the optical contrast ratio. (Reprinted with permission from Elsevier, 2008.)

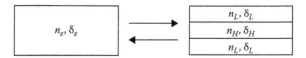

FIGURE 2.5
Three-layer symmetrical system and its equivalent single layer. (Reprinted with permission from Elsevier, 2008.)

According to Equations 2.21 and 2.22 for the simple three symmetrical layers (aL) (bH) (aL), which consist of dielectric materials free from absorption, the characteristic matrix of the combination can be written as

$$M = \begin{pmatrix} M_{11} & M_{12} \\ M_{21} & M_{22} \end{pmatrix}$$

$$= \begin{pmatrix} \cos\delta_L & i\sin\dfrac{\delta_L}{n_L} \\ in_L\sin\delta_L & \cos\delta_L \end{pmatrix} \cdot \begin{pmatrix} \cos\delta_H & i\sin\dfrac{\delta_H}{n_H} \\ in_H\sin\delta_H & \cos\delta_H \end{pmatrix} \cdot \begin{pmatrix} \cos\delta_L & i\sin\dfrac{\delta_L}{n_L} \\ in_L\sin\delta_L & \cos\delta_L \end{pmatrix} \quad (2.28)$$

where the elements of the matrix M can be derived as

$$M_{11} = \cos 2\delta_L \cos\delta_H - \frac{1}{2}\left(\frac{n_L}{n_H} + \frac{n_H}{n_L}\right)\sin 2\delta_L \sin\delta_H \quad (2.29)$$

$$M_{12} = \frac{i}{n_L}\left[\sin 2\delta_L \cos\delta_H + \frac{1}{2}\left(\frac{n_L}{n_H} + \frac{n_H}{n_L}\right)\cos 2\delta_L \sin\delta_H + \frac{1}{2}\left(\frac{n_L}{n_H} - \frac{n_H}{n_L}\right)\sin\delta_H\right] \quad (2.30)$$

$$M_{21} = inL \left[\sin 2\delta_L \cos \delta_H + \frac{1}{2} \left(\frac{n_L}{n_H} + \frac{n_H}{n_L} \right) \cos 2\delta_L \sin \delta_H - \frac{1}{2} \left(\frac{n_L}{n_H} - \frac{n_H}{n_L} \right) \sin \delta_H \right]$$

and (2.31)

$$M_{22} = M_{11}$$

Since the characteristic matrix of the symmetrical layers is exactly in the same form as a single layer, it can be represented as a special equivalent single layer as

$$M = \begin{pmatrix} \cos \delta_e & i \sin \dfrac{\delta_e}{n_e} \\ in_e \sin \delta_e & \cos \delta_e \end{pmatrix}$$

(2.32)

where the equivalent refractive index n_e and equivalent phase thickness de can be determined by the following relationships:

$$n_e = n_L \left(\frac{\sin 2\delta_L \cos \delta_H + \rho^+ \cos 2\delta_L \sin \delta_H + \rho^- \sin \delta_H}{\sin 2\delta_L \cos \delta_H + \rho^+ \cos 2\delta_L \sin \delta_H - \rho^- \sin \delta_H} \right)^{1/2}$$

(2.33)

$$\delta_e = \cos^{-1} \left[\cos 2\delta_L \cos \delta_H - \rho^+ \sin 2\delta_L \sin \delta_H \right]$$

(2.34)

where
$\rho^\pm = 1/2(x \pm 1/x)$

x is the optical contrast ratio

Equations 2.33 and 2.34 show frequency-dependent refractive index ne and phase thickness de; as a result, this equivalent layer acts somewhat different from normal layers that comprise a single material. If we assume the equivalent phase thickness $\delta_e = \pi/2$, $(2a+b=1)$, at angular frequency ω_0, we get

$$\cos 2\delta_L = \frac{\left(n^2_L + n^2_H \right) \left(n^2_e - n^2_L \right)}{\left(n^2_H - n^2_L \right) \left(n^2_e - n^2_L \right)} \pm$$

(2.35)

and

$$\tan \delta_H = \frac{2n_L n_H}{\left(n^2_L + n^2_H \right)} \frac{1}{\tan 2\delta_L}$$

(2.36)

Using Equations 2.35 and 2.36 and setting the equivalent optical contrast ratio $x_e = n_H/n_{e0}$ to desired values lower than n_H/n_L, where n_{e0} is the equivalent refractive index at ω_0, the corresponding phase thicknesses (d_H, d_L) and consequently the physical thicknesses d_H and d_L of the fractional layers can be determined. The frequency dependent of the equivalent layers may result in a whole difference in the reflectance spectrum with respect to the ordinary layers. Therefore, replacing each of the low-index layers by a combination of three symmetrical layers with proper thicknesses that are defined by Equations 2.35 and 2.36 in basic 1D-PCs structure $(M_L M_H)^q$, we obtain $(M_e M_H)^q$. Similar to this way, we get Equation 2.27, if we set 2X equals the trace of the unit period matrix $M_e M_H$, IXI = 1 corresponds to the edge of bandgap in this structure. So we numerically calculate the relative bandwidth for a different effective optical contrast $x_e = n_H / n_{e0}$; by replacing the equivalent layers, the index profile of our structure is changed, and for the different values of effective optical contrast ratio, $x_e = n_H / n_{e0}$ has a different configuration. Equation 2.27 gives an approximate method for engineering the PBG in 1D PCs with high-index and equivalent layers, and this bandwidth can be changed continuously to any desired value below the ordinary value.

2.9 Application of Nonlinear Optical Polymers in Nanophotonics

Nonlinear optical effects arise from the interaction of electromagnetic fields in various dielectric media to produce new fields altered in phase, frequency, amplitude, or other propagation characteristics relative to the incident optical fields. When a beam of light propagates through a material, the interaction of the optical field with organic molecules in the material induces charge variation and displacement of associated atoms. The amount of charge displacement in the molecules is proportional to the plain amplitude of the electric field. The oscillating frequency of charges in the molecules and the incident light become in phase. Under the small intensity of incident light, the displacement of charge from the equilibrium position, polarization (p), in a microscopic regime is proportional to the applied field, E. With sufficiently strong intensity of laser radiation, the linear relation is modified to be a nonlinear one and expanded to include nonlinear terms in the description of field-induced polarization. In polymers or macromolecules that contain acentric molecular constituents, the dielectric polarization varies nonlinearly to an applied oscillating electric field, and this asymmetrical response can be influenced by an external DC field. This perturbation of the electron distribution gives rise to the harmonic generation and the linear EO effect. EO activity can be thought of as control of the index of refraction of a material by application of a DC or AC voltage. The application of an electric field acts to change the charge distribution of the asymmetrical NLO chromophore. This change in charge distribution and its interaction will, in turn, alter the velocity of light propagating through the material. Although microscopic nonlinearity can be defined in an asymmetrically substituted chromophore exhibiting ground-state charge-density asymmetry, the bulk matrix bearing them exhibits bulk NLO activity of a magnitude that is proportional to the degree of net poled order. Randomization of dipolar order leads to an isotropic medium, and it results in a virtual diminishment of the second-order NLO effect in the bulk material. At the molecular level, asymmetrical molecular design to induce a high microscopic molecular hyperpolarizability (β) is quite easy. Asymmetrical ground-state electron density is related to molecular first hyperpolarizability (β). In order to achieve asymmetrical electron distribution in the ground state, the chromophore can be designed by virtue of a π-conjugated bridge, such as an aromatic ring or polyene. Typical rigid rod structure of the chromophore contains an electron donating and withdrawing group at both ends. In contrast to the ease by which asymmetry is induced at the molecular level, the induction and the stabilization of dipolar ordering of these individual chromophores in the polymer or dendritic material are the critical issues for practical applications.

The increase of the overall dipolar order and its stability became very important subjects of many research projects that have been performed in the area of materials chemistry. Nonlinear optical polymers and dendrimers show great promise for device applications in the communications and photonics industries. At present, there are a number of NLO applications that utilize inorganic crystals. A few of these processes include optical switches, EO modulators, optical data storage, and optical fiber–based devices. The study of the organic NLO chromophore–doped polymers allows them to preferentially replace their inorganic counterparts. A number of chromophores utilizing modified structural components and designs have been shown to produce considerably high nonlinearities despite potentially detrimental intermolecular electrostatic interactions due to the increases in molecular dipole. The large numbers of recently reported chromophores utilizing new heterocyclic components as acceptor moieties, either tricyanofuran (TCF) or TCP derivatives, underscore the importance of this new class of acceptor groups. These chromophores have achieved outstanding optical nonlinearities and have been shown to be effective for the design and molecular construction of supra-chromophoric architectures [157].

2.9.1 Nanoscale Nonlinear Optical Imaging Technique

Nonlinear optical imaging technique has the advantages of effective rejection of background, reduced volume of photo bleaching, and depth discrimination [158]. Some of these advantages can benefit NSOM and photon-scanning tunneling microscopy (PSTM). The schematic diagrams of two-photon NSOM and PSTM are shown in Figure 2.6a and b.

A self-mode-locked Ti/sapphire laser is used for two-photon NSOM, as an excitation source at 800 nm with an average power of 1.5 mW. The laser light coupled into a 1 m single-mode optical fiber illuminates the sample through an aluminum-coated probe with an aperture of 50 nm. Dispersion induced by the optical fiber is compensated by a 350-line/mm grating pair. ZnS: Mn nanoparticles with a diameter of 150–250 nm and encapsulated with 2-{4-{(ε)-2- {4-(sulfonyl) phenyl}-1-ethenyl} (methyl) anilino}-1-ethanethiol (APSS-SH) have been imaged. The APSS-SH contains a thiol-terminated group that chemically bonds to the surface of a ZnS: Mn particle. The improved resolution of two-photon excitation is due to the quadratic dependence of two-photon excitation, which limits the effective excitation to a small volume at the aperture of the probe and enhances the light confinement. Due to the enhanced light confinement, the two-photon NSOM can also effectively reject the background from the illuminated zone and improve the image contrast. However, it is found that the nonlinear emission in the fiber overlaps with the fluorescence from the specimen and contributes to the background [159–161]. Although a pulse compressor can compensate for the group-velocity dispersion induced by the fiber, the nonlinear effect such as self-phase modulation is not easily corrected [160]. To suppress the background signal and improve the image contrast, a metal tip illuminated with a femtosecond laser has been utilized as a localized excitation source to generate a strongly enhanced electric field at the metal tip for two-photon imaging in NSOM. The two-photon PSTM can provide advantages in signal-to-noise improvement and in system alignment compared to two-photon NSOM. The same self-mode-locked Ti/sapphire laser is used as an excitation source with an average power of 12 mW for two-photon PSTM. The laser beam is focused by a lens and illuminates the sample that is mounted with index-matching oil on a fused silica prism under total internal reflection. Since the incident light is linearly polarized, once excited, the molecules act as electric dipoles, and dipole orientations approximately follow the excitation polarization. Therefore, the spatial fluorescence feature indicates the high degree of molecular order in the isolated nanoparticles. With two-photon excitation, the quadratic dependence

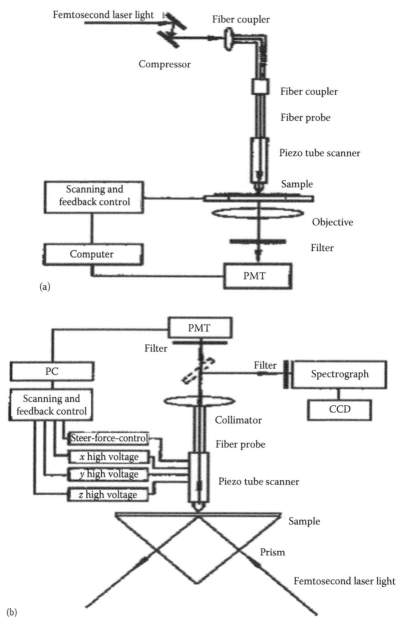

FIGURE 2.6
Schematic of (a) two-photon NSOM and (b) PSTM. (Reprinted with permission from American Chemical Society, 2000.)

of two-photon excitation on light intensity enables a light confinement at the aperture of the probe and enhances the image contrast with a coated probe. Since the evanescent field decays exponentially, the depth of penetration for PSTM is limited [162]. However, two-photon PSTM still provides some advantages in reduced alignment constraints and minimized background interference and shows potential benefit in thin-film analysis.

2.10 Computational Modeling of Nano- and Biophotonics Using FDTD Design

Optical simulations are the only way to get a deeper understanding of light propagation and scattering in advanced nanophotonic structures. Optical wave phenomenon at the nanoscale requires the application of rigorous numerical modeling. The situation is very similar in nanobiophotonics diagnostics and imaging research studies where the optical scattering phenomena are initiated at a comparable scale of dimensions. The methods for the numerical modeling of light scattering from single or multiple biological cells are of particular interest as they provide information about the fundamental light-cell interaction phenomena that are highly relevant for the practical interpretation of cell images by pathologists. The FDTD simulation and modeling of the light interaction with single and multiple, normal, and pathological–biological cells and subcellular structures has attracted attention since 1996 [163]. The emerging relevance of nanobiophotonics imaging research has established the FDTD method as one of the powerful tools for studying the nature of light-cell interactions. The main advantages of the FDTD method are (i) numerical simplicity and physical basis since it is a numerical solution of Maxwell equations; (ii) ability to simulate electromagnetic wave phenomena in complex geometries, including advanced models; and (iii) ability to make accurate predictions such that the results can be used for real-world photonic designs. The FDTD potential for advanced visualization of the electromagnetic fields is associated with a greater capability for numerical simulation control that has greatly contributed to its broader adoption as a research tool in many photonics research. The sub-wavelength grating (SWG) design exploits the effective medium principle, which states that different optical materials, combined at sub-wavelength scales, can be approximated by an effective homogeneous material [164,165]. Within this approximation, an effective medium can be characterized by an effective refractive index defined by a power series of the homogenization parameter $\chi = \Lambda/\lambda$, where Λ is the grating pitch and λ is the wavelength of light. Provided that the pitch Λ is less than the first-order Bragg period $\Lambda_{Bragg} = \lambda/(2_{neff})$, the grating operates in a sub-wavelength regime and the diffraction effects are frustrated. An example of a basic structure that exemplifies the use of refractive-index engineering in a waveguide is shown in Figure 2.7a [2].

It is a nonresonant photonic structure formed by etching a linear periodic array of rectangular segments into a 260 nm thick single-crystal silicon layer of silicon-on-insulator wafer. A 2 mm thick bottom oxide (SiO_2) layer separates the waveguide from the underlying silicon substrate. The waveguide core is a composite medium formed by interlacing the high-refractive-index segments with a material of a lower refractive index, which at the same time is used as the cladding material. The refractive index of the core is controlled lithographically by changing the volume fractions of the two materials. By intermixing Si and SU-8 materials at the sub-wavelength scale, the refractive-index range of ≈ 1.6–3.5 can be obtained. In order to avoid the formation of standing waves due to Bragg scattering and the opening of a bandgap near 1550 nm wavelength, a nominal structural period $d = 300$ nm is chosen, which is less than a half of the effective wavelength of the waveguide mode λ_{eff}. Figure 2.7b shows the dispersion diagrams of the periodic SWG waveguide and of an equivalent strip waveguide with a core index of 2.65 [166]. The comparison of the two dispersion curves shows that the dispersion away from the bandgap resonance matches that of an equivalent strip waveguide. SWG design optimization issues. The adoption of a FDTD simulation approach includes the opportunity to investigate the effect of variation

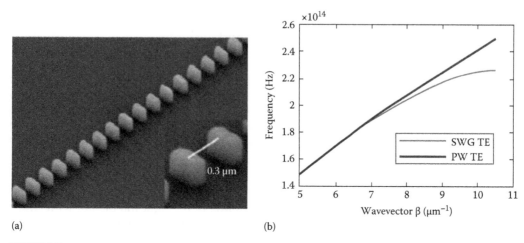

(a)   (b)

FIGURE 2.7
(a) SWG waveguide (SEM image) and (b) dispersion diagrams for SWG and an equivalent strip waveguide with
an engineered core refractive index of 2.65 (TE polarization). (Reprinted with permission from Elsevier, 2011.)

of the grating period and/or chirping. The limitation arises when the spatial numerical
resolution becomes comparable to the design parameter variation step. A guided periodic
Bloch mode is lossless in an ideal waveguide fabricated with materials without absorption.
Provided the pitch is outside the Bragg condition, a loss mechanism arises from the imperfections in fabrication resulting in the segments not being perfectly periodic as well as
introducing some additional random variations of the periodicity coming from the roughness itself.

Conventional waveguide loss theory suggests that to minimize waveguide loss, one
must reduce the scattering efficiency of the imperfections by reducing the field intensity
at the rough boundary. In the aforementioned case, the SWG parameters width/pitch/
duty cycle are designed to maximize delocalization (minimize loss) and yet minimize
light leakage to the substrate (the buried oxide layer was 2 mm thick), which created an
upper limit to the degree of delocalization. The pitch of 300 nm is chosen to minimize the
fusing of segments by keeping the dimensions far away from the minimum feature size.
Other studies on SWG couplers [167] have shown that an increase of the grating pitch from
200 to 300 nm leads to only a 0.03 dB difference in coupling efficiency and that a variation
of the grating pitch between 300 and 400 nm leads to very similar experimental results.
It was therefore logical to use those parameters for designing the SWG waveguides. The
best-performing SWG couplers are a good indicator of grating parameters because the
structure comprises of a 50 mm coupler and a 150 mm SWG waveguide with constant
grating parameters. A thorough optimization of the design parameter space of a SWG
requires a model of the effect of roughness on the periodic mode profiles and total waveguide loss. The FDTD approach reaches one of its natural limitations when the minimum
possible mesh size becomes comparable to the random variations coming from to the surface roughness of the grating segments. This limitation is particularly relevant in the field
of nanophotonics. It, however, does not diminish the key advantages of using the FDTD
method. The FDTD method lends itself very well to visualizing the transverse evolution
of the mode. This is particularly important and revealing when assessing the steady state
of a particular simulation. Apart from visualizing the field profile, the phase fronts can
easily be seen, which is particularly illuminating when designing gratings. The simulation

example demonstrates both the advantages and the limitations of the FDTD approach to the design and modeling of a new waveguide-crossing principle based on SWG waveguides. An important advantage of this SWG structure is that it can be fabricated with a single etch step. SWG crossings have the potential to facilitate massive interconnectivity and minimize the device footprint for future complex planar waveguide circuits. The 3D FDTD formulation is based on a modified version [168–170] of TFSF FDTD formulation [123]. It uses a TFSF region that contains the biological cell and extends beyond the limits of the simulation domain. The extension of the transverse dimension of the input field beyond the limits of the computational domain through the perfectly matched boundaries leads to distortions of the ideal plane-wave shape and eventually leads to a numerical distortion of the simulation results. To avoid these distortions, Bloch periodic boundary conditions (Figure 2.8a and b) are used in the lateral x and y directions that are perpendicular to the direction of propagation $-z$ [4–6].

Phase-contrast microscopy produces high-contrast images of transparent specimens such as living cells and subcellular components. In a conventional flow cytometry configuration, a beam of light of a single wavelength is directed onto a hydro-dynamically focused stream of fluid driving a periodical array of cells to flow through it. The optical phase-contrast microscopy (OPCM) simulation model requires the explicit availability of the forward-scattered transverse distribution of the fields. The phase of the scattered field accumulated by a plane wave propagating through a biological cell within a cytometric cell flow is shown in Figure 2.9a and b, where an image with a strong image contrast ratio is created by coherently interfering a reference (R) with a beam (D) that is diffracted from one particular cell in the cell flow. The phase-contrast microscope uses incoherent annular illumination that is modeled by adding up the results of eight different simulations using ideal input plane waves incident at a given polar angle (308), an azimuthal angle (0, 90, 180, or 2708), and a specific light polarization (parallel or perpendicular to the plane of the graph). Every single FDTD simulation provides the near-field components in

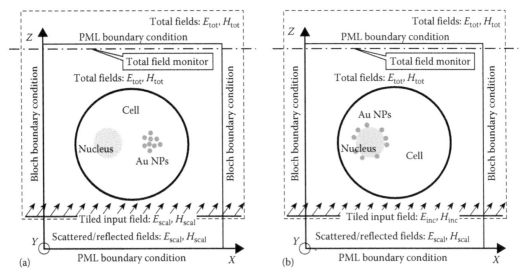

FIGURE 2.8
Schematic representation of the 3D FDTD formulation including (a) a cell with a nucleus and a cluster of gold nanoparticles in the cytoplasm and (b) a cell with gold nanoparticles randomly distributed on the nucleus surface. (Reprinted with permission from Elsevier, 2011.)

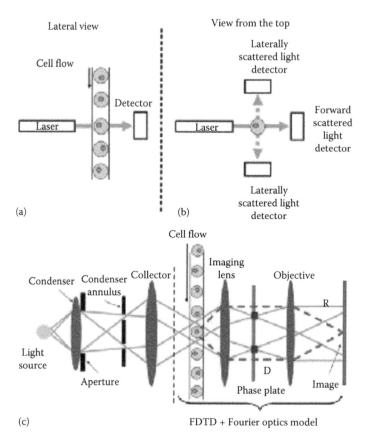

FIGURE 2.9
OPCM cytometer (a and b) with a visual representation of the FDTD OPCM model (c) including the propagation of the reference (R) beam without the cell flow and the propagation of the beam diffracted (D) from one of the cells. (Reprinted with permission from Elsevier, 2011.)

a transverse monitoring plane located right behind the cell (see Figure 2.8a and b). The far-field transformations use the calculated near fields behind the cell and return the three complex components of the electromagnetic fields far enough from the location of the near fields, that is, in the far field [123]. The amplitudes and the phases of the calculated far-field components can be used then to do Fourier optics with both the scattered and the reference beams. The magnification factor of the optical lens system was implemented by merely modifying the angle of light propagation—it was applied to the far fields before the interference of the diffracted (D) and reference (R) beams (Figure 2.9c) at the image plane [168–170].

The effect of the numerical aperture (NA) is to clip any light that has too steep an angle and is not collected by the lens system. The OPCM images at the image plane are calculated by adding up the scattered and the reference beam at any desired phase offset C. The 3D FDTD modeling of OPCM imaging of a single biological cell uses optical magnification factor M-10 and NA = 0.8. The cell is modeled as a sphere with a radius $R_c = 5$ mm with membrane thickness $d = 20$ nm. It corresponds to an effective (numerical) thickness of approximately 10 nm. The cell nucleus is also spherical with a radius $R_n = 1.5$ mm centered at a position 2.0 mm away from the cell center in a direction perpendicular to the direction of light propagation. The refractive index of the cytoplasm is $n_{cyto} = 1.36$,

$n_{nuc} = 1.4$ of the nucleus, $n_{mem} = 1.47$ of the membrane, and $n_{ext} = 1.33$ of the extracellular material (no refractive-index matching (RIM)). The case of $n_{ext} = 1.36$ corresponds to RIM that through optical clearing ensures a better contrast of the cell image [171]. Figure 4a shows the schematic positioning of a cluster of 42 NPs in the cytoplasm used in simulations. The cell center is located in the middle ($x = y = z = 0$) of the computational domain with dimensions 15 μm × 12 μm × 15 μm (Figure 1a). The nucleus' center is located at $x = 2$ μm, $y = z = 0$ μm. The cluster of gold NPs is located at $x = -2$ μm, $y = z = 0$ μm. Figure 4b shows another simulation scenario where the cluster of 42 NPs is randomly distributed on the surface of the cell nucleus. One of the main goals of this section is to illustrate the ability of the FDTD approach to model generate OPCM cell images including the imaging effects of optical clearing and gold NP resonance. The nanobiophotonics simulation example considered here opens a number of questions related to the potential optimization of the specific numerical approach. The first one is about the choice of using periodic boundary conditions in the lateral direction. In order to proceed with a brief justification of this specific choice, it is important to point out that numerical excitation of the OPCM needed a tilted incident plane wave. Bloch boundary conditions are periodic boundary conditions that appeared to be the only choice in dealing with the spurious phase effects due to the tilting of the input plane waves incoming at lateral simulation boundaries. Employing periodic boundary conditions means that what we are actually modeling is a periodic row of biological cells. The near-scattered fields, however, are calculated in the transverse planes located in the close proximity to the cell where the coupling effect due to waves scattered from the virtually adjacent cells is negligible. This effect can be further minimized or completely removed by controlling the lateral dimension of computational domain by using a large enough period of the periodic cell structure. The larger is this period, the smaller is the coupling effect. The FDTD technique averages out the refractive-index values at the interface of two different materials effectively reducing the membrane thickness value and eventually, due to the staircase approximation error, destroying the continuity of the membrane. In the case of the numerical design of nanoparticles, the challenges were slightly different in nature. The OPCM simulation example demonstrates both the advantages and the limitations of the FDTD approach in addressing a specific nanobiophotonics problem. The key achievement consists in the adequate construction of OPCM images of biological cells with a realistic size including clusters of nanoparticles. It is an example that points out the potential of the FDTD approach in modeling advanced optical nano-bio-imaging instrumentation. FDTD shows the ability to help in addressing a critical issue in the design and fabrication of microphotonic waveguide structures—the optimal choice for the most suitable refractive-index contrast of the waveguides. The real challenge in making this choice consists in the requirement for the refractive-index values to be sufficiently high in order to guarantee a proper light confinement and, at the same time, dealing with all the consequent side effects, such as higher propagation loss, higher sensitivity to fabrication imperfections, and sidewall roughness. Usually, the refractive index cannot be chosen at will but must be selected within a limited set of optical platforms. In the new design [166,172], the waveguide is longitudinally patterned with a SWG, consisting of segments of a high-refractive-index core material interlaced with a lower-refractive-index cladding material. Since the refractive-index contrast can be changed by simply controlling the grating period, SWG waveguides with different optical parameters (mode confinement, effective index, chromatic dispersion, and so on) can be realized on the same chip. This approach nicely fits the fabrication processes of planar light wave circuits and represents a radical step forward with respect to the existing methods.

The extremely low loss of the experimentally fabricated waveguides [166,172] shows that SWG waveguide applications are now able to compete with the ones based on conventional waveguides. It is important to point out that all the results correspond to the case when there is a RIM between the cytoplasm and the extracellular medium that leads to the optical clearing of the cell images. The refractive index of the extracellular fluid can be externally controlled by the administration of an appropriate chemical agent [171] leading to increased light transmission through the cell due to the matching of the refractive indexes of some of its organelles to that of the extracellular medium [169,123]. Due to optical clearing, the image contrast of the cell cytoplasm can be drastically reduced to zero levels, and it is only the image of the nucleus that will remain sharply visible. Such scenario shows an unprecedented opportunity to use the optical clearing effect for the analysis of pathological changes in the eccentricity and the chromatin texture of cell nuclei within the context of OPCM configurations. This opportunity is associated with the fact that at RIM conditions, the cell image is efficiently transformed into a high-contrast image of the nucleus. In such conditions, the imaging effect of the NPs is significantly enhanced. The presented results did not allow analyzing the scaling of the NP imaging effect as a function of the number of the NPs. However, the validation of the model provides a basis for future research in this direction. The typical light scatterers in cellular and subcellular biomedical optics are of nanoscale dimensions that only enhance the potential for the application of the FDTD approach in newly emerging research areas involving more sophisticated combinations of nano- and biophotonics materials. The simulation challenges in the two different situations might be different but are usually numerical in nature and emerge from some of the natural limitations of the FDTD methodology. The FDTD approach reaches the natural limits of its resolution that makes the numerical design very challenging and questions the value of the optimization efforts. Part of the problem with optimization comes from the fact that the FDTD method always models a (or one) specific device and cannot be easily integrated with statistical models that could take into account the random variation of some of the design parameters due to fabrication tolerances or imperfections. The transverse distribution of the propagating field is critically important—in the SWG waveguide–crossing case, it is a valuable monitoring tool facilitating the overall design of the nanophotonics structure; in the OPCM nanobioimaging case, it is the final goal of the simulation efforts leading to the construction of the OPCM images. In both cases, there is a component involving the purposeful manipulation of the refractive-index values—in the SWG waveguide–crossing case, the refractive-index engineering represents the most innovative part of the design; in the OPCM nanobioimaging case, the RIM ensures the maximum contrast of the nanoparticle cluster images. This point may be considered in favor of the adoption commercially available software simulation tools; however, it should be interpreted within a completely scientific context since (i) powerful GUI and visualization capabilities lead to better photonics designs and (ii) powerful GUI and visualization capabilities help the replicability of research results that is a key component of the scientific enterprise in general. The value of in-house simulation tools is unquestionable, and in many cases, the degree of their numerical sophistication is far beyond the one provided by commercial tools. However, photonics research groups should cooperate with optical software simulation tool developers in order to appropriate as much as possible the value of existing tools as well as to challenge them in promoting more innovative open design strategies aiming at the development of better simulation tools and new application areas. We believe that the unique features of the FDTD technique provide a very good opportunity for such development.

2.11 Future Outlook

Majority of CI's applications are genetic algorithms, where there is only a little activity on the application of fuzzy logic control. As the seamless interconnection of optical and existed technologies like IP is progressing, through the GMPLS protocol, the functionality is also enhanced. This fact together with the appearance of more demanding applications will require efficient management of the network. At the stage where maturity and operability of optical networks' is attained, it is expected that new design and management optimization issues will emerge, as seen in the case of the well-established ATM technology where the use of fuzzy, neuro-fuzzy, or evolutionary fuzzy techniques is quite extended [71]. Based in the described current research work in optical networks, we discussed and proposed also future trends in design and optimization. However, optical networks are a rapidly growing area with continuously emerging new issues. When optical technology reaches a mature state and components like optical buffers or all-optical packet header recognition chips will become cost-effective solutions, then OPS will become a reality. We should expect then new design and optimization issues for packet-switched networks, employing again heuristics and genetic algorithms [71] and following also similar methodologies as in the previously existed communications technology [16,18]. The complexity of large communication networks like the Internet requires highly efficient and adaptive routing algorithms. Recent studies are examining novel methods for the consideration of adaptive routing, such as EGT, as described in [72]. Optical networks are evolved also toward the direction of all-optical networks and consequently to the realization of all-optical Internet. In this direction, the application of game theory and EGT in the large-scale noncooperative all-optical networks, as first described recently in [173,174], could help our better understanding of future global all-optical networks.

References

1. Weisbuch, C., Benisty, H., Houdre, R. 2000. Overview of fundamentals and applications of electrons, excitons and photons in confined structures. *J. Lumin.* 85: 271–293.
2. Kumar, N. D., Lal, M., Prasad, P. N. 1998. *Science and Technology of Polymers and Processing of Multifunctional Polymers and Composites for Photonics*, Plenum Press, New York.
3. Romanov, S. G., Maka, T., Torres, C. M. S., Muller, M., Zentel, R. 1999. Photonic band-gap effects upon the light emission from a dye–polymer–opal composite. *Appl. Phys. Lett.* 75: 1057–1059.
4. Paspalakis, E., Keitel, C. H., Knight, P. L. 1998. Fluorescence control through multiple interference mechanisms. *Phys. Rev. A* 58: 4868–4877.
5. Lin, S. Y., Fleming, J. G., Sigalas, M. M., Biswas, R., Ho, K. M. 1999. Photonic band-gap microcavities in three dimensions. *Phys. Rev. B* 59: 15579–15582.
6. Miyazak, H., Ohtaka, K. 1998. Near-field images of a monolayer of periodically arrayed dielectric spheres. *Phys. Rev. B* 58: 6920–6937.
7. Chan, Y. S., Chan, C. T., Liu, Z. Y. 1998. Photonic band gaps in two dimensional photonic quasicrystals. *Phys. Rev. Lett.* 80: 956–959.
8. Russell, P. S., Roberts, P. J., Allan, D. C. 1999. Single-mode photonic band gap guidance of light in air. *Science* 285: 1537–1539.
9. Herndon, M. K., Collins, R. T., Hollingsworth, R. E., Larson, P. R., Johnson, M. B. 1999. Near-field scanning optical nanolithography using amorphous silicon photoresists. *Appl. Phys. Lett.* 74: 141–143.

10. Nolte, S., Chichkov, B. N., Welling, H., Shani, Y., Lieberman, K., Terkal, H. 1999. Nanostructuring with spatially localized femtosecond laser pulses. *Opt. Lett.* 24: 914–916.
11. Betzig, E., Trautman, J. K., Wolfe, R., Gyorgy, E. M., Finn, P. L. 1992. Near-field magneto-optics and high density data storage. *Appl. Phys. Lett.* 61: 142–144.
12. Terris, B. D., Mamin, H. J., Rugar, D. 1996. Near-field optical data storage. *Appl. Phys. Lett.* 68: 141–143.
13. Hosaka, S., Shintani, T., Miyamoto, M., Hirotsune, A., Terao, M., Yoshida, M., Fujita, K., Kammer, S. 1996. Nanometer-sized phase-change recording using a scanning near-field optical microscope with a laser diode. *Jpn. J. Appl. Phys.* 35: 443–447.
14. Tan, W., Shi, Z. Y., Smith, S., Birnbaum, D., Kopelman, R. 1992. Submicrometer intracellular chemical optical fiber sensors. *Science* 258: 778–781.
15. Tancevski, L. 2003. Intelligent next generation WDM optical networks, Information. *Sciences* 149: 211–217.
16. Pedrycz, W., Vasilakos, A. V. 2000. *Computational Intelligence in Telecommunications Networks*, CRC Press, Boca Raton, FL.
17. Vasilakos, A. V., Pedrycz, W. 2006. *Ambient Intelligence, Wireless Networking, Ubiquitous Computing*, Art House, Boston, MA.
18. Sekercioglu, Y. A., Pitsilides, A., Vasilakos, A. 2001. Computational intelligence in management of ATM networks: A survey of current state of research. *Soft Comput.* 5(4): 257–263.
19. Shokooh-Saremi, M., Nourian, M., Mirsalehi, M. M., Keshmiri, S. H. 2004. Design of multilayer polarizing beam splitters using genetic algorithm. *Opt. Commun.* 233(1–3): 57–65.
20. Alcantara, L. D. S., Lima, M. A. C., Cesar, A. C., Borges, B. H. V., Teixeira, F. L. 2000. Design of a multifunctional integrated optical isolator switch based on nonlinear and nonreciprocal effects. *Opt. Eng.* 44(12): 124002–124004.
21. Moosburger, R., Kostrzewa, C., Fischbeck, G., Petermann, K. 1997. Shaping the digital optical switch using evolution strategies and BPM. *IEEE Phot. Technol. Lett.* 9(11): 1484–1486.
22. Riziotis, C., Zervas, M. N. 2001. Design considerations of optical add-drop filters based on grating assisted mode conversion in null couplers. *IEEE/OSA J. Ligt. Wav. Technol.* 19(1): 92–104.
23. West, B. R., Honkanen, S. 2004. MMI devices with weak guiding designed in three dimensions using a genetic algorithm. *Opt. Exp.* 12(12): 2716–2722.
24. Cheng, C. 2004. A global design of an erbium-doped fiber and an erbium-doped fiber amplifier. *Opt. Laser. Technol.* 36(8): 607–612.
25. Wei, H., Tong, Z., Jian, S. S. 2004. Use of a genetic algorithm to optimize multistage erbium-doped fiber-amplifier systems with complex structures. *Opt. Exp.* 12(4): 531–544.
26. Gao, M. Y., Jiang, C., Hu, W. S., Wang, J. Y. 2006. Two-pump fiber optical parametric amplifiers with three-section fibers allocation. *Opt. Laser. Technol.* 38(3): 186–191.
27. Xin, Y. F., Rouskas, G. N., Perros, H. G. 2003. On the physical and logical topology design of large-scale optical networks. *J. Ligt. Wave Technol.* 21(4): 904–915.
28. Rouskas, G. N., Perros, H. G. 2002. A tutorial on optical networks, networking 2002 tutorials. *Lecture Notes in Computer Science* 2497: 155–193.
29. Willner, A. E., Gurkan, D., Sahin, A. B., McGeehan, J. E., Hauer, M. C. 2003. All-optical address recognition for optically-assisted routing in next-generation optical networks. *IEEE Opt. Commun.* 41: S38–S44.
30. Teh, P. C., Thomsen, B. C., Ibsen, M., Richardson, D. J. 2003. Multi-wavelength (40 WDM 10 Gbit/s) optical packet router based on superstructure fiber Bragg gratings, *IEICE Trans. Special Issue Recent Prog. Optoelectron. Commun.* E86B(5): 1487–1492.
31. Thomsen, B. C., Teh, P. C., Ibsen, M., Lee, J. H., Richardson, D. J. 2002. A multihop optical packet switching demonstration employing all optical grating based header generation and recognition, in: *Proceedings of the European Conference Optical Communications, ECOC*, Copenhagen, Denmark, September 8–12, 2002.
32. Xu, L. S., Perros, H. G., Rouskas, G. N. 2003. Access protocols for optical burst-switched ring networks. *Inform. Sci.* 149(1–3): 75–81.

33. Zang, H., Jue, J. P., Mukherjee, B. 2000. A review of routing and wavelength assignment approaches for wavelength-routed optical WDM networks. *Opt. Networks Mag.* 1: 47–60.

34. Ramaswami, R., Sivarajan, K. N. 1995. Routing and wavelength assignment in all-optical networks, *IEEE-ACM Trans. Network.* 3(5): 489–500.

35. Chu, X. W., Li, B., Chlamtac, I. 2003. Wavelength converter placement under different RWA algorithms in wavelength-routed all-optical networks. *IEEE Trans. Commun.* 51(4): 607–617.

36. Datta, R., Sengupta, I. 2005. Static and dynamic connection establishment WDM optical networks: A review. *IETE J. Res.* 51(3): 209–222.

37. Ali, R., Ramamurthy, B., Deogun, J. S. 2000. Routing and wavelength assignment with power considerations in optical networks. Computer networks. *Int. J. Comp. Telecommun. Network.* 32(5): 539–555.

38. Talay, A. C., Oktug, S. 2004. A GA/heuristic hybrid technique for routing and wavelength assignment in WDM networks, Applications of evolutionary computing. *Lecture Notes in Computer Science* 3005: 150–159.

39. Talay, A. C. 2004. Parallel genetic algorithm/heuristic based hybrid technique for routing and wavelength assignment in WDM networks, *Computer and Information Sciences, ISCIS 2004, Proceedings 2004, Lecture Notes in Computer Science* 3280: 819–826.

40. Chu, X. W., Li, B. 2005. Dynamic routing and wavelength assignment in the presence of wavelength conversion for all-optical networks. *IEEE-ACM Trans. Network.* 13(3): 704–715.

41. Le, V. T., Ngo, S. H., Jiang, X. H., Horiguchi, S., Guo, M. Y. 2004. A genetic algorithm for dynamic routing and wavelength assignment in WDM networks, in: *Parallel and Distributed Processing and Applications, Proceedings 2004, Lecture Notes in Computer Science* 3358: 893–902.

42. Le, V. T., Jiang, X. H., Ngo, S. H., Horiguchi, S. 2005. Dynamic RWA based on the combination of mobile agents technique and genetic algorithms in WDM networks with sparse wavelength conversion. *IEICE Trans. Info. Sys.* E88D(9): 2067–2078.

43. Hwang, I. S., Huang, I. F., Yu, S. C. 2005. Dynamic fuzzy controlled RWA algorithm for IP/GMPLS over WDM networks. *J. Comp. Sci. Technol.* 20(5): 717–727.

44. Durand, F. R., Lima, M. A. C., Cesar, A. C., Moschim, E. 2005. Impact of PMD on hybrid WDM/OCDM networks. *IEEE Phot. Technol. Lett.* 17(12): 2787–2789.

45. Skorin-Kapov, N. 2006. Multicast routing and wavelength assignment in WDM networks: A bin packing approach. *J. Opt. Network.* 5(4): 266–279.

46. Skorin-Kapov, N. 2006. Heuristic algorithms for virtual topology design and routing and wavelength assignment in WDM networks, Doctoral Thesis, University of Zagreb, Zagreb, Croatia.

47. Din, D. R. 2004. Genetic algorithms for multiple multicasts on WDM ring network, *Comp. Commun.* 27(9): 840–856.

48. Chen, M. T., Tseng, S. S. 2005. A genetic algorithm for multicast routing under delay constraint in WDM network with different light splitting. *J. Info. Sci. Eng.* 21(1): 843–847.

49. Siregar, J. H., Zhang, Y. B., Takagi, H. 2005. Optimal multicast routing using genetic algorithm for WDM optical networks. *IEICE Trans. Commun.* E88B(1): 219–226.

50. Din, D. R. 2005. Anycast routing and wavelength assignment problem on WDM network, *IEICE Trans. Commun.* E88B(10): 3941–3951.

51. Borne, S., Gourdin, E., Liao, B., Mahjoub, A. R. 2006. Design of survivable IP-over-optical networks. *Annal. Oper. Res.* 146: 41–73.

52. Lee, H. J., Choi, H., Subramaniam, S., Choi, H. A. 2003. Survivable embedding of logical topologies in WDM ring networks. *Info. Sci.* 149: 151–160.

53. Sahasrabuddhe, L., Ramamurthy, S., Mukherjee, B. 2002. Fault management in IP-over-WDM networks: WDM protection versus IP restoration. *IEEE J. Select. Area. Commun.* 20(1): 21–33.

54. Koo, S., Subramaniam, S. 2003. Performance evaluation of optical mesh restoration schemes. *Info. Sci.* 149(1–3): 183–195.

55. Modiano, E., Narula-Tam, A. 2002. Survivable lightpath routing: a new approach to the design of WDM-based networks. *IEEE J. Select. Area. Commun.* 20(4): 800–809.

56. Narula-Tam, A., Modiano, E., Brzezinski, A. 2004. Physical topology design for survivable routing of logical rings in wdm-based networks. *IEEE J. Select. Area. Commun.* 22(8): 1525–1538.

57. Xin, Y. F., Rouskas, G. N. 2004. A study of path protection in large-scale optical networks. *Phot. Network. Commun.* 7(3): 267–278.

58. Wuttisittikulkij, L., Mahony, M. J. O'. 1997. Use of spare wavelengths for traffic restoration in a multi wavelength transport network. *Fiber. Integrat. Opt.* 16(4): 343–354.

59. Chong, H. W., Kwong, S. 2003. Optimization of spare capacity in survivable WDM networks, in: *Genetic and Evolutionary Computation, GECCO 2003, PT II, Proceedings 2003, Lecture Notes in Computer Science* 2724: 2396–2397.

60. Gokisik, D., Bilgen, S. 2003. Neural network based optical network restoration with multiple classes of traffic, in: *Computer and Information Sciences, ISCIS 2003, Lecture Notes in Computer Science* 2869: 771–778.

61. Goncalves, C. H. R., Oliveira, M., Andrade, R. M. C., de Castro, M. F. 2004. Applying artificial neural networks for fault prediction in optical network links, in: *Telecommunications and Networking, ICT 2004, Lecture Notes in Computer Science* 3124: 654–659.

62. Pickavet, M., Demeester, P. 2000. Multi-period planning of survivable WDM networks. *Eur. Trans. Telecommun.* 11: 7–16.

63. Fumagalli, F., Balestra, G., Valcarenghi, L. 1998. Optimal amplifier placement in multi-wavelength optical networks based on simulated annealing, in: J. M. Senior, C. Qiao, (Eds.), *Proceedings of SPIE, All-Optical Networking: Architecture, Control, and Management Issues*, Boston, MA, Vol. 3531, pp. 268–279.

64. Jukan, A., Franzl, G. 2004. Path selection methods with multiple constraints in service-guaranteed WDM networks. *IEEE/ACM Trans. Network.* 12(1): 59–72.

65. Wang, Z. X., Chen, Z. Q., Yuan, Z. Z. 2004. QoS routing optimization strategy using genetic algorithm in optical fiber communication networks. *J. Comp. Sci. Technol.* 19(2): 213–217.

66. Jia, X. H., Du, D. Z., Hu, X. D., Lee, M. K., Gu, J. 2001. Optimization of wavelength assignment for QoS multicast in WDM networks. *IEEE Trans. Commun.* 49(2): 341–350.

67. Ravikumar, C. P., Sharma, M., Jain, P. 2002. Optimal design of delay-bounded WDM networks using a genetic algorithm. *IETE J. Res.* 48(5): 423–428.

68. Wang, X. W., Cheng, H., Li, J., Huang, M., Zheng, L. D. 2004. A QoS-based multicast algorithm for CSCW in IP/DWDM optical Internet, Grid and Cooperative Computing. *Lecture Notes in Computer Science* 3033: 1059–1062.

69. Yang, H. S., Maier, M., Reisslein, M., Carlyle, W. M. 2003. A genetic algorithm-based methodology for optimizing multi service convergence in a metro WDM network. *J. Light Wave Technol.* 21(5): 1114–1133.

70. Shen, S. T., Chuang, Y. R., Chen, Y. H., Lai, E. 2003. An optimization solution for packet scheduling: A pipeline-based genetic algorithm accelerator, *Genetic and Evolutionary Computation—GECCO 2003, Proceedings 2003, Lecture Notes in Computer Science* 2723: 681–692.

71. Fischer, S., Vocking, B. 2005. Evolutionary game theory with applications to adaptive routing, in: *Proceedings of the European Conference on Complex Systems (ECCS)*, Paris France, November.

72. Wagner, R. S., Ellis, W. C. 1964. Vapor-liquid-solid mechanism of single crystal growth. *Appl. Phys. Lett.* 4: 89–91.

73. Stach, E. A., Pauzauskie, P. J., Kuykendall, T., Goldberger, J., He, R., Stach, P. Y. 2003. Watching GaN nanowires grow. *Nano Lett.* 3: 867–869.

74. Wu, Y. Y., Yang, P. D. 2001. Direct observation of vapor-liquid-solid nanowire growth. *J. Am. Chem. Soc.* 123: 3165–3166.

75. Colas, E., Simhony, S., Kapon, E., Bhat, R., Hwang, D. M., Colas, P. S. D. L. 1990. Growth of GaAs quantum wire arrays by organometallic chemical vapor deposition on submicron gratings. *Appl. Phys. Lett.* 57: 914–917.

76. Morales, A. M., Lieber, C. M. 1998. A laser ablation method for the synthesis of crystalline semiconductor nanowires. *Science* 279: 208–211.

77. Hu, J., Ouyang, M., Yang, P., Lieber, C. M. 1999. Controlled growth and electrical properties of heterojunctions of carbon nanotubes and silicon nanowires. *Nature* 399: 48–45.

78. Duan, X. F., Lieber, C. M. 2000. General synthesis of compound semiconductor nanowires. *Adv. Mater.* 12: 298–302.
79. Hochbaum, A.I., Fan, R., He, R., Yang, P. 2005. Controlled growth of Si nanowire Arrays for device integration. *Nano Lett.* 5: 457–460.
80. Wu, Y., Yan, H., Huang, M., Messer, B., Song, J. H., Yang, P. 2002. Inorganic semiconductor nanowires: Rational growth, assembly, and novel properties. *Chem. Eur. J.* 8: 1260–1268.
81. He, R. R., Gao, D., Fan, R., Hochbaum, A. I., Carraro, C., Maboudian, R., Yang, P. 2005. Si nanowire bridges in micro trenches: Integration of growth into device fabrication. *Adv. Mater.* 17: 2098–2102.
82. Huang, M. H., Mao, S., Feick, H., Yan, H., Wu, Y., Kind, H., Weber, E., Russo, R., Yang, P. 2001. Room-temperature ultraviolet nanowire nanolasers. *Science* 292: 1897–1899.
83. Johnson, J. C., Yan, H., Schaller, D. R., Haber, L. H., Saykally, R. J., Yang, P. 2001. Single nanowire lasers. *J. Phys. Chem. B* 105: 11387–11390.
84. Yang, P., Yan, H., Mao, S., Russo, R., Johnson, J., Saykally, R., Morris, N., Pham, J., He, R., Choi, H.-J. 2002. Controlled growth of ZnO nanowires and their optical properties. *Adv. Funct. Mater.* 12: 323–331.
85. Law, M., Green, L. E., Johnson, J. C., Saykally, R., Yang, P. 2005. Nanowire dye-sensitized solar cells. *Nat. Mater.* 4: 455–459.
86. Johnson, J. C., Yan, H., Schaller, R. D., Petersen, P. B., Yang, P., Saykally, R. J. 2002. *Nano Lett.* 2: 279–283.
87. Liu, L., Edgar, J. H. 2002. Substrates for gallium nitride epitaxy. *Mater. Sci. Eng. R* 37: 61–127.
88. Kuykendall, T., Pauzauskie, P., Lee, S., Zhang, Y., Goldberger, J., Yang, P. 2003. Metal organic chemical vapor deposition route to GaN nanowires with triangular cross sections. *Nano Lett.* 3: 1063–1066.
89. Kuykendall, T., Pauzauskie, P., Lee, S., Zhang, Y., Goldberger, J., Sirbuly, D., Denlinger, J., Yang, P. 2004. Crystallographic alignment of high-density gallium nitride nanowire arrays. *Nat. Mater.* 3: 524–528.
90. Choi, H.-J., Johnson, J. C., He, R., Lee, S.-K., Kim, F., Pauzauskie, P., Goldberger, J., Saykally, R. J., Yang, P., Choi, H. 2003. Self-organized GaN quantum wire UV lasers. *J. Phys. Chem. B* 107: 8721–8725.
91. Kim, H.-M., Cho, Y.-H., Lee, H., Kim, S., Ryu, Sung, R. D., Kim, Y., Kang, T. W., Chung, K. S. 2004. High-brightness light emitting diodes using dislocation-free indium gallium nitride/gallium nitride multi quantum-well nanorod arrays. *Nano Lett.* 4: 1059–1062.
92. Nakamura, S. 1998. The roles of structural imperfections in InGaN-based blue light-emitting diodes and laser diodes. *Science* 281: 956–961.
93. Foster, M. A., Turner, A. C., Sharping, J. E., Schmidt, B. S., Lipson, M., Gaeta A. L. 2006. Broadband optical parametric gain on a silicon photonic chip. *Nature* 441: 960–963.
94. Jacobsen, R. S., Andersen, K. N., Borel, P. I., Pedersen, J. F., Frandsen, L. H., Hansen, O., Kristensen, M., Lavrinenko, A. V., Moulin, G., Ou, H., Peucheret, C., Zsigri, B., Bjarklev, A. 2006. Strained silicon as a new electro-optic material. *Nature* 441: 199–202.
95. Xu, Q. F., Schmidt, B., Pradhan, S., Xu, M. L. 2005. Micrometre-scale silicon electro-optic modulator. *Nature* 435: 325–327.
96. Rong, H., Liu, A., Jones, R., Cohen, O., Hak, D., Nicolaescu, R., Fang, A., Paniccia, M. 2005. An all-silicon Raman laser. *Nature* 433: 292–294.
97. Rong, H., Jones, R., Liu, A., Cohen, O., Hak, D., Fang, A., Paniccia, M. 2005. A continuous-wave Raman silicon laser. *Nature* 433: 725–728.
98. Goldberger, J., Hochbaum, A. I., Fan, R., Yang, P. 2006. Silicon vertically integrated nanowire field effect transistors. *Nano Lett.* 6: 973–977.
99. Vashaee, D., Shakouri, A., Goldberger, J., Kuykendall, T., Pauzauskie, P., Yang, P. 2006. Electrostatics of nanowire transistors with triangular cross sections. *J. Appl. Phys.* 99: 054310–054315.
100. Goldberger, J., Sirbuly, D. J., Law, M., Yang, P. 2005. ZnO nanowire transistors. *J. Phys. Chem. B* 109: 9–14.

101. Qian, F., Li, Y., Gradečak, S., Wang, D., Barrelet, C. J., Lieber, C. M. 2004. Gallium nitride-based nanowire radial hetero structures for nanophotonics. *Nano Lett.* 4: 1975–1979.

102. Hayden, O. 2006. Nanoscale avalanche photodiodes for highly sensitive and spatially resolved photon detection. *Nat. Mater.* 5: 352–356.

103. Law, M., Sirbuly, D. J., Johnson, J. C., Goldberger, J., Saykally, R. J., Yang, P. 2004. Nanoribbon wave guides for sub wavelength photonics integration. *Science* 305: 1269–1273.

104. Riziotis, C., Vasilakos, A. V. 2007. Computational intelligence in photonics technology and optical networks: A survey and future perspectives. *Info. Sci.* 177: 5292–5315.

105. Sirbuly, D. J., Law, M., Pauzauskle, P., Yan, H., Maslov, A. V., Knutsen, K., Ning, C. Z., Saykally, R. J., Yang, P. D. 2005. Optical routing and sensing with nanowire assemblies. *Proc. Natl. Acad. Sci. USA* 102: 7800–7805.

106. Tao, A., Kim, F., Hess, C., Goldberger, J., He, R., Sun, Y., Xia, Y., Yan, P. 2003. Langmuir–blodgett silver nanowire monolayers for molecular sensing using surface-enhanced Raman spectroscopy. *Nano Lett.* 3: 1229–1233.

107. Yang, P. 2003. Wires on water. *Nature* 425: 243–244.

108. Yang, P., Kim, F. 2002. Langmuir–blodgett assembly of one-dimensional nanostructures. *Chem. Phys. Chem.* 3: 503–506.

109. Huang, J. X., Kim, F., Tao, A. R., Connoer, S., Yang, P. 2005. Spontaneous formation of nanoparticle stripe patterns through dewetting. *Nat. Mater.* 4: 896–900.

110. Grier, D. G. 2003. A revolution in optical manipulation. *Nature* 424: 810–816.

111. Glandorf, N. L., Perkins, T. T. 2004. Measuring 0.1-nm motion in 1 ms in an optical microscope with differential back-focal-plane detection. *Opt. Lett.* 29: 2611–2613.

112. Korda, P. T., Spalding, G. C., Dufresne, E. R., Grier, D. G. 2002. Nanofabrication with holographic optical tweezers. *Rev. Sci. Instrum.* 73: 1956–1957.

113. Ashkin, A., Dziedzic, J. M., Bjorkholm, J. E., Chu, S. 1986. Observation of a single-beam gradient force optical trap for dielectric particles. *Opt. Lett.* 11: 288–290.

114. Sato, S., Harada, Y., Waseda, Y. 1994. Optical trapping of microscopic metal particles. *Opt. Lett.* 19: 1807–1809.

115. Svoboda, K., Block, S. M. 1994. Optical trapping of metallic Rayleigh particles. *Opt. Lett.* 19: 930–932.

116. Friese, M. E. J., Nieminen, T. A., Heckenberg, N. R., Dunlop, H. R. 1998. Optical alignment and spinning of laser-trapped microscopic particles. *Nature* 394: 348–350.

117. Yu, T., Cheong, F.-C., Sow, C.-H. 2004. The manipulation of CuO nanorods by optical tweezers. *Nanotechnology* 15: 1732–1738.

118. Pauzauskie, P. J., Radenovic, A., Trepagnier, E., Shroff, H., Yang, P., Liphardt, J. 2006. Optical trapping and integration of semiconductor nanowire assemblies in water. *Nat. Mater.* 5: 97–101.

119. Ozbay, E. 2006. Plasmonics: merging photonics and electronics at nanoscale dimensions. *Science* 311: 189–193.

120. Busch, K., Freymann, G., von Linden, S., Mingaleev, S. F., Tkeshelashvili, L., Wegener, M. 2007. Periodic nanostructures for photonics. *Phys. Rep.* 444: 101–202.

121. García de Abajo, F. J. 2007. Light scattering by particle and hole arrays. *Rev. Modern Phys.* 79: 1267–1290.

122. Taflove, A., Hagness, S. C. 2005. *Computational Electrodynamics: The Finite-Difference Time-Domain Method*, 3rd edn., Artech House Publishers, Norwood, MA.

123. Farjadpour, A., Roundy, D., Rodriguez, A., Ibanescu, M., Bermel, P., Joannopoulos, J. D., Johnson, S. G. 2006. Improving accuracy by sub pixel smoothing in the finite-difference time domain. *Opt. Lett.* 31: 2972–2974.

124. Deinega, A., Valuev, I. 2007. Sub pixel smoothing for conductive and dispersive media in the finite-difference time-domain method. *Opt. Lett.* 32: 3429–3431.

125. Mohammadi, A., Jalali, T., Agio, M. 2008. Dispersive contour-path algorithm for the two-dimensional finite-difference time-domain method. *Opt. Express* 16: 7397–7406.

126. Tornberg, A.-K., Engquist, B. 2008. Consistent boundary conditions for the Yee scheme. *J. Comp. Phy.* 227: 6922–6943.

127. Jin, J. 2002. *Computational Electrodynamics: The Finite Element Method in Electromagnetics*, 2nd edn., John Wiley & Sons, New York.

128. Hesthaven, J., Warburton, T. 2002. Nodal high-order methods on unstructured grids—I. Time-domain solution of Maxwell's equations. *J. Comput. Phys.* 181(1): 186–221.

129. Niegemann, J., Konig, M., Stannige, K., Busch, K. 2009. Higher-order time-domain methods for the analysis of nano-photonic systems. *Phot. Struct. Fund. Appl.* 7: 2–11.

130. Warburton, T. 2006. An explicit construction of interpolation nodes on the simplex. *J. Eng. Math.* 56(3): 247–262.

131. Carpenter, M. H., Kennedy, C. A. 1994. Fourth-Order 2N-Storage Runge–Kutta Schemes, Tech. Re NASA-TM-109112, NASA Langley Research Center, VA, June.

132. Kanevsky, A., Carpenter, M. H., Gottlieb, D., Hesthaven, J. S. 2007. Application of implicit–explicit high order Runge–Kutta methods to discontinuous-Galerkin schemes. *J. Comput. Phys.* 225(2): 1753–1781.

133. Niegemann, J., Tkeshelashvii, L., Busch, K. 2007. Higher-order time-domain simulations of Maxwell's equations using Krylov-subspace methods. *J. Comput. Theor. Nanos.* 4(3): 627–634.

134. Busch, K., Niegemann, J., Pototschnig, M., Tkeshelashvili, L., Krylov, A. 2007. A subspace based solver for the linear and nonlinear Maxwell equations. *Phys. Status Solidi B* 244(10): 3479–3496.

135. Lu, T., Zhang, P., Cai, W. 2004. Discontinuous Galerkin methods for dispersive and lossy Maxwell's equations and PML boundary conditions. *J. Comput. Phys.* 200(2): 549–580.

136. Sacks, Z., Kingsland, D., Lee, R., Lee, J. 1995. A perfectly matched anisotropic absorber for use as an absorbing boundary condition. *IEEE Trans. Antennas Propagat.* 43(12): 1460–1463.

137. Nevels, R., Jeong, J. 2004. The time domain Green's function and propagator for Maxwell's equations. *IEEE Trans. Antennas Propagat.* 52: 3012–3018.

138. Press, W., Teukolsky, S., Vetterling, W., Flannery, B. 1992. *Numerical Recipes in C*, 2nd edn., Cambridge University Press, Cambridge, U.K.

139. Johnson, P. B., Christy, R. W. 1972. Optical constants of the noble metals. *Phys. Rev. B* 6(12): 4370–4379.

140. Joannopoulos, J. D., Meade, R. D., Winn, J. N. 1995. *Photonic Crystals: Molding the Flow of Light*, Princeton University Press, Princeton, NJ.

141. Sakoda, K. 2001. *Optical Properties of Photonic Crystals. Springer Series in Optical Sciences*, Springer-Verlag, Berlin, Germany.

142. Fan, S., Johnson, S. G., Joannopoulos, J. D., Manolatou, C., Haus, H. A. 2001. Waveguide branches in photonic crystals. *J. Opt. Soc. Am. B* 18: 162–165.

143. Costa, R., Melloni, A., Martinelli, M. 2003. Band pass resonant filters in photonic-crystal waveguides. *IEEE Photon. Tech. Lett.* 15: 401–403.

144. Sharkawy, A., Shi, S., Prather, D. W. 2001. Multichannel wave- length division multiplexing using photonic crystals. *Appl. Opt.* 40: 2247–2252.

145. Dowling, J. P., Scalora, M., Bloemer, M. J., Bowden, C. M. 1994. The photonic band edge laser: A new approach to gain enhancement. *J. Appl. Phys.* 75(4): 1896–1899.

146. Aguanno, G. D., Centini, M., Scalora, M., Sibilia, C., Bloemer, M. J., Bowden, C. M., Haus, J. W., Bertolotti, M. 2001. Group velocity, energy velocity, and superluminal propagation in finite photonic band-gap structures. *Phys. Rev. E* 63: 036610–036615.

147. Hache, A., Bourgeois, M. 2000. Ultrafast all-optical switching in a silicon-based photonic crystal. *Appl. Phys. Lett.* 77(25): 4089–4051.

148. Mingaleev, S. F., Miroshnichenko, A. E., Kivshar, Y. S., Busch, K. 2006. All-optical switching, bistability, and slow-light transmission in photonic crystal waveguide-resonator structures. *Phys. Rev. E* 74: 046603–046618.

149. Fukamachi, T., Hosomi, K., Sugawara, T., Kikuchi, N., Katsuyama, T., Arakawa, Y. 2005. Dispersion compensation in 40 Gb/s non-return-to-zero optical transmission system using coupled-cavity photonic crystals. *Appl. Phys.* 44(41): L1283–L1284.

150. Tsurumachi, N., Yamashita, S., Muroi, N., Fuji, T., Hattori, T., Nakatsuka, H. 1999. Enhancement of optical effect in one dimensional photonic crystals structures. *Jpn. J. Appl. Phys.* 38: 6302–6308.

151. Schiek, R. 1993. Nonlinear refraction caused by cascaded second-order nonlinearity in optical waveguide structures. *J. Opt. Soc. Am. B* 10: 1848–1855.

152. Inoue, M., Arai, K., Fujii, T., Abe, M. 1998. Magneto-optical properties of one-dimensional photonic crystals composed of magnetic and dielectric layers. *J. Appl. Phys.* 83(11): 6768–6771.

153. Wang, X., Hu, X., Li, Y., Jia, W., Xu, C., Liu, X., Zia, J. 2002. Enlargement of omnidirectional total reflection frequency range in one-dimensional photonic crystals by using photonic heterostructures. *Appl. Phys. Lett.* 80: 4292–4294.

154. Xi, J.-Q., Ojha, M., Plawsky, J. L., Gill, W. N., Kim, J. K., Schuberta, E. F. 2005. Internal high-reflectivity omni-directional reflectors. *Appl. Phys. Lett.* 87: 031111–031113.

155. Lee, H.-Y., Yao, T. 2003. Design and evaluation of omnidirectional one-dimensional photonic crystals. *J. Appl. Phys.* 93(2): 819–830.

156. Armenta, A. G., Mendieta, F. R., Villa, F. 2003. One-dimensional photonic crystals: Equivalent systems to single layers with a classical oscillator like dielectric function. *Opt. Commun.* 216: 361–367.

157. Choa, M. J., Choia, D. H., Sullivan, P. A., Akelaitis, A. J. P., Dalton, L. R. 2008. Recent progress in second order nonlinear optical polymers and dendrimers. *Prog. Polym. Sci.* 33: 1013–1058.

158. Denk, W., Strickler, J. H., Webb, W. W. 1990. Two-photon laser scanning fluorescence microscopy. *Science* 248: 73–76.

159. Lewis, M. K., Wolanin, P., Gafni, A., Steel, D. G. 1998. Near-field scanning optical microscopy of single molecules by femtosecond two-photon excitation. *Opt. Lett.* 23: 1111–1113.

160. Hell, S. W., Booth, M., Wilms, S., Schnetter, C. M., Kirsch, A. K., Arndt-Jovin, D. J., Jovin, T. M. 1998. Two-photon near- and far-field fluorescence microscopy with continuous-wave excitation. *Opt. Lett.* 23: 1238–1240.

161. Jakubczyk, D., Shen, Y., Lal, M., Friend, C., Kim, S. K., Swiatkiewicz, J., Prasad, P. N. 1999. Near-field probing of nanoscale nonlinear optical processes. *Opt. Lett.* 16: 1151–1153.

162. Fornel, F., Salomon, L., Adam, P., Bourillot, E., Goudonnet, J. P. 1992. Resolution of the photon scanning tunneling microscope: Influence of physical parameters. *Ultramicroscopy* 42: 422–429.

163. Dunn, A., Kortum, R. R. 1996. Three-dimensional computation of light scattering from cells. *IEEE J. Select. Top. Quant. Elect.* 2: 898–905.

164. Rytov, S. 1956. Electromagnetic properties of a finely stratified medium. *Soviet Phys. JETP* 2: 466–475.

165. Lalanne, P., Hugonin, J.-P. 1998. High-order effective-medium theory of sub wavelength gratings in classical mounting: Application to volume holograms. *J. Opt. Soc. Amer.* A15: 1843–1851.

166. Cheben, P., Bock, P., Schmid, J., Lapointe, J., Janz, S., Xu, D.-X., Densmore, A., Delâge, A., Lamontagne, B., Hall, T. 2010. *Opt. Lett.* 35(15): 2526–2528.

167. Cheben, P., Xu, D.-X., Janz, S., Densmore, A. 1999. Subwavelength waveguide grating for mode conversion and light coupling in integrated optics. *Opt. Exp.* 14(11): 4695–4702.

168. Tanev, S., Sun, W., Pond, J., Tuchin, V., Zharov, V. 2010. DTD simulation of light interaction with cells for diagnostics and imaging in nanobiophotonics, in: V. Tuchin (Ed.), *Handbook of Photonics for Biomedical Science, Series in Medical Physics and Biomedical Engineering*, CRC Press, Taylor & Francis Ltd., 2010 (Chapter 1).

169. Tanev, S., Sun, W., Pond, J., Tuchin, V., Zharov, V. 2009. Flow cytometry with gold nanoparticles and their clusters as scattering contrast agents: FDTD simulation of light-cell interaction. *J. Biophot.* 2(8–9): 505–520.

170. Tanev, S., Pond, J., Paddon, P., Tuchin, V. 2008. A new 3D simulation method for the construction of optical phase contrast images of gold nanoparticle clusters in biological cells. *Adv. Opt. Technol.* 727418–727427.

171. Tuchin, V. V. 2006. *Optical Clearing of Tissues and Blood, PM 154*, SPIE Press, Bellingham, WA.

172. Bock, P., Cheben, P., Schmid, J., Lapointe, J., Delâge, A., Xu, D.-X., Janz, S., Densmore, A., Hall, T. 2010. Sub wavelength grating crossings for silicon wire waveguides. *Opt. Express* 18(15): 16146–16155.

173. Bilo, V., Moscardelli, L. 2004. The price of anarchy in all-optical networks. *Structural Information and Communication Complexity, Proceedings 2004, Lecture Notes in Computer Science*, 3104: 13–22.

174. Bilo, V., Flammini, M., Moscardelli, L. 2005. On Nash equilibria in non-cooperative all-optical networks. *STACS 2005, Proceedings, Lecture Notes in Computer Science* 3404: 448–459.

3

Nanowire Photonics and Their Applications

Sun-Kyung Kim, Thomas J. Kempa, Charles M. Lieber, and Hong-Gyu Park

CONTENTS

3.1 Introduction

Nanostructures have enabled numerous novel and multifunctional devices with impact in many areas of science and technology [1–33]. Precise synthetic control of nanomaterial parameters, including chemical composition, size, and morphology, is key to eliciting unique device properties and thus applications [1–11,25–33]. Semiconductor nanowires are particularly attractive building blocks for integrated nanosystems because they can function as both device elements and interconnects [1–11,29,33]. This concept has been demonstrated in nanoelectronics with the assembly of devices such as field-effect transistors and integrated nanowire logic gates [8–11] and in nanophotonics with the assembly of individual light-emitting diodes and laser diodes [4–7,29,33].

One attractive feature of nanowires is that small structural changes can lead to dramatic changes in their optical properties [3–7,29–40]. As a result, such nanostructures allow for significant modulation of optical properties as compared to conventional (e.g., bulk material) structures and thus can enable efficient photonic devices. Recent studies have demonstrated that semiconductor nanowires can function as photonic waveguides [7,33]. Semiconductor nanowires can act as high-bandwidth, high-speed, and ultrasmall optoelectronic components in a compact and high-density integrated circuit [4–7,29,33]. In addition, semiconductor nanowires have been implemented as nanolasers characterized by a single cavity mode, low lasing thresholds, and low power consumption [5,6,34]. Furthermore, nanowire photovoltaic devices exhibit enhanced light absorption compared to conventional planar devices owing to highly confined resonant modes and large absorption cross sections [3,30–32,35–40]. Nanowire photovoltaic devices also demonstrate an ability to selectively tune several or select resonant modes through morphological changes [3,30,36].

The photonic properties of nanowires can be accurately predicted by analytical solutions or numerical computations [3,30–40]. Simulations have successfully supported various experiments in nanomaterial photonics and contributed to deeper understanding of light–matter interaction at length scales of approximately 20–300 nm [3,30–33,35,36]. While analytical solutions provide fast asymptotic results, they typically assume circular symmetry and a homogeneous environment and are therefore limited for analysis of relatively simple nanowire systems [30,31,35,38–41]. On the other hand, numerical full-field electromagnetic simulations are less constrained, having the ability to represent more complex structures and to provide quantitative as well as qualitative analyses. Both analytical and numerical computational methods are used extensively to design nanostructures.

This chapter discusses 1D nanowire materials, their optical properties, and their applications in practical photonic devices. Section 3.2 will focus on nanowire photovoltaic devices and their unique optical properties. We will illustrate that light absorption in a nanowire depends significantly on its morphology. Section 3.3 will investigate modal characteristics such as quality and confinement factors and nonlinear properties in nanowire cavities. Light confinement in sub-wavelength-scale plasmonic cavities will also be discussed.

3.2 Nanowires as Optical Absorbers

Semiconductor nanowires can be regarded as sub-wavelength optical cavities and thus can possess optical properties distinct from bulk structures [3,30]. For example, in contrast to its bulk counterpart, a nanowire with the same absorption coefficient will show morphology-dependent absorption spectra [30]. Light absorption in nanowires is dictated by several parameters: (1) the absorption coefficient of the nanowire material, (2) the coupling efficiency between incident light and a nanowire, (3) the spatial profile of resonant modes excited in a nanowire cavity, and (4) the nanowire size-dependent optical antenna effect, which means that a nanowire can interact with incident photons beyond its projected area. Together, these factors can lead to an overall enhancement in the absorption efficiency of a nanowire.

3.2.1 Basic Computational Methods

An absorption spectrum illustrates how nanostructures interact with incident photons. To understand the optical properties of nanowires and their assembled structures, it is crucial to calculate an absorption spectrum that accurately reproduces the frequency and amplitude of resonant features.

Analytical Lorenz–Mie scattering solutions or numerical full-field electromagnetic simulations are widely used to verify experiments and understand light–matter interaction in nanostructures. In the Lorenz–Mie approach, both incident plane wave and scattered field are expanded into spherical vector wave functions, and the expansion coefficients of the scattered field are calculated using boundary conditions on the spherical surface [41]. The Lorenz–Mie approach can be applied to optimize nanowire optical cavities to first order and thus is useful for quickly surveying a wide parameter space prior to more refined numerical simulation. Lorenz–Mie accurately predicts properties for highly confined optical systems wherein the field is marginally perturbed by an ambient medium [30]. However, the Lorenz–Mie solution is restricted to symmetric structures contained within a homogeneous medium (Figure 3.1a). Moreover, as the size of a nanowire becomes smaller or the wavelength of incident light becomes longer, the effective length of the evanescent field, which is associated with resonant modes supported within the nanowire, will extend appreciably outside of the nanowire. Due to this scenario, there can be substantial discrepancy between experimental results and the result predicted by Lorenz–Mie.

While the Lorenz–Mie solutions provide qualitative agreement with experimental results, full-field electromagnetic simulations including finite-difference time-domain (FDTD) [3,34] or finite-difference frequency-domain (FDFD) methods [31,32] are not limited by any geometrical constraints and can simulate an experimental system accurately. For example, field distributions can be described exactly for a nanowire positioned on a substrate (Figure 3.1b). In full-field electromagnetic simulations, the Maxwell equations are solved with simulation precision and accuracy determined by grid size in the calculation domain and appropriate choice of dispersion properties. It is thus crucial to accurately model the dispersive properties of materials used in the simulations [3,32,42].

In the FDTD method, the absorption efficiency and current density of a nanowire are calculated as follows [3]. First, the absorption cross section of the nanowire is calculated by integrating J·E at each grid point within a nanowire, while a normally incident plane wave with transverse electric (TE) or transverse magnetic (TM) polarization states

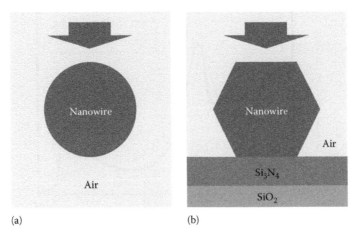

(a) (b)

FIGURE 3.1
Schematics of two simulation methods. (a) Lorenz–Mie scattering analytical solution and (b) full-field electromagnetic method. A nanowire interacts with a normally incident plane wave.

interacts with a nanowire cavity; J and E are the polarization current density and electric field, respectively. The absorption cross section is averaged over one optical cycle. Next, the absorption efficiency is obtained by the ratio of the absorption cross section to the projected area of the nanowire. The wavelength of the incident plane wave is iterated with a finite step size, covering the whole spectral range of nanowire material. The short-circuit current density is obtained by multiplying the absorption spectrum by solar irradiance (AM 1.5 G spectrum) over the spectral range under consideration in experiment. To describe a single nanowire placed on an indefinite substrate, periodic boundary conditions are applied along the axis of a nanowire and the total-field scattered-field (TFSF) method is used [43]. The TFSF allows a single nanowire to experience an infinite plane wave.

To demonstrate the accuracy of the two computational methods, we consider the external quantum efficiency (EQE) spectra for a single Ge nanowire photodetector (Figure 3.2a) [31]. Measured EQE is compared to the calculation results obtained from FDFD simulation and the Lorenz–Mie scattering solution (Figure 3.2b). The measured resonant frequencies and amplitudes are in better quantitative agreement with the full-field electromagnetic

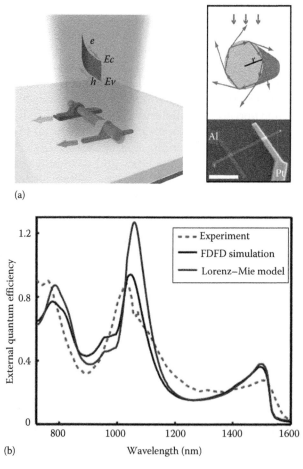

FIGURE 3.2
(a) Schematic of a Ge nanowire photodetector. (b) Experimental and calculated spectra of EQE from a Ge nano-wire device with a diameter of 280 nm. (Adapted from Cao, L. et al., *Nano Lett.*, 10, 1229, 2010.)

simulation, which more faithfully represents the cross-sectional shape of and substrate supporting the nanowire.

3.2.2 Single Horizontally Oriented Nanowires

Single semiconductor nanowires can sustain localized resonant modes and optical antenna effects, which, when verified and tailored, have important implications for optoelectronic applications [3,30,32]. In particular, nanowire photovoltaics are emerging as promising candidates for third-generation solar cell technology [2,3,44–53]. To understand the absorption properties of a nanowire cavity, several experiments have been performed to measure the wavelength-dependent light absorption [3,30,35,46,54,55] and scattering cross sections [32] of these structures. Concurrently, a number of theoretical analyses based on numerical simulations [3,30–32] or analytical methods [31,35,38–40] have been reported.

Resonant modes and optical antenna effects differentiate nanowires from their bulk counterparts (Figure 3.3). Localized resonant modes in a nanowire cavity are easily perturbed by subtle changes in structure, and this fact can be exploited to tune or amplify specific resonant modes. Also, optical antenna effects allow a nanowire cavity to interact with incident light beyond its projected area, which results in extended absorption and scattering cross sections for some wavelengths [3,21,30,32].

One way to quantify accurately the absorption properties of a nanowire in experiment is to make a functional single-nanowire photovoltaic device with high internal quantum efficiency. Recently, p/in core–shell crystalline Si nanowire photovoltaics were successfully demonstrated [3]. A coaxial Si photovoltaic device is fabricated by depositing lithographically defined metal contacts to the p-type core (exposed after brief wet etching) and n-type shell of a previously synthesized Si nanowire (Figure 3.4). In addition to enhanced light absorption resulting from the sub-wavelength cavity, the coaxial nanowire structure also exhibits high-quality electrical performance involving efficient charge collection and low recombination loss. To support and understand the measured photocurrent spectra of nanowire photovoltaics or photodiodes, computational methods such as full-field electromagnetic simulations or Lorenz–Mie scattering solutions are used.

3.2.2.1 Optical Antenna Effects and Size- and Morphology-Dependent Resonant Modes

In a solar absorber, one can readily enhance light absorption or tune specific resonant modes by increasing the thickness of the absorbing material. For example, absorption

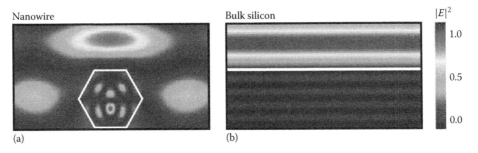

FIGURE 3.3
Electric field intensity distribution of a nanowire (a) and a bulk structure (b) when they interact with normally incident plane wave. (Adapted from Kempa, T.J. et al., *Proc. Natl. Acad. Sci. USA*, 109, 1407, 2012.)

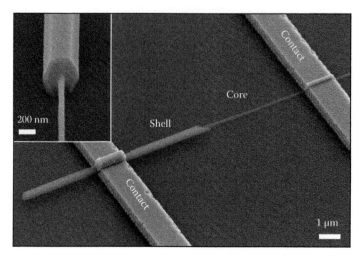

FIGURE 3.4
SEM image of a p/in core–shell Si nanowire photovoltaic device. (Adapted from Kempa, T.J. et al., *Proc. Natl. Acad. Sci. USA*, 109, 1407, 2012.)

efficiency increases with increasing thickness of a bulk Si structure, particularly for long wavelengths where the efficiency is not saturated (Figure 3.5).

In contrast to a bulk Si structure, it is less straightforward to predict the wavelength-dependent absorption behavior of a nanowire due to its optical antenna effects and localized resonant modes. The optical antenna effect from a nanowire cavity is one of the primary determinants of absorption efficiency. The scattering cross section of a nanowire becomes larger than its physical cross section due to the optical antenna effect, and absorption efficiency is enhanced as a result. The optical antenna effect in a nanowire can be quantified by measuring [32] or calculating its scattering cross section [38].

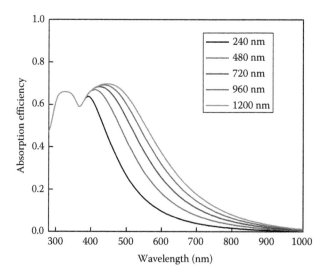

FIGURE 3.5
Calculated absorption spectra of bulk Si structures with Si thicknesses of 240, 480, 720, 960, and 1200 nm. A 40 nm thick SiO_2 is coated on the top.

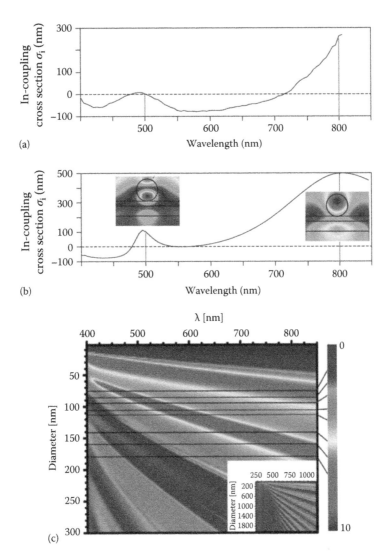

FIGURE 3.6
Measured (a) and calculated (b) scattering spectrum for a Si nanowire with a height of 90 nm. The absorption cross section increases on resonance. (Adapted from Barnard, E.S. et al., *Nat. Nanotechnol.*, 6, 588, 2011.) (c) Calculated scattering spectrum of a Si nanowire as a function of the size of nanowire. The absorption cross section increases at a smaller nanowire diameter. (Adapted from Bronstrup, G. et al., *ACS Nano*, 4, 7113, 2010.)

In Figure 3.6, a nanowire is placed above a metal–Si–metal photodiode with Schottky contacts, and the generated photocurrent is directly proportional to the square of the local electric field near the nanowire antenna. To understand the measured resonant behavior, FDFD simulation is carried out (Figure 3.6a and b). In Figure 3.6c, a Lorenz–Mie calculation is used to generate a 2D map of the Si nanowire scattering cross section as a function of incident wavelength and nanowire size. Such results reveal two important features. First, the optical antenna effects are maximized at each resonance and the scattering spectrum is highly structured with multiple discrete peaks. In Figure 3.6a and b, the full-field electromagnetic simulation shows good agreement with the measured result. A sharp peak at ~500 nm and a broad peak at ~800 nm correspond to monopolar and dipolar resonances, respectively. Second, optical antenna effects are enhanced with decreasing nanowire size. As shown in Figure 3.6c, when the nanowire size is smaller or the wavelength of incident light is longer, the scattering cross section increases. The 2D dispersion map can be useful when designing nanowires with enormously large scattering cross sections [21].

The second important factor affecting the absorption efficiency of a nanowire is the presence of localized resonant modes [3,30,34,35]. Since a nanowire can be regarded as a waveguide with a high refractive index, resonant modes in a nanowire cavity can be identified using standard nomenclature (l, m) and according to their polarization states, TE or TM (Figure 3.7a) [30,35]. In order to couple effectively to resonant modes of a cavity, incident photons must satisfy stringent conditions including momentum matching as well as frequency matching. For this reason, it is important to note that not all calculated modes will necessarily appear in a measured absorption spectrum. In the 2D dispersion map showing the absorption cross section (Figure 3.7b), all discrete bands above ~400 nm are shifted monotonically with increasing nanowire size.

More interesting features can be monitored by comparing the absorption spectrum of a smaller nanowire with that of a larger nanowire (Figure 3.8). The measured photocurrent spectra are acquired from Si nanowire photovoltaic devices and simulated spectra are performed by FDTD computations. The absorption spectra of Figure 3.8a illustrate several key features. First, the number of peaks in the absorption spectrum increases with nanowire size. Each peak corresponds to resonant modes excited in the nanowire cavity and is classified as 2D whispering-gallery (WG) or 1D Fabry–Perot modes (Figure 3.8b). Second, in a larger nanowire, all absorption peaks are redshifted, while the mode profiles corresponding to those peaks are preserved. In addition to increasing the density of modes sustained by a nanowire cavity, one can increase the current density of a nanowire photovoltaic device by superimposing some resonant peaks onto maxima of the solar spectral irradiance [3]. Third, the maximum EQE amplitude for the larger nanowire is somewhat decreased as compared to the smaller nanowire owing to a reduced optical antenna effect from the larger nanowire [3,32,38].

The current density of a nanowire is calculated by integrating the product of its absorption spectrum with the solar spectrum. Figure 3.9 shows calculated current density as a function of nanowire size. As a result of the competition between optical antenna effects and the number of absorption peaks, the resultant current density exhibits a nonmonotonic trend as a function of nanowire size. As the nanowire size increases, an increased number of absorption peaks emerge in the spectrum, while optical antenna effects are weakened. Consequently, a local maximum in current density is observed at a nanowire diameter of <100 nm (Figure 3.9). Such local absorption maxima underscore a general advantage of nanomaterial devices, namely, the possibility of achieving superior absorption, and therefore device performance, compared to planar or bulk systems, which would consume more material.

3.2.2.2 Cross-Sectional Dependence

The aspect ratio and cross-sectional shape of an optical cavity can dictate the type of resonant modes, which can be efficiently excited. For example, as an optical cavity approaches a semi-infinite bulk structure, 1D Fabry–Perot modes predominate [56]. On the other hand, as one critical dimension of an optical cavity approaches a length scale comparable to the incident wavelength, 2D resonant modes can emerge. In such wavelength-scale optical cavities, even subtle changes in the cavity cross section may influence absorption

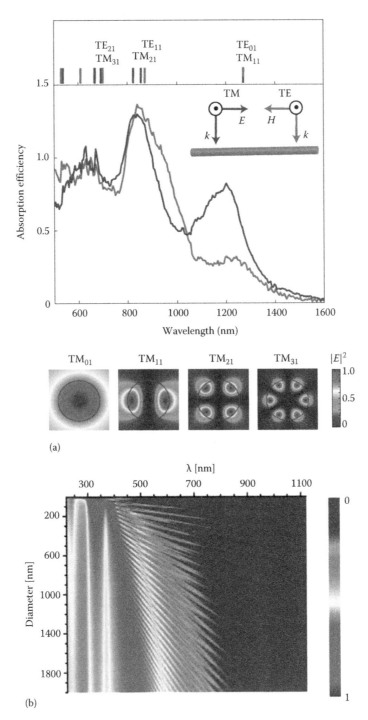

FIGURE 3.7
(See color insert.) (a) Measured absorption spectrum of a Ge nanowire with a diameter of 220 nm and the electric field intensity profiles for several resonant modes. (Adapted from Cao, L. et al., *Nat. Mater.*, 8, 643, 2009.) (b) Calculated absorption spectrum of a Si nanowire as a function of nanowire size. (Adapted from Bronstrup, G. et al., *ACS Nano*, 4, 7113, 2010.)

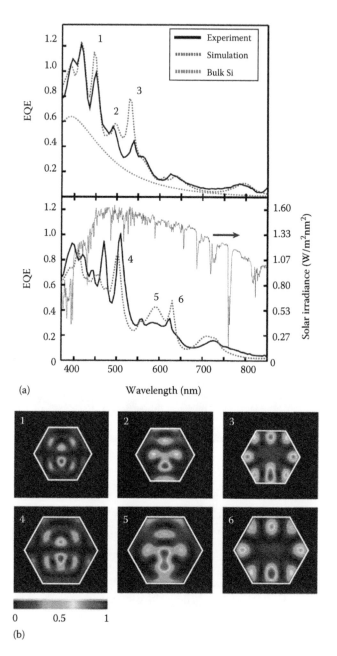

FIGURE 3.8
(a) Measured and calculated absorption spectra of Si nanowires with a height of 240 nm (top) and 305 nm (bottom). (b) Absorption intensity of resonant modes marked in the spectra shown in (a). (Adapted from Kempa, T.J. et al., *Proc. Natl. Acad. Sci. USA*, 109, 1407, 2012.)

properties. Gas-phase synthesis affords precise control over the morphology of semiconductor nanowires and enables a greater diversity of crystal structure and shape (e.g., cross section), which might otherwise not be feasible through conventional top-down fabrication approaches [2,3,50]. In principle, it is possible to manipulate the semiconductor nanowire structure by exploiting its inherent crystal facets during a specially tailored synthesis.

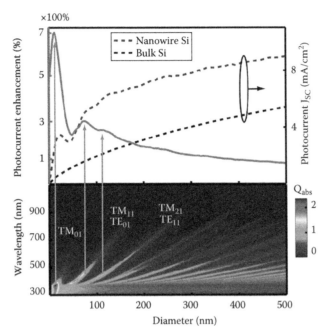

FIGURE 3.9
Top: calculated current density for Si nanowires (gray dashed) and for bulk silicon (black dashed) as a function of their size. The solid line indicates the photocurrent enhancement of the nanowire per unit volume of material. Bottom: 2D plot of the calculated absorption efficiency of a Si nanowire as a function of wavelength and diameter. All the calculations are performed by Lorenz–Mie scattering solution. (Adapted from Cao, L. et al., *Nano Lett.*, 10, 439, 2010.)

The current density and absorption spectra of nanowires with various cross sections (i.e., square, circle, hexagon, and triangle) can be quantified accurately only by performing full-field electromagnetic simulations. If a nanowire structure is entirely composed of highly absorptive material such as amorphous Si [30], the short attenuation depth of photons will reduce optical feedback within the cavity leading to weak excitation of resonant modes. In such a scenario, one can anticipate that a designed morphological change in the highly absorptive nanowire will not significantly influence absorption efficiency. As shown in the simulation results (Figure 3.10), amorphous Si nanowires with different cross sections do not exhibit significantly distinct absorption features in current density, absorption spectrum, or resonant mode profile. The difference in the current densities results only from the difference of cross-sectional area. It is also noted that optical antenna effects are retained in any cross-sectional shape.

On the other hand, when a nanowire is composed of less absorptive material, such as single-crystalline Si, the attenuation depth becomes much larger than the nanowire size, allowing extensive optical feedback within the cavity and thus the excitation of well-defined resonant modes. As a result, the optical properties of a single-crystalline Si nanowire could be more readily influenced by its cross-sectional shape with the concomitant enhancement or suppression of specific absorption peaks [36]. Figure 3.11 shows the absorption spectra of rectangular nanowires whose size is varied while maintaining an aspect ratio of 1:1.13. The width and height of the $Si_3N_4/SiO_2/Si$ substrate are infinitely large. The thicknesses of Si_3N_4, SiO_2, and Si are 200, 100 nm, and infinite, respectively. Resonant peaks are tuned gradually by changing the size of the rectangular cross section,

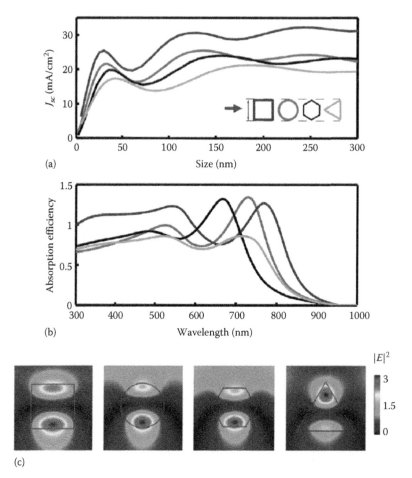

FIGURE 3.10
(a) Calculated current densities of amorphous Si nanowires with various cross-sectional shapes as a function of the size of the nanowire. (b) Calculated absorption spectra of each nanowire considered in (a). The size of each nanowire corresponds to that associated with the second maximum peaks shown in (a). (c) Electric field distribution from resonant peaks at ~700 nm shown in (b). (Adapted from Cao, L. et al., *Nano Lett.*, 10, 439, 2010.)

presenting the possibility of creating a photodetector with broad bandwidth through integration of multiple rectangular nanowires of different sizes. Although the current density of a nanowire is not improved significantly by a change in its cross-sectional shape, enhanced absorption at specific wavelengths could be useful for designing an efficient photodetector, which would operate in a narrow range of wavelengths.

3.2.2.3 External Effects: Substrate and Dielectric Shell

The absorption efficiency of a photovoltaic device is proportional to the interaction time of incident light with the absorbing material, provided that absorption is not completely saturated at the wavelengths under consideration. Therefore, it is important to design an optical cavity, which effectively traps photons inside a photovoltaic device. The introduction of an appropriately designed substrate [57] and/or dielectric shell on a nanowire [37,40,58] can increase its absorption efficiency.

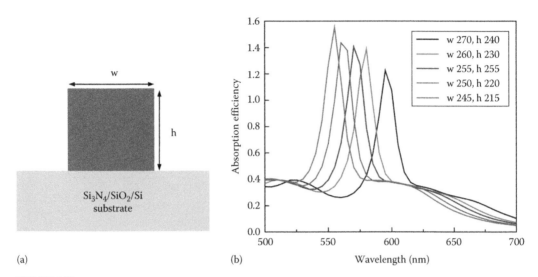

(a)   (b)   Wavelength (nm)

FIGURE 3.11
(a) Schematic of a nanowire with a rectangular cross section. (b) Absorption spectra of rectangular nanowires with different sizes, (w, h) = (270, 240 nm), (260, 230 nm), (255, 225 nm), (250, 220 nm), and (245, 215 nm). The aspect ratio is fixed at ~0.87. The TFSF method was used in the simulation.

In general, a bottom substrate reduces optical cavity loss and thus allows photons to interact with the cavity for longer times (Figure 3.12a). For a bulk Si structure, resonant peaks appear only in the presence of a bottom reflecting substrate as a result of multiple optical feedbacks (Figure 3.12b). The amplitude of a resonant peak depends on the reflectivity of the substrate and increases significantly at every resonance.

On the other hand, the substrate effect is more pronounced for a nanowire because of its good resonator properties. A substrate underneath a nanowire enhances the

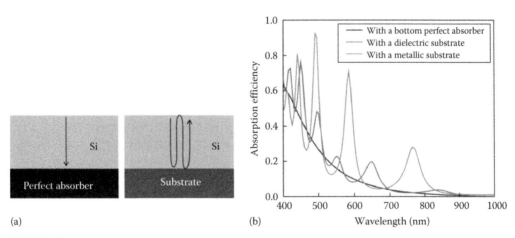

(a)   (b)   Wavelength (nm)

FIGURE 3.12
(a) Schematics of a bulk Si structure with and without bottom substrate. (b) Calculated absorption efficiency of a bulk Si structure with and without bottom substrate. The dielectric substrate is composed of 200 nm thick Si_3N_4, 100 nm thick SiO_2, and Si. Silver is used as a metallic substrate. The thickness of Si is 240 nm for all cases. The width of Si, Si_3N_4, and SiO_2 are all infinite. In simulation, we used periodic boundary condition in the horizontal direction.

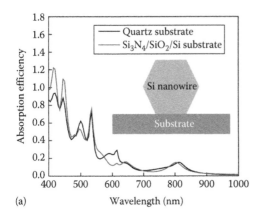
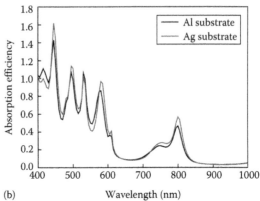

(a)   Wavelength (nm)   (b)   Wavelength (nm)

FIGURE 3.13

Calculated absorption efficiency of a Si nanowire on dielectric substrate (a) or metal substrate (b). The height of a nanowire is 240 nm. The substrate is infinitely large. The TFSF method was used in the simulation.

amplitude of all cavity modes. In particular, the absorption efficiency of a nanowire increases dramatically when it is positioned on metal substrate compared to its efficiency on a dielectric substrate (Figure 3.13). Significantly, the difference in absorption efficiency cannot be simply attributed to the difference in reflectance of the two substrates. The enhancement of absorption efficiency is primarily due to amplification of resonant modes and optical antenna effects within a nanowire. In general, a nanowire cavity sustains both 1D and 2D resonant modes such as Fabry–Perot and WG modes, whereas a bulk structure sustains only 1D Fabry–Perot modes. A metal substrate acting as an omnidirectional mirror will be able to effectively increase the interaction time between 2D modes in a nanowire cavity and incident light. Consequently, the absorption efficiency of a nanowire on a metal substrate (silver or aluminum) can exceed unity for several wavelengths. On the other hand, a dielectric substrate consisting of a single or multiple layers, which can be introduced to tailor the reflectance at normal incidence, only marginally influences the absorption efficiency of a nanowire [3].

While a dielectric substrate has nominal influence on the absorption efficiency of a nanowire resonator, a conformal dielectric coating presents a useful route toward increasing light absorption. Reflection losses can be reduced by introducing a dielectric layer at the interface between a photovoltaic device and its ambient medium. In a bulk structure, a dielectric coating induces destructive interference of successively reflected waves (Figure 3.14a). In contrast, the conformal dielectric coating on a nanowire assists trapping of photons inside the nanowire while also reducing the backscattering losses (Figure 3.14b). Essentially, a dielectric shell on a nanowire increases the probability for photons to couple to the cavity leading to an enhancement of current density.

A theoretical study based on Lorenz–Mie scattering was performed to investigate the enhancement of light absorption and its dependence on the thickness and refractive index of a dielectric shell [40,58]. In a cylindrical nanowire with a SiO_2 dielectric shell, the calculated absorption spectrum reveals two key features (Figure 3.15). First, the absorption efficiency is greatly enhanced at off-resonance wavelengths as compared to an uncoated nanowire, which is also observed in a bulk structure with an antireflection coating. However, the absorption efficiency in a nanowire increases steadily with increasing the thickness of the dielectric shell when the size of a nanowire is larger than ~200 nm [58]. In this case, the dielectric shell enhances the optical antenna effect that

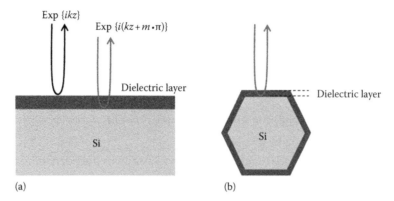

FIGURE 3.14
(a) Schematic of antireflection dielectric coating on a bulk structure. (b) Schematic of a nanowire with a conformal dielectric coating.

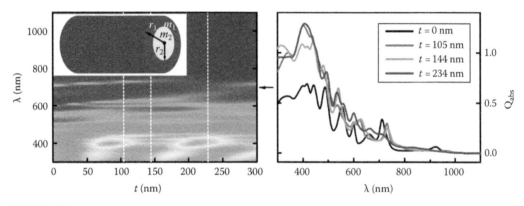

FIGURE 3.15
(See color insert.) The absorption spectrum of a Si/SiO_2 core/shell nanowire (inset) calculated using Lorenz–Mie theory. The diameter of Si core is 300 nm and the thickness of SiO_2 shell is t. (Adapted from Liu, W. and Sun, F., *Adv. Mater. Res.*, 391–392, 264, 2012.)

additionally augments the light absorption of a nanowire. Second, all resonant modes are redshifted with the increasing thickness of the dielectric shell, while their mode profiles are unchanged [37,40]. The redshift behavior is clearly seen from the modes located at ~700 nm in the absorption spectrum (Figure 3.15). The substantial shift in wavelength stems from extension of the evanescent field beyond the physical dimensions of the nanowire, one of the unique characteristics of such a sub-wavelength cavity. In general, tailoring substrate and dielectric shell environments represents a simple but powerful route toward enhancing light absorption in nanowire optical systems. Furthermore, these effects can be robustly applied to large-scale nanowire array systems.

3.2.2.4 Vertical Stacks and Horizontal Arrays

A single nanowire shows many intriguing absorption properties that depend on localized resonant modes and optical antenna effects. However, it is crucial to integrate multiple nanowires in parallel in order to achieve scaled power outputs sufficient to drive electronic circuits at the nano-, micro-, and macroscale [3,25,57,59]. Nanowire subunits can be

(a)   (b)   (c)

FIGURE 3.16
Schematics of horizontal array (a) and vertical stack (b) composed of two nanowires and nanowire woodpile (c).

stacked vertically one by one [3,60] or arranged periodically in the horizontal plane to yield multicomponent assembled structures (Figure 3.16). In vertical nanowire stacks, nanowires with different materials or morphologies can form tandem cells showing broadband absorption. Also, in conjunction with the vertical stack and horizontal array, 3D nanowire woodpile structures [15] can be proposed as an ultimate solar absorber with high light absorption efficiency.

The absorption spectrum of a vertical nanowire stack is compared to that of a single nanowire by FDTD simulations and photocurrent measurements (Figure 3.17). As shown in Figure 3.17a, the absorption efficiency of a vertical nanowire stack is significantly increased relative to a single nanowire over the entire spectral range, and this increase continues as additional nanowires are stacked. For a broad range of blue wavelengths, the absorption efficiency is larger than unity. In addition, absorption profiles at resonance show that modes are well preserved in the top and bottom nanowires. The broadband enhancement observed in the spectrum is due to optical antenna effects that mitigate screening by allowing nanowires deeper within the stack to absorb a substantial amount of photons. Ultimately, this effect supports more balanced absorption intensity in each stacked nanowire and thus helps mitigate the typical attenuation observed in bulk film absorption (Figure 3.17b).

For two Si nanowires vertically stacked and integrated to function as a photovoltaic device, measurements show that the short-circuit current density is as large as 25 mA/cm^2 with a silver backside reflector. Numerical simulations show that efficiencies of >15% can

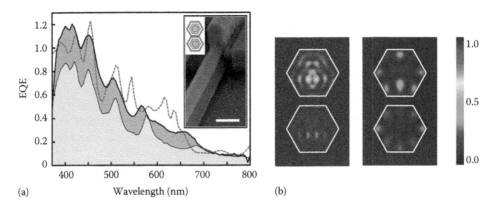

(a)   (b)

FIGURE 3.17
(a) Measured (solid) and simulated (dashed) absorption spectra of a single-nanowire (a bottom curve) and a two-nanowire vertical stack (top curves). Each Si nanowire is ~240 nm thick. (b) Absorption mode profiles (TM) of the two-nanowire stack shown in (a) at the wavelengths of 460 nm (left) and 540 nm (right). (Adapted from Kempa, T.J. et al., *Proc. Natl. Acad. Sci. USA*, 109, 1407, 2012.)

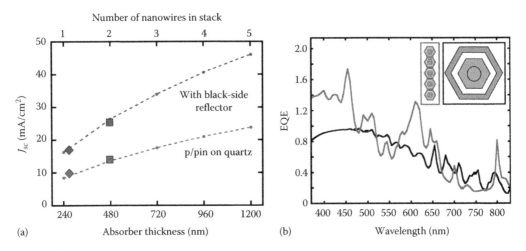

FIGURE 3.18
(a) Dashed curves denote simulated current densities for stacked nanowires with increasing number of nanowires in the stack (1, 2, 3, 4, 5) without and with a backside reflector, respectively. (b) Simulated absorption spectra of a five-nanowire vertical stack (gray) and a micron wire (black). For both structures, the total thickness of Si is 1200 nm. (Adapted from Kempa, T.J. et al., *Proc. Natl. Acad. Sci. USA*, 109, 1407, 2012.)

be attained for ~1 µm thick stacks of 5 nanowires (Figure 3.18a). Unlike a bulk structure or a micron wire, the current density of a nanowire stack retains linearity with respect to the number of added nanowires, and this once again is due to broadband enhancement in the absorption spectrum (Figure 3.18b).

Horizontally oriented nanowire arrays are another important structure and can be composed of closely packed or periodically arranged nanowires with arbitrary filling fraction [61]. A fundamental question is whether the current density of a single nanowire can be preserved in a nanowire array. The optical antenna effects that are partially responsible for enhanced light absorption in a single nanowire are either lost or severely attenuated in a nanowire array. Fortunately, however, the amplitudes of absorption peaks can be maintained to nearly the same values as for a single nanowire, thanks to a diffraction effect of the periodic nanowire array [37]. Periodically modulated surfaces generally yield broadband antireflection, and in this case, the current density of a closely packed nanowire array should be nearly identical to that of the single nanowire. Most importantly, an important design rule for assembled multicomponent nanowire solar cells emerges. The current density of any type of nanowire can be preserved in an array of that nanowire. While individual nanowire photovoltaics allow for optimization of fundamental factors affecting device efficiency, another important step will be to create large-area arrays of these nanowires to enable a new of generation solar cells.

3.2.2.5 Plasmonic Photovoltaic Devices

We have discussed several single-nanowire photovoltaic structures with enhanced light absorption. In this section, we will introduce a new type of photovoltaic device, which exploits surface plasmon polaritons (SPPs). SPPs are electromagnetic waves localized at the interface between metals and dielectrics [24,62,63]. The field intensity of SPPs is maximal at metal–dielectric interfaces and this field exponentially decays in both directions. In general, excitation of SPPs at metal–dielectric interfaces depends upon satisfaction of a phase-matching condition, which can be achieved through incorporation of a groove, textured

FIGURE 3.19
Schematic illustrations showing the interaction between normally incident plane wave and a bulk structure with metallic particles or metallic grating. (Adapted from Atwater, H.A. and Polman, A., *Nature Mater.*, 9, 205, 2010.)

surface, or grating. SPPs can be exploited to excite new resonant modes in photovoltaic devices. For example, when interfaced with a bulk structure, metallic structures function either as scattering elements to couple and trap incident plane waves (left, Figure 3.19) or as optical antennas that result in increased absorption cross section (middle, Figure 3.19). Metallic gratings or corrugated films can also convert sunlight into SPP modes (right, Figure 3.19).

The absorption enhancement due to SPPs can be calculated using full-field electromagnetic simulations [64–67]. When a 1D silver grating coupler is placed on a Si film (Figure 3.20a), incident photons with particular frequencies can be coupled to SPP resonant modes. The evanescent field of the SPP mode increases the absorption efficiency at that wavelength (Figure 3.20b). Absorption enhancement due to coupled SPP modes is predicted to be much greater than that due to any conventional dielectric structure, because SPP fields are highly concentrated at a metal–dielectric interface. In addition, because SPP modes can be tuned by the pitch size of the grating (Figure 3.20c), light absorption can be readily enhanced at specific wavelengths. However, it is worth noting that there is a broad range of wavelengths for which light absorption in Si is reduced due to plasmonic absorption loss in the metal [68,69]. Therefore, in order to achieve efficient light absorption, it is crucial to reduce overlap of the SPP fields with the metal [63,70].

Recently, a related approach was used to excite SPPs in single-nanowire devices [66,67]. Gold nanoparticles were dispersed on a single nanowire (Figure 3.21a) and were shown to induce SPP modes at specific wavelengths so that the absorption could be enhanced (Figure 3.21b). The enhancement in light absorption was shown to depend exponentially on the gap between the gold nanoparticle and nanowire. Although semiconductor nanowires combined with metallic nanostructures may offer a new opportunity to increase light absorption at discrete or a broad range of wavelengths, further study will be necessary to achieve an overall enhancement in current density while maintaining good device electrical properties.

3.3 Nanowire Active Devices

3.3.1 Nanowire Optical Cavities

Chemically synthesized semiconductor nanowires typically have diameters and lengths that can be controlled over a range of 20–200 nm and 2–40 µm, respectively. A large refractive index contrast, necessary for strong light confinement, is a prerequisite to forming a

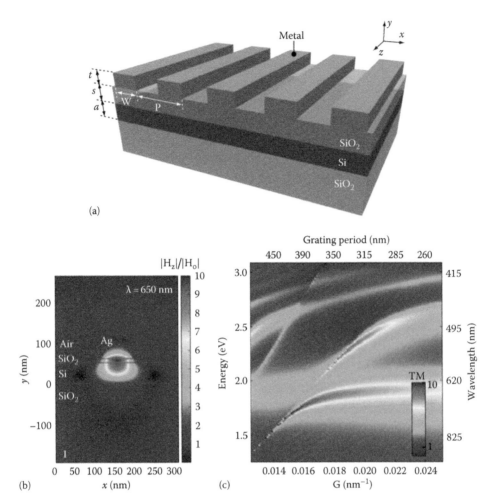

FIGURE 3.20
(a) Schematic of thin Si film with metallic grating as an SPP-boosted photovoltaic device. (b) Calculated time-averaged magnetic field intensity (TM) of a Si film with a metallic grating having a pitch size of 312 nm. (c) Calculated map of the absorption enhancements in a 50 nm thick Si film versus the incident photon energy and reciprocal lattice vector of the grating for TE illumination. All the calculations are performed using FDFD method. (Adapted from Pala, R.A. et al., *Adv. Mater.*, 21, 3504, 2009.)

high Q-factor cavity [71]. A 1D Fabry–Perot cavity can be formed along the axis of a single nanowire. In this section, we investigate the modal characteristics of nanowire optical cavities and their implementations as nanowire lasers and nonlinear mixers.

3.3.1.1 Modal Characteristics: Quality Factors and Confinement Factors

Semiconductor nanowires have various geometric cross sections that depend, in part, on the nanowire crystallographic growth axis. For example, the cross section of a GaN nanowire can be triangular or hexagonal when the nanowire is grown along ⟨11–20⟩ or ⟨0001⟩ directions, respectively [5,72,73]. In this section, we study the optical characteristics of a single GaN nanowire with a triangular cross section (Figure 3.22a). Lasing has been previously demonstrated from a triangular GaN nanowire [6,72]. Quality (Q) factors,

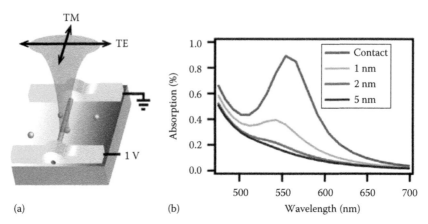

(a)   (b)

FIGURE 3.21
(a) Schematic of a Si nanowire decorated by gold nanoparticles. (b) Calculated absorption spectrum from the structure shown in (a) for different gap sizes between gold particle and Si nanowire, using FDTD method. (Adapted from Hyun, J.K. and Lauhon, L.J., *Nano Lett.*, 11, 2731, 2011.)

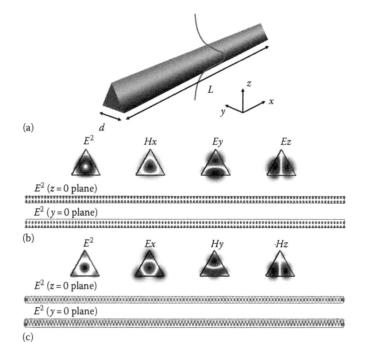

FIGURE 3.22
(a) Schematic of a single GaN nanowire cavity with a triangular cross section. Electric and magnetic field components of the TE mode (b) and the TM mode (c) at the nanowire size of 300 nm. (Adapted from Seo, M.-K. et al., *Nano Lett.*, 8, 4534, 2008.)

confinement factors, and single-mode conditions are calculated using 3D FDTD simulations, while the nanowire size at its base, d, is varied from 175 to 300 nm [34].

Two transverse modes, TE and TM modes, are excited in the nanowire as shown in Figure 3.22b and c [34]. The FDTD simulations show that H_x, E_y, and E_z field components in the TE mode (Figure 3.22b) or E_x, H_y, and H_z field components in the TM mode (Figure 3.22c) are

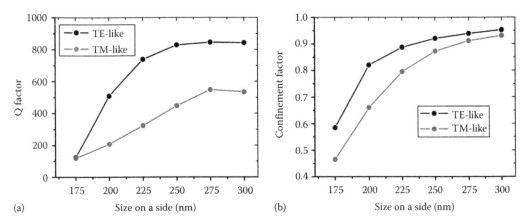

FIGURE 3.23
Q and confinement factors of the TE and the TM modes in a free-standing GaN nanowire cavity. (a) Q factor calculated as a function of nanowire size. (b) Confinement factor calculated as a function of nanowire size. (Adapted from Seo, M.-K. et al., *Nano Lett.*, 8, 4534, 2008.)

excited separately. Both modes have distinguishable electric field intensity distributions as seen by the presence of an intensity node at the center of the TE mode and a central intensity antinode in the case of the TM mode.

Q factors of the TE and TM resonant modes are calculated as a function of the nanowire size [34]. Figure 3.23a shows that the highest Q factors of the TE and TM modes are 840 and 530, respectively. The TM mode profile with intensity antinodes at three sharp corners of the cross section suffers more optical losses (Figure 3.22c), while the TE mode is well confined by the index guiding (Figure 3.22b). These calculated Q factors are relatively smaller compared to those in other nanocavities such as photonic crystals [74] and microdisks [75], and this is due to the smaller size of the nanowire cavity. Nevertheless, despite a nanowire cavity Q factor, which is <800, lasing action has been achieved in experiment [5,72] and this is largely due to a large confinement factor. The confinement factor in a free-standing GaN nanowire cavity is calculated as a function of the nanowire size using the FDTD method (Figure 3.23b). It is worth noting that large confinement factors of >0.9 are obtained for both transverse modes at nanowire sizes >275 nm.

To examine single-mode conditions in the nanowire cavity, dispersion curves of the TE and the TM modes are calculated using 3D FDTD simulations with a periodic boundary condition in the x direction (Figure 3.24) [34]. A dispersion curve is a useful tool when determining the proper structural parameters for a nanowire cavity that is intended to support a single transverse mode. In Figure 3.24a, the normalized frequency, d/λ, of the TE mode in a free-standing GaN nanowire cavity is plotted as a function of the wave vector. As shown in the red region, the single TE mode is observed when d/λ varies from 0.5 to 0.9, whereas the single TM mode occurs at slightly smaller d/λ.

3.3.1.2 Nanowire Laser

The FDTD simulations performed in Section 3.3.1.1 provide optimized structural parameters for the development of low-threshold single GaN nanowire lasers. By inserting the calculated Q and confinement factors into the rate equations, lasing thresholds in the

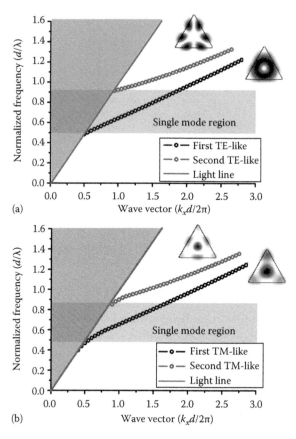

FIGURE 3.24
Dispersion curves of a free-standing GaN nanowire cavity for the TE mode (a) and the TM mode (b). (Adapted from Seo, M.-K. et al., *Nano Lett.*, 8, 4534, 2008.)

nanowire cavities can be estimated. The carrier density, N, and photon density, P, in the cavity are described by the following rate equations in the case of optical pumping [76,77]:

$$\frac{dN}{dt} = \eta_i \frac{L_{in}}{h\omega_p V} - \left(AN + BN^2 + CN^3\right) - \Gamma g(N)P$$

$$\frac{dP}{dt} = \Gamma g(N)P - \frac{P}{\tau_p} + \beta BN^2$$

where
   η_i is the absorbed ratio in active materials
   ω_p is the frequency of a pumping laser
   A is the surface nonradiative recombination coefficient
   B is the radiative recombination coefficient
   C is the Auger nonradiative recombination coefficient
   Γ is the confinement factor
   V is the active volume
   β is the spontaneous emission factor

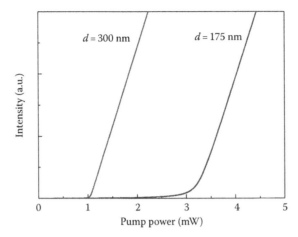

FIGURE 3.25
Light-in versus light-out curves obtained from the rate equation analysis, for the TE transverse modes at the nanowire sizes of 300 and 175 nm. (Adapted from Seo, M.-K. et al., *Nano Lett.*, 8, 4534, 2008.)

The photon lifetime, τ_p, is related to the Q factor as follows: $\tau_p = \lambda Q/2\pi c$, where λ is the wavelength of the output laser. The optical gain is represented by $g(N) = g_0 (c/n_{eff})$ $(N - N_{tr})$, and Q and confinement factors calculated in Figure 3.23b and c are inserted into the rate equations. The calculated light-in versus light-out curves are then plotted for the TE transverse modes for nanowire sizes of 300 and 175 nm (Figure 3.25). The lasing threshold for a 175 nm nanowire is approximately 3.1 times larger than the threshold for a 300 nm nanowire. In conclusion, FDTD simulations and rate equation analysis can be used to design optimized nanowire laser cavities that support low lasing thresholds.

3.3.2 Nonlinear Mixing in Nanowires

Scaling down tabletop optics implementations of nonlinear mixing to integrated photonics is necessary to realize functional integrated photonic circuits [78,79]. Some semiconductor nanowires can provide a noncentrosymmetric crystal with large second-order nonlinear response. This means that nonlinear-mixing processes such as second-harmonic generation could be demonstrated in an ultracompact photonic integrated system by using such nanowires with a waveguide geometry defined by the nanowire axis. The basic structure consists of a CdS nanowire embedded in a lithographically defined poly(methyl methacrylate) (PMMA) cladding (Figure 3.26) [33]. CdS functions as the nonlinear medium with the goal being to funnel as much light as possible into this core. CdS nanowires can be synthesized with widths of ~80 nm and length of ~5 μm.

In simulations, a CdS nanowire core with a square cross section is surrounded by a PMMA cladding. Figure 3.27a shows the electric energy distribution ($\varepsilon|E|^2$) at wavelengths of 1064 nm (bottom, ω) and 532 nm (top, 2ω). It is important to note that most of the field distribution for the fundamental-transverse mode at the wavelength of 1064 nm is present in the PMMA cladding due to the sub-wavelength-scale nanowire diameter.

Coherence length is a measure of the maximum length within which the nonlinear process is efficient. The coherence length for a second-harmonic generation

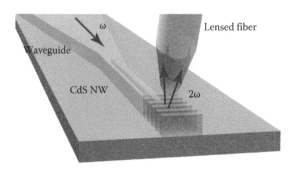

FIGURE 3.26
Schematic of the nonlinear-mixing process in a nanowire sub-wavelength waveguide. (Adapted from Barrelet, C.J. et al., *Nano Lett.*, 11, 3022, 2011.)

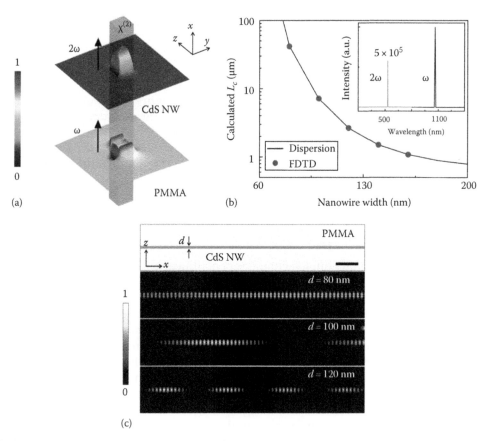

FIGURE 3.27
(a) Electric energy distributions ($\varepsilon|E|^2$) for the fundamental mode excited within an 80 nm wide CdS waveguide core. The wavelength is 1064 nm (bottom, ω) and 532 nm (top, 2ω). (b) Coherence length for the second-harmonic generation as a function of nanowire width. In the inset, the second-harmonic generation signal is observed at 2ω (gray) by injecting the pump beam at ω (black) with an intensity of 1 W/μm^2. (c) Electric energy distributions ($\varepsilon|E|^2$) at 532 nm calculated using nonlinear FDTD for nanowire widths (d) of 80, 100, and 120 nm, respectively. Scale bar is 1 μm. (Adapted from Barrelet, C.J. et al., *Nano Lett.*, 11, 3022, 2011.)

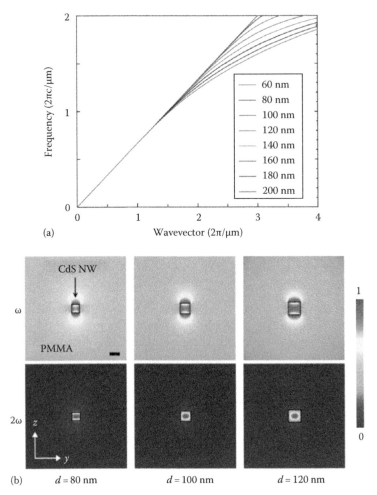

FIGURE 3.28

(a) Dispersion curves at nanowire widths ranging from 60 to 200 nm. (b) Cross-sectional views of electric energy distribution ($\varepsilon|E|^2$) for wavelengths of 1064 nm (ω) and 532 nm (2ω) in nanowires with widths d of 80 nm (left), 100 nm (middle), and 120 nm (right), respectively. Scale bar is 100 nm. (Adapted from Barrelet, C.J. et al., *Nano Lett.*, 11, 3022, 2011.)

process, L_c, can be calculated in a nanowire waveguide either from a dispersion curve or from nonlinear FDTD [33]. A simulation of the dispersion curve is performed using conventional FDTD with Bloch and perfectly matched layer boundary conditions along the nanowire axis and in the plane perpendicular to the nanowire, respectively (Figure 3.28). The dispersive permittivity of CdS is modeled using the Drude model. Figure 3.28a shows the calculated dispersion curves for nanowire widths ranging from 60 to 200 nm. The coherence length for second-harmonic generation can be calculated from these dispersion curves using the equation, $L_c = 2\pi/\Delta k$, where $\Delta k = k_{2\omega} - 2k_\omega$. The coherence length (Figure 3.27b) increases significantly as the nanowire width decreases. The CdS nanowire with a width of 80 nm yields $L_c = 42$ µm, whereas for a bulk CdS crystal, $L_c = 1.6$ µm, confirming that the sub-wavelength waveguide can provide a longer effective length over which the nonlinear process can take place.

3D nonlinear FDTD simulation is also used to calculate coherence length [80]. In nonlinear FDTD simulations (Figure 3.27b and c), the full second-order susceptibility tensor of bulk CdS is employed and one end of the nanowire cavity is pumped, while the second-harmonic generation signal is detected along the nanowire. In Figure 3.27c, the simulation outputs a second-harmonic generation signal arising from nonlinear mixing in the sub-wavelength waveguide. Significantly, the electric energy distribution produces a clear second-harmonic generation signal even in a nanowire as small as 80 nm. The coherence length is calculated by measuring the distance between nodes in this electric energy distribution (red dots, Figure 3.27b), and this result agrees well with the length calculated from the dispersion curve (black line, Figure 3.27b) and confirms the trend that nanowires with smaller width exhibit longer coherence length. A spectrum measured at the end of the 80 nm nanowire (inset, Figure 3.27b) clearly identifies a second-harmonic generation signal at 2ω for a distal pump beam at ω. This 3D nonlinear FDTD simulation strongly supports the observation of second-harmonic generation in the nanowire waveguide.

3.3.3 Sub-Wavelength Plasmonic Cavities

Wavelength-scale dielectric optical cavities can show high Q factors and small mode volumes and could potentially enable low-power consumer optical devices and sensitive optical sensors [74,75]. However, the diffraction-limited size of these cavities restricts further device miniaturization. In contrast, SPP cavities do not face such a size limitation and can support sub-wavelength-scale mode volumes [81]. The resonant wavelength of the SPPs excited at the dielectric–metal interface can be shorter than the wavelength of resonances in a dielectric material.

Several SPP cavities and lasers have recently been demonstrated. The electrically pumped gold-finger laser represents the first attempt to present plasmonic characteristics [82] and 2D SPP confinement was achieved using a CdS nanowire on a silver substrate (Figure 3.29a) [70]. Also, gold/dye-doped silica core/shell nanospheres allowed the

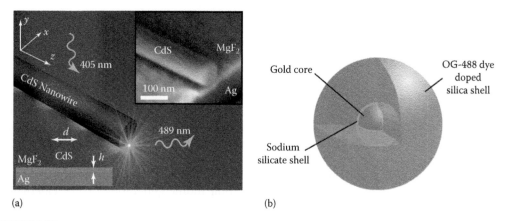

(a)   (b)

FIGURE 3.29
(See color insert.) Schematics of plasmonic cavities. (a) A CdS nanowire on silver substrate. (Adapted from Oulton, R.F. et al., *Nature*, 461, 629, 2009.) (b) A gold/dye-doped silica core/shell nanosphere. (Adapted from Noginov, M.A. et al., *Nature*, 460, 1110, 2009.)

demonstration of an optically pumped surface plasmon amplification by stimulated emission of radiation (SPASER) (Figure 3.29b) [83]. More recently, optically pumped plasmonic lasing was demonstrated from a semiconductor nanodisk covered with a silver nanopan structure (Figure 3.33) [84].

Lasing from such an extremely small sub-wavelength cavity represents significant progress toward realizing a highly miniaturized coherent light source. To demonstrate an SPP laser, it is necessary to implement an optical design, which can reduce losses due to optical and metallic absorption. In this section, we propose two 3D SPP cavities, a dielectric-core/metal-shell nanowire structure [85] and dielectric nanodisk/metal nanopan structure [84,86]. We characterize the unique optical properties of these plasmonic cavities, which bear a mode volume that is considerably smaller than the physical limit for a conventional optical cavity, $\sim(\lambda/2n)^3$.

3.3.3.1 Metal-Coated Dielectric Nanowires

In a metal-coated dielectric nanowire with an axial heterostructure (Figure 3.30a), SPPs are confined at the interface between a nanowire core and metal shell [85]. The cavity, with a length L, consists of a nanowire core sheathed by a metal shell and is flanked by SPP mirrors, which are composed of a nanowire core of the same material, a low-index dielectric shell, and the metal shell. To understand the confinement mechanism of SPPs in this cavity, we first consider an infinitely long waveguide that consists of a high-index-core/low-index-shell/metal-shell structure with a square cross section (inset, Figure 3.30b) and calculate the dispersion curves of the fundamental-transverse SPP waveguide mode (Figure 3.30b). In the dispersion curves, the frequency of the SPP mode is significantly altered by the presence of the low-index dielectric shell. In particular, there is a cutoff frequency at a wave vector of 0, which is a unique property of a 2D SPP waveguide [87]. Therefore, for a frequency range of 3000–4000 THz, SPP modes excited in the part of the waveguide without the low-index dielectric shell are not coupled to the SPP modes in the part of the waveguide with the shell. This entails 3D confinement of SPPs. In addition, Figure 3.30c shows the electric field profiles of the fundamental-transverse SPP mode appearing in the metal-coated nanowire waveguide without the low-index dielectric shell.

The (nanowire core)/(silver shell) structure with a length L is sandwiched by two (nanowire core)/(low-index dielectric shell)/(silver shell) structures (Figure 3.30a). The SPPs are strongly confined in the ultrasmall cavities having L of 40, 120, and 200 nm, without remarkable scattering (Figure 3.31a through c). In the cavity with $L=40$ nm (Figure 3.31a), a longitudinal mode with an antinode number of 1 is observed. On the other hand, in an SPP control structure that has no axial heterostructure along the nanowire axis (Figure 3.31d), the SPPs are spread over the nanowire–metal interface along the nanowire axis and concomitant scattering of SPPs is observed at the nanowire end facets.

The resonant wavelengths, mode volumes, Q factors, and confinement factors of the SPP modes are calculated as a function of L, using 3D FDTD simulation (Figure 3.32) [85]. The thickness of the silver shell and the total length along the nanowire axis are 125 nm and $L+500$ nm, respectively. Figure 3.32a shows that the resonant wavelength can be tuned by varying L and can be well adjusted to the emission spectrum of a dielectric nanowire. The mode volume is reduced to $\sim 10^{-5}$ μm³, which is approximately 100 times smaller than that of nanoscale optical cavities previously reported [74,75]. At $L=40$ nm,

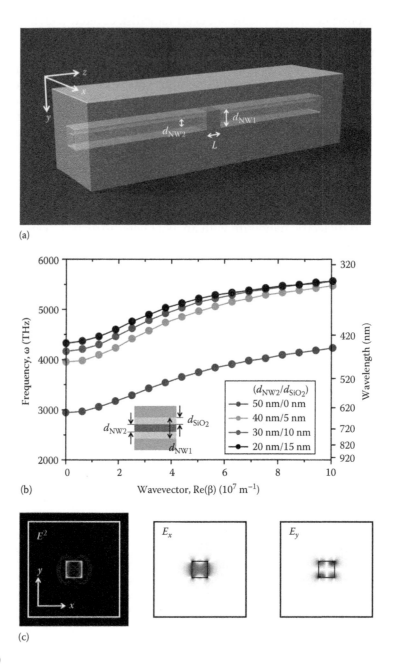

(a)

(b)

(c)

FIGURE 3.30

(a) Schematic of the metal-coated dielectric nanowire SPP cavity. (b) The dispersion curves of the fundamental-transverse SPP modes in an infinitely long waveguide with square cross section (inset). (c) The electric field profiles (E^2, E_x, and E_y) of the fundamental-transverse SPP mode. (Adapted from Seo, M.-K. et al., *Nano Lett.*, 9, 4078, 2009.)

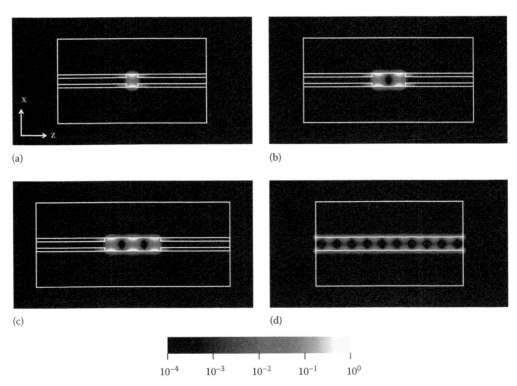

FIGURE 3.31
(a) The electric field intensities (log E^2) in the SPP cavities with L of 40 nm (a), 120 nm (b), and 200 nm (c). (d) The SPP mode in the cavity that has no axial heterostructure in the nanowire core and a length of 540 nm. (Adapted from Seo, M.-K. et al., *Nano Lett.*, 9, 4078, 2009.)

the mode volume is calculated to be ~0.020 $(\lambda/2n_{NW})^3$, where n_{NW} is 2.6. This 3D sub-wavelength SPP cavity overcomes the diffraction limit of conventional optics. The Q factors of the SPP cavities are calculated at 20 K (Figure 3.32b) because a lower temperature substantially reduces absorption loss due to the metal [82,88,89]. Under these conditions, a high Q factor of ~38,000 is obtained, which approaches the metal-loss-limited value. The confinement factor, which is defined as the ratio of the electric field energy confined in the dielectric nanowire core of the cavity to the total electric field energy, is also calculated in Figure 3.32b. A large confinement factor of >~0.45 is obtained for this sub-wavelength-scale SPP cavity.

To further investigate the effect of the metal absorption loss, the Q factor of the $L = 40$ nm cavity was calculated as a function of temperature (Figure 3.32c). To represent a range of temperatures in the FDTD simulation, the damping collision frequency of silver was scaled by a factor of the room-temperature conductivity divided by the low-temperature conductivity in the Drude model [82,88,89]. FDTD simulation shows that the Q factor of the SPP mode increases exponentially with decreasing temperature. The fact that the Q factor is inversely proportional to the damping collision frequency suggests that the Q factor is metal-loss limited and that the intrinsic optical loss of the SPP cavity mode is sufficiently low.

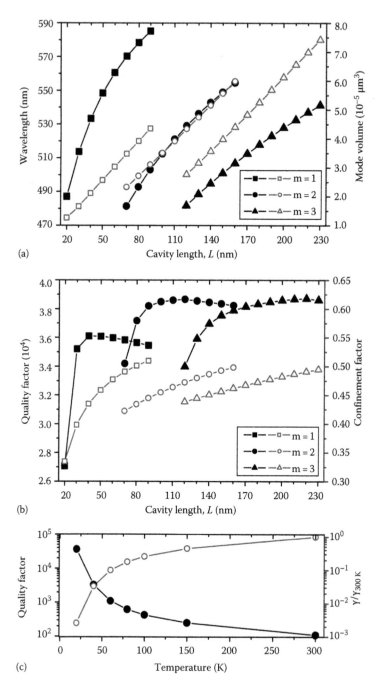

FIGURE 3.32
(a) The resonant wavelengths and mode volumes are calculated as a function of L. (b) The Q factors and confinement factors are calculated as a function of L. The Q factors are calculated at 20 K. (c) The Q factor of the SPP mode in the cavity with $L = 40$ nm is calculated as a function of temperature. (Adapted from Seo, M.-K. et al., *Nano Lett.*, 9, 4078, 2009.)

3.3.3.2 Temperature-Dependent Quality Factors

Another full 3D SPP cavity consists of an InP nanodisk with a diameter of <1000 nm and a silver nanopan (Figure 3.33) [84]. SPP lasing action was demonstrated from this plasmonic cavity. The top of the nanodisk is bonded to a transparent glass substrate for optical pumping and the bottom and sidewall of the disk are covered with silver for the excitation of SPP modes. The optical properties of the resonant modes excited in the nanodisk/nanopan structure are calculated using 3D FDTD simulation.

To understand how SPP modes are excited in the nanodisk/nanopan structure, optical modes are calculated in a nanodisk on a glass substrate without a nanopan structure. The conventional TM-like and TE-like WG optical modes, whose electric field directions are perpendicular to the bottom surface and sidewall of the nanodisk, respectively, are excited inside the nanodisk (Figure 3.34a). If the metal nanopan structure is introduced, these optical modes are converted into SPP modes (Figure 3.34b) [86]. Since the electric field direction of an SPP mode is perpendicular to the metal surface, the TM-like and TE-like WG optical modes are changed to TM-like and TE-like WG SPP modes, respectively. The TM-like WG SPP mode is confined to the bottom disk–silver interface, whereas the TE-like WG SPP mode is at the sidewall interface, as shown in Figure 3.34b.

Figure 3.35 shows all resonant modes excited in the nanodisk/nanopan structure with a disk radius of 500 nm [86]. There are three SPP modes (Figure 3.35a) and two optical modes (Figure 3.35b). The SPP modes with an electric field maximum at the disk–silver interface are classified by TM-like radial, TM-like WG, and TE-like WG SPP modes. The optical modes with an electric field maximum at the center of the disk slab are classified by dipole and monopole modes. The TM-like and TE-like WG SPP modes and dipole optical modes are doubly degenerate, whereas the TM-like radial SPP mode and monopole optical mode are nondegenerate. The mode volume of each mode is presented in Table 3.1. Indeed, the mode volumes of the TM-like radial and WG SPP modes are smaller than $(\lambda/2n)^3$, while the mode volumes of the optical modes are larger.

Finally, the Q factors are calculated as a function of temperature (Figure 3.36) [86]. In the TM-like WG SPP mode, a high Q factor of 4200 is obtained at 4 K. In addition, the Q

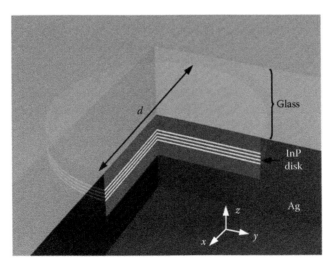

FIGURE 3.33
Schematic of the nanodisk/nanopan structure. (Adapted from Kwon, S.-H. et al., *Nano Lett.*, 10, 3679, 2010.)

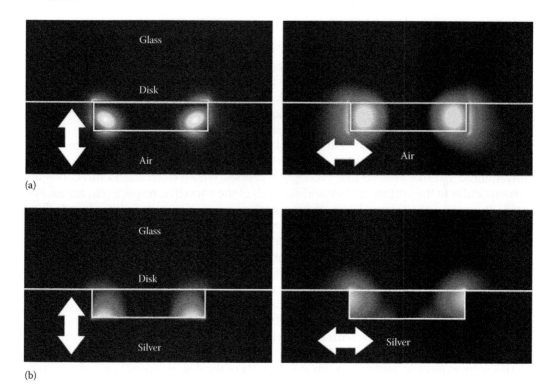

(a)

(b)

FIGURE 3.34

Side views of electric field intensity profiles (log scale) of (a) TM-like (left) and TE-like (right) WG optical modes and (b) TM-like (left) and TE-like (right) WG SPP modes. The InP disks on glass substrates are surrounded by (a) air and (b) silver. White arrows show each polarization direction. The disk radii are 500 nm for (a) and (b).

factor of the SPP modes exhibits significant temperature dependence, contrary to the optical modes. The Q factors of the SPP modes are higher than those of the optical modes at low temperature due to low metallic absorption loss but are seriously deteriorated with increasing temperature. On the other hand, the optical modes with relatively small overlap with the silver nanopan show almost constant Q factors regardless of temperature. A temperature-dependent Q factor is a key distinguishing feature of SPP modes versus conventional optical modes [84–86].

It would be vital to obtain a high Q-factor SPP mode at room temperature. Inserting a low-index dielectric material between the nanodisk and silver nanopan could be useful to increase the Q factor, although confinement of the SPP modes at the disk–nanopan interface would suffer slightly from this modification [70]. Further study will be necessary to demonstrate low-loss room-temperature SPP cavities and lasers.

3.4 Conclusions

Over the past decade, semiconductor nanowires have enabled a diverse range of promising electronic devices. More recently, they have attracted considerable interest as ultrasmall nanophotonic devices such as photovoltaic devices, nanoscale cavities, and plasmonic

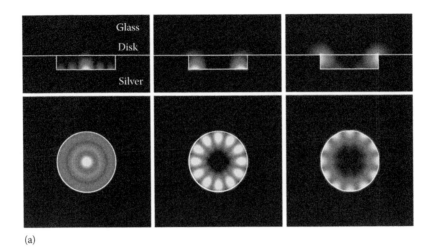

(a)

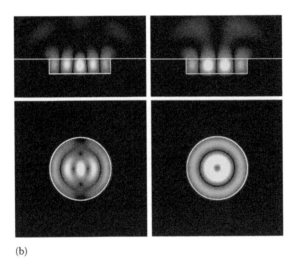

(b)

FIGURE 3.35
Side (upper panels) and top (lower panels) views of electric field intensity profiles (log scale) of the cavity modes in the nanodisk/nanopan structure with a radius of 500 nm. (a) TM-like radial (left), TM-like WG (middle), and TE-like WG (right) SPP modes and (b) dipole (left) and monopole (right) optical modes. (Adapted from Kwon, S.-H. et al., *IEEE J. Quantum Electron.*, 47, 1346, 2011.)

TABLE 3.1

Calculated Mode Volumes of the Cavity Modes in the Nanodisk/Nanopan Structure with a Radius of 500 nm

Mode	Mode Volume
TM-like radial SPP	$0.65 \ (\lambda/2n)^3$
TM-like WG SPP	$0.56 \ (\lambda/2n)^3$
TE-like WG SPP	$1.2 \ (\lambda/2n)^3$
Dipole optical	$2.7 \ (\lambda/2n)^3$
Monopole optical	$3.2 \ (\lambda/2n)^3$

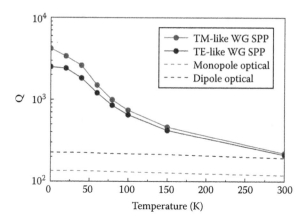

FIGURE 3.36
Calculated Q factors as a function of temperature. The disk radii are all 500 nm in the simulation.

lasers. In this chapter, we studied the practical photonic applications of nanomaterials and nanostructures, focusing mainly on semiconductor nanowires. Enhanced light absorption and tunability of resonant modes in Si nanowires were investigated by analytical solutions and numerical simulations. Modal characteristics and nonlinear properties of nanowires were also examined using full-field electromagnetic simulations. In addition, the plasmonic resonant modes supported by sub-wavelength-scale nanocavities were investigated. Numerical simulations and analytical solutions demonstrated the ability to accurately predict the unique optical characteristics of these nanomaterials and nanostructures.

Miniaturization of photonic structures could enable enhanced device performance, but this might come at the expense of optical loss. A variety of computational methods can be used to design optimized nanostructures for practical applications. We believe that nanomaterials and nanostructures are a versatile platform for the next generation of highly efficient photonic devices with extremely small thermal overhead.

References

1. C. M. Lieber, Semiconductor nanowires: A platform for nanoscience and nanotechnology, *MRS Bull.* **36**, 1052–1063 (2011).
2. B. Tian, X. Zheng, T. J. Kempa, Y. Fang, N. Yu, G. Yu, J. Huang, and C. M. Lieber, Coaxial silicon nanowires as solar cells and nanoelectronic power sources, *Nature* **449**, 885–890 (2007).
3. T. J. Kempa, J. F. Cahoon, S.-K. Kim, R. W. Day, D. C. Bell, H.-G. Park, and C. M. Lieber, Coaxial multishell nanowires with high-quality electronic interfaces and tunable optical cavities for ultrathin photovoltaics, *Proc. Natl. Acad. Sci. USA* **109**, 1407–1412 (2012).
4. J. Wang, M. S. Gudiksen, X. Duan, Y. Cui, and C.M. Lieber, Highly polarized photoluminescence and photodetection from single indium phosphide nanowires, *Science* **293**, 1455–1457 (2001).
5. J. C. Johnson, H. J. Choi, K. P. Knutsen, R. D. Schaller, P. D. Yang, and R. J. Saykally, Single gallium nitride nanowire lasers, *Nature Mater.* **1**, 106–110 (2002).
6. F. Qian, Y. Li, S. Gradecak, H.-G. Park, Y. Dong, Y. Ding, Z. L. Wang, and C. M. Lieber, Multiquantum-well nanowire heterostructures for wavelength-controlled lasers, *Nature Mater.* **7**, 701–706 (2008).

7. C. J. Barrelet, J. Bao, M. Loncar, H.-G. Park, F. Capasso, and C.M. Lieber, Hybrid single-nanowire photonic crystal and microresonator structures, *Nano Lett.* **6**, 11–15 (2006).

8. B. Tian, T. Cohen-Karni, Q. Qing, X. Duan, P. Xie, and C.M. Lieber, Three-dimensional, flexible nanoscale field-effect transistors as localized bioprobes, *Science* **329**, 831–834 (2010).

9. X. Duan, R. Gao, P. Xie, T. Cohen-Karni, Q. Qing, H. S. Choe, B. Tian, X. Jiang, and C. M. Lieber, Intracellular recordings of action potentials by an extracellular nanoscale field-effect transistor, *Nat. Nanotechnol.* **7**, 174–179 (2012).

10. P. Xie, Q. Xiong, Y. Fang, Q. Qing, and C. M. Lieber, Local electrical potential detection of DNA by nanowire-nanopore sensors, *Nat. Nanotechnol.* **7**, 119–125 (2012).

11. H. Yan, H. S. Choe, S.W. Nam, Y. Hu, S. Das, J. F. Klemic, J. C. Ellenbogen, and C. M. Lieber, Programmable nanowire circuits for nanoprocessors, *Nature* **470**, 240–244 (2011).

12. J. P. Reithmaier, G. Sek, A. Löffler, C. Hofmann, S. Kuhn, S. Reitzenstein, L. V. Keldysh, V. D. Kulakovskii, T. L. Reinecke, and A. Forchel, Strong coupling in a single quantum dot–semiconductor microcavity system, *Nature* **432**, 197–200 (2004).

13. J. D. Joannopoulos, P. R. Villeneuve, and S. Fan, Photonic crystals: Putting a new twist on light, *Nature* **386**, 143–149 (1997).

14. O. Painter, R. K. Lee, A. Scherer, A. Yariv, J. D. O'Brien, P. D. Dapkus, and I. Kim, Two-dimensional photonic band-gap defect mode laser, *Science* **11**, 1819–1821 (1999).

15. S. Noda, K. Tomoda, N. Yamamoto, and A. Chutinan, Full three-dimensional photonic bandgap crystals at near-infrared wavelengths, *Science* **289**, 604–606 (2000).

16. Y. Xia, B. Gates, Y. Yin, and Y. Lu, Monodispersed colloidal spheres: Old materials with new applications, *Adv. Mater.* **12**, 693–713 (2000).

17. S. Noda and M. Fujita, Light-emitting diodes: Photonic crystal efficiency boost, *Nat. Photon.* **3**, 129–130 (2009).

18. S. Fan, P. R. Villeneuve, J. D. Joannopoulos, and E. F. Schubert, High extraction efficiency of spontaneous emission from slabs of photonic crystals, *Phys. Rev. Lett.* **78**, 3294–3297 (1997).

19. T. Fujii, Y. Gao, R. Sharma, E. L. Hu, S. P. DenBaars, and S. Nakamura, Increase in the extraction efficiency of GaN-based light-emitting diodes via surface roughening, *Appl. Phys. Lett.* **84**, 855–857 (2004).

20. J. J. Wierer, A. David, and M. M. Mergens, III-nitride photonic-crystal light-emitting diodes with high extraction efficiency, *Nat. Photon.* **3**, 163–169 (2009).

21. L. Novotny and N. van Hulst, Antennas for light, *Nat. Photon.* **5**, 83–90 (2011).

22. J. N. Anker, W. P. Hall, O. Lyandres, N. C. Shah, J. Zhao and R. P. Van Duyne Anker, Biosensing with plasmonic nanosensors, *Nat. Mater.* **7**, 442–453 (2008).

23. A. Kinkhabwala, Z. Yu, S. Fan, Y. Avlasevich, K. Müllen, and W. E. Moerner, Large single-molecule fluorescence enhancements produced by a bowtie nanoantenna, *Nat. Photon.* **3**, 654–657 (2009).

24. D. K. Gramotnev and S. I. Bozhevolnyi, Plasmonics beyond the diffraction limit, *Nat. Photon.* **4**, 83–91 (2010).

25. S. W. Boettcher, J. M. Spurgeon, M. C. Putnam, E. L. Warren, D. B. Turner-Evans, M. D. Kelzenberg, J. R. Maiolo, H. A. Atwater, and N. S. Lewis, Energy-conversion properties of vapor-liquid-solid–grown silicon wire-array photocathodes, *Science* **8**, 185–187 (2010).

26. T. Thurn-Albrecht, J. Schotter, G. A. Kästle, N. Emley, T. Shibauchi, L. Krusin-Elbaum, K. Guarini, C. T. Black, M. T. Tuominen, and T. P. Russell, Ultrahigh-density nanowire arrays grown in self-assembled diblock copolymer templates, *Science* **15**, 2126–2129 (2000).

27. F. Bonaccorso, Z. Sun, T. Hasan, and A. C. Ferrari, Graphene photonics and optoelectronics, *Nat. Photon.* **4**, 611–612 (2010).

28. K. S. Kim, Y. Zhao, H. Jang, S. Y. Lee, J. M. Kim, K. S. Kim, J.-H. Ahn, P. Kim, J.-Y. Choi, and B. H. Hong, Large-scale pattern growth of graphene films for stretchable transparent electrodes, *Nature* **457**, 706–710 (2009).

29. H.-G. Park, C. J. Barrelet, Y. Wu, B. Tian, F. Qian, and C. M. Lieber, A wavelength-selective photonic-crystal waveguide coupled to a nanowire light source, *Nat. Photon.* **2**, 622–626 (2008).

30. L. Cao, P. Fan, A. P. Vasudev, J. S. White, Z. Yu, W. Cai, J. A. Schuller, S. Fan, and M. L. Brongersma, Semiconductor nanowire optical antenna solar absorbers, *Nano Lett.* **10**, 439–445 (2010).

31. L. Cao, J.-S. Park, P. Fan, B. Clemens, and M. L. Brongersma, Resonant germanium nanoantenna photodetectors, *Nano Lett.* **10**, 1229–1233 (2010).

32. E. S. Barnard, R. A. Pala, and M. L. Brongersma, Photocurrent mapping of near-field optical antenna resonances, *Nat. Nanotechnol.* **6**, 588–593 (2011).

33. C. J. Barrelet, H.-S. Ee, S.-H. Kwon, and H.-G. Park, Nonlinear mixing in nanowire subwavelength waveguides, *Nano Lett.* **11**, 3022–3025 (2011).

34. M.-K. Seo, J.-K. Yang, K.-Y. Jeong, H.-G. Park, F. Qian, H.-S. Ee, Y.-S. No, and Y.-H. Lee, Modal characteristics in a single-nanowire cavity with a triangular cross section, *Nano Lett.* **8**, 4534–4538 (2008).

35. L. Cao, J. S. White, J.-S. Park, J. A. Schuller, B. M. Clemens, and M. L. Brongersma, Engineering light absorption in semiconductor nanowire devices, *Nat. Mater.* **8**, 643–647 (2009).

36. S.-K. Kim, R. W. Day, J. F. Cahoon, T. J. Kempa, K.-D. Song, H.-G. Park, and C. M. Lieber, Tuning light absorption in core/shell silicon nanowires photovoltaic devices through morphological design, *Nano Lett.* **12**, 4971–4976 (2012).

37. T. J. Kempa, R. W. Day, S.-K. Kim, H.-G. Park, and C. M. Lieber, Semiconductor nanowires: A platform for exploring limits and concepts for nano-enabled solar cells, *Energy Environ. Sci.* **6**, 719–733 (2013).

38. G. Bronstrup, N. Jahr, C. Leiterer, A. Csaki, W. Fritzsche, and S. Christiansen, Optical properties of individual silicon nanowires for photonic devices, *ACS Nano* **4**, 7113–7122 (2010).

39. W. F. Liu, J. I. Oh, and W. Z. Shen, Light absorption mechanism in single c-Si (core)/a-Si (shell) coaxial nanowires, *Nanotechnology* **22**, 125705–125708 (2011).

40. W. F. Liu, J. I. Oh, and W. Z. Shen, Light trapping in single coaxial nanowires for photovoltaic applications, *IEEE Electron Device Lett.* **32**, 45–47 (2011).

41. C. F. Bohren and D. R. Huffman, *Absorption and Scattering of Light by Small Particles*, Wiley: New York (1998).

42. A. Vial and T. Laroche, Comparison of gold and silver dispersion laws suitable for FDTD simulations, *Appl. Phys. B: Lasers Opt.* **93**, 139–143 (2008).

43. A. Taflove and S. C. Hagness, *Computational Electrodynamics: The Finite-Difference Time-Domain Method*, Artech House: Norwood, MA (2005).

44. A. I. Hochbaum and P. Yang, Semiconductor nanowires for energy conversion, *Chem. Rev.* **110**, 527–546 (2010).

45. J. Zhu, C. M. Hsu, Z. Yu, S. Fan, and Y. Cui, Nanodome solar cells with efficient light management and self-cleaning, *Nano Lett.* **9**, 1979–1984 (2009).

46. J. Tang, Z. Huo, S. Brittman, H. Gao, and P. Yang, Solution-processed core-shell nanowires for efficient photovoltaic cells, *Nat. Nanotechnol.* **6**, 568–572 (2011).

47. K. Q. Peng and S. T. Lee, Silicon nanowires for photovoltaic solar energy conversion, *Adv. Mater.* **23**, 198–215 (2011).

48. T. J. Kempa, B. Tian, D.R. Kim, J. Hu, X. Zheng, and C. M. Lieber, Single and tandem axial p-i-n nanowire photovoltaic devices, *Nano Lett.* **8**, 3456–3460 (2008).

49. B. Tian, T. J. Kempa, and C. M. Lieber, Single nanowire photovoltaics, *Chem. Soc. Rev.* **38**, 16–24 (2009).

50. Y. Dong, B. Tian, T. Kempa, and C. M. Lieber, Coaxial group III-nitride nanowire photovoltaics, *Nano Lett.* **9**, 2183–2187 (2009).

51. C. Colombo, M. Heibeta, M. Gratzel, A. Fontcuberta i Morral, Gallium arsenide p-i-n radial structures for photovoltaic applications, *Appl. Phys. Lett.* **94**, 173103–173108 (2009).

52. M. Heurlin, P. Wickert, S. Falt, M. T. Borgstrom, K. Deppert, L. Samuelson, and M. H. Magnusson, Axial InP nanowire tandem junction grown on a silicon substrate, *Nano Lett.* **11**, 2028–2031 (2011).

53. M. D. Kelzenberg, S. W. Boettcher, J. A. Petykiewicz, D. B. Turner-Evans, M. C. Putnam, E. L. Warren, J. M. Spurgeon, R. M. Briggs, N. S. Lewis, and H. A. Atwater, Enhanced absorption and carrier collection in Si wire arrays for photovoltaic applications, *Nat. Mater.* **9**, 239–244 (2010).

54. M. D. Kelzenberg, D. B. Turner-Evans, B. M. Kayes, M. A. Filler, M. C. Putnam, N. S. Lewis, and H. A. Atwater, Photovoltaic measurements in single-nanowire silicon solar cells, *Nano Lett.* **8**, 710–714 (2008).

55. J. Grandidier, D. M. Callahan, J. N. Munday, and H. A. Atwater, Light absorption enhancement in thin-film solar cells using whispering gallery modes in dielectric nanospheres, *Adv. Mater.* **23**, 1272–1276 (2011).

56. Z. Yu, A. Raman, and S. Fan, Fundamental limit of nanophotonic light trapping in solar cells, *Proc. Natl. Acad. Sci. USA* **107**, 17491–17496 (2010).

57. M. D. Kelzenberg, D. B. Turner-Evans, M. C. Putnam, S. W. Boettcher, R. M. Briggs, J. Y. Baek, N. S. Lewis, and H. A. Atwater, High-performance Si microwire photovoltaics, *Energy Environ. Sci.* **4**, 866–871 (2011).

58. W. Liu and F. Sun, Light absorption enhancement in single Si(core)/SiO₂(shell) coaxial nanow-ires for photovoltaic applications, *Adv. Mater. Res.* **391–392**, 264–268 (2012).

59. G. Mariani, P.-S. Wong, A. M. Katzenmeyer, F. Leonard, J. Shapiro, and D. L. Huffaker, Patterned radial GaAs nanopillar solar cells, *Nano Lett.* **11**, 2490–2494 (2011).

60. L. Cao, P. Fan, and M. L. Brongersma, Optical coupling of deep-subwavelength semiconductor nanowires, *Nano Lett.* **11**, 1463–1468 (2011).

61. A. Javey, S. Nam, R. S. Friedman, H. Yan, and C. M. Lieber, Layer-by-layer assembly of nanow-ires for three-dimensional, multifunctional electronics, *Nano Lett.* **7**, 773–777 (2007).

62. M. L. Brongersma and P. G. Kik, *Surface Plasmon Nanophotonics, Springer Series in Optical Sciences*, Springer: Dordrecht, the Netherlands (2007).

63. H. A. Atwater and A. Polman, Plasmonics for improved photovoltaic devices, *Nat. Mater.* **9**, 205–213 (2010).

64. R. A. Pala, J. White, E. Barnard, J. Liu, and M. L. Brongersma, Design of plasmonic thin-film solar cells with broadband absorption enhancements, *Adv. Mater.* **21**, 3504–3509 (2009).

65. J. N. Munday and H. A. Atwater, Large integrated absorption enhancement in plasmonic solar cells by combining metallic gratings and antireflection coatings, *Nano Lett.* **11**, 2195–2201 (2011).

66. S. Brittman, H. Gao, E. C. Garnett, and P. Yang, Absorption of light in a single-nanowire silicon solar cell decorated with an octahedral silver nanocrystal, *Nano Lett.* **11**, 5189–5195 (2011).

67. J. K. Hyun and L. J. Lauhon, Spatially resolved plasmonically enhanced photocurrent from Au nanoparticles on a Si nanowire, *Nano Lett.* **11**, 2731–2734 (2011).

68. S.-K. Kim, H.-S. Ee, W. Choi, S.-H. Kwon, J.-H. Kang, Y.-H. Kim, H. Kwon, and H.-G. Park, Surface-plasmon-induced light absorption on a rough silver surface, *Appl. Phys. Lett.* **98**, 011109 (2011).

69. Z. Xu, Y. Chen, M. R. Gartia, J. Jiang, and G. L. Liu, Surface plasmon enhanced broadband spectrophotometry on black silver substrates, *Appl. Phys. Lett.* **98**, 241904 (2011).

70. R. F. Oulton, V. J. Sorger, T. Zentgraf, R.-M. Ma, C. Gladden, L. Dai, G. Bartal, and X. Zhang, Plasmon lasers at deep subwavelength scale, *Nature* **461**, 629–632 (2009).

71. A. V. Maslov and C. Z. Ning, Far-field emission of a semiconductor nanowire laser, *Opt. Lett.* **29**, 572–574 (2004).

72. S. Gradecak, F. Qian, Y. Li, H.-G. Park, and C. M. Lieber, GaN nanowire lasers with low lasing thresholds, *Appl. Phys. Lett.* **87**, 173111 (2005).

73. F. Qian, S. Gradecak, Y. Li, C. Wen, and C. M. Lieber, Core/multishell nanowire heterostruc-tures as multicolor, high-efficiency light-emitting diodes, *Nano Lett.* **5**, 2287–2291 (2005).

74. H.-G. Park, S.-H. Kim, S.-H. Kwon, Y.-G. Ju, J.-K. Yang, J.-H. Baek, S.-B. Kim, and Y.-H. Lee, Electrically driven single-cell photonic crystal laser, *Science* **305**, 1444–1447 (2004).

75. A. C. Tamboli, E. D. Haberer, R. Sharma, K. H. Lee, S. Nakamura, and E. L. Hu, Room-temperature continuous-wave lasing in GaN/InGaN microdisks, *Nat. Photon.* **1**, 61–64 (2006).

76. H.-G. Park, J.-K. Hwang, J. Huh, H.-Y. Ryu, S.-H. Kim, J.-S. Kim, and Y.-H. Lee, Characteristics of modified single-defect two-dimensional photonic crystal lasers, *IEEE J. Quantum Electron.* **38**, 1353–1365 (2002).

77. L. A. Coldren and S. W. Corzine, *Diode Lasers and Photonic Integrated Circuits*, John Wiley & Sons, Inc.: New York (1995).

78. D. Cotter, R. J. Manning, K. J. Blow, A. D. Ellis, A. E. Kelly, D. Nesset, I. D. Phillips, A. J. Poustie, and D. C. Rogers, Nonlinear optics for high-speed digital information processing, *Science* **286**, 1523–1528 (1999).

79. R. W. Boyd, *Nonlinear Optics*, Academic Press: New York (2008).

80. M. A. Alsunaidi, H. M. Al-Mudhaffar, and H. M. Masoudi, Vectorial FDTD technique for the analysis of optical second-harmonic generation, *IEEE Photon. Technol. Lett.* **21**, 310–312 (2009).

81. S. A. Maier, *Plasmonics: Fundamentals and Applications*, Springer: New York (2007).

82. M. T. Hill, Y.-S. Oei, B. Smalbrugge, Y. Zhu, T. de Vries, P. J. van Veldhoven, F. W. M. van Otten et al., Lasing in metallic-coated nanocavities, *Nat. Photon.* **1**, 589–594 (2007).

83. M. A. Noginov, G. Zhu, A. M. Belgrave, R. Bakker, V. M. Shalaev, E. E. Narimanov, S. Stout, E. Herz, T. Suteewong, and U. Wiesner, Demonstration of a spaser-based nanolaser, *Nature* **460**, 1110–1112 (2009).

84. S.-H. Kwon, J.-H. Kang, C. Seassal, S.-K. Kim, P. Regreny, Y.-H. Lee, C. M. Lieber, and H.-G. Park, Subwavelength plasmonic lasing from a semiconductor nanodisk with silver nanopan cavity, *Nano Lett.* **10**, 3679–3683 (2010).

85. M.-K. Seo, S.-H. Kwon, H.-S. Ee, and H.-G. Park, Full three-dimensional subwavelength high-Q surface-plasmon-polariton cavity, *Nano Lett.* **9**, 4078–4082 (2009).

86. S.-H. Kwon, J.-H. Kang, S.-K. Kim, and H.-G. Park, Surface plasmonic nanodisk/nanopan lasers, *IEEE J. Quantum Electron.* **47**, 1346–1353 (2011).

87. J. A. Dionne, H. J. Lezec, and H. A. Atwater, Highly confined photon transport in subwavelength metallic slot waveguides, *Nano Lett.* **6**, 1928–1932 (2006).

88. Y. Y. Gong and J. Vuckovic, Design of plasmon cavities for solid-state cavity quantum electrodynamics applications, *Appl. Phys. Lett.* **90**, 033113 (2007).

89. D. R. Lide (Ed.), *CRC Handbook of Chemistry and Physics*, 87th edn., Internet Version 2007, Taylor & Francis: Boca Raton, FL (2007).

4

Modeling and Characterization of Nonlinear Optical Effects in Photonic Nanowires

Rim Cherif, Mourad Zghal, and Amine Ben Salem

CONTENTS

4.1 Introduction

Highly nonlinear optical waveguides have great interest for compact, low-power, and all optical nonlinear devices. Nonlinearity in waveguides can be enhanced either by modifications in the structure to reduce the effective mode area (A_{eff}) or by using materials with higher Kerr nonlinearity (n_2). In fact, photonic nanowires with diameters smaller than the wavelength of the guided light attract considerable interest due to their unique properties and wide range of applications [1]. Tapering is a commonly used method for reducing optical fiber dimensions and engineering the waveguide dispersion. Low-loss fiber tapers with sub-wavelength waist diameters have been fabricated from both standard single-mode fibers (SMFs) [1–3] and photonic crystal fibers (PCFs) [3–5]. These structures not only are platforms for microphotonic devices but also enable nonlinear process such as soliton-effect compression and supercontinuum (SC) generation at low input pulse energies [5,6]. The reduction in the effective mode area and the subsequent enhancement of the nonlinearity are ultimately limited by the index contrast of air and material (e.g., silica) in the case of SMFs [7], as well as the structural dimensions in the case of PCFs [8]. The pronounced evanescent field surrounding the fiber and the strong radial confinement of the light [7] make nanofibers well suited for an efficient and controlled interaction of the guided light with matter.

In this chapter, we will conduct detailed studies to linearly and nonlinearly characterize photonic nanowires in order to design devices generating broadband SC in the blue and/ or red wavelength regions. In fact, we will investigate the modal optical properties of photonic nanowires, composed from various glass materials, in order to determine the optimal nanowire dimension in which nonlinear interactions are highly exhibited. Then, generating compressed temporal pulses and efficient broadband SC from blue to mid-infrared (IR) regions could be achieved.

The analysis of the group velocity dispersion (GVD) profiles depicts that by reducing the size of the air–silica nanowire, the region of the anomalous dispersion exhibited between two zero dispersion wavelengths (ZDWs) can be shifted toward the blue wavelength region [9–11]. We demonstrate the generation of 1.4 fs-compressed pulse by pumping with low input pulse energy of 2.5 nJ, an 800 nm air–silica nanowire yielding the generation of more than one-octave spanning SC from 260 to 1800 nm [10,11]. For highly nonlinear glasses, different from silica, based on chalcogenide glasses and providing high IR transmission quality, we find that increasing the nanowire diameter shifts the first ZDW toward mid-IR wavelengths [12,13]. Moreover, fs-compressed pulses in an 800 nm core chalcogenide nanowire are generated with very low pulse energy of pJ levels, and two-octave spanning SC was generated at 1550 nm [12]. Aiming to design mid-IR devices operating beyond 2 μm wavelength, the design of a tapered chalcogenide PCF is proposed with a highly redshifted ZDW, and more than three-octave spanning spectrum was generated with 100 pJ energy at 4.7 μm [13].

4.2 Photonic Nanowires

4.2.1 Definitions and Concepts

In the last decade, a new class of small-core fibers has been introduced. These nanofibers, which have also been referred to as optical nanowires, photonic nanowires, are obtained from tapering, under high constant temperature, an SMF [1–3] or a PCF [3–5] or even drown directly from bulk glass [9]. Figure 4.1 summarizes the different types of photonic nanowires. They are characterized by tight mode confinement, high effective nonlinearity, and GVD controlling due to their small core diameters, which are below 1 μm. These photonic nanowires are able to guide light at visible and IR wavelengths with low bend loss.

They have been manufactured from a range of amorphous materials, which include phosphate [14], tellurite [14], lead silicate [15], bismuthate [15], and chalcogenide glasses [16] and a variety of polymers [17–20].

Photonic nanowires have attracted so much attention because of the numerous extraordinary optical and mechanical properties that they offer [21]:

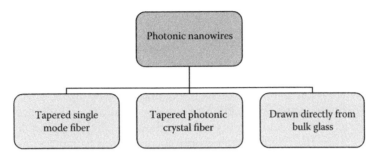

FIGURE 4.1
Different types of photonic nanowires.

1. *Large evanescent field*: With a small core diameter, a considerable fraction of the transmitted power can propagate in the evanescent field outside the physical boundaries of photonic nanowire.

2. *Strong mode confinement*: When the photonic nanowire has a diameter comparable to the wavelength of the injected light, the propagating beam is confined by diffraction to its minimum waist diameter. Thus, strong mode confinement enhances the effective nonlinearities and yields to high soliton-effect compression and broadband SC generation.

3. *Flexibility*: Due to its small rigidity, photonic nanowire can support micrometric bending radii, providing the ultimate device compactness.

4. *Configurability*: Photonic nanowire preserves the original optical fiber dimensions at its input and output ends; this allows for low-loss splicing to standard fibers and easy interconnection to other optical fibers and fiberized components and detectors.

5. *Robustness*: Photonic nanowires have a superior mechanical strength. This allows for a relatively easy handling of a nanowire with macroscopic tools and equipment typical of the macroscopic world.

4.2.2 Fabrication

The "flame-brushing" technique (FBT) [3,22] has been used to manufacture the majority of photonic nanowires reported in the literature. FBT was originally developed for the fabrication of optical fiber tapers and couplers [23,24] and relies on a small flame moving under an optical fiber, which is being stretched. The taper shape can be established with an extreme accuracy by finely controlling the flame movement and the fiber elongation.

It consists of flame pulling a commercial single-mode optical fiber resulting in a tapered optical fiber with a nanofiber waist. In order to efficiently couple light into and out of the nanofiber waist, the transformation of the fiber mode within the taper sections has to be adiabatic, thereby ensuring a high transmission of the overall structure. At the same time, it is desirable for technical reasons to keep the tapered fiber as short as possible. The combination of the two requirements makes it essential to carefully design the taper profile.

The lack of a cladding layer causes these nanowires to be susceptible to both physical contact leakage and scattering-induced environmental losses, as a significant proportion of the intensity of the guided mode travels outside the wire. This environmental sensitivity can be desirable for fiber-optic sensing [1,2,25] but is likely to be detrimental for applications in nonlinear optics or in mode coupling.

Leon-Saval et al. [3] reported a novel approach to fabricating waveguides that possess similar optical properties to the nanowires reported in [26]. It has been shown that PCFs can be tapered while retaining the cross-sectional profile [8]. By tapering a PCF, they scaled the core diameter by a factor of 10, down to 300 nm, resulting in enhanced optical non-linearities and modified dispersion properties without the high coupling losses typical of small-core PCFs. They then demonstrated optical continuum generation in this nanowire using lower pump intensities than previously reported using conventional untapered microstructured fibers [27].

Three other different fabrication techniques have been used to manufacture photonic nanowires:

1. The "self-modulated taper drawing"
2. The modified FBT
3. Micropipette puller

The "self-modulated taper drawing" is a two-step process. Firstly, a taper with a diameter of few micrometers is drawn from an optical fiber using the conventional FBT; subsequently, the taper is separated into two pigtailed halves, one of which is wrapped onto a hot sapphire rod and pulled to sub-micrometric diameters. The sapphire rod is heated by a flame positioned at a distance from the fiber, and it conveys the heat in regular manner into a small volume, helping to stabilize the temperature distribution.

The "modified FBT" has emerged in the last few years as the technique of choice to manufacture photonic nanowires from compound glass optical fibers. It is derived from the FBT, and it replaces the flame with a different heat source such as a microheater or a sapphire capillary tube heated by a CO_2 laser beam.

"Micropipette puller" is a multistage heating–pulling process originally developed to manufacture micropipettes. The fiber is heated by a CO_2 laser source while it is stretched by two spring-loaded clamps. Since the heat source is static and the stages movement cannot be accurately controlled, this technique cannot achieve a good taper shape control, and the final profile is somewhat stochastic. Still, micropipette pullers can provide tips with diameter as small as 50 nm and very steep profiles.

4.2.3 Applications

Applications include, among others, optical sensing [28–30], nonlinear optics [31–34], nanowire-based dye lasers [35,36], nanowire-based evanescent wave spectroscopy [37,38], and cold atom physics [39–43]. These applications demand high-quality photonic nanowires with high diameter uniformity and a low surface roughness.

Photonic nanowires exploit the strong dependence of the evanescent field on the surrounding environment. Additionally, photonic nanowires exhibit all the benefits characteristic of optical fiber sensors: They are unaffected by electromagnetic interference from static electricity, strong magnetic fields, or surface potentials, and these factors are extremely important for nanosensing, where forces exerted by electrostatic charges become comparable to forces experienced by macroscopic sensors (like gravity or friction).

4.3 Optical Characterization of Photonic Nanowires

In order to determine the optical properties of the fundamental mode including the effective index and the chromatic dispersion, we solve the 2D scalar-wave equation of the electric field $E(x,y)$ given by

$$\nabla^2 E + \left(\frac{2\pi}{\lambda}\right)^2 \left(n^2 - n_{eff}^2\right)E = 0 \qquad (4.1)$$

where
  ∇ is the transverse Laplacian operator in the (x,y) plane
  λ is the optical wavelength in vacuum
  n is the refractive index of the silica glass
  $n_{eff} = \beta\lambda/(2\pi)$ is the effective index of the optical mode

The resolution of Equation 4.1 provides not only the effective index but also the spatial pattern of the optical electric field $E(x,y)$. By applying the finite element method (FEM) with a mesh made up of $10^4 \sim 10^5$ elements according to the dimensions of the photonic nanowire, the transverse electric field equation can be solved. In fact, by dividing the fiber cross section into curvilinear hybrid edge/nodal elements and applying the variational finite element procedure, we obtain the following eigenvalue equation:

$$[K]\{E\} - k_0^2 n_{eff}^2 [M]\{E\} = \{0\} \qquad (4.2)$$

where
  $[K]$ and $[M]$ are the finite element matrices
  $\{E\}$ is the discretized electric field vector consisting of the edge and nodal variables
  k_0 is the free-space wave number

4.3.1 Optical Modal Analysis

4.3.1.1 Air–Silica Nanowires

In this section, we determine the optical properties of air–silica nanowires, which are produced from tapering conventional SMFs. Air–silica nanowires with core diameters less than one micron are modeled as circular silica rods in air since they exhibit large refractive-index difference between the core and the air cladding.

4.3.1.1.1 Chromatic Dispersion

The GVD referred to as chromatic dispersion (D_c) can be determined from the second derivative of the mode effective index as a function of the wavelength. It is given by

$$D_c = -\frac{\lambda}{c}\frac{d^2 n_{eff}}{d\lambda^2} \qquad (4.3)$$

where
  c is the velocity of light in vacuum
  λ is the wavelength

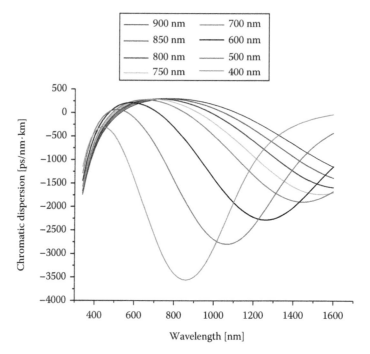

FIGURE 4.2
Calculated chromatic dispersion of air–silica nanowires with core diameters ranging from 400 to 900 nm.

We calculate the chromatic dispersion as a function of the wavelength for different core diameter air–silica nanowires ranging from 400 to 900 nm, as seen in Figure 4.2. These calculations are made by means of FEM, which assures high solution accuracy. Good agreement is found with chromatic dispersions calculated for the same range of core diameter nanowires and published in Ref. [3].

By adjusting the size of the nanowire and reducing its core diameter, the region of the anomalous dispersion can be shifted toward the blue wavelength region. This behavior is as a result of the dominance of the waveguide dispersion over the material dispersion for air–silica nanowires. These photonic nanowires provide exotic dispersion profiles and high effective nonlinearities.

4.3.1.1.2 Effective Nonlinearity and Evanescent Field

Due to the ultrasmall core diameter dimensions and the large core-cladding refractive-index difference, the photonic nanowires exhibit tighter light confinement and higher nonlinear coefficient than standard fibers. The nonlinear coefficient γ is defined as

$$\gamma = \frac{2\pi}{\lambda} \frac{\int n_2 \cdot S_z^2 d^2r}{\left(\int S_z d^2r \right)^2} \tag{4.4}$$

where
  S_z is the longitudinal component of the Poynting vector
  n_2 is the nonlinear index

Because a significant fraction of the optical mode propagates in the air outside the core and n_2 of the cladding (air) is negligible compared to that of the core glass, the integrals in the numerator are evaluated only over the glass core region. The integrals in the denominator are evaluated over the total transverse profile of the electric filed. In addition, to obtain complete information about the power residing in the evanescent field, we evaluate the fractional power η_{EF} outside the core. It is given by $\eta_{EF}=1-\eta$, where η is the fractional power inside the core and is expressed by

$$\eta = \frac{\int_0^{d/2} S_{z1}\cdot rdr \cdot \int_0^{2\pi} d\varphi}{\int_0^{d/2} S_{z1}\cdot rdr \cdot \int_0^{2\pi} d\varphi + \int_{d/2}^{\infty} S_{z2}\cdot rdr \cdot \int_0^{2\pi} d\varphi} \qquad (4.5)$$

As mentioned before, by resolving Equation 4.1, we can calculate the spatial pattern of the optical field $E(x,y)$, so that the effective mode area and the longitudinal component of the power flow density, which is proportional to the electric field intensity, can be determined.

Figure 4.3 shows that the power flow density of the 800 nm core diameter is not totally confined in the core region and presents a significant optical portion residing in the evanescent field outside the core field region. This evanescent field interacts directly with the cladding (air)–glass interfaces.

Figure 4.4 depicts the effective mode area A_{eff} and the effective nonlinear coefficient γ calculated at a pump wavelength of $\lambda p = 800$ nm for different air–silica nanowires with core diameters ranging from 200 to 1000 nm. It illustrates that the mode confinement determined by the effective mode area becomes stronger when reducing the core diameter of the air–silica nanowire. This behavior continues linearly to evolve and reaches a minimum effective mode area/maximum nonlinear coefficient value for a core diameter of about 560 nm, and then it diverges rapidly. This is justified by the fact that the

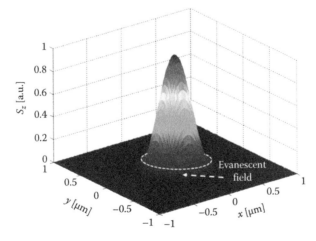

FIGURE 4.3
Power flow density (S_z) of the 800 nm core diameter nanowire. Dashed white contour indicates the limits of the core field region from the evanescent field.

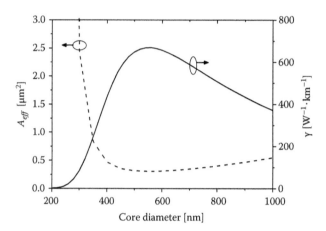

FIGURE 4.4
Effective mode area (A_{eff}) and nonlinear coefficient (γ) of air–silica nanowires at $\lambda_p = 800$ nm.

air–silica nanowire no longer tightly confines the light so that the evanescent field starts to dominate. For the 800 nm core diameter air–silica nanowire, we calculate an effective mode area A_{eff} equal to 0.403 μm². Thus, this nanowire exhibits high nonlinear coefficient of 507 (W km)$^{-1}$ by using $n_2 = 2.6 \times 10^{-20}$ m² W^{-1}.

4.3.1.2 Tapered Silica Photonic Crystal Fibers (TSPCFs)

In this subsection, we show the calculated chromatic dispersion of a tapered silica photonic crystal fiber (TSPCF) with a core diameter of 800 nm. We set the pitch $\Lambda = 0.5$ μm and the air hole diameter $d = 0.2$ μm.

As we can see from Figure 4.5, the chromatic dispersion profile of the 800 nm TSPCF is different from the 800 nm air–silica nanowire and presents a unique ZDW λ_{ZDW} around 1.25 μm comparing to that of the 800 nm air–silica nanowire having two ZDWs.

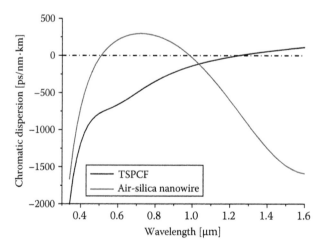

FIGURE 4.5
Chromatic dispersions of 800 nm TSPCF and 800 nm air–silica nanowire.

4.3.1.3 Tapered Chalcogenide Fibers

We have carried out a rigorous calculation of the chromatic dispersion of the different air–As$_2$Se$_3$ photonic nanowires by using the Sellmeier equation of the As$_2$Se$_3$ glass defined as $n(\lambda) = \sqrt{A + B\lambda^2/(\lambda^2 - D) + C\lambda^2}$, where $A = -4.5102$, $B = 12.0582$, $C = 0.0018\ \mu m^{-2}$, and $D = 0.0878\ \mu m^2$ [21]. The unit of λ is μm. Figure 4.1 depicts the chromatic dispersion as a function of the wavelength of air–As$_2$Se$_3$ nanowires with core diameters ranging from 600 to 1000 nm.

Figure 4.6 depicts the chromatic dispersion calculated for different air–As$_2$Se$_3$ nanowires with core diameters ranging from 600 to 1000 nm with a step size of 100 nm. We notice that the ZDW is redshifted when the core diameter increases. As the core diameter is reduced to submicron dimensions, the waveguide dispersion becomes dominant over the material dispersion (bulk As$_2$Se$_3$) allowing the overall GVD to be highly engineered. We find that the 900 nm air–As$_2$Se$_3$ has a ZDW around 1550 nm. The 800 nm air–As$_2$Se$_3$ presents a negative chromatic dispersion value at 1550 nm and low third-order dispersion ($\beta_2 = -0.48$ ps^2 m^{-1} and $\beta_3 = 7.79 \times 10^{-3}$ ps^3 m^{-1}). Consequently, pumping at this wavelength in the anomalous dispersion regime is attractive to simulate soliton-effect compression. Before investigating the nonlinear propagation in such air–As$_2$Se$_3$ nanowires, one should characterize the optical properties of the air–As$_2$Se$_3$ nanowires under investigation.

Figure 4.7 shows the nonlinear coefficient of different air–As$_2$Se$_3$ nanowires with core diameters ranging from 200 to 1000 nm at a pump wavelength of 1.55 μm. We notice that the mode confinement determined by the effective mode area becomes stronger when reducing the core diameter of the air–silica nanowire. This behavior continues linearly to evolve and shows an optimal nanowire size exhibiting the minimum effective mode area/the highest effective nonlinearity with a core diameter of about 450 nm. After this optimum value, the mode confinement diverges rapidly. This is justified by the fact that the air–As$_2$Se$_3$ nanowire no longer tightly confines the light so that the evanescent field starts to dominate.

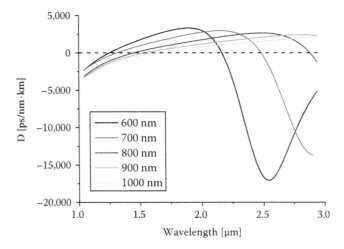

FIGURE 4.6
Calculated chromatic dispersion of air–As$_2$Se$_3$ nanowires with core diameters ranging from 600 to 1000 nm.

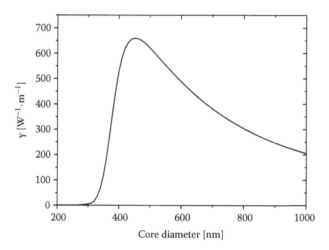

FIGURE 4.7
Nonlinear coefficient (γ) of air–As_2Se_3 nanowires at $\lambda p = 1550$ nm.

4.4 Nonlinear Propagation

Considering the optical waveguide properties of the photonic nanowire, the investigation of the nonlinear propagation is based on the resolution of the generalized nonlinear Schrödinger equation (GNLSE) [46]:

$$\frac{\partial U}{\partial z} = -\frac{\alpha}{2}U + \sum_{m \geq 0} \frac{j^{m+1}\beta_m}{m!}\frac{\partial^m U}{\partial t^m} + j\gamma\left(1 + \frac{j}{\omega_0}\frac{\partial}{\partial t}\right) \times \left(U(z,t)\int_{-\infty}^{+\infty} R(t')|U(z,t-t')|^2\,dt'\right) \quad (4.6)$$

where
    $U(z,t)$ is the slowly varying envelope
    α is the attenuation coefficient
    β_m is the higher-order dispersion coefficients of the propagation constant β at the input pulse's central frequency ω_0

The nonlinear response function $R(t)$ includes the instantaneous and the delayed Raman contributions. It is given by [44]

$$R(t) = (1 - f_R)\delta(t) + f_R h_R(t) \quad (4.7)$$

where f_R is the fractional contribution of the material; for instance, it is typically taken as $f_R = 0.18$ for the fused silica. The delayed Raman response $h_R(t)$ is expressed through the Green's function of the damped harmonic oscillator [45]:

$$h_R(t) = \frac{\tau_1^2 + \tau_2^2}{\tau_1\tau_2^2}\exp\left(-\frac{t}{\tau_2}\right)\sin\left(\frac{t}{\tau_1}\right) \quad (4.8)$$

where the parameters τ_1 and τ_2 correspond respectively to the inverse of the phonon oscillation frequency and the bandwidth of the Raman gain spectrum.

The resolution of the GNLSE is performed by the symmetrized split-step Fourier method. To start the resolution, we need to set the input pulse shape and duration. For example, the excitation by an input soliton order N has an envelope field expression given by [44]

$$U(0,t) = \sqrt{P_0} \sec h\left(\frac{t}{T_0}\right) \tag{4.9}$$

where

P_0 is the peak power
T_0 is the input soliton duration defined as $T_{FWHM}/1.763$
T_{FWHM} is the input pulse full width at half maximum (FWHM) duration

The soliton order N is defined as [46]

$$N^2 = \frac{L_D}{L_{NL}} = \frac{T_0^2 \gamma P_0}{|\beta_2|} \tag{4.10}$$

where

$L_D = T_0^2 / |\beta_2|$ is the dispersion length
$L_{NL} = 1/\gamma P_0$ is the nonlinear length

4.4.1 Soliton Self-Compression and Supercontinuum Generation in Air–Silica Nanowires

4.4.1.1 Study of the Effect of the Input Pulse Shape and Duration

Before starting the study of the soliton self-compression in air–silica nanowires, we need to define some parameters in order to characterize the efficiency of the temporal compression. We denote by z_{opt} the optimal length at which a maximum compressed pulse can be extracted and $F_c = T_{FWHM}/T_{comp}$ the compression factor defined as the ratio of the FWHM pulse durations at the input (T_{FWHM}) and the output of the nanowire (T_{comp}). The quality factor $Q_c = P_{comp}/(P_0 F_c)$ is defined as the ratio of the peak power (P_{comp}) of the compressed pulse with the input peak power P_0 and the compression factor F_c.

In order to validate our results, we consider the 800 nm core diameter air–silica nanowire studied by Foster et al. [46]. We inject a 500 pJ and initially 30 fs Gaussian input pulse in the 800 nm air–silica nanowire. We extract a 1.78 fs-compressed pulse at an optimal length of 645 μm, as shown in Figure 4.8.

The selection of the optimum compression length z_{opt} is done when we identify the presence of pulse intensity fluctuations and peak perturbations so that we decide to set the optimal length z_{opt} when a maximum compressed and preserved soliton shape is reached.

Consider now an input pulse assumed to be a nonchirped hyperbolic-secant pulse centered at 800 nm and having FWHM duration of 100 fs.

Figure 4.9 shows the temporal evolution of an initial 100 fs pulse with an input pulse energy of 0.5 nJ as a function of the propagation distance. We notice that a maximum compressed pulse is found at an optimal length $z_{opt} = 2.97$ mm. We extract a 3.26 fs-compressed pulse corresponding to the excitation of an input soliton order of $N = 8.8$.

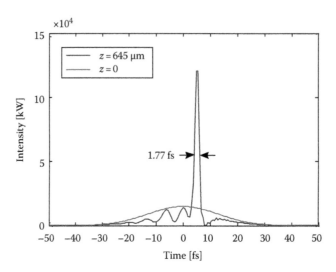

FIGURE 4.8
Temporal evolution of a 30 fs input Gaussian pulse with an input energy of 500 pJ in the 800 nm core diameter air–silica nanowire at 800 nm central wavelength.

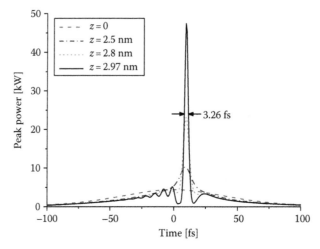

FIGURE 4.9
Temporal evolution of a 100 fs input pulse with an energy of 0.5 nJ in the 800 nm air–silica nanowire at $\lambda p = 800$ nm. The maximum compression is obtained at $z = 2.97$ mm.

This interaction corresponds to a compression factor $F_c = 30.68$ and a quality factor $Q_c = 0.35$. We recall that a compression factor $F_c = 16.67$ was achieved at the same input pulse energy of 0.5 nJ in the 800 nm core diameter air–silica nanowire starting from an input 30 fs pulses.

4.4.1.2 Study of the Effect of the Photonic Nanowire Size

In order to study the effect of the size of the air–silica nanowire on the soliton self-compression, we excite the five air–silica nanowires operating in the anomalous dispersion regime at a pump wavelength of 800 nm with the same soliton order N. We inject a 100 fs FWHM input soliton order of $N = 9$, which corresponds to a low input energy

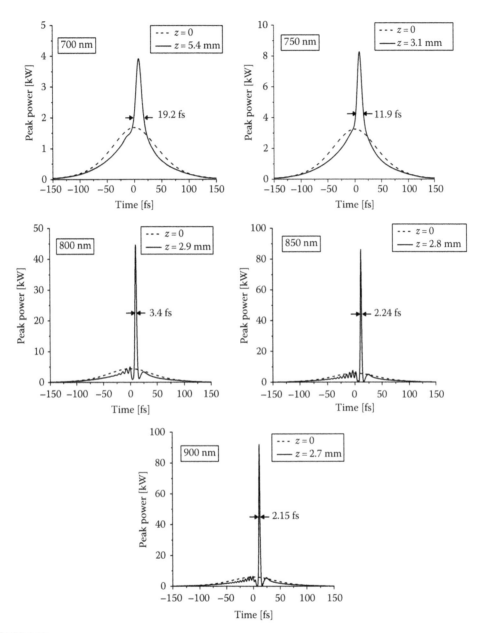

FIGURE 4.10
Soliton self-compression enhanced when the nanowire diameter increases from 700 to 900 nm for an input soliton order ($N=9$).

of 0.2 nJ for the 700 nm core diameter nanowire. Figure 4.10 depicts the evolution of the temporal compressed pulse obtained at the optimal lengths for various silica nanowire diameters ranging from 700 to 900 nm with a step size of 50 nm. We clearly see that the optimal length z_{opt} at which the output pulse is compressed to its maximum decreases as the core diameter of the air–silica nanowire increases. This is explained by the fact that the chromatic dispersion of the nanowire, which shifts toward the longest wavelengths, becomes higher, so that the dispersion length L_D becomes smaller.

We also find that the compressed pulse depicts a temporal FWHM, which decreases as the nanowire diameter increases. This behavior is justified by the enhancement of the peak power when exciting all the nanowires by the same soliton order N expressed by Equation 4.10. The increase of the peak power P_0 is explained by the decrease of the nonlinear coefficient γ and the increase of the GVD when the core diameter of the nanowire is enlarged. Consequently, the compression factor F_c increases from 5.21 for the 700 nm nanowire to 46.51 for the 900 nm nanowire. Furthermore, the correspondent optimal nanowire length z_{opt} decreases from 5.4 to 2.7 mm.

4.4.1.3 Study of the Effect of the Input Pulse Energy

In order to investigate the pulse compression in the anomalous dispersion regime in the 800 nm core diameter nanowire, we study the effect of the variation of the input pulse energy on the temporal compression factor F_c as well as the optimal nanowire length z_{opt}.

Figure 4.11 depicts the decrease of the temporal width of the generated compressed pulses when increasing the input pulse energy from 0.5 to 2.5 nJ. The enhancement of the temporal compression is absolutely accompanied by a decrease in the nanowire length. In fact, we extract maximum compressed pulses at an optimal length z_{opt} equal to 2.97, 1.95, and 1.15 mm for input pulse energy of 0.5, 1, and 2.5 nJ, respectively. We show that we can reach a very high compression factor of $F_c = 65.36$ for an input pulse energy of 2.5 nJ. This corresponds to the excitation of a soliton order of 19.8, which undergoes temporal compression down to 1.53 fs.

Figure 4.12 shows the spectral evolution of 100 fs input soliton, with 2.5 nJ energy, leading to the maximum soliton compression achieved in the 800 nm core diameter air–silica

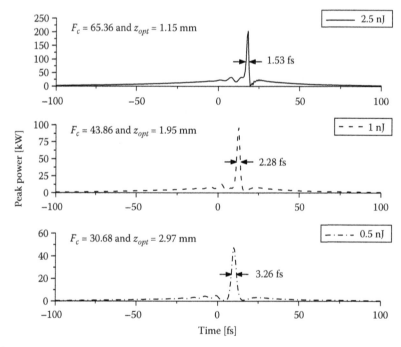

FIGURE 4.11
Effect of the input pulse energy on the temporal soliton compression in the 800 nm air–silica nanowire. A 1.53 fs pulse is generated with an input energy of 2.5 nJ.

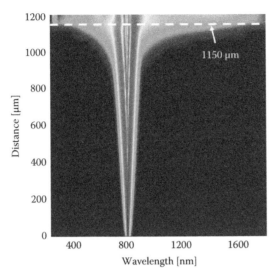

FIGURE 4.12
Maximum spectral broadening obtained of 100 fs input pulse with energy of 2.5 nJ after 1.15 mm propagation distance.

nanowire. We obtain the generation of more than 1-octave spanning SC extending from 260 to 1800 nm. The maximum spectral broadening is achieved after a 1.15 mm propagation distance.

Increasing the input pulse energy considerably affects the spectral broadening, which can shift toward extreme silica wavelength regions. This proves that small core diameter air–silica nanowires are good candidates for generating SC from ultraviolet to near-IR regions.

4.4.2 Soliton Self-Compression and Supercontinuum Generation in As$_2$Se$_3$ Nanowires

After determining the optical properties of air–As$_2$Se$_3$ photonic nanowires, we introduce them in the GNLSE as inputs to investigate the nonlinear propagation in the 800 nm air–As$_2$Se$_3$ nanowire. As we found that the 800 nm air–As$_2$Se$_3$ performs in the anomalous dispersion regime around a pump wavelength of 1550 nm, we select, based on the availability, a femtosecond laser delivering 250 fs FWHM at 1550 nm. The Raman response parameters of the As$_2$Se$_3$ glass are set to be $\tau_1 = 23$ fs, $\tau_2 = 164.5$ fs, and $f_R = 0.148$. The coefficient loss $\alpha = 1$ dB m^{-1}.

Taking into account the high two-photon absorption coefficient ($\alpha_2 = 2.5 \times 10^{-12}$ m W^{-1}) and the high nonlinearity exhibited in As$_2$Se$_3$ glass, we select a very low input pulse energy to characterize the soliton self-compression in the 800 nm air–As$_2$Se$_3$ nanowire. We set the input pulse energy to 10 pJ at a pump wavelength of 1550 nm and inject an input hyperbolic-secant pulse initially chirped with $C = -0.2$, which is found to optimize the temporal compression. This optimal value of the chirp is determined by carrying out several calculations, at a fixed input pulse energy, in which we vary the initial chirp over the interval [−2, 2] with a step size of 0.1 and register the shortest FWHM duration of the generated compressed pulse. Figure 4.13a shows a maximum generated compressed pulse with a duration of 25.4 fs in 2.1 mm nanowire length. This nonlinear interaction corresponds to the injection of a soliton order $N = 14.5$. We characterize the compression

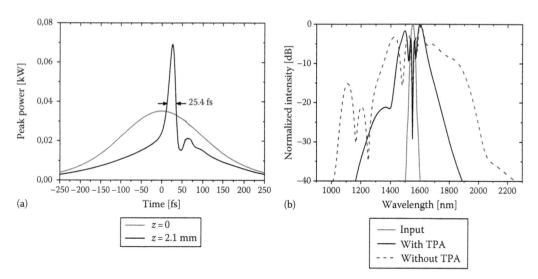

FIGURE 4.13

(a) Temporal and (b) spectral evolution of 250 fs input pulse with 10 pJ input energy in the 800 nm air–As$_2$Se$_3$ nanowire at $\lambda p = 1550$ nm. Supercontinua are generated with and without TPA.

efficiency by the evaluation of two parameters, namely, the compression factor F_c and the quality factor Q_c. The compression factor F_c, defined as the ratio of the FWHM pulse duration at the beginning and the end of the nanowire, is found to be 9.84. The correspondent quality factor Q_c, which is equal to the peak power of the compressed pulse normalized to the input peak power and the compression factor, is evaluated to be 0.21. Spectral evolution of the initial pre-chirped hyperbolic-secant pulse is depicted in Figure 4.13b. The SC is generated taking into account the effect of TPA (for which we extracted the maximum compressed pulse) and then without TPA, and we notice its big effect on the spectral generated bandwidth. In fact, an extra 500 nm bandwidth is generated when one neglects the TPA. Thus, a perfect modeling has to consider the effect of TPA in order to obtain correct results simulating the nonlinear propagation in highly nonlinear air–As$_2$Se$_3$ nanowires. We notice the generation of over 700 nm bandwidth of near-IR SC showing a symmetrical broadening, which confirms that self-phase modulation is the dominant effect giving such broadening and leading to temporal soliton self-compression.

4.5 Conclusion

In this chapter, we present photonic nanowires and their fabrication techniques. These structures with submicron core diameters have attracted so much attention because of the numerous extraordinary optical properties that they offer. They have been the subject of intense investigations. We showed that with air–silica nanowires, we can generate few optical cycles in few millimeters length. We found that less than single optical cycle can be generated in 800 nm air–silica nanowire at low input pulse energy and for short interaction distance. This corresponds to the excitation of higher-order soliton, which undergoes soliton self-compression in the anomalous dispersion regime. Consequently, the generation of more than one-octave spanning SC from ultraviolet to near IR is shown with low input pulse energy of 2.5 nJ. Air–silica nanowires present suitable waveguides for high soliton compression, which is very promising for designing attosecond-pulse sources. Detailed studies of the optical properties have been achieved in chalcogenide photonic nanowires based on As_2Se_3 glass. Efficient soliton self-compression requires large anomalous GVD region with low third-order dispersion. This soliton-effect compression results from the interplay between the dispersion and the self-phase modulation effects. This fact is due to the exotic dispersion profiles and high nonlinearities provided by photonic nanowires. In addition, the generation of mid-IR two octaves spanning SC around 1550 nm in only 0.84 mm air–As_2S_3 nanowire is shown. Chalcogenide photonic nanowires are very promising devices for designing mid-IR light coherent sources with short lengths and low powers. They are suitable waveguides for high soliton-effect compression, which opens new horizon toward extreme nonlinear optics and attosecond physics.

References

1. L. M. Tong, R. R. Gattass, J. B. Ashcom, S. L. He, J. Y. Lou, M. Y. Shen, I. Maxwell, and E. Mazur, Subwavelength-diameter silica wires for low-loss optical wave guiding, *Nature* **426**, 816–819 (2003).

2. G. Brambilla, V. Finazzi, and D. J. Richardson, Ultra-low-loss optical fiber nanotapers, *Opt. Express* **12**, 2258–2263 (2004).

3. S. G. Leon-Saval, T. A. Birks, W. J. Wadsworth, P. S. J. Russell, and M. W. Mason, Supercontinuum generation in submicron fibre waveguides, *Opt. Express* **12**, 2864–2869 (2004).

4. E. C. Mägi, P. Steinvurzel, and B. J. Eggleton, Tapered photonic crystal fibers, *Opt. Express* **12**(5), 776–784 (2004).

5. Y. K. Lize, E. C. Magi, V. G. Ta'eed, J. A. Bolger, P. Steinvurzel, and B. J. Eggleton, Microstructured optical fiber photonic wires with subwavelength core diameter, *Opt. Express* **12**, 3209–3217 (2004).

6. R. R. Gattass, G. T. Svacha, L. M. Tong, and E. Mazur, Supercontinuum generation in submicrometer diameter silica fibers, *Opt. Express* **14**, 9408–9414 (2006).

7. M. A. Foster, K. D. Moll, and A. L. Gaeta, Optimal waveguide dimensions for nonlinear interactions, *Opt. Express* **12**, 2880–2887 (2004).

8. H. C. Nguyen, B. T. Kuhlmey, M. J. Steel, C. L. Smith, E. C. Magi, R. C. McPhedran, and B. J. Eggleton, Leakage of the fundamental mode in photonic crystal fiber tapers, *Opt. Lett.* **30**, 1123–1125 (2005).

9. A. Ben Salem, R. Cherif, and M. Zghal, Generation of few optical cycles in air-silica nanowires, *Proc. SPIE* **8001**, 80011J (2011)

10. A. Ben Salem, R. Cherif, and M. Zghal, Study of soliton-self compression in photonic nanowires, *Proc. SPIE* **8011**, 80119N (2011).

11. A. Ben Salem, R. Cherif, and M. Zghal, Low-energy single-optical-cycle soliton self-compression in air-silica nanowires, *SPIE J. Nanophoton.* **5**, 059506 (2011).

12. A. Ben Salem, R. Cherif, and M. Zghal, Soliton-self compression in highly nonlinear chalcogenide photonic nanowires with ultralow pulse energy, *OSA Opt. Express* **19**, 19955–19966 (2011).

13. A. Ben Salem, R. Cherif, and M. Zghal, Tapered As_2S_3 chalcogenide photonic crystal fiber for broadband mid-infrared supercontinuum generation, *OSA FIO* 2011, San Jose, CA.

14. L. Tong, L. Hu, J. Zhang, J. Qiu, Q. Yang, J. Lou, Y. Shen, J. He, and Z. Ye, Photonic nanowires directly drawn from bulk glasses, *Opt. Express* **14**(1), 82–87 (2006).

15. G. Brambilla, F. Koizumi, X. Feng, D.J. Richardson, Compound-glass optical nanowires, *Electron Lett.* **4**, 1400–1402 (2005).

16. E. C. Mägi, L. B. Fu, H. C. Nguyen, M. R. E. Lamont, D. I. Yeom, B. J. Eggleton, Enhanced Kerr nonlinearity in sub-wavelength diameter As2Se3 chalcogenide fiber tapers, *Opt. Express* **15**, 10324–10329 (2007).

17. S. A. Harfenist, S. D. Cambron, E. W. Nelson, S. M. Berry, A. W. Isham, M. M. Crain, K. M. Walsh, R. S. Keynton and R. W. Cohn, *Nano Lett.* **4**, 1931–1937 (2004).

18. A. S. Nain, J. C. Wong, C. Amon and M. Sitti, *Appl. Phys. Lett.* **89**, 183105–183107 (2006).

19. F. X. Gu, L. Zhang, X. F. Yin, and L. M. Tong, *Nano Lett.* **8**, 2757–2761 (2008).

20. X. Xing, Y. Wang, and B. Li, *Opt. Express* **16**, 10815–10822 (2008).

21. G. Brambilla, Optical fibre nanowires and microwires: A review, *J. Opt.* **12**(4), 043001 (2010).

22. G. Brambilla, F. Xu, and X. Feng, Fabrication of optical fibre nanowires and their optical and mechanical characterization, *Electron. Lett.* **42**, 517–518 (2006).

23. F. Bilodeau, K. O. Hill, S. Faucher, and D. C. Johnson, Low-loss highly overcoupled fused couplers—Fabrication and sensitivity to external pressure, *J. Lightw. Technol.* **6**, 1476–1482 (1988).

24. T. A. Birks and Y. W. Li, The shape of fiber tapers, *J. Lightw. Technol.* **10**, 432–438 (1992).

25. D. A. Akimov, M. Schmitt, R. Maksimenka, K. V. Dukel'skii, Y. N. Kondrat'ev, A. V. Khokhlov, V. S. Shevandin, W. Kiefer, and A. M. Zheltikov, Supercontinuum generation in a multiple-submicron-core microstructure fiber: Toward limiting waveguide enhancement of nonlinear-optical processes, *Appl. Phys.* **B77**, 299–305 (2003).

26. L. Tong, J. Lou, and E. Mazur, Single-mode guiding properties of subwavelength-diameter silica and silicon wire waveguides, *Opt. Express* **12** (6), 1025–1035 (2004),

27. J. K. Ranka, R. S. Windeler, and A. J. Stentz, Visible continuum generation in air silica micro-structure optical fibers with anomalous dispersion at 800 nm, *Opt. Lett.* **25**(1), 25–27 (2000).

28. J. Villatoro and D. Monz'on-Hern'andez, Fast detection of hydrogen with nano fiber tapers coated with ultra thin palladium layers, *Opt. Express* **13**, 5087–5092 (2005).

29. P. Polynkin, A. Polynkin, N. Peyghambarian, and M. Mansuripur, Evanescent field-based optical fiber sensing device for measuring the refractive index of liquids in microfluidic channels, *Opt. Lett.* **30**, 1273–1275 (2005).

30. L. Zhang, F. Gu, J. Lou, X. Yin, and L. Tong, Fast detection of humidity with a subwavelength-diameter fiber taper coated with gelatin film, *Opt. Express* **16**, 13349–13353 (2008).

31. W. J. Wadsworth, A. Ortigosa-Blanch, J. C. Knight, T. A. Birks, T.-P. Martin Man, and P. St. J. Russell, Supercontinuum generation in photonic crystal fibers and optical fiber tapers: A novel light source, *J. Opt. Soc. Am.* **B19**, 2148–2155 (2002).

32. J. Teipel, K. Franke, D. Turke, F. Warken, D. Meiser, M. Leuschner, and H. Giessen, Characteristics of supercontinuum generation in tapered fibers using femtosecond laser pulses, *Appl. Phys.* **B 77**, 245–251 (2003).

33. V. Grubsky and A. Savchenko, Glass micro-fibers for efficient third harmonic generation, *Opt. Express* **13**, 6798–6806 (2005).

34. U. Wiedemann, K. Karapetyan, C. Dan, D. Pritzkau, W. Alt, S. Irsen, and D. Meschede, Measurement of submicrometre diameters of tapered optical fibres using harmonic generation, *Opt. Express* **18**, 7693–7704 (2010).

35. G. J. Pendock, H. S. MacKenzie, and F. P. Payne, Dye lasers using tapered optical fibers, *Appl. Opt.* **32**, 5236–5242 (1993).

36. X. S. Jiang, Q. H. Song, L. Xu, J. Fu, and L. M. Tong, Microfiber knot dye laser based on the evanescent-wavecoupled gain, *Appl. Phys. Lett.* **90**, 233501 (2007).

37. F. Warken, E. Vetsch, D. Meschede, M. Sokolowski, and A. Rauschenbeutel, Ultra-sensitive surface absorption spectroscopy using sub-wavelength diameter optical fibers, *Opt. Express* **15**, 11952–11958 (2007).

38. A. Stiebeiner, O. Rehband, R. Garcia-Fernandez, and A. Rauschenbeutel, Ultra-sensitive fluorescence spectroscopy of isolated surface-adsorbed molecules using an optical nanofiber, *Opt. Express* **17**, 21704–21711 (2009).

39. G. Sague, E. Vetsch, W. Alt, D. Meschede, and A. Rauschenbeutel, Cold-atom physics using ultrathin optical fibres: Light-induced dipole forces and surface interactions, *Phys. Rev. Lett.* **99,** 163602 (2007).
40. K. P. Nayak, P. N. Melentiev, M. Morinaga, F. Le Kien, V. I. Balykin, and K. Hakuta, Optical nanofiber as an efficient tool for manipulating and probing atomic fluorescence, *Opt. Express* **15**, 5431–5438 (2007).
41. K. P. Nayak and K. Hakuta, Single atoms on an optical nanofibre, *New J. Phys.* **10**, 053003 (2008).
42. M. J. Morrissey, K. Deasy, Y. Q. Wu, S. Chakrabarti, and S. N. Chormaic, Tapered optical fibers as tools for probing magneto-optical trap characteristics, *Rev. Sci. Instrum.* **80**, 053102 (2009).
43. E. Vetsch, D. Reitz, G. Sagu'e, R. Schmidt, S. T. Dawkins, and A. Rauschenbeutel, Optical interface created by laser-cooled atoms trapped in the evanescent field surrounding an optical nanofiber, *Phys. Rev. Lett.* **104**, 203603 (2010).
44. G. P. Agrawal, *Nonlinear Fiber Optics*, 3rd edn., Academic Press, San Diego, CA (2001).
45. K. J. Blow and D. Wood, Theoretical description of transient stimulated Raman scattering in optical fibers, *IEEE J. Quantum Electron.* **25**, 2665–2673 (1989).
46. M. Foster, A. L. Gaeta, Q. Cao, and R. Trebino, Soliton-effect compression of supercontinuum to few-cycle durations in photonic nanowires, *Opt. Express* **13**(18), 6848–6855 (2005).

5

Modeling Optical Applications of Nanofibers/Nanowires

Yipei Wang and Limin Tong

CONTENTS

5.1 Introduction

As typical 1D wire-type waveguides with diameter or width close to or below the wavelength of the light, optical nanofibers/nanowires offer interesting properties including tight optical confinement, high fractional evanescent waves, steep field gradient, and abnormal dispersion, which open opportunities for developing nanophotonic components and devices ranging from filters, interferometers, resonators, and lasers to sensors.

The nanofiber/nanowire discussed here consists of two layers of structures: a solid cylindrical core surrounded by a clad with a relatively low refractive index. In contrast to the theoretical modeling for conventional waveguides, the weak guidance approximation is not valid for high-index-contrast nanofibers/nanowires. The rigorous method is to solve Maxwell equations analytically. However, in many practical cases, analytical solution is difficult to obtain and we must restore to numerical method. For waveguiding structures like nanofibers/nanowires with relatively small calculation dimension, finite-difference time-domain (FDTD) simulation or finite-element method (FEM) is one of the best numerical methods regarding accuracy and efficiency. This chapter introduces modeling optical properties of nanofibers/nanowires, as well as applications of these 1D structures for photonic components or devices.

5.2 Single-Mode Waveguiding Properties of Dielectric Nanofibers/Nanowires

In the first section, we introduce basic waveguiding properties of air-clad sub-wavelength-diameter nanofibers, by solving Maxwell equations with boundary conditions.

5.2.1 Mathematical Model

In basic models, a straight nanofiber is assumed to have a circular cross section, a uniform diameter, a smooth surface, and an infinite air clad with a step-index profile. The length of the nanofiber is assumed to be large enough (e.g., >10 μm) to establish the spatial steady state. The mathematical model with the aforementioned assumptions is shown in Figure 5.1, in which a is the radius of the nanofiber and n_1 and n_2 are the refractive indexes of the fiber material and the air, respectively.

We assume the refractive index of air (n_1) is 1.0, and the materials of the nanofiber are silica and silicon with the refractive indexes obtained from the Sellmeier-type dispersion formula.

For silica [2],

$$n^2 - 1 = \frac{0.6961663\lambda^2}{\lambda^2 - (0.0684043)^2} + \frac{0.4079426\lambda^2}{\lambda^2 - (0.1162414)^2} + \frac{0.8974794\lambda^2}{\lambda^2 - (9.896161)^2} \tag{5.1}$$

and for silicon [3],

$$n^2 = 11.6858 + \frac{0.939816}{\lambda^2} + \frac{0.000993358}{\lambda^2 - 1.22567} \tag{5.2}$$

In most practical applications, the sub-wavelength-diameter nanofibers can be regarded as non-dissipative and source-free. Therefore, Maxwell equations can be reduced to the following Helmholtz equations:

$$\begin{aligned}
\left(\nabla^2 + n^2 k^2 - \beta^2\right) \quad \vec{e} = 0, \\
\left(\nabla^2 + n^2 k^2 - \beta^2\right) \quad \vec{h} = 0
\end{aligned} \tag{5.3}$$

where
  k represents the wavevector
  β is the propagation constant

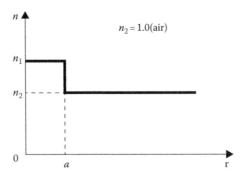

FIGURE 5.1
Mathematical model of an air-clad cylindrical nanofiber. (Adapted from Tong, L.M. et al., *Opt. Express*, 12, 1025, 2004.)

Exact solutions of Equation 5.3 can be found in Ref. [4]. The eigenvalue equations of Equation 5.3 for different modes are obtained as follows:

For HE_{vm} and EH_{vm} modes,

$$\left\{ \frac{J_v'(U)}{UJ_v(U)} + \frac{K_v'(W)}{WK_v(W)} \right\} \left\{ \frac{J_v'(U)}{UJ_v(U)} + \frac{n_2^2 K_v'(W)}{n_1^2 WK_v(W)} \right\} = \left(\frac{v\beta}{kn_1} \right)^2 \left(\frac{V}{UW} \right)^4 \tag{5.4}$$

for TE_{0m} modes,

$$\frac{J_1(U)}{UJ_0(U)} + \frac{K_1(W)}{WK_0(W)} = 0 \tag{5.5}$$

and for TM_{0m} modes,

$$\frac{n_1^2 J_1(U)}{UJ_0(U)} + \frac{n_2^2 K_1(W)}{WK_0(W)} = 0 \tag{5.6}$$

where

J_v is the Bessel function of the first kind

K_v is the modified Bessel function of the second kind

$U = D\sqrt{\left(k_0^2 n_1^2 - \beta^2\right)}, W = D\sqrt{\left(\beta^2 - k_0^2 n_2^2\right)}, V = k_0 a \sqrt{\left(n_1^2 - n_2^2\right)},$ and $D = 2a$

5.2.2 Propagation Constants and Single-Mode Condition

In most cases, searching propagation constants (β) is the starting point for theoretical investigations of guided modes in nanofibers/nanowires. By numerically solving Equations 5.4 through 5.6, the diameter (D)-dependent β of silica and silicon nanofibers at 633 and 1550 nm wavelength are obtained and plotted in Figures 5.2 and 5.3, respectively, in which V number is defined as $V = 2\pi \cdot \dfrac{a}{\lambda_0} \sqrt{\left(n_1^2 - n_2^2\right)}$. It shows that when D is reduced to a certain value (denoted as D_{SM}), all the modes exhibit cutoffs except the HE_{11} mode, corresponding to the single-mode condition that can be retrieved from Equations 5.5 and 5.6 and expressed as

$$V = 2\pi \cdot \frac{a}{\lambda_0} \sqrt{\left(n_1^2 - n_2^2\right)} \approx 2.405 \tag{5.7}$$

Figure 5.3 shows single-mode conditions of the air-clad silica and silicon nanofibers with respect to the wavelengths and diameters (plotted as solid lines). The region beneath the solid line corresponds to the single-mode region. We can see that compared with the wavelength of the propagating light inside the nanofiber (plotted as dashed line), a sub-wavelength-diameter silica nanofiber is always single mode, while a sub-wavelength-diameter silicon nanofiber is single mode only when its diameter lies below the solid line.

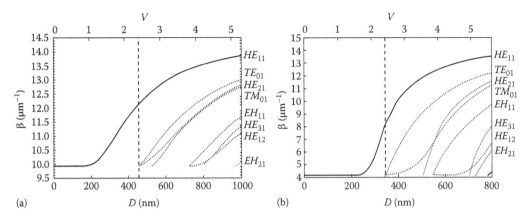

FIGURE 5.2
Calculated propagation constant (β) of (a) air-clad silica and (b) silicon nanofibers at 633 nm wavelength. Solid line, fundamental mode. Dotted lines, high-order modes. Dashed line, critical diameter for single-mode operation (D_{SM}). (Adapted from Tong, L.M. et al., *Opt. Express*, 12, 1025, 2004.)

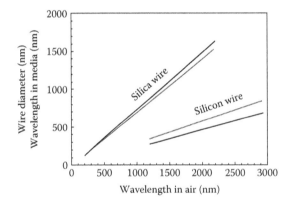

FIGURE 5.3
Single-mode condition of an air-clad silica and silicon nanofibers. Solid line, critical diameter for single-mode operation. Dotted line, wavelength in media. (Adapted from Tong, L.M. et al., *Opt. Express*, 12, 1025, 2004.)

5.2.3 Modal Fields of Fundamental Modes

For practical applications, the single-mode operation is usually favorable or required. So now we only focus on the fundamental mode (HE_{11}) and take a deeper investigation. In single-mode condition, Equation 5.4 becomes

$$\left\{ \frac{J_1'(U)}{UJ_1(U)} + \frac{K_1'(W)}{WK_1(W)} \right\} \left\{ \frac{J_1'(U)}{UJ_1(U)} + \frac{n_2^2 K_1'(W)}{n_1^2 WK_1(W)} \right\} = \left(\frac{v\beta}{kn_1} \right)^2 \left(\frac{V}{UW} \right)^4 \tag{5.8}$$

When we express the electromagnetic fields as

$$\begin{cases} \vec{E}(r,\phi,z) = (e_r \hat{r} + e_\phi \hat{\phi} + e_z \hat{z})e^{i\beta z}e^{-i\omega t} \\ \vec{H}(r,\phi,z) = (h_r \hat{r} + h_\phi \hat{\phi} + h_z \hat{z})e^{i\beta z}e^{-i\omega t} \end{cases} \tag{5.9}$$

electric fields of the fundamental mode can be obtained as follows [4]:

inside the core ($0 < r < a$),

$$e_r = -\frac{a_1 J_0(UR) + a_2 J_2(UR)}{J_1(U)} \cdot f_1(\phi) \tag{5.10}$$

$$e_\phi = -\frac{a_1 J_0(UR) + a_2 J_2(UR)}{J_1(U)} \cdot g_1(\phi) \tag{5.11}$$

$$e_z = \frac{-iU}{\alpha\beta} \frac{J_1(UR)}{J_1(U)} \cdot f_1(\phi) \tag{5.12}$$

and outside the core ($r > a$),

$$e_r = -\frac{U}{W} \frac{a_1 K_0(WR) - a_2 K_2(WR)}{K_1(W)} \cdot f_1(\phi) \tag{5.13}$$

$$e_\phi = -\frac{U}{W} \frac{a_1 K_0(WR) - a_2 K_2(WR)}{K_1(W)} \cdot g_1(\phi) \tag{5.14}$$

$$e_z = \frac{-iU}{\alpha\beta} \frac{K_1(UR)}{K_1(U)} \cdot f_1(\phi) \tag{5.15}$$

where

$f_1(\phi) = \sin(\phi),\, g_1(\phi) = \cos(\phi)$

$$a_1 = \frac{F_2 - 1}{2}, \quad a_3 = \frac{F_1 - 1}{2}, \quad a_5 = \frac{F_1 - 1 + 2\Delta}{2}, \quad a_2 = \frac{F_2 + 1}{2}, \quad a_4 = \frac{F_1 + 1}{2}, \quad a_6 = \frac{F_1 + 1 - 2\Delta}{2}$$

$$F_1 = \left(\frac{UW}{V}\right)^2 [b_1 + (1 - 2\Delta)b_2], \quad F_2 = \left(\frac{V}{UW}\right)^2 \frac{1}{b_1 + b_2}$$

$$b_1 = \frac{1}{2U}\left\{\frac{J_0(U)}{J_1(U)} - \frac{J_2(U)}{J_1(U)}\right\}, \quad b_2 = -\frac{1}{2W}\left\{\frac{K_0(W)}{K_1(W)} + \frac{K_2(W)}{K_1(W)}\right\}$$

By substituting the propagation constant (β) from Figure 5.2 into Equations 5.10 through 5.15, the normalized electric components of the fundamental modes in cylindrical coordinates are obtained and shown in Figure 5.4. For comparison, Gaussian profiles (dashed line) are plotted in the radial distributions, and the electric field of a nanofiber with a diameter of D_{SM} is also provided as a dotted line. Compared to the Gaussian profile, air-clad silica nanofiber shows much tighter field confinement within a certain diameter range (e.g., around 400 nm), which is attributed to the high-index contrast between the air and silica. Besides, when D reduces to a certain degree (e.g., 200 nm), the field begins to extend to a far distance with considerable amplitude, indicating that the majority of the field is no longer tightly confined inside or around the nanofiber.

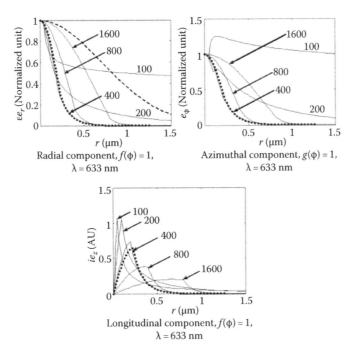

FIGURE 5.4

Electric components of HE_{11} modes of silica nanofibers at 633 nm wavelength with different diameters in cylindrical coordination. Normalizations are applied as $\varepsilon e_r(r=0)=1$ and $e_\varphi(r=0)=1$. Nanofiber diameters are arrowed to each curve in unit of nm. (Adapted from Tong, L.M. et al., *Opt. Express*, 12, 1025, 2004.)

5.2.4 Power Distributions and Effective Diameters

In order to characterize the confinement of the waveguide, here we discuss the power distributions and effective diameters of the waveguides. These parameters are important especially for optic circuits and evanescent wave-based sensors. Tight confinement is helpful for reducing the modal width and increasing the integrated density of the optical circuits with less cross talk [4,5], while weaker confinement will be favorable for energy exchanging between nanofibers within a short interaction length [6], as well as for improving the sensitivity of evanescent wave-based sensors [7].

For an ideal wire waveguide considered here, the average energy flows in the radial (r) and azimuthal (φ) directions are zero. The z-components of Poynting vectors are obtained as follows [4]:

inside the core ($0 < r < a$),

$$S_{z1} = \frac{1}{2}\left(\frac{\varepsilon_0}{\mu_0}\right)^{1/2} \frac{kn_1^2}{\beta J_1^2(U)}\left[a_1a_3J_0^2(UR) + a_2a_4J_2^2(UR) + \frac{1-F_1F_2}{2}J_0(UR)J_2(UR)\cos(2\phi)\right] \quad (5.16)$$

and outside the core ($r > a$),

$$S_{z2} = \frac{1}{2}\left(\frac{\varepsilon_0}{\mu_0}\right)^{1/2} \frac{kn_1^2}{\beta K_1^2(W)} \frac{U^2}{W^2}\left[a_1a_5K_0^2(WR) + a_2a_6K_2^2(WR) - \frac{1-2\Delta-F_1F_2}{2}K_0(WR)K_2(WR)\cos(2\phi)\right]$$

$$(5.17)$$

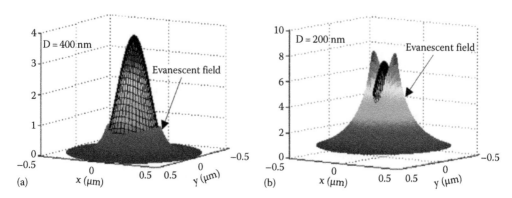

FIGURE 5.5

Z-direction Poynting vectors of silica nanofibers at 633 nm wavelength with diameters of (a) 400 and (b) 200 nm. Mesh, field inside the core. Gradient, field outside the core. (Adapted from Tong, L.M. et al., *Opt. Express*, 12, 1025, 2004.)

Figure 5.5 are the profiles of Poynting vectors for 200 and 400 nm diameter silica nanofibers at 633 nm wavelength, in which the mesh profile represents propagating fields inside the nanofiber and the gradient profile stands for evanescent fields in air. It shows that for a 400 nm diameter nanofiber, the major power is confined inside the wire; for a 200 nm diameter nanofiber, a large amount of light is guided outside as evanescent waves.

With the Poynting vectors obtained in the preceding text, we define the fractional power inside the core (n) as

$$\eta = \frac{\int_0^a S_{z1} dA}{\int_0^a S_{z1} dA + \int_a^\infty S_{z2} dA},$$ (5.18)

where $dA = r \cdot dr \cdot d\theta$ is the cross-sectional surface element of the nanowire in the cylindrical coordinates. Figure 5.6 gives the calculated D-dependent n for silica nanofiber at 633 nm

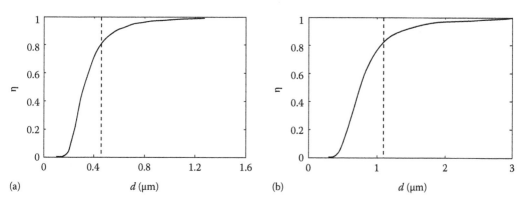

FIGURE 5.6

Fractional power of the fundamental modes inside the core of an air-clad silica nanofiber operated at (a) 633 nm and (b) 1.5 μm wavelength. Dashed line, critical diameter for single-mode operation. (Adapted from Tong, L.M. et al., *Opt. Express*, 12, 1025, 2004.)

and 1.5 μm wavelengths. It shows that at D_{SM} (dashed line), n is about 81% (633 nm and 1.5 μm wavelength). The diameter for confining 90% energy inside the core is about 566 nm (633 nm wavelength) and 1342 nm (1.5 μm wavelength), while the diameter for confining 10% energy is 216 nm (633 nm wavelength) and 513 nm (1.5 μm wavelength).

To obtain more straightforward information of the confinement, the effective diameter (D_{eff}) of the optical field is defined as a hypothetic diameter inside which 86.5% (i.e., $1 - e^2$) of the total power is included and it can be expressed as

$$
\begin{cases}
\dfrac{\displaystyle\int_0^{D_{eff}} S_{z1}dA}{\displaystyle\int_0^a S_{z1}dA + \int_a^\infty S_{z2}dA} = 86.5\%, \quad (if\ D_{eff} \le a) \\[4mm]
\dfrac{\displaystyle\int_0^a S_{z1}dA + \int_a^{D_{eff}} S_{z1}dA}{\displaystyle\int_0^a S_{Z1}dA + \int_a^\infty S_{z2}dA} = 86.5\%, \quad (if\ D_{eff} > a)
\end{cases}
\tag{5.19}
$$

Figure 5.7 gives the D_{eff} of the fundamental mode of silica nanofibers at 633 nm and 1.5 μm wavelengths. For comparison, the real diameters of nanofibers (D_{real}) are provided as dotted lines. Result shows that when D is very small, D_{eff} is large. For example, for a 200 nm diameter silica nanofiber at 633 nm wavelength, D_{eff} is about 2.3 μm, which is over 10 times larger than the fiber diameter. Meanwhile, with the increasing D, D_{eff} decreases quickly until it reaches a minimum value. For a nanofiber with a large D (e.g., 450 nm), it is able to confine major power within a sub-wavelength scale. Moreover, in Figure 5.7, the critical diameters (D_{SM}) are shown as dashed line. It shows that two curves (D-dependent D_{real} and D-dependent D_{eff}) intersect near the D_{SM} and the D_{real} exceeds the D_{eff} thereafter. For practical applications, the intersection point in Figure 5.7 represents the minimum usable nanofiber diameter at the nominated wavelength when using an 86.5% confinement for estimation.

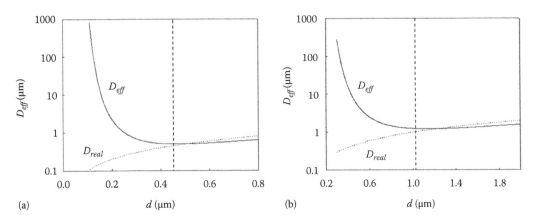

(a) (b)

FIGURE 5.7
Effective diameters (D_{eff}) of the light fields of the fundamental modes of an air-clad silica nanofiber operated at (a) 633 nm wavelength and (b) 1.5 μm wavelength. Solid line, D_{eff}; dotted line, real diameter; dashed line, D_{SM}. (Adapted from Tong, L.M. et al., *Opt. Express*, 12, 1025, 2004.)

5.2.5 Group Velocity and Waveguide Dispersion

Group velocity of fundamental mode can be obtained as [4]

$$v_g = \frac{c}{n_1^2} \cdot \frac{\beta}{k} \cdot \frac{1}{1 - 2\Delta(1-\eta)} \tag{5.20}$$

from which D-dependent group velocities (v_g) of HE_{11} modes of air-clad silica and silicon nanofibers are obtained and shown in Figure 5.8. We can see that when D is very small, v_g approaches the light speed (c) in vacuum since most of the light energy is propagated in air. With increasing D, more and more energy is confined inside the core, and v_g decreases until it reduces to a minimum value that is smaller than c/n_1 (the group velocity of a plane wave in the nanofiber material). After this point, v_g increases gradually with D and finally approaches c/n_1 when the D is large enough.

Figure 5.9 gives the wavelength dependence of v_g for nanofibers with different diameters. It exhibits a similar behavior to the diameter-dependent situation. For a given D,

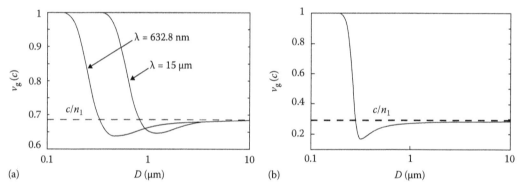

(a)   (b)

FIGURE 5.8
Diameter-dependent group velocities of the fundamental modes of air-clad (a) silica nanofiber at 633 nm and 1.5 μm wavelengths and (b) silicon nanofiber at 1.5 μm wavelength. (Adapted from Tong, L.M. et al., *Opt. Express*, 12, 1025, 2004.)

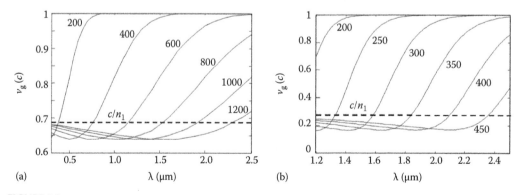

(a)   (b)

FIGURE 5.9
Wavelength-dependent group velocities of the fundamental modes of air-clad (a) silica nanofiber and (b) silicon nanofiber with different diameters (nanofiber diameters are labeled on each curve in unit of nm). (Adapted from Tong, L.M. et al., *Opt. Express*, 12, 1025, 2004.)

when the wavelength is large, v_g approaches c. Conversely, when the wavelength is very small, v_g approaches c/n_1 with a minimum value smaller than c/n_1.

With group velocity obtained in the preceding text, the waveguide dispersion (D_w) is obtained as [8]

$$D_w = \frac{d(v_g^{-1})}{d\lambda} \tag{5.21}$$

where λ is the wavelength.

Diameter- and wavelength-dependent waveguide dispersions (D_w) of air-clad silica and silicon nanofibers are illustrated in Figures 5.10 and 5.11, respectively. For comparison, material dispersions of fused silica and single-crystal silicon calculated from

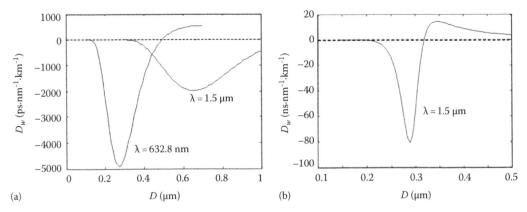

FIGURE 5.10
Diameter-dependent waveguide dispersion of fundamental modes of air-clad (a) silica nanofiber at 633 nm and 1.5 μm wavelengths and (b) silicon nanofiber at 1.5 μm wavelengths. (Adapted from Tong, L.M. et al., *Opt. Express*, 12, 1025, 2004.)

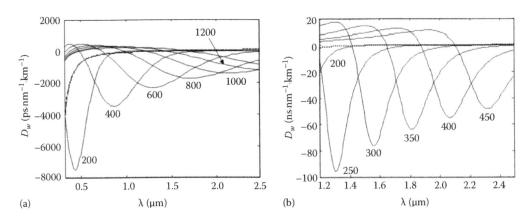

FIGURE 5.11
Wavelength-dependent waveguide dispersion of fundamental modes of air-clad (a) silica nanofiber and (b) silicon nanofiber with different wire diameters (the diameter is labeled on each curve in unit of nm). Material dispersion is plotted in dotted line. (Adapted from Tong, L.M. et al., *Opt. Express*, 12, 1025, 2004.)

Equations 5.1 and 5.2 are also provided in Figure 5.11. Compared to those of weakly guiding fibers and bulk material, D_w of calculated nanofibers can be very large. For example, D_w of an 800 nm diameter silica nanofiber at 1.5 μm wavelength is as high as −1400 ps·nm⁻¹·km⁻¹, which is about 70 times larger than that of the material dispersion. In contrast to those of ps·nm⁻¹·km⁻¹ for weakly guided glass fibers [9], D_w of silica and silicon wires can reach ns·nm⁻¹·km⁻¹ level. Moreover, the total dispersion (combined material and waveguide dispersions) of a wire can be made zero, positive, or considerably negative when the diameter is chosen properly, which provides opportunities for achieving enhanced dispersions with reduced sizes.

5.3 Dispersion in Optical Nanowires with Thin Dielectric Coatings

Controlling light-propagation properties by tailoring waveguide dispersion is of special interest for applications such as optical communication, optical sensing, nonlinear optics, and atom trapping [10–13]. As shown in the previous section, we have already studied the dispersion of an air-clad nanowire, in which the dispersion depends on the parameters including the diameter and refractive index of the nanowire and the wavelength of the guided light. While in some applications, a large dispersion or dispersion shift is required or favorable. Here we introduce a three-layer-structured cylindrical waveguide (a nanowire coated with a thin non-dissipative dielectric material). Compared to air-clad nanowires, nanowires with dielectric coatings may lead to considerable dispersion shift with minimum change in their geometric dimensions.

5.3.1 Mathematical Model

The mathematical model is schematically illustrated in Figure 5.12. A long, straight nanowire with coat and air clad is assumed to be a cylindrical structure of translation symmetry, which involves three regions in the cross section (Figure 5.12a): a circular dielectric core with radius ρ, a dielectric coat with thickness d_c, and an infinite air cladding. Refractive indexes of the core, coat, and air are assumed to be n_s, n_c, and n_a, respectively (Figure 5.12b). For numerical calculations, silica is selected as core material of nanowires ($n_s = 1.45$) since it is commonly used at visible and near-infrared spectral ranges. Dielectric materials with relatively low to

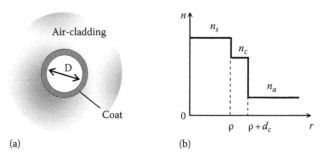

(a)   (b)

FIGURE 5.12
Mathematical model of an optical nanowire with a cylinder coat. (a) Cross-sectional view and (b) refractive-index profile of the nanowire. (Adapted from Lou, J.Y. et al., *Opt. Express*, 14, 6993, 2006.)

high refractive index including SiO_2 ($n_c = 1.45$), $PbCl_2$ ($n_c = 2.20$), and TiO_2 ($n_c = 2.70$) are used as the cylinder coats. The refractive index of air clad (n_a) is assumed to be 1.0.

By solving Maxwell equations in cylindrical coordinates (r, θ, z), we obtain the components of the electromagnetic field for the mth mode as [15,16]

$$E_r = \left(\frac{i\beta}{k_j} Z_m^{j\prime}(k_j r)a_m^j - \frac{m\omega}{k_j^2 r} Z_m^j(k_j r)b_m^j \right) F_m, \qquad H_r = \left(\frac{mn_j^2 k_0^2}{\omega k_j^2 r} Z_m^j(k_j r)a_m^j + \frac{i\beta}{k_j} Z_m^{j\prime}(k_j r)b_m^j \right) F_m,$$

$$E_\theta = \left(-\frac{m\beta}{k_j^2 r} Z_m^j(k_j r)a_m^j - \frac{i\omega}{k_j} Z_m^{j\prime}(k_j r)b_m^j \right) F_m, \qquad H_\theta = \left(\frac{in_j^2 k_0^2}{\omega k_j} Z_m^{j\prime}(k_j r)a_m^j - \frac{m\beta}{k_j^2 r} Z_m^j(k_j r)b_m^j \right) F_m,$$

$$E_z = Z_m^j(k_j r)a_m^j F_m, \qquad\qquad\qquad H_z = Z_m^j(k_j r)b_m^j F_m, \qquad\qquad (5.22)$$

where

the index $j = s$ denotes the components inside the dielectric core ($r < \rho$)
$j = c$, the cylinder coat ($\rho < r < \rho + d_c$)
$j = a$, the infinite air ($r > \rho + d_c$)
$Z_m^s(x) \equiv J_m(x)$, the Bessel function of order m
$Z_m^a(x) \equiv H_m^{(1)}(x)$, the Hankel function of the first kind of order m
$Z_m^c(x) \equiv c_1 J_m(x) + c_2 H_m^{(1)}(x)$, the linear combination of the Bessel function and the Hankel function
The prime denotes differentiation with respect to the argument $x \equiv k_j r$
β is the propagation constant and k_0 is the free-space wave number
$k_0 = \omega/c$. a_m^j and b_m^j are complex coefficients determined from the boundary conditions

F_m is the exponential factor:

$$F_m = \exp(im\theta + i\beta z - i\omega t) \qquad\qquad (5.23)$$

and k_j is the transverse wave number in the respective medium:

$$k_j^2 = n_j^2 k_0^2 - \beta^2 \qquad\qquad (5.24)$$

For single-mode operation, here we consider the fundamental modes and thus set $m = 1$ in Equation 5.22. Additionally, the diameter (D) of the nanowire is assumed to be 400 nm to ensure that only the fundamental mode exists.

By applying the boundary conditions (i.e., the tangential components of the electromagnetic field E and H must be continuous at the inner and outer cylinder surfaces ($r = \rho$ and $r = \rho + d_c$)), a system of eight linear homogeneous equations is obtained. The system admits a nontrivial solution only in case its determinant is zero. Thus, the propagation constant (β) is determined by the condition that the determinant of the system of linear equations shall vanish:

$$det[M(\beta)] = 0 \qquad\qquad (5.25)$$

where M is the resulting matrix of the system of equations.

5.3.2 Dispersion Shifts in Silica Nanowires with High-Index Coat

The group velocity (v_g) can be expressed as [8]

$$v_g = \frac{d\omega}{d\beta} = -\frac{2\pi c}{\lambda^2}\frac{d\lambda}{d\beta},$$

(5.26)

where
   λ is the wavelength
   c is the light speed in vacuum

The waveguide dispersion (D_w) can be obtained from Equation 5.21. By substituting the given ρ, n_s, n_c, n_a, and d_c into Equation 5.25, the propagation constants (β) can be obtained numerically. For reference, when the thickness of the coat (d_c) is assumed to be 5 nm, wavelength-dependent dispersion of 400 nm diameter silica nanowires with several typical coatings obtained from Equations 5.21 and 5.26 are plotted in Figure 5.13. Dispersion of a freestanding 410 nm diameter silica nanowire (i.e., a 5 nm thickness silica coat) is also provided for comparison. We can see that starting from the short wavelength side, D_w increases quickly until it reaches a maximum around 500 nm wavelength, and then it decreases to a minimum around 900 nm wavelength. After that, D_w goes up and approaches zero at the IR edge. Within the broad spectral range (e.g., from 500 to 1500 nm wavelength), considerable dispersion shift is produced by adding a thin high-index coat compared to that of the freestanding nanowire. Moreover, the larger the coating index, the larger the D_w. For example, at 500 nm wavelength, D_w of the fundamental modes of a freestanding silica nanowire is about −480 ps·nm⁻¹·km⁻¹. When a 5 nm thickness coating with index of 2.7 is added, a positive 485 ps·nm⁻¹·km⁻¹ shift is generated, resulting in the positive D_w beyond the zero-dispersion point. The maximum dispersion shift is about

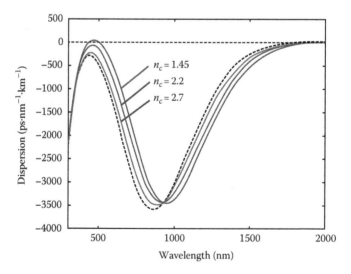

FIGURE 5.13
Modified dispersion of 400 nm diameter silica nanowire with different coatings (d_c = 5 nm). Dashed line: dispersion of 400 nm diameter silica nanowire without coat. (Adapted from Lou, J.Y. et al., *Opt. Express*, 14, 6993, 2006.)

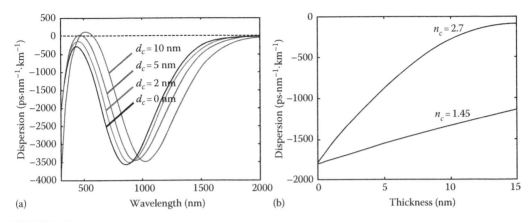

FIGURE 5.14
(a) Modified dispersion of a 400 nm diameter silica nanowire with different coating thicknesses. (b) Coatthickness-dependent dispersion of a 400 nm diameter silica wire at the wavelength of 633 nm. Refractive indexes of the coats are assumed to be 2.7 and 1.45, respectively. (Adapted from Lou, J.Y. et al., *Opt. Express*, 14, 6993, 2006.)

700 ps·nm^{-1}·km^{-1}(observed within 560–800 nm wavelength range) and the minimum is zero at about 920 nm wavelength.

Now we turn to discuss the dependence of dispersion shift with respect to the coat thickness (d_c). For calculation, silica nanowires are assumed to be 400 nm in diameters coated with a high-index (TiO$_2$, n_c=2.7) film with d_c of 2, 5, and 10 nm, respectively. Calculated wavelength-dependent dispersion is given in Figure 5.14a. One can see that a very thin coating may produce considerable dispersion shift. For example, a 2 nm thickness coat on a 400 nm diameter wire (i.e., 1% increase in wire diameter) leads to as large as 415 ps·nm^{-1}·km^{-1} shift in dispersion at 633 nm wavelength. Besides, Figure 5.14b shows the d_c-dependent D_w of a 400 nm diameter nanowire coated with TiO$_2$ at the wavelength of 633 nm. D_w increases continuously and smoothly with the increasing d_c, indicating great potentials for fine modification in D_w of nanowires. For comparison, the dispersion shift caused by solely increasing the diameter of the same nanowire (i.e., 400 nm diameter silica nanowire coated with silica layers) is also provided. It shows that for dispersion modification, coating a high-index layer is much more efficient than increasing the nanowire diameter. For example, adding a 5 nm thickness high-index coat leads to a dispersion shift of about 950 ps·nm^{-1}·km^{-1}, whereas increasing the diameter to the same thickness brings a dispersion shift of only 250 ps·nm^{-1}·km^{-1}.

5.4 Roughness-Induced Radiation Losses

So far, all the nanofiber/nanowire discussed previously is assumed to have a smooth surface, while practically, the sidewalls of nanofibers/nanowires are not ideally perfect. Take the typical dielectric nanofiber fabricated by taper drawing of molten glasses [17–19] for example; during the solidification of the glass, the thermally excited surface capillary waves freeze onto the surface at the glass transition temperature [20], which will inevitably contribute to surface roughness and lead to a certain degree of scattering (or radiation) loss. Practically, for the sub-wavelength-diameter nanowire, surface roughness–induced scattering loss is an important loss mechanism. To investigate the surface roughness–induced loss, we introduce an induced current model, in which an induced current is assumed to represent the nonuniformities [4], and the perturbation is treated as induced current sources on the surface of an unperturbed nanofiber. Since the diameter fluctuation of a taper-drawn nanofiber can go down to 10^{-7} [21], treating optical loss of a nanofiber solely based on sidewall roughness should represent one of the typical situations as has been applied in photonic crystal fibers [22,23].

5.4.1 Theoretical Analysis

The mathematical model of nanofibers with rough surface is shown in Figure 5.15. To model the roughness, we assume the capillary wave as a sine deform along its length L. This is because any function can be expressed in sine waves through Fourier transformation. The refractive-index profile of a real (with surface perturbation as shown in Figure 5.15a) and an ideal (without perturbation as shown in Figure 5.15b) nanofiber is assumed to be $n(r, \phi, z)$ and $\bar{n}(r, \phi, z)$, respectively; the sinusoidal perturbation can be represented as current sources induced by an electric field, with an induced current density given by [4]

$$\mathbf{J} = i\sqrt{\frac{\varepsilon_0}{\mu_0}}\,k\left(\bar{n}^2 - n^2\right)\mathbf{E} \tag{5.27}$$

where
ε_0 and μ_0 are permittivity and magnetic permittivity in vacuum
k represents the wavevector
$\mathbf{E}(r, \phi, z)$ is the total electric field of the perturbed nanofiber

Previous works have pointed out that the sidewall roughness of taper-drawn nanofiber is usually lower than 0.5 nm [18,19,25]. Meanwhile, when the nanofiber is operated at visible and near-infrared spectral, its diameter is usually larger than 100 nm. Therefore, the perturbation is very small, and it is reasonable to assume that $\mathbf{E}(r, \phi, z) \cong \bar{\mathbf{E}}(r, \phi, z)$ in Equation 5.27, where $\bar{\mathbf{E}}(r, \phi, z)$ is the electric field of an unperturbed nanofiber [4]. Since single-mode operation is usually favorable, we only consider the fundamental mode, in which the $\bar{\mathbf{E}}(r, \phi, z)$ can be obtained from Equations 5.9 through 5.15.

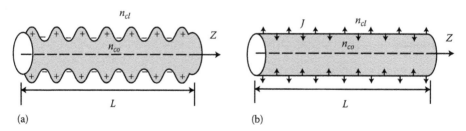

(a)   (b)

FIGURE 5.15
Mathematical model of (a) a real nanofiber and (b) an ideal nanofiber with induced currents on the surface. The length of the nanofiber is L, and the refractive indexes of the core and cladding of the nanofiber are n_{co} and n_{cl}, respectively. (Adapted from Zhai, G.Y. and Tong, L.M., *Opt. Express*, 15, 13805, 2007.)

For a slightly perturbed nanofiber, the induced current can be well approximated by [4]

$$\mathbf{J} = i\sqrt{\frac{\varepsilon_0}{\mu_0}}k\left(\bar{n}^2 - n^2\right)\bar{\mathbf{E}} = i\sqrt{\frac{\varepsilon_0}{\mu_0}}k\left(\bar{n}^2 - n^2\right)\bar{a}_1\mathbf{e}_1\exp\left(i\beta z\right) \tag{5.28}$$

where
$\bar{\mathbf{E}} = \bar{a}_1\mathbf{e}_1\exp(i\beta z)$ is the electric field of the HE_{11} mode of an unperturbed MNF
$\bar{a}_1$ and β are the amplitude and propagation constant, respectively

With a sine-deformed surface, the radius $\rho(z)$ of a real nanofiber (see Figure 5.15a) can be expressed as

$$\rho(z) = \rho_0 + \xi\sin\omega z \tag{5.29}$$

where
ρ_0 is the radius of an unperturbed nanofiber
ξ is the amplitude of the surface roughness
ω the is spatial frequency of the perturbation

Since $\xi \ll \rho_0$, we assume that the currents are localized on the surface of the ideal fiber, as shown in Figure 5.15b. Therefore, the item $\bar{n}^2 - n_2$ in Equation 5.28 can be approximated as [4]

$$\bar{n}^2 - n^2 \cong -\left(n_{co}^2 - n_{cl}^2\right)\xi\sin\left(\omega z\right)\bar{\delta}\left(r - \rho_0\right) \tag{5.30}$$

where
$\bar{\delta}\left(r - \rho_0\right)$ is the Dirac delta function
n_{co} and n_{cl} are refractive indexes of the core and cladding of the nanofiber, respectively

By substituting Equation 5.30 into 5.28, the current on the surface of the nanofiber in Figure 5.15b is obtained as

$$\mathbf{J} = -i\sqrt{\frac{\varepsilon_0}{\mu_0}}k\left(n_{co}^2 - n_{cl}^2\right)\bar{\delta}\left(r - \rho_0\right)\xi\left(\sin\omega z\right)\bar{a}_1\mathbf{e}_1\exp\left(i\beta z\right) \tag{5.31}$$

Then, we can express that the amplitudes of the radiation modes excited by the surface current are as follows [4]:

For forward-propagating radiation mode (i.e., $z \geq L$),

$$a_j^r(Q) = -\frac{1}{4N_j^r(Q)}\int_0^L\int_{A_\infty}\mathbf{e}_j^*(Q)\cdot\mathbf{J}\exp\left(-i\beta(Q)z\right)dAdz$$

$$= \frac{ik\bar{a}}{4N_j^r(Q)}\left(\frac{\varepsilon_0}{\mu_0}\right)^{1/2}\left(n_{co}^2 - n_{cl}^2\right) \tag{5.32a}$$

$$\times\int_0^L\int_{A_\infty}\xi\sin\left(\omega z\right)\bar{\delta}\left(r - \rho_0\right)\mathbf{e}_1\cdot\mathbf{e}_j^{*r}(Q)\exp\left(i\left(\beta_1 - \beta(Q)\right)z\right)dAdz$$

and for backward-propagating radiation mode (i.e., $z \leq 0$),

$$a^r_{-j}(Q) = -\frac{1}{4N^r_j(Q)} \int_0^L \int_{A_\infty} \mathbf{e}^{*r}_{-j}(Q) \cdot \mathbf{J} \exp(i\beta(Q)z) \, dA dz$$

$$= \frac{ik\bar{a}}{4N^r_j(Q)} \left(\frac{\varepsilon_0}{\mu_0}\right)^{1/2} \left(n^2_{co} - n^2_{cl}\right)$$

$$\times \int_0^L \int_{A_\infty} \xi \sin(\omega z) \bar{\delta}(r - \rho_0) \mathbf{e}_1 \cdot \mathbf{e}^{*r}_{-j}(Q) \exp\left(i\left(\beta_1 + \beta(Q)\right)z\right) dA dz \qquad (5.32b)$$

where

    $\mathbf{e}^r_j(Q)$ and $\mathbf{e}^r_{-j}(Q)$ are electric fields of forward- and backward-propagating radiation modes

    A_∞ is the infinite cross section

    * denotes complex conjugate

    $\beta(Q)$ is the propagation constant of the radiation mode

$N^r_j(Q) = \dfrac{1}{2}\left|\displaystyle\int_{A_\infty} \mathbf{e}^r_j(Q) \times \mathbf{h}^{*r}_j(Q) \cdot \hat{\mathbf{z}} dA\right|$ is the normalization factor of the power of the radiation mode, in which $\hat{\mathbf{z}}$ is the unit vector parallel to the waveguide axis

It should be noticed that in Equation 5.32, when the sinusoidal function (i.e., $sin(\omega z)$) is negative, we use the inside-fiber (core) electric fields, and when $sin(\omega z)$ is positive, we use the outside-fiber (cladding) electric fields.

In right-hand parts of Equation 5.32, all modes with $j \neq 1$ vanish after the spatial integration. Thus, the radiation field of the perturbed nanofiber is obtained as [4]

$$\mathbf{E}_{rad} = a^{r(ITE)}_1(Q)\mathbf{e}^{r(ITE)}_1(Q)\exp(-i\beta(Q)z) + a^{r(ITE)}_{-1}(Q)\mathbf{e}^{r(ITE)}_{-1}(Q)\exp(i\beta(Q)z)$$

$$+ a^{r(ITM)}_1(Q)\mathbf{e}^{r(ITM)}_1(Q)\exp(-i\beta(Q)z) + a^{r(ITM)}_{-1}(Q)\mathbf{e}^{r(ITM)}_{-1}(Q)\exp(i\beta(Q)z) \qquad (5.33)$$

where

    superscripts *ITE* and *ITM* denote the *ITE* (free-space TE) and *ITM* (free-space TM) radiation modes, respectively

    the positive and negative subscripts denote the forward- and backward-propagating modes, respectively

Therefore, the total power radiated by the surface current is [4]

$$P_{rad} = \frac{1}{2}\mathrm{Re}\left\{\int_{A_\infty} \mathbf{E}_{rad} \times \mathbf{H}^*_{rad} \cdot \hat{\mathbf{z}} dA\right\}$$

$$= \int_0^{kpn_{cl}} \left(\left|a^{r(ITE)}_1(Q)\right|^2 + \left|a^{r(ITE)}_{-1}(Q)\right|^2\right) N^{r(ITE)}_1(Q) dQ$$

$$+ \int_0^{kpn_{cl}} \left(\left|a^{r(ITM)}_1(Q)\right|^2 + \left|a^{r(ITM)}_{-1}(Q)\right|^2\right) N^{r(ITM)}_1(Q) dQ \qquad (5.34)$$

The loss coefficient is then obtained as

$$\alpha = 10 \lg \frac{\left(\bar{P} - P_{rad}/\bar{P}\right)}{L},\qquad(5.35)$$

where

$\bar{P} = |\bar{\mathbf{E}}|^2 = |\bar{a}_1|^2 N$ is the incident power of the nanofiber, and $N = \dfrac{1}{2}\left|\displaystyle\int_{A\infty} \mathbf{e}_1 \times \mathbf{h}_1^* \cdot \hat{\mathbf{z}} dA\right|$ is the normalization factor of the power of the HE_{11} mode, in which $\mathbf{h}_1$ is the magnetic field of the HE_{11} mode, $\hat{\mathbf{z}}$ is the unit vector parallel to the waveguide axis.

5.4.2 Dependence of Roughness-Induced Loss on the Perturbation Period

Using the model set up earlier, we now begin to investigate the roughness-induced loss of nanofibers numerically. We first consider the dependence of the loss coefficient (α) on the perturbation period (Γ) ($\Gamma = 2\pi/\omega$). The amplitude of the surface roughness (ξ) is assumed to be 0.2 nm. Three typical nanofibers including silica ($\rho_0 = 350$ nm, $n_{c0} = 1.44$), phosphate ($\rho_0 = 350$ nm, $n_{c0} = 1.54$), and tellurite ($\rho_0 = 350$ nm, $n_{c0} = 2.05$) nanofibers are selected. The wavelength of light used here is 1550 nm.

With the aforementioned assumptions, calculated Γ-dependent α is shown in Figure 5.16. For reference, the loss behavior of a silicon nanofiber ($\rho_0 = 200$ nm, $n_{c0} = 3.48$) is also provided. The result shows that with the increasing Γ, α shows an oscillation dependence with a series of minima (α_{min}), indicating that the roughness-induced loss can be very weak when Γ goes to some specific values (denoted as Γ_{min}). The oscillation behavior can be explained by destructive interference between radiation waves originated at the opposite side of the sine-deformed surface, which is similar to the behaviors in weakly guiding waveguides [26,27]. For example, for a 350 nm radius phosphate nanofiber plotted as light grey lines, the maximum α is about 5.35 dB/mm with $\Gamma = 16$ μm, and the corresponding α_{min} is about

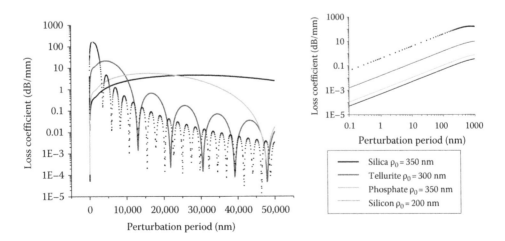

FIGURE 5.16
Loss coefficients (α) of air-clad nanofibers as a function of the perturbation periods Γ. The four nanofibers are assumed to have the same roughness amplitude ξ of 0.2 nm and be operated at 1550 nm wavelength. Inset, Γ-dependent with relatively small Γ. (Adapted from Zhai, G.Y. and Tong, L.M., *Opt. Express*, 15, 13805, 2007.)

0.00043 dB/mm (over four orders of magnitude lower) with $\Gamma = 47.8$ μm. Moreover, the higher the refractive index of the nanofiber, the more the minima with a given length. This can also be attributed to the destructive interference, in which the nanofiber with higher effective index (or equivalently, larger propagation constant) provides higher spatial frequency (or equivalently, smaller spatial period) for offering more minima with a given length. We can also see from Figure 5.16 that the maximum and the minimum α decrease with the increasing of Γ. In addition, for the same diameter nanofiber, the first Γ_{min} required for the first α_{min} of a higher-index nanofiber (e.g., phosphate, light grey line) is smaller than that of a lower-index nanofiber (e.g., silica, black line), which can be explained as larger propagation constant in a higher-index nanofiber.

For relatively small Γ, due to stronger losses in higher core-cladding index contrast, with the increasing of the refractive index of nanofiber, radiation loss increases monotonously, that is, $\alpha_{silicon} > \alpha_{tellurite} > \alpha_{phosphate} > \alpha_{silica}$ (see inset of Figure 5.16).

5.4.3 Dependence of Roughness-Induced Loss on the Amplitude of the Surface Roughness

Now we turn to investigate the influence of the amplitude of the surface roughness (ξ) on Γ_{min}. Γ-dependent α of 600 nm diameter tellurite nanofibers with ξ of 0.1, 0.2, and 0.4 nm are plotted in Figure 5.17, respectively. Rationally, for a given Γ, the larger the ξ, the larger the α. We can also see that Γ_{min} is almost independent on ξ, due to the fact that the propagation constant is almost independent on ξ when ξ is small.

To further study the relations between ξ and α, Figure 5.18 gives calculated α of silica, phosphate, tellurite, and silicon nanofibers as a function of ξ, with a given perturbation period $\Gamma = 10$ nm. The result shows that for a given nanofiber, α increases monotonously with ξ, while for a given ξ, the higher refractive index of the nanofiber results in the larger α. For example, the α of a 700 nm diameter phosphate nanofiber with $\xi = 0.2$ nm is 0.011 dB/mm, which is over two times higher than that of a 700 nm diameter silica nanofiber (0.0051 dB/mm). This is reasonable due to the fact that larger index contrast leads to higher radiation loss.

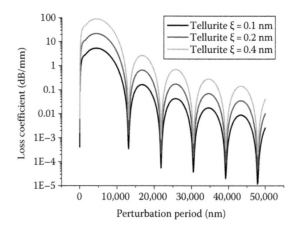

FIGURE 5.17

Loss coefficient (α) of 600 nm diameter tellurite nanofibers as a function of the perturbation period Γ with a 1550 nm wavelength guided light. (Adapted from Zhai, G.Y. and Tong, L.M., *Opt. Express*, 15, 13805, 2007.)

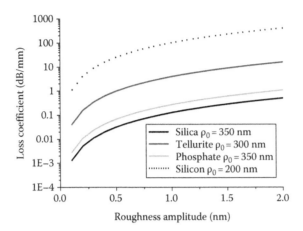

FIGURE 5.18
Loss coefficient (α) of nanofibers as a function of the roughness amplitude ξ, with the light of 1550 nm wavelength and $\Gamma = 10$ nm. (Adapted from Zhai, G.Y. and Tong, L.M., *Opt. Express*, 15, 13805, 2007.)

5.4.4 Dependence of Roughness-Induced Loss on the Diameters of Nanofibers

To investigate the loss coefficients α of nanofibers with different diameters, we assumed that all the nanofibers here have the same roughness (ξ) of 0.2 nm and the wavelength of light used is 1550 nm. Figure 5.19 shows Γ-dependent α of silica nanofibers with diameters of 700, 800, and 900 nm, respectively. It shows that with the increasing diameter, the radiation loss increases. For example, at $\Gamma = 10$ μm and $\xi = 0.2$ nm, α of a 700 nm diameter nanofiber is 2.41 dB/mm, which is over four times higher than that of a 900 nm diameter one (0.58 dB/mm). The explanation is as follows: in thinner nanofibers, the field intensity is much higher than that of the thicker one as discussed in Section 5.2.3. In this case, the interaction between guided light and the surface of nanofibers is enhanced. Besides these, Figure 5.19 also shows that the thinner nanofiber provides larger first Γ_{min} compared to the thicker one, due to the smaller propagation constant in thinner nanofibers.

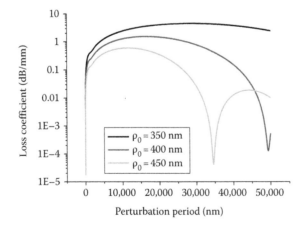

FIGURE 5.19
Γ-dependent loss coefficient (α) of silica nanofibers of different diameters, with the light of 1550 nm wavelength and $\xi = 0.2$ nm. (Adapted from Zhai, G.Y. and Tong, L.M., *Opt. Express*, 15, 13805, 2007.)

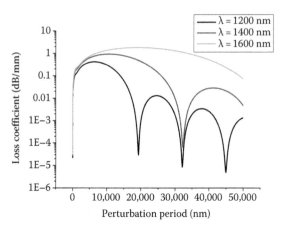

FIGURE 5.20
Γ-dependent loss coefficient (α) of silica nanofibers operated at different wavelength. Nanofibers are assumed to have a diameter of 800 nm diameter and a ξ of 0.2 nm. (Adapted from Zhai, G.Y. and Tong, L.M., *Opt. Express*, 15, 13805, 2007.)

5.4.5 Dependence of Roughness-Induced Loss on the Wavelength of Light

It is also interesting to investigate the loss coefficients α of nanofibers with different wavelengths. For calculation, the surface roughness (ξ) is assumed to be 0.2 nm, and the diameter of nanofiber is assumed to be 800 nm. Figure 5.20 gives Γ-dependent α of the nanofiber assumed earlier operated at 1200, 1400, and 1600 nm wavelengths, respectively. It shows that the larger the wavelength of light, the larger the α, as a consequence of higher fractional power around the surface with larger wavelength of guided light. It is also reasonable to see that the guided light with longer wavelength provides the larger first Γ_{min} due to its smaller propagation constant.

Now we turn to discuss the surface roughness–induced loss within a broad spectral range, since supercontinuum generation in nanofibers is of special interest [11,28,29]. All the nanofibers here are assumed to have the same ξ (0.2 nm) and Γ (100 nm). Figure 5.21 shows wavelength-dependent α of the 300 and 500 nm diameter silica nanofibers, respectively. We can see that α increases with the increasing wavelength, and the thinner nanofiber suffers much higher loss than the thicker one, which agrees with those shown in Figures 5.19 and 5.20. In addition, when the ratio of the wavelength and the fiber diameter (defined as $\lambda/(2\rho_0)$) exceeds a certain value (here ~2, e.g., 600 nm wavelength for the 300 nm diameter nanofiber), the increasing of α becomes very slow.

5.5 Bending Losses

The aforementioned mathematical models are solved by the exact solutions of Maxwell equations analytically or numerically. Here, we introduce an FDTD method, one of the best numerical methods to investigate structures like nanowires/nanofibers with relatively small calculation dimensions, to investigate the bending loss of nanowires with circular 90° bends. In practical applications, bent nanowires serve as important building blocks for components and devices ranging from highly compact photonic integrated circuits (PICs),

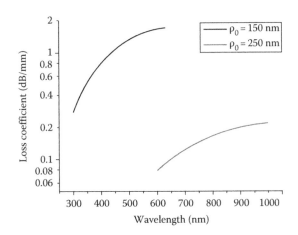

FIGURE 5.21
Wavelength-dependent loss coefficient (α) of silica nanofibers with diameters of 300 and 500 nm, respectively. Nanofibers are assumed to have same roughness parameters with amplitude ξ of 0.2 nm and period Γ of 100 nm. (Adapted from Zhai, G.Y. and Tong, L.M., *Opt. Express*, 15, 13805, 2007.)

resonators, and interferometers to lasers [30–33]. The high-index contrast between the core and cladding allows the nanowire to guide light through sharp bends (micrometer level) with low optical loss [18,30]. In this section, dependences of bending losses on bending radii, diameters, refractive indexes of the nanowires, wavelengths, and polarizations of guided light are discussed using FDTD simulations.

5.5.1 Mathematical Model

The mathematical model for a bent nanowire is illustrated in Figure 5.22, in which the nanowire is assumed to have a circular cross section with a diameter D, a bending radius R with circular 90° bends, and a step-index profile with an infinite air clad. For generality, typical materials with relatively low to high refractive index (n), including silica (SiO_2, $n = 1.46$), polystyrene (PS, $n = 1.59$), and zinc oxide (ZnO, $n = 2.0$) are selected as materials for nanowire waveguide. In order to observe the cross-plane energy distribution, a vertical transverse cross-plane P1 is set. Since single-mode operation is favorable in most applications, here we only investigate the fundamental mode (HE_{11}) in the simulation. The excite source is set at the start of a 2 μm length straight part of the nanowire, which

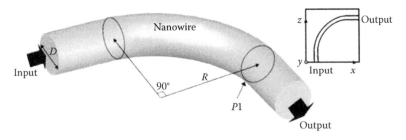

FIGURE 5.22
Mathematical model for 3D FDTD simulation of a circular 90° bent nanowire. Inset, topography profile of the bent nanowire. (Adapted from Yu, H.K. et al., *Appl. Opt.*, 48, 4365, 2009.)

evolves to steady state before entering the bent part of the nanowire. Additionally, since the HE_{11} mode propagates with two orthogonal polarizations (x- and y-polarizations) [34], both of the two orthogonal polarizations are excited in the simulation. The bending losses (α) are obtained as

$$\alpha = -10\log_{10}\left(\frac{I_{out}}{I_{in}}\right) \tag{5.36}$$

where
  I_{out} is the steady output power (at the output end of the nanowire shown in the inset of Figure 5.22)
  I_{in} is steady input power (before entering the bent part).

It should be noticed that two types of bending losses, that is, the pure bending loss and transition loss, are included in the calculation.

The computational domain is discretized into a uniform orthogonal 3D mesh with cell size of one-twelfth of the wavelength in medium and terminated by the perfectly matched layer (PML) boundary condition [35]. The FDTD simulations are performed by *Meep* [36].

5.5.2 Electric Field Energy Distribution

To give a general view of optical losses in bent nanowires, electric field intensity distributions of typical silica nanowires with low and high losses are shown in Figure 5.23, respectively. The wavelength of the x-polarized light used here is 633 nm and the diameter of the nanowire is assumed to be 450 nm. Figure 5.23a and b gives the electric field intensity distribution in the x–z plane and the output modal profiles of the 5 μm bent wire, respectively. We can see that no power leakage (0.14 dB/90° calculated bending loss) is virtually observed, except for a slight shift (positive z direction) of output modal profiles.

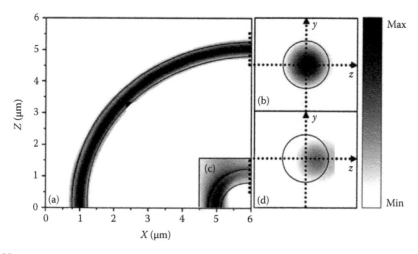

FIGURE 5.23
Electric field intensity distributions in x–z plane ($y = 0$) of (a) 5 μm and (c) 1 μm bent nanowires. The wavelength of the x-polarized light is 633 nm and the diameter of the nanowires is 450 nm. Output modal profiles at the P1 transverse cross section of 5 and 1 μm bent nanowires are shown in (b) and (d), respectively. (Adapted from Yu, H.K. et al., *Appl. Opt.*, 48, 4365, 2009.)

While the bending radius decreases to 1 µm (Figure 5.23c and d), serious energy leakage (4.8 dB/90° calculated bending loss) occurs around the bending region with an obvious lateral shift of output modal profiles.

5.5.3 Dependence of Bending Loss on the Bending Radii and Polarizations

For calculation, the wavelength of the guided light is assumed to be 633 nm with x- and y-polarizations, respectively. R-dependent α of a 350 nm diameter SiO_2 nanowire, a 350 nm diameter PS nanowire, and a 270 nm diameter ZnO nanowire are obtained and shown in Figure 5.24. It shows that α decreases monotonically with the increasing R. For example, for PS nanowire with x-polarization light, when $R = 1$ µm, α is as high as 3.9 dB/90°, while R increases to 1 µm, α reduces to an acceptable 0.5 dB/90°. Meanwhile, for the same R, the higher the refractive index of the nanowire, the lower the bending loss.

For the dependence of bending losses with light polarizations, as shown in Figure 5.24, the light of x-polarization suffers higher losses than that of y-polarization (the nanowire is curved in the $x–z$ plane). Moreover, the loss difference between the two orthogonal polarizations decreases with the increasing R and eventually becomes imperceptible when the R exceeds 3 µm.

5.5.4 Dependence of Bending Loss on Wavelength of the Guided Light

The light wavelength is another important factor that directly determines bending loss (α). For the same bending radius ($R = 2$ µm) and light polarization (x-polarization), Figure 5.25 gives wavelength-dependent bending losses of a 450 nm diameter silica nanowire and a 250 nm diameter ZnO nanowire. It shows that with decreasing wavelength, the bending loss (α) decreases monotonously due to the tighter confinement of optical field (defined as the ratio of fractional power inside the nanowire, see in Section 5.2.4). In addition, compared to the SiO_2 nanowire, α of the ZnO nanowire decreases at a faster rate with respect to the decreasing wavelength, as a consequence of stronger

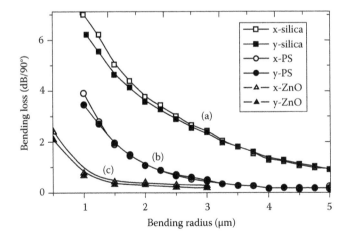

FIGURE 5.24
Bending radius (R)-dependent bending losses (α) of (a) a 350 nm diameter silica nanowire, (b) a 350 nm diameter PS nanowire, and (c) a 270 nm diameter ZnO nanowire at 633 nm wavelength. (Adapted from Yu, H.K. et al., *Appl. Opt.*, 48, 4365, 2009.)

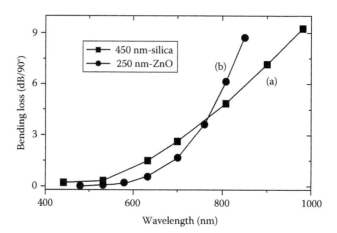

FIGURE 5.25
Wavelength-dependent bending losses (α) of (a) 450 nm diameter silica nanowire and (b) 250 nm diameter ZnO nanowire with x-polarized sources. The bending radius is 2 μm. (Adapted from Yu, H.K. et al., *Appl. Opt.*, 48, 4365, 2009.)

waveguide dispersion of the wavelength-dependent effective index induced by the higher-index contrast of the ZnO nanowire.

5.5.5 Dependence of Bending Loss on Diameter of the Nanowire

Similar to the wavelength factor, the nanowire diameter (D) is an equivalent parameter that determines the bending loss (α). Using an x-polarized source with 633 nm wavelength, D-dependent α of SiO_2 and ZnO nanowires with 2 μm radius 90° bends are shown in Figure 5.26. It is reasonable to see that with an increasing D, α decreases due to the increasing of optical confinement. Moreover, for the SiO_2 nanowire, α reduces to an acceptable level of 1 dB/90° when the D exceeds 500 nm, and for the high-index-contrast ZnO nanowire, the same α (1 dB/90°) can be achieved at a much smaller D of 240 nm.

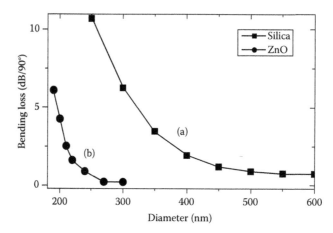

FIGURE 5.26
Diameter (D)-dependent bending losses (α) of (a) 450 nm diameter silica nanowire and (b) 250 nm diameter ZnO nanowire with a 2 μm bending radius. The wavelength of guided light is 633 nm with x-polarization. (Adapted from Yu, H.K. et al., *Appl. Opt.*, 48, 4365, 2009.)

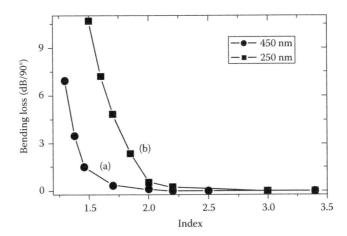

FIGURE 5.27
Refractive-index-dependent bending losses of (a) 450 nm diameter nanowire and (b) 250 nm diameter nanowire with a 2 μm bending radius. The wavelength of guided light is 633 nm with *x*-polarization. (Adapted from Yu, H.K. et al., *Appl. Opt.*, 48, 4365, 2009.)

5.5.6 Dependence of Bending Loss on Refractive Index of the Nanowire

Figure 5.27 shows the refractive-index-dependent bending loss (α) of a 250 and 450 nm diameter nanowire with 2 μm radius 90° bends. The wavelength of guided light is assumed to be 633 nm with *x*-polarization. As shown, α decreases with the increasing refractive index, which is attributed to stronger optical confinement. For example, with a refractive index of 1.38, the α of the 450 nm diameter nanowire is as high as 3.5 dB/90°; when the refractive index increases to 2.0, the bending loss decreases to a negligible value of 0.1 dB/90°. Moreover, the loss difference between 250 and 450 nm diameter nanowires decreases with the increasing refractive index. Also, results shown in Figure 5.27 may provide a valuable reference for selecting nanowire parameters for particular applications. For example, to reduce the bending loss of a 450 nm diameter nanowire below 1 dB/90°, the refractive index of the nanowire should be higher than 1.5 (e.g., PS, ZnO).

5.6 End-Face Output Patterns of Nanowires/Nanofibers

End-face output patterns (EOPs) of nanowires/nanofibers are of great importance for sub-wavelength-dimension illuminations, photon-momentum-induced effects, nanoprobe sensing, laser trapping, and laser scalpel for nanosurgery [38–41]. In this subsection, using a 3D FDTD simulation, the intensity distribution and beam widths of near- or far-field output patterns of freestanding silica and tellurite nanofibers with flat, spherical, and tapered end faces in air and/or water are investigated.

5.6.1 Mathematical Model

Figure 5.28 shows the basic model for numerical simulation, in which the nanofiber is assumed to have a cylindrical core with a diameter *D*, a length *L*, an infinite cladding

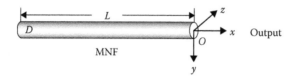

FIGURE 5.28
Mathematical model for investigation of the EOPs of nanofibers. (Adapted from Wang, S.S. et al., *Opt. Express,* 16, 8887, 2008.)

(air or water), and a step-index profile. Silica and tellurite are selected as typical core materials for nanofibers. The wavelength of light used here is 633 nm, and the indexes of air, water, silica, and tellurite are 1.0, 1.33, 1.46, and 2.02, respectively. As shown in Figure 5.28, a Cartesian coordinate with its origin located at the center of the output end face is applied to model the nanofiber. Due to the cylindrical symmetry, the source is assumed to be z-polarized without losing generality. The computational domain is discretized into a uniform orthogonal 3D mesh with cell size of 40 nm for silica nanofibers and 20 nm for tellurite nanofibers, terminated by PML boundaries [35]. The FDTD simulations are performed by *Meep* [36].

5.6.2 Nanofibers with Flat End Faces

At first, we investigate the EOPs of single-mode freestanding silica and tellurite nanofibers with flat end face. Here, nanofibers are assumed to be embedded in air and water, and the wavelength of light used is assumed to be 633 nm wavelength. To ensure the single-mode operation, the diameters of silica and tellurite nanofibers are assumed to be 400 and 250 nm, respectively. Figure 5.29a through d shows the calculated output patterns in x–y plane ($z=0$) of a 4.2 μm length silica and tellurite nanofiber with the aforementioned assumptions, respectively. We can clearly see the standing wave patterns along the lengths of nanofibers, due to the interference between the forward-propagating light and backward reflection generated at the output end face. Moreover, the nanofibers embedded in water show higher fraction of evanescent fields, smaller divergence in EOPs and weaker standing wave patterns than those of air-clad nanofibers embedded in air, due to the fact that the tighter confinement and stronger end-face reflection in air-clad nanofibers caused by the higher-index contrast.

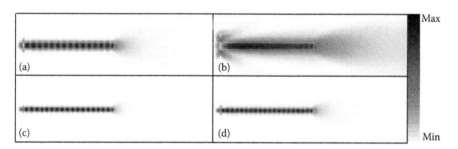

FIGURE 5.29
Output patterns in x–y plane ($z=0$) of (a) a 4.2 μm length, 400 nm diameter silica nanofiber in air; (b) a 4.2 μm length, 400 nm diameter silica nanofiber in water; (c) a 4.2 μm length, 250 nm diameter tellurite nanofiber in air; and (d) a 4.2 μm length, 250 nm diameter nanofiber in water. (Adapted from Wang, S.S. et al., *Opt. Express,* 16, 8887, 2008.)

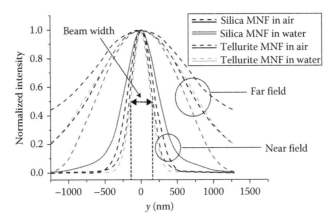

FIGURE 5.30

Normalized intensity distributions along the y-axis in x–y plane ($z=0$) with distances of 100 nm ($x=100$ nm, near field, in solid lines) and 3000 nm ($x=3000$ nm, far field, in dashed lines) departed from the end facets of the silica nanowire in air, silica nanowire in water, tellurite nanowire in air, and tellurite in water. (Adapted from Wang, S.S. et al., *Opt. Express*, 16, 8887, 2008.)

Then we investigate the optical confinement of the near- and far-field output patterns. For calculation, the wavelength of guided light used here is 633 nm. Figure 5.30 gives the normalized intensity distributions along the y-axis in x–y plane ($z=0$) with distances of 100 nm ($x=100$ nm, near field) and 3000 nm ($x=3000$ nm, far field) departed from the end facets of nanofibers. As shown, the intensity peaks locates at the central axis of the nanofiber ($y=0$) and decreases monotonously when departing from the central axis. We define a "beam width" as the full widths at half maximum (*FWHMs*) of the intensity distributions to characterize the spatial concentration of the end-face output. It shows that in the near-field region, the majority of energy is confined within a sub-wavelength scale (e.g., the near-field *FWHM* of a 250 nm diameter tellurite nanowire in water is ~290 nm); in a relatively far field, although the *FWHM* is significantly broadened (e.g., the far-field *FWHM* of a 250 nm diameter tellurite nanowire in water is ~1.6 µm), the beam divergence is not large.

To search the minimum beam width of the near-field output in nanowires, we calculate the *FWHM* ($x=100$ nm, near field) with respect to the normalized fiber diameter (D/λ), as plotted in Figure 5.31. It shows that the *FWHM* of near-field output can be tuned by the ratio of fiber diameter and light wavelength. And more importantly, there always exists a minimum *FWHM* smaller than the wavelength. For example, for a silica nanowire embedded in water, when $D/\lambda=0.822$ (i.e., $D=520$ nm with 633 nm wavelength light), the *FWHM* reduces to a minimum (480 nm in water). For a high-index tellurite nanofiber, much smaller *FWHM* (320 nm in water) can be obtained at a smaller fiber diameter (367 nm in water).

5.6.3 Nanofibers with Spherical and Tapered End Faces

Spherical or tapered fiber tips are usually used for focusing or dispersing light and have been intensively investigated in conventional fibers [43,44]. However, when the tip size goes down to wavelength or sub-wavelength level, the light beam behaves differently at the end face due to the large fraction of diffraction and evanescent fields.

Figure 5.32a and b gives the output patterns in x–y plane ($z=0$) of 400 nm diameter nanofibers with spherical tip, in which the diameters of the tip is 400 and 800 nm, respectively. The wavelength of guided light used here is assumed to be 633 nm. As shown, the output

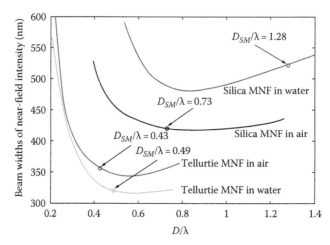

FIGURE 5.31
Beam widths (FWHMs) of near-field outputs ($x = 100$ nm) with respect to the normalized fiber diameter (D/λ) at the wavelength of 633 nm for silica nanofibers in air or water and tellurite nanofibers in air or water. Open circles denote the critical diameters for single-mode operation. (Adapted from Wang, S.S. et al., *Opt. Express*, 16, 8887, 2008.)

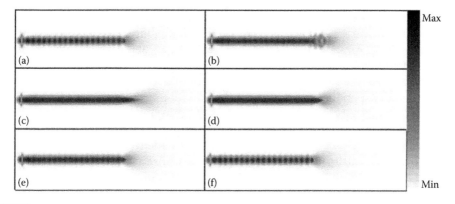

FIGURE 5.32
Output patterns of nanofibers in x–y plane ($z = 0$) at the wavelength of 633 nm. The nanofibers have spherical tips with sphere diameters of (a) 400 and (b) 800 nm and tapered tips with tapering angles of (c) 15°, (d) 30°, (e) 60°, and (f) 120°, respectively. (Adapted from Wang, S.S. et al., *Opt. Express*, 16, 8887, 2008.)

light from the tip spreads out along the x direction, indicating that the spherical tip has no evident focusing effect. Moreover, for the nanofiber with a 400 nm spherical tip as shown in Figure 5.32a, a standing wave pattern can be clearly observed. Figure 5.32c through f shows the output patterns in x–y plane ($z = 0$) of nanofibers with tapered tips, in which the shape of the taper is assumed to be an isosceles triangle with vertex angle of 15°, 30°, 60°, and 120°, respectively. It shows that output light from the tapered tip exhibits a similar behavior as that from the spherical tip, without evident focusing effect. In addition, for tips with large vertex angle (e.g., 120°), evident standing wave patterns are formed in the nanofiber.

For better understanding, Figure 5.33 provides intensity distribution of the nanofiber output along the y-axis in x–y plane ($z = 0$) with a distance of 100 nm ($x = 100$ nm, near field) and 3000 nm ($x = 3000$ nm, far field) from the apex of the tip. It shows that with

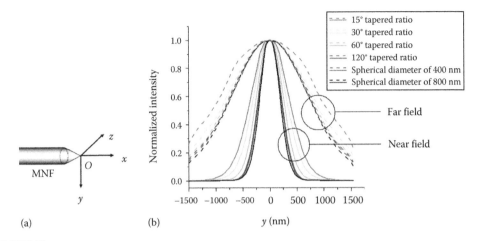

FIGURE 5.33
(a) Coordinate system for nanofibers with tapered or spherical tips. (b) Normalized intensity distributions along the *y*-axis in *x–y* plane (*z* = 0) with distances of 100 nm (*x* = 100 nm, near field, in solid lines) and 3000 nm (*x* = 3000 nm, far field, in dashed lines) departed from the apex of tapered or spherical nanofibers in air. (Adapted from Wang, S.S. et al., *Opt. Express*, 16, 8887, 2008.)

the increasing vertex angles of tapered ends, the divergence of output patterns decreases. And compared to the spherical tips, the tapered tips produce slightly larger divergence.

5.7 End-Face Reflectivities of Nanowires

End-face reflectivities (ERs) are critical to some applications of waveguiding nanowires. For example, in nanowire lasers, ERs determine the quality factor Q of the Fabry–Perot cavity for lasing oscillation, as well as the lasing threshold. Unlike the ERs of conventional fibers that can be estimated by the Fresnel formula, this is not valid for nanowires with tight confinement and high fractional evanescent waves. Based on 3D FDTD simulation, here we investigate ERs of both freestanding and substrate-supported silica, tellurite, PMMA, and semiconductor nanowires/nanofibers.

5.7.1 Mathematical Model

The mathematical model is illustrated in Figure 5.34. We assume that the nanowire has a circular cross section, and a step-index profile with a diameter D and a length L. In this time, we will consider two cases, freestanding nanowire and nanowire supported by a low-index substrate. As shown, Cartesian coordinate is applied with its origin located at the center of the left end face of the nanowire. The nanowires we investigated are operated in single mode. For the freestanding nanowire, the excite source is assumed to be *x*-polarized, and for the nanowire with a substrate, sources of two polarization states (*x*- and *y*-polarization) are used separately. The computational domain is divided into a uniform orthogonal 3D mesh with the length of cell size less than one-twentieth of the light wavelength and terminated by PML boundaries [35]. And the FDTD simulations are performed by *Meep* [36].

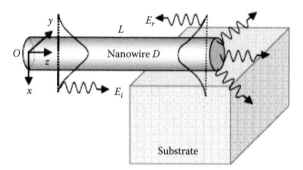

FIGURE 5.34
Schematic of a nanowire with substrate. (Adapted from Wang, S.S. et al., *Opt. Express*, 17, 10881, 2009.)

To obtain the ERs of nanowires, we use two steps to run the simulation. In the first step, we set the left and right PML boundaries next to the left and right end faces of the nanowire, respectively. In this way, only incident wave is propagating along the fiber with no reflections at both end faces. A vertical plane one wavelength away from the right side of the source plane is chosen to calculate the total incident energy flux. For the second step, the right side PML is set ten wavelengths away from the right end face of nanowire so that ER exists at the right end face. Energy flux is calculated again at the same plane as the first step. Finally, the ER is calculated based on the two energy fluxes from these two steps.

5.7.2 End-Face Reflectivities of Freestanding Nanowires

At first, we investigate the dependence of ERs on the refractive indexes of nanowires with typical diameters and working wavelengths. The indexes of nanowires span over a range of 1.38–2.65. As shown in Figure 5.35, the ERs demonstrate monotonous increasing with the indexes, due to the tighter optical confinement provided by higher-index nanowires.

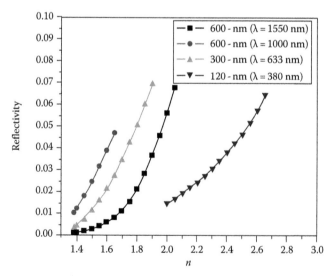

FIGURE 5.35
ERs versus indexes of nanowires with typical diameters and wavelengths. (Adapted from Wang, S.S. et al., *Opt. Express*, 17, 10881, 2009.)

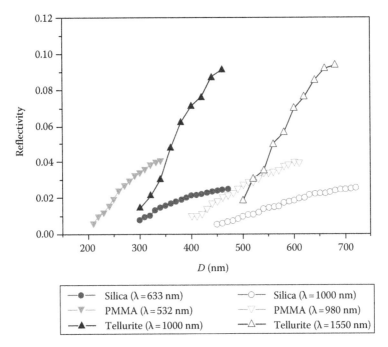

FIGURE 5.36
ERs versus diameters of nanowires with typical wavelengths. (Adapted from Wang, S.S. et al., *Opt. Express,* 17, 10881, 2009.)

For example, when the index is 1.45 (silica), the ER of 600 nm diameter nanowire at 1550 nm wavelength is as low as 0.002, indicating ER from silica nanowire of the same parameters is negligibly small. When the index increases to 2.05 (tellurite), ER increases to 0.068, which cannot be ignored. Additionally, for the same diameter (e.g., 600 nm) and refractive index, the nanowire worked at 1000 nm wavelength has relatively higher ER than the one at 1550 nm wavelength, mainly due to its relatively higher effective index.

The diameter of the nanowire is another factor that directly affects the ER. Figure 5.36 gives the calculated ERs of freestanding silica ($n = 1.45$), tellurite ($n = 2.05$), and PMMA ($n = 1.59$) nanowire with their typical working wavelengths. It is reasonable to see that the ERs increase with the diameters due to the stronger optical confinement and weaker end-face diffraction.

5.7.3 End-Face Reflectivities of Nanowires with Substrates

In experimental cases, a supporting substrate is usually indispensable for nanowire manipulation and characterization. To investigate the influence of substrates, we take the typical dielectric material MgF_2 ($n = 1.38$) for example. Figure 5.37 shows the calculated ERs of silica and tellurite nanowires supported by MgF_2 at wavelengths of 633 and 1550 nm, respectively. The supporting length between the nanowire and the substrate are assumed to be 3.5 μm for silica and 9.5 μm for tellurite. Two polarization states have been taken into account due to the breaking of cylindrical symmetry. It shows that ERs of the nanowires with substrates are considerably lower than those of freestanding nanowires, due to the fact that the presence of the substrate shifts the mode fields further toward the substrate.

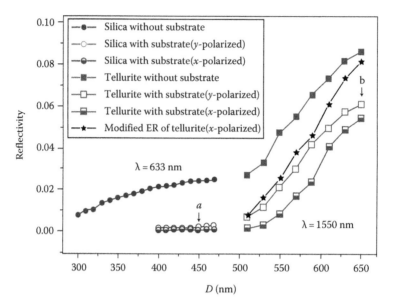

FIGURE 5.37
ERs of MgF$_2$-supported silica nanowires at 633 nm wavelength and tellurite nanowires at 1550 nm with x- and y-polarizations. The interaction lengths are 3.5 µm for silica nanowire and 9.5 µm for tellurite nanowire. (Adapted from Wang, S.S. et al., *Opt. Express*, 17, 10881, 2009.)

For better understanding, the intensity distributions of 450 nm diameter silica and 650 nm diameter tellurite nanowires on y=0 plane, as well as the output patterns on z=L plane are plotted in Figure 5.38, respectively. For a MgF$_2$-supported 450 nm diameter silica nanowire (Figure 5.38a), no guided mode exists, as has been demonstrated experimentally [46]. For comparison, in a higher-index MgF$_2$-supported 650 nm diameter tellurite nanowire, no obvious mode profile shift are observed on z=L plane, as shown in Figure 5.38b. As a result, ERs of tellurite nanowires are considerably large (open and half-open squares in Figure 5.37). Besides, ERs of x-polarization are smaller than those of the y-polarization, due to better coupling of the x-polarized fields to the substrate.

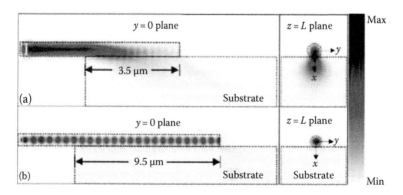

FIGURE 5.38
Intensity distributions in y=0 plane and output patterns in z=L plane of (a) 450 nm diameter silica nanowire (point a in Figure 5.37) and (b) 650 nm diameter tellurite nanowire (point b in Figure 5.37) on y=0 and z=L planes. The light source is y-polarized. (Adapted from Wang, S.S. et al., *Opt. Express*, 17, 10881, 2009.)

From Figure 5.38, we can see that ER is strongly influenced by the energy leakage induced by the substrate. To reduce or eliminate the influence of the leakage loss in the calculation, we set the energy flux calculation plane one wavelength away on the left side of the right end face to calculate the modified ERs. The optical mode is stable close to the flux plan, and thus the energy leakage effect is eliminated. In this case, the dependence of modified ER as a function of wire diameter with x-polarized guided light is plotted in Figure 5.37 (pentagrams).

5.8 Evanescent Coupling between Two Parallel Nanowires

Evanescent coupling between two nanowires is of special interest for nanophotonic passive components and active devices. Investigating the coupling efficiency and energy exchange between two nanowires is important for applications such as all fiber interferometric sensors, wavelength filters, and ring resonator. Usually, evanescent coupling between adjacent weakly guiding waveguides can be described by perturbation theory, in which the weakly coupled system is usually made of low-index-contrast waveguides. However, when they are closely contacted, evanescently coupled high-index-contrast nanowires are not weakly coupled systems, in which perturbation theory is not valid. The rigorous method to investigate the mode coupling between two parallel nanowires is to solve Maxwell equations in the different regions and to apply the boundary conditions to determine the modes of the overall system. However, it is difficult to perform the calculation analytically. Here we introduce a 3D FDTD method to investigate the coupling behaviors of two parallel nanowires. Basic properties including the coupling efficiency and the transfer length for light exchange between the two nanowires are discussed.

5.8.1 Mathematical Model

Figure 5.39 shows the mathematical model for numerical analysis, in which two cylindrical nanowires with diameters of D_1 and D_2 are placed in parallel with a separation (H) and an overlap (L). The coupling system is divided into three regions. In region I, eigenvalue modes are guided along the input nanowire; in region II, energy exchanges between the two wires; and in region III, the output is collected for calculating the coupling efficiency. For simplicity, we assume that the nanowire has an intrinsically smooth surface so that

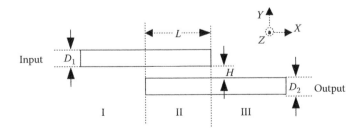

FIGURE 5.39
Mathematical model for coupling of two parallel nanowires. (Adapted from Huang, K.J. et al., *Appl. Opt.*, 46, 1429, 2007.)

TABLE 5.1

Refractive Indexes of Silica, Silicon,
and Tellurite Nanowires

Material	Wavelength (nm)	Refractive Index
Silica	633	1.46
Silica	1550	1.44
Silicon	1550	3.48
Tellurite	633	2.08

Source: Adapted from Huang, K.J. et al., *Appl. Opt.*, 46, 1429, 2007.

the roughness-induced scattering is negligible. The wavelength of guided light used here is 633 and 1550 nm. Silica, silicon, and tellurite with refractive indexes shown in Table 5.1 are selected as the materials of nanowires. The *V* number (see in Equation 5.7) is kept below 2.405 to ensure single-mode operation. Since the coupling efficiency is polarization dependent due to the asymmetry of the structure, sources of two orthogonal lights with *y*- and *z*-polarization at the input plane are used separately. The computational domain is discretized into a uniform orthogonal 3D mesh with cell size of one-tenth of the nanowire diameter, terminated by PML boundaries [35]. The 3D FDTD simulations are performed by *Meep* [36].

5.8.2 Evanescent Coupling between Two Identical Nanowires

At first, we investigate the coupling efficiency (η) of two identical nanowires with respect to the overlapping length (*L*). Typical 350, 400, and 425 nm diameter silica nanowires and 633 nm wavelength of light with *y*- and *z*-polarizations are selected. Usually, when two nanowires are brought in close proximity in air or in vacuum, they attract each other into tight contact, due to Vander Waals or electrostatic forces. Thus, the separation (*H*) between two nanowires is assumed to be zero.

Using the parameters set previously, the calculated *L*-dependent η relations are plotted in Figure 5.40. The hollow and solid circle points represent the *y*- and *z*-polarizations, respectively. Results show that the η oscillates with the increasing *L*, exhibiting a similar behavior as that of weakly coupled systems [8]. However, because of the strong coupling between the two nanowires, the minimum transfer length (L_T) for energy exchange is much shorter than that in weakly coupled waveguides. Meanwhile, unlike the minimum coupling efficiency (η_{min}) in the weakly coupled systems, which is usually close to zero, here η_{min} of the strongly coupled nanowires is considerably higher than zero (e.g., ~34% for two 350 nm diameter silica nanowires). To get better understanding of this, Figure 5.41 gives intensity distributions of evanescent coupling between two parallel 350 nm diameter silica nanowires with typical *L* of 0, 2.4, and 4.8 μm, respectively. The considerably larger η_{min} behavior is clearly illustrated in Figure 5.41a and c, in which *L* is assumed to be 0 and 4.8 μm to achieve the lowest coupling efficiency, respectively. We can see that even with the lowest η, a certain amount of energy is still able to transfer from the upper nanowire to the bottom one. Moreover, Figure 5.41a and c also shows that there is an obvious radiation loss at the right end of the upper nanowire. In addition, the maximum coupling efficiency (η_{max}) is lower than 100%, due to the fact that a part of the guided mode transfers to radiation modes in overlapping areas caused by the breaking of the symmetrical structure. However, such loss is relatively small, as shown in Figure 5.41b, where no obvious radiation loss is

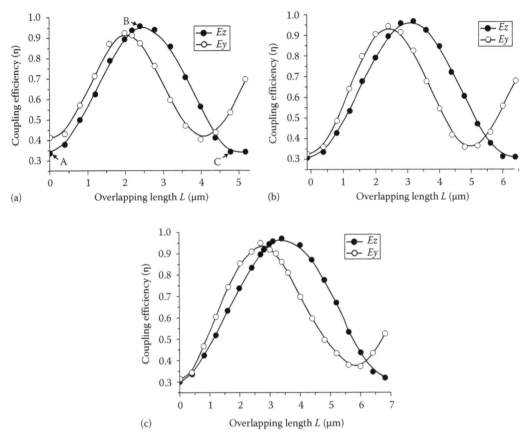

FIGURE 5.40
Overlapping-length (L)-dependent coupling efficiency (η) of two identical nanowires with a diameter of (a) 350, (b) 400, and (c) 425 nm. The wavelength of light source is 633 nm with two polarizations (z and y), respectively. (Adapted from Huang, K.J. et al., *Appl. Opt.*, 46, 1429, 2007.)

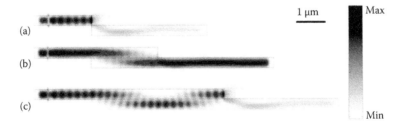

FIGURE 5.41
Intensity distributions of evanescent coupling between two parallel 350 nm diameter silica nanowires with overlapping length (L) of (a) 0, (b) 2.4 μm, and (c) 4.8 μm. The source is z-polarized with a wavelength of 633 nm. (Adapted from Huang, K.J. et al., *Appl. Opt.*, 46, 1429, 2007.)

observed. Interestingly, for nanowires with diameters around the critical value (D_{SM}) for single-mode operation, the η_{max} can approach 100% (e.g., $\eta_{max} > 97\%$, for 425 nm diameter silica nanowires with z-polarization).

Besides, back to Figure 5.40, lines with hollow and solid circle points also show that the coupling behaviors is affected by the polarizations of guided light. Compared with

TABLE 5.2

Calculated L_T, η_{max}, and η_{min} of Two Parallel Nanowires with $H=0$

Material	Wavelength (nm)	D (nm)	L_T (μm)		η_{max}		η_{min}	
			z Pol.	y Pol.	z Pol. (%)	y Pol. (%)	z Pol. (%)	y Pol. (%)
Silica	633	350	2.4	2	96	92	34	41
Silica	633	400	3.2	2.4	96	94	31	33
Silica	633	425	3.4	2.7	97	94	30	31
Silica	1550	900	6.5	5.1	96	94	38	43
Silica	1550	1000	7.7	6	97	94	36	38
Silicon	1550	350	1.4	1	96	96	29	30
Tellurite	633	250	1.4	1	96	95	30	34

Source: Adapted from Huang, K.J. et al., *Appl. Opt.*, 46, 1429, 2007.

the z-polarization, the y-polarization has a smaller L_T due to stronger overlapping of the optical fields.

For further details, calculated L_T, η_{max}, and η_{min} of evanescently coupled silica nanowires with typical diameters are listed in Table 5.2. It shows that at the given wavelength of 633 or 1550 nm, the thicker silica nanowire shows higher η_{max} and lower η_{min} than the thinner one. For example, the η_{max} and η_{min} of a 425 nm silica nanowire (at a 633 nm wavelength with z-polarization) is 97% and 30%, respectively, while the η_{max} and η_{min} of a 350 nm silica nanowire is 96% and 34%, respectively. In addition, typical results of single-mode high-index silicon and tellurite nanowires are also provided in Table 5.2. Compared to that of the silica nanowire, L_T of the high-index nanowires is much smaller. For example, at a wavelength of 633 nm, L_T of the tellurite nanowire is as large as 1.4 μm, which is about 40% of that of the silica nanowire (1.4 μm).

The small transfer length (L_T) shown in Table 5.2 suggests potentials for developing ultracompact optical devices based on the strong coupling of these optical nanowires. Assuming a 3-dB coupler assembled by two 350 nm diameter silica nanowires and operated at 633 nm wavelength, the interaction length required is about 0.8 μm (directly read from Figure 7.2a), much smaller than that of most other couplers [6,48].

Table 5.2 also shows that at the given wavelength, L_T decreases with the decreasing of the wire diameter. For example, at 633 nm wavelength, L_T of silica nanowires decreases from 3.4 to 2.4 μm when the wire diameter decreases from 425 to 350 nm, indicating the opportunity for achieving smaller L_T by using thinner wires. However, when the wire diameter decreases to a certain value, a behavior similar to antisymmetric supermode cutoff in weakly coupling systems is observed [49]. For better understanding, intensity distributions of evanescent coupling between two parallel 250 nm diameter silica nanowires with 633 nm wavelength are shown in Figure 5.42. We can see that the oscillating behavior disappears, and the explanation can be as follows: for nanowires with very small diameters, the coupled system (which can be regarded as one waveguide in region II in Figure 5.39) works as a single-mode waveguide and only supports the fundamental mode, resulting in the disappearance of oscillation.

The dependence of the coupling efficiency (η) on the lateral separation (H) is also investigated. Figure 5.43a gives the calculated η for two 350 nm diameter silica nanowires at 633 nm wavelength with z-polarization. The overlapping length (L) is set to be 3.6 μm and a smoothing B-spline fit is applied to the calculated results (marked as the square points). It shows that η increases with the increasing H until it reaches the maximum value

FIGURE 5.42
Intensity distributions of evanescent coupling between two parallel 250 nm diameter silica nanowires with separation of (a) $H=0$ and (b) $H=70$ nm. The light source is z-polarized with wavelength of 633 nm, the overlapping length (L) is 6 μm. (Adapted from Huang, K.J. et al., *Appl. Opt.*, 46, 1429, 2007.)

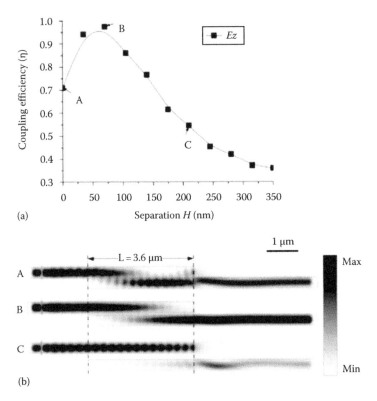

FIGURE 5.43
Evanescent coupling behavior of two 350 nm diameter silica nanowires with a z-polarized 633 nm wavelength source. (a) H-dependent η and (b) power maps with lateral separation (H) of 0 (point A), 70 (point B), and 210 nm (point C). The overlapping length (L) is 3.6 μm. (Adapted from Huang, K.J. et al., *Appl. Opt.*, 46, 1429, 2007.)

($\eta_{max}=97\%$, $H=70$ nm) and decreases thereafter. For better understanding, intensity distributions with respect to $H=0$ nm (point A), $H=70$ nm (point B), and $H=210$ nm (point C) are illustrated in Figure 5.43b. We can see that when $H=0$, due to $L_T<L$, two nanowires are overcoupled with η ~ 70%; when H increases to 70 nm, $L_T=L=3.6$ μm, and n reaches its maximum; further increasing H leads to larger L_T (i.e., $L_T>L$), resulting in the decreasing n with respect to the increasing H.

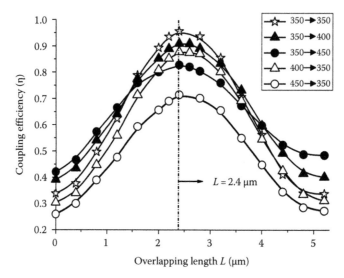

FIGURE 5.44

Overlapping-length (*L*)-dependent coupling efficiency (η) of two silica nanofibers with a *z*-polarized 633 nm wavelength source. Diameters of the nanofibers are denoted as $x \rightarrow y$, in which x and y stand for the diameters of the input and output nanofiber, respectively. (Adapted from Huang, K.J. et al., *Appl. Opt.*, 46, 1429, 2007.)

5.8.3 Evanescent Coupling between Two Silica Nanowires with Different Diameters

For evanescent coupling between nanowires with different diameters, we use silica nanowires with typical diameters of 350, 400, and 450 nm. We assume that the two nanowires are in tight contact (H = 0) and operated at 633 nm wavelength with z-polarization.

Figure 5.44 gives the calculated *L*-dependent η with the aforementioned assumptions, in which *L*-dependent η of two identical 350 nm diameter silica nanowires is also provided for reference. Rationally, η_{max} of two identical 350 nm diameter nanowires (denoted as $350 \rightarrow 350$) is larger than that of nanowires with different diameters. For example, η_{max} of $350 \rightarrow 350$, $350 \rightarrow 400$ (i.e., the input nanowire is 350 nm and output nanowire is 400 nm), and $350 \rightarrow 450$ is about 96%, 91%, and 81%, respectively. This can be explained as effective coupling between identical nanowires with matched propagation constants; with the increasing diameter difference, the mismatch of propagation constant increases, leading to the reduction of η_{max}. Moreover, Figure 5.44 also shows that the coupling efficiency of nanowires with different materials is directional dependent: coupling light from a thinner wire to a thicker one shows higher efficiency than in the opposite direction, due to the fact that the thicker nanowire has a higher effective index and exhibits a stronger capability to confine and attract the light field than the thinner one. For example, η_{max} of $350 \rightarrow 450$ is as large as 82%, while the η_{max} of $450 \rightarrow 350$ is about 71%. Additionally, despite the diameter difference in five coupling conditions (i.e., $350 \rightarrow 350$, $350 \leftrightarrow 400$, $350 \leftrightarrow 450$), the minimum transfer lengths (L_T) required for obtaining the η_{max} are almost the same (about 2.4 μm).

5.8.4 Evanescent Coupling between Silica and Tellurite Nanowires

Now we turn attention to discuss the evanescent coupling between silica and tellurite nanowires. For calculation, we assume that the two nanowires are in tight contact (H = 0) and operated at 633 nm wavelength with z-polarization. Typical diameters of

TABLE 5.3

Calculated L_T and η_{max} of Evanescent Coupling
between Silica and Tellurite Nanofibers with $H=0$
at 633 nm Wavelength

Diameter (nm)			η_{max}	
Tellurite (T)	Silica (S)	L_T (µm)	T→S (%)	S→T (%)
200	325	1.3	78	81
200	350	1.5	83	86
200	400	1.8	92	91
200	425	2.0	95	92
200	450	2.2	96	92

silica (from 325 to 450 nm) and tellurite (200 nm) nanowires are selected to ensure the single-mode operation.

The calculated results with the aforementioned assumptions are summarized in Table 5.3. For reference, intensity distribution of evanescent coupling between a 200 nm diameter tellurite nanowire and a 450 nm diameter silica nanowire with $L = L_T = 2.2$ µm is also provided in Figure 5.45. It shows that despite the large index difference between silica and tellurite nanowires, high coupling efficiency (η) can still be obtained when the diameters of nanowires are properly chosen. For example, when a 450 nm diameter silica nanowire is used to couple light out of a 200 nm diameter tellurite nanowire (Figure 5.45a), the η can go up to 96%. Moreover, for the given tellurite nanowire shown in Table 5.3, L_T and η_{max} increase monotonously with the increasing diameter of the silica nanowire. This can be explained as the weaker interaction (therefore, lower scattering loss) between nanowires due to the increasing diameters. In addition, similar to the coupling behavior discussed in Section 5.8, η_{max} is also directional dependent. For example, η_{max} of coupling light from a 200 nm diameter tellurite nanowire to a 450 nm diameter silica nanowire is 96%, while it decreases to 92% when the light propagates in the opposite direction.

Generally, matching propagation constants is favorable to achieve high coupling efficiency between two different nanowires, which means that the nanowire with higher refractive index should have a smaller diameter. However, other factors (e.g., optical confinement) may also affect the coupling loss, and sometimes a slight difference in propagation constants of the two nanowires may provide the maximum coupling efficiency.

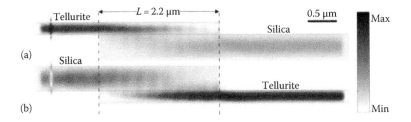

FIGURE 5.45
Intensity distributions of evanescent coupling between a 200 nm diameter tellurite nanowire and a 450 nm diameter silica nanowire with a z-polarized 633 nm wavelength source. The overlapping length $L = L_T = 2.2$ µm. (a) Coupling light from the tellurite nanowire to the silica nanowire. (b) Coupling light from the silica nanowire to the tellurite nanowire. (Adapted from Huang, K.J. et al., *Appl. Opt.*, 46, 1429, 2007.)

5.9 Longitudinal Lorentz Force on Nanofibers

Theoretically studying the longitudinal Lorentz force on nanofibers is helpful to solve the Minkowski–Abraham dilemma (a brief review on this problem can be found in Refs. [50,51]), in which the form of Maxwell stress tensor inside media has been debated over the past century. A clear understanding of these topics is also of great interest for both optics research and optomechanical applications [52–55]. Here we use analytical calculations and numerical simulations to investigate longitudinal Lorentz forces that a propagating light exerts on a sub-wavelength-diameter optical nanofiber. The method presented here provides us with a comprehensive example, which combines the rigorous computational method with the FDTD simulation.

5.9.1 Longitudinal Lorentz Forces on an Infinitely Long Nanofiber

At first, we investigate the longitudinal Lorentz forces with an analytical method, in which the nanofiber discussed is assumed to have a cylindrical silica ($n = 1.45$) core with a uniform diameter (D) of 450 nm, an air clad, a smooth sidewall, and an infinite length. To ensure the single-mode operation, the light source is assumed to be a 980 nm wavelength CW (continuous wave) laser. For simplicity, the nonlinear optical effects have not been taken into account in the simulation.

During the propagation of light, the instantaneous Lorentz force density that electromagnetic field acts on the media is obtained as [56]

$$f = \rho_b E + J_b \times \mu_0 H = -(\nabla \cdot P)E + \frac{\partial P}{\partial t} \times \mu_0 H, \qquad (5.37)$$

where
E and H are the electric and magnetic fields of the optical mode in the fiber, respectively
J_b is the bound current density
μ_0 is the permeability of vacuum
P is the electric polarization density

Here, since the longitudinal (i.e., z direction) force is of special interest, we only focus on Lorentz force density along longitudinal direction, and it can be expressed as

$$f_z = \rho_b E_z + J_b \times \mu_0 H_z \qquad (5.38)$$

As mentioned earlier, we only consider the fundamental mode, in which the electric and magnetic field components are obtained as [19]

$$E_t = \math{Re} E_t \cos(\beta z - \omega t) = (\hat{r} E_r + \hat{\phi} E_\phi) \cos(\beta z - \omega t) \qquad (5.39)$$

$$H_t = \math{Re} H_t \cos(\beta z - \omega t) = (\hat{r} H_r + \hat{\phi} H_\phi) \cos(\beta z - \omega t) \qquad (5.40)$$

$$E_z = \math{Re} E_z \sin(\beta z - \omega t) \qquad (5.41)$$

$$H_z = \cancel{2} H_z \sin(\beta z - \omega t) \tag{5.42}$$

$$P = \varepsilon_0 (\varepsilon_r - 1) E \tag{5.43}$$

where

$\hat{t}$, $\hat{z}$, $\hat{r}$, and $\hat{\phi}$ are unit vector along transverse, longitudinal, radial, and azimuthal directions, respectively

E_t (H_t), $E_z(H_z)$, E_r (H_r), and E_ϕ (H_ϕ) are the transverse, longitudinal, radial, and azimuthal electric (magnetic) fields, respectively

β is the propagation constant

ω is the angular frequency of the field

ε_0 and ε_r are the permittivity of vacuum and relative permittivity, respectively

By substituting Equations 5.39 through 5.43 into Equation 5.38, the instantaneous longitudinal Lorentz force density is obtained as follows:

$$f_z = \cancel{2} \frac{1}{2} \varepsilon_0 (\varepsilon_r - 1) \left[\left(E_z \frac{\partial}{\partial r} E_r + E_z \frac{1}{r} \frac{\partial}{\partial \phi} E_\phi + \beta E_z^2 \right) + \mu_0 \omega \left(E_r H_\phi - H_r E_\phi \right) \right] \sin 2(\beta z - \omega t) \tag{5.44}$$

where $E_r(H_r)$, $E_\phi(H_\phi)$, and $E_z(H_z)$ are the electric(magnetic) components of the fundamental mode in the cylindrical coordination [4]. Then we substitute the primittivity of the silica nanofiber and the angular frequency of the incident light into Equation 5.44 and integrate the f_z over the transverse cross section

$$F_z = \iint f_z d\sigma \tag{5.45}$$

where $d\sigma$ is the cross-sectional surface element of the nanofiber in the cylindrical coordination. Finally, the instantaneous longitudinal force distribution as a function of z for a 450 nm diameter silica nanofiber at 980 nm wavelength are obtained and plotted as shown in Figure 5.46 (solid line). We can see that the longitudinal Lorentz forces exhibit a spatial oscillation with a period of $2\pi/2\beta$, in which the calculated peak amplitude ($\sim$9.7 pN/(μm·mW)) is considerably large in micro-/nanometer scale [57].

The total time-averaged longitudinal force is obtained by integrating the F_z over an optical period T:

$$\langle F_z \rangle = \frac{1}{T} \int_0^T dt \iint f_z d\sigma. \tag{5.46}$$

Obviously, for the calculated fundamental mode, the total time-averaged longitudinal force $\langle F_z \rangle$ is zero inside the nanofiber (dashed line in Figure 5.46). Also, same results can be obtained for higher-order modes propagating inside the fiber.

As we know, while the light propagates inside a medium, the interaction between the light and the medium is accompanied by momentum exchange, and the part obtained by the medium is mechanical momentum. As to the nanofiber, by integrating F_z over one period,

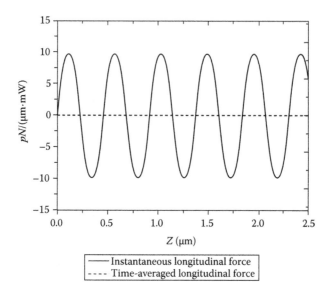

FIGURE 5.46
Instantaneous (solid line) and time-averaged (dashed line) longitudinal force distribution as a function of z for a 450 nm diameter silica nanofiber at 980 nm wavelength. (Adapted from Yu, H.K. et al., *Phys. Rev. A*, 83, 053830, 2011.)

we can see that the net impulse that medium gained is zero, indicating no momentum or energy loss for CW light of a stationary mode in a homogeneous and transparent medium.

5.9.2 Longitudinal Lorentz Forces on a Nanofiber End

For practical applications, the nanofiber cannot be infinitely long as assumed earlier, and it should have an end. In this case, when light arrives at the free end face of the nanofiber, end-face scattering occurs (i.e., reflection and diffraction, due to the abrupt interface). As schematically illustrated in Figure 5.47a, the end-face-scattering output can be roughly divided into two parts: the light reflected back into the upper half-space or diffracted

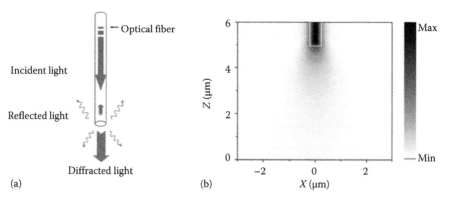

FIGURE 5.47
(a) Schematic diagram of the reflected and diffracted light at the end face of the nanofiber. (b) Calculated z-direction Poynting vector distribution in x–z($y=0$) cross section for a 450 nm diameter silica fiber with refractive index 1.45 at wavelength of 980 nm (x-polarized). (Adapted from Yu, H.K. et al., *Phys. Rev. A*, 83, 053830, 2011.)

forward into the lower half-space. Note that the guided light here cannot be simply treated as a plane wave, so that the analytical solution is hard to obtain and we have to resort to the FDTD simulation.

In the simulation, a 450 nm diameter silica nanofiber with flat end face and a 980 nm wavelength CW light source with x-polarization are selected. The length of nanofiber is assumed to be 9.5 μm, and the computational domain is 10 μm × 10 μm × 8 μm, which is divided into a uniform orthogonal 3D mesh with the length of cell size of 20 nm and terminated by PML boundaries [35]. The 3D FDTD simulations are performed by *Meep* [36].

Figure 5.47b gives the calculated distribution of z-direction Poynting vector in x-z ($y = 0$) cross section, in which the measured fractional power of light in the z direction is about 92% across a plane 5.5 μm away from the output end face due to the diffraction. Meanwhile, as schematically illustrated in Figure 5.47a, a small fractional power (~1% of the total incident power) reflected back into the upper half-space at the end face. In general, part of the reflected light propagates along the opposite direction of incident light as guided mode, while the rest transmit to the free space as radiation mode.

To obtain the longitudinal Lorentz forces on the nanofiber, we set the following steps to perform the calculation: In the first step, we obtain the electric and magnetic fields (e.g., E_x, as plotted in Figure 5.48a) through the FDTD simulation. For the second step, we calculate the bound charge and bound current densities by the method indicated in Ref. [59]. Then, by substituting the results obtained from the first and second steps into Equation 5.38, Lorentz force density along longitudinal direction (f_z) is obtained. Finally, using Equation 5.46 with the aforementioned f_z, time-averaged Lorentz force over the transverse cross section ($\langle F_z \rangle$) as a function of z is obtained and shown in Figure 5.48b. Unlike that of the infinitely long nanofiber, in which $\langle F_z \rangle = 0$, $\langle F_z \rangle$ of the real nanofiber exhibits an oscillation with a period of π/β, as a consequence of the interference between the incident wave and the reflected wave. Moreover, we can see that the amplitude of the oscillation decreases with distance away from the end face and stabilizes after several periods. This is because a fraction of the reflected light escapes from the fiber into the free space as the radiation mode, due to the finite size of the fiber end face.

Figure 5.48b also shows that $\langle F_z \rangle$ is discontinuous at the fiber end face. As is known, the Lorentz forces on the nanofiber consist of electric and magnetic components, stemming from interaction of the electric and magnetic fields with the bound charge and bound current densities. In homogeneous media, the bound charge exists only on the surface, while the end face is treated as a layer with a finite thickness of 20 nm in the FDTD simulation. For the given end-face layer, the time-averaged pull force contributed by the bound current is calculated to be 96 fN/mW, while the contribution from the bound current is much smaller (8.8 fN/mW). To further investigate this surface effect, we divided the total force by the thickness of the surface layer to get an average force density. After integrating $\langle F_z \rangle$ along the z-axis from the fiber end face ($z = 9.5$ μm) into the fiber (i.e., $\int_{z=9.5\mu m}^{z=Z(Z<9.5\mu m)} \langle F_z \rangle dz$), we obtain an oscillating overall pull force, shown in Figure 5.48c. While the oscillation stabilizes ($z < 8$ μm), an average value of 0.4 pN/mW (the dashed line in Figure 5.48c) exerted on the fiber tip is obtained. This overall pull force is obtained not solely from the force caused by the surface charge or current as discussed earlier (shown as a small gap at $z = 9.5$ μm) but also by the force accumulating over the near-field region close to the end face, where the radiated mode escapes from the fiber. With an incident light power of 1 mW, the magnitude of the pull force is comparable with the weight of a 10 μm long silica nanofiber with 450 nm in diameter.

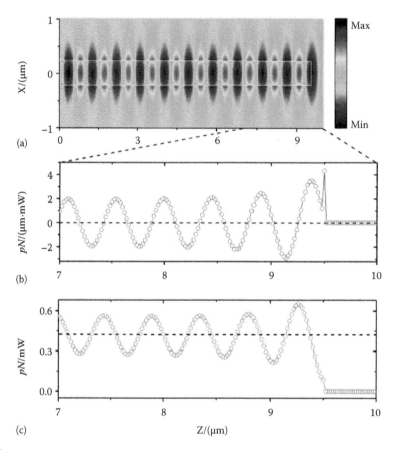

FIGURE 5.48
(a) Calculated instantaneous E_x distribution on $x-z$ ($y=0$) cross section. The white line represents the profile of the fiber. (b) Time-averaged Lorentz force over the transverse cross-sectional plane (F_z) as a function of z. (c) Integrated F_z from the fiber end face to position where $z < 9.5$ μm. (Adapted from Yu, H.K. et al., *Phys. Rev. A*, 83, 053830, 2011.)

5.10 Plasmonic Waveguiding Properties of Metal Nanowires

Surface plasmon polaritons (SPPs) are electromagnetic waves coupled with collective electron oscillations at a metal and dielectric interface. Owing to their capacity of photonics and miniaturization of electronics, plasmonic metal nanowires open opportunities for confining light on the deep-sub-wavelength scale, which inspired great potentials ranging from ultracompact optoelectronic circuits [60,61] and optical sensors [62,63] to quantum electrodynamics research [64–67]. Plasmonic waveguiding properties of a freestanding metal nanowire can be obtained from exact solutions of Maxwell equations [16,68–70]. However, for experimental manipulations and characterizations of the nanowire, a supporting substrate is usually indispensable. Unfortunately, an analytic calculation regarding this nanowire–substrate system is hard to perform, so that we have to restore to the numerical method, here an FEM for this purpose. Single-mode waveguiding properties of Au nanowires including modal profiles, propagation constants, effective mode areas, propagation lengths, and losses are obtained using a COMSOL Multiphysics FEM.

5.10.1 Mathematical Model

The mathematical model is illustrated in Figure 5.49, in which a straight Au nanowire with a cylindrical core, a diameter of D, and a smooth sidewall is placed on a dielectric substrate. Dielectric materials with relatively low to high refractive index including MgF_2 ($n = 1.38$), SiO_2 ($n = 1.45$), ITO ($n = 1.84$), and TiO_2 ($n = 2.7$) are selected as supporting substrates. The wavelength of light used in the simulation is assumed to be 660 nm, which is the typical plasmonic resonance wavelength of Au.

The propagation constant (β) and propagation length (L_m) of the nanowire are obtained as [72]

$$\beta = \mathrm{Re}(\beta) + i\,\mathrm{Im}(\beta) \tag{5.47}$$

$$L_m = \frac{1}{2\,\mathrm{Im}(\beta)} \tag{5.48}$$

The propagation loss (α) is inversely proportional to L_m as

$$\alpha = \frac{-10\log(1/e)}{L_m} \approx \frac{4.343}{L_m} \tag{5.49}$$

The effective mode area (A_m) is defined as [73]

$$A_m = \frac{W_m}{\max\{W(r)\}} \tag{5.50}$$

where
W_m is the total mode energy
$W(r)$ is the energy density (per unit length flowed along the direction of propagation)

For dispersive and lossy materials, the $W(r)$ inside can be calculated as [73]

$$W(r) = \frac{1}{2}\left(\frac{d(\varepsilon(r)\omega)}{d\omega}|E(r)|^2 + \mu_0|H(r)|^2\right) \tag{5.51}$$

where $\varepsilon(r)$ is the complex dielectric function. At the wavelength of 660 nm used in this work, the electric permittivity of the Au (ε_{Au}) is assumed as $\varepsilon_{Au} = -13.6815 - 1.0356i$ [74] and $d(\varepsilon(r)\omega)/d\omega$ is obtained from analytical models from Ref. [75].

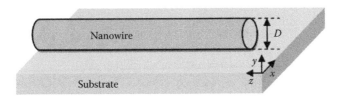

FIGURE 5.49
Mathematical model for a metal nanowire–substrate waveguiding system. (Adapted from Wang, Y.P. et al., *Opt. Express*, 20, 19006, 2012.)

The calculations are performed from the study type of "boundary mode analysis" combined with "frequency domain" provided by the software COMSOL. In calculation, we only focus on the fundamental mode, due to the fact that the fundamental mode dominates the waveguiding properties for thin nanowires [76]. The computational domain is discretized into a triangular mesh with an element size of one-tenth of the nanowire diameter (e.g., 10 nm for a 100 nm diameter nanowire), terminated by PML boundaries.

5.10.2 Modal Profiles and Propagation Constants

Using the mathematical model set previously, Figure 5.50 gives the calculated modal profiles in terms of energy density distributions for Au nanowires with different diameters (50, 100, 200 nm) and substrates (air, MgF$_2$, SiO$_2$, ITO, and TiO$_2$), respectively. We can see that unlike that of dielectric nanowires, in which the fundamental mode is HE$_{11}$ mode with two orthogonal polarizations, the fundamental mode of freestanding metal nanowires is axially symmetric TM$_0$ mode resulting from electrons oscillating parallel to the wire axis (Figure 5.50a). Moreover, as shown, with the presence of the substrate, the plasmon mode of Au nanowires becomes asymmetric with the modal profiles shifting toward the substrate (Figure 5.50b through d). Meanwhile, with the increasing of nanowire diameter, fractional energy confined around the nanowire–substrate interface increases (e.g., Figure 5.50d vs. 5.50c). The energy density enhancement is due to the polarized substrate (inset of Figure 5.50), which is similar to nanoparticle–substrate system [77,78]. In addition, for the given diameters of Au nanowires (e.g., 100 nm, Figure 5.50e through h), the higher the refractive indexes of substrates, the tighter the energy confinement around the interface (e.g., modal profiles in Figure 5.50h vs. 5.50g). It can be explained as follows: when the refractive index of the substrate increases, the polarizability in the substrate increases.

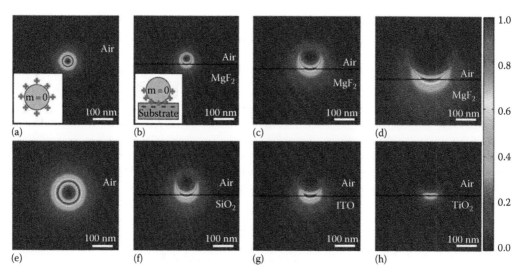

FIGURE 5.50

Energy density distribution on the cross section of Au nanowires. (a) $D = 50$ nm, no substrate; inset, schematic illustration of the polarized fields. (b) $D = 50$ nm, MgF$_2$ substrate; inset, schematic illustration of the polarized fields. (c) $D = 100$ nm, MgF$_2$ substrate. (d) $D = 200$ nm, MgF$_2$ substrate. (e) $D = 100$ nm, no substrate. (f) $D = 100$ nm, SiO$_2$ substrate. (g) $D = 100$ nm, ITO substrate. (h) $D = 100$ nm, TiO$_2$ substrate. The wavelength of light used here is 660 nm. (Adapted from Wang, Y.P. et al., *Opt. Express*, 20, 19006, 2012.)

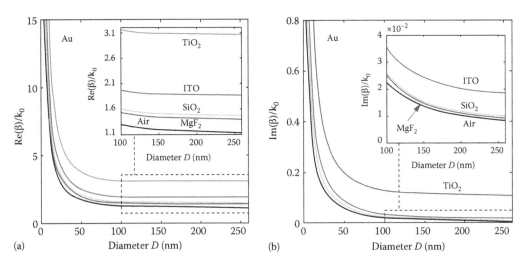

FIGURE 5.51
Numerical solutions of propagation constants of Au nanowires. (a) Re(β) and (b) Im(β). Insets, propagation constants of the nanowire diameter larger than 100 nm. The wavelength of light used here is 660 nm. (Adapted from Wang, Y.P. et al., *Opt. Express*, 20, 19006, 2012.)

The propagation constants (β) of Au nanowires are obtained from the eigenvalue equations by the COMSOL software. The real part (Re(β)) and the imaginary part (Im(β)) of the calculated propagation constants are illustrated in Figure 5.51, respectively. It shows that for a large D (e.g., >100 nm), Re(β) approaches to a constant; however, when D goes below 100 nm, Re(β) starts to increase with respect to the decreasing D. It is noticeable that for a very small D (e.g., $kD \ll 1$, k represents the wave vector), Re(β) exhibits a $1/D$ behavior, which can be treated as quasistatic [79]. In this case, the substrate exerts little impact on the SPP mode. Meanwhile, Re(β) is also influenced by the substrate refractive index (n_{sub}). For a given diameter of a nanowire, the higher the n_{sub}, the larger the Re(β), due to the lower light-propagation velocity in the higher-n_{sub} medium.

5.10.3 Optical Confinement and Effective Mode Areas

To quantify the confinement, effective mode areas (A_m) of nanowires are obtained with Equation 5.50. Figure 5.52 shows calculated effective mode areas of Au nanowires with dielectric substrates of MgF$_2$, SiO$_2$, ITO, and TiO$_2$, respectively. For comparison, mode areas of freestanding Au nanowires are also provided. It shows that with decreasing D, A_m of the plasmonic nanowire decreases monotonously, exhibiting a contrary behavior to that of a tightly confined dielectric waveguide (e.g., a silicon nanowire), in which A_m increases with decreasing D. Such behaviors of metal nanowires provide possibilities to achieve ultra-tightly confined guiding modes. For example, A_m of a 50 nm diameter Au nanowire is as small as 0.0026 μm², which is about 0.6% of $λ^2$, while the minimum A_m of a silicon nanowire is about 3% of $λ^2$ (shown in Section 5.2). Furthermore, compared to that of the freestanding nanowire, A_m of the nanowire with substrates is significantly reduced. For example, the A_m of a 100 nm diameter freestanding nanowire is about 0.01 μm², while the A_m of the 100 nm diameter nanowire supported by SiO$_2$ is as small as 0.003 μm². It is also noticed that with increasing substrate index, A_m decreases considerably.

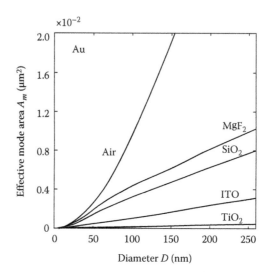

FIGURE 5.52
Effective mode area (A_m) of Au nanowires. The wavelength of light used here is 660 nm. (Adapted from Wang, Y.P. et al., *Opt. Express*, 20, 19006, 2012.)

5.10.4 Propagation Lengths and Losses

Metal nanowires with tight confinement usually suffer from high ohmic losses when operated around plasmonic resonance frequency. To study this behavior, the imaginary part of propagation constants ($Im(\beta)$) of Au nanowires with diameters ranging from 5 to 260 nm are calculated and shown in Figure 5.53.

By substituting the $Im(\beta)$ in the preceding text into the definition in Equations 5.48 and 5.49, we obtained the propagation lengths (L_m) and losses (*a*) of Au nanowires with diameters ranging from 5 to 260 nm, as is plotted in Figure 5.53. It is reasonable to see that L_m increases monotonously with the increasing D, reflecting the trade-off relations between the confinement and the loss of the plasmonic waveguide. Also, L_m of the freestanding

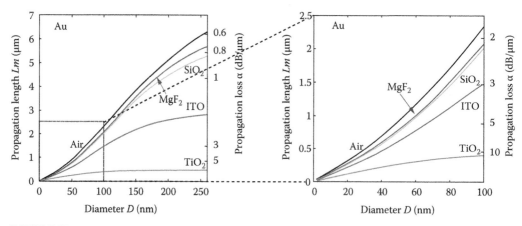

FIGURE 5.53
Propagation lengths (L_m) and losses (α) of Au nanowires. The wavelength of light used here is 660 nm. (Adapted from Wang, Y.P. et al., *Opt. Express*, 20, 19006, 2012.)

nanowire is always larger than those with the substrate, and L_m decreases monotonously with n_{sub} increasing from MgF_2, SiO_2, and ITO to TiO_2. For example, the L_m for a 100 nm diameter freestanding Au nanowire is about 2.3 μm, while the L_m for a SiO_2-supported nanowire is about 2.1 μm. Comparing the difference in L_m with A_m obtained in Section 5.10, the nanowire–substrate structure can simultaneously provide much smaller A_m and a relatively large L_m, when the parameters of the nanowire and the substrate are properly chosen.

5.11 Conclusion

In this chapter, we have introduced the modeling of nanofibers/nanowires using the commercially available computational softwares for nanophotonics. We started from the modeling of basic waveguiding properties of sub-wavelength-diameter air-clad nanofibers. Based on exact solutions of Maxwell equations and numerical calculations, single-mode guiding properties of nanofibers including single-mode conditions, modal fields, power distributions, group velocities, and waveguide dispersions were presented. These concepts and properties have been widely accepted and adopted in photonic applications of optical nanofibers/nanowires. Meanwhile, we have also reviewed the modeling of dispersion shifts in dielectric-coated nanowires and roughness-induced losses by solving Maxwell equations analytically. Then we introduced the numerical methods for calculating more complicated cases, in which the analytical solutions are difficult or impossible to obtain. Based on the FDTD methods, bending losses, EOPs, ERs, evanescent couplings, and longitudinal Lorentz forces in waveguiding nanofibers/nanowires were investigated. Finally, we concluded this chapter with a brief introduction of waveguiding properties of plasmonic metal nanowires using FEM.

References

1. L. M. Tong, J. Y. Lou, and E. Mazur, Single-mode guiding properties of subwavelength-diameter silica and silicon wire waveguides, *Opt. Express* **12**, 1025 (2004).
2. P. Klocek, *Handbook of Infrared Optical Materials*, Marcel Dekker, New York (1991).
3. E. D. Palik, *Handbook of Optical Constants of Solids*, Academic Press, New York (1998).
4. A. W. Snyder and J. D. Love, *Optical Waveguide Theory*, Chapman & Hall, New York (1983).
5. C. Manolatou, S. G. Johnson, S. Fan, P. R. Villeneuve, H. A. Haus, and J. D. Joannopoulos, High-density integrated optics, *J. Lightwave Technol.* **17**, 1682–1692 (1999).
6. G. Kakarantzas, T. E. Dimmick, T. A. Birks, R. Le Roux, and P. S. Rusell, Miniature all-fiber devices based on CO_2 laser microstructuring of tapered fibers, *Opt. Lett.* **26**, 1137–1139 (2001).
7. Z. M. Qi, N. Matsuda, K. Itoh, M. Murabayashi, and C. R. Lavers, A design for improving the sensitivity of a Mach-Zehnder interferometer to chemical and biological measurands, *Sens. Actuat.* **B81**, 254–258 (2002).
8. B. E. A. Saleh and M. C. Teich, *Fundamentals of Photonics*, John Wiley & Sons, New York (1991).
9. A. Ghatak and K. Thyagarajan, *Introduction to Fiber Optics*, Cambridge University Press, Cambridge, U.K. (1998).
10. P. P. Bishnu, *Fundamentals of Fibre Optics in Telecommunication and Sensor Systems*, John Wiley & Sons, New York (1993).

11. T. A. Birks, W. J. Wadsworth, and P. S. Russell, Supercontinuum generation in tapered fibers, *Opt. Lett.* **25**, 1415–1417 (2000).

12. L. F. Mollenauer, Nonlinear optics in fibers, *Science* **302**, 996–997 (2003).

13. V. I. Balykin, K. Hakuta, F. Le Kien, J. Q. Liang, and M. Morinaga, Atom trapping and guiding with a subwavelength-diameter optical fiber, *Phys. Rev. A* **70**, 011401 (2004).

14. J. Y. Lou, L. M. Tong, and Z. Z. Ye, Dispersion shifts in optical nanowires with thin dielectric coatings, *Opt. Express* **14**, 6993–6998 (2006).

15. J. A. Stratton, *Electromagnetic Theory*, McGraw-Hill, New York (1941).

16. U. Schroter and A. Dereux, Surface plasmon polaritons on metal cylinders with dielectric core, *Phys. Rev. B* **64**, 125420 (2001).

17. J. Bures and R. Ghosh, Power density of the evanescent field in the vicinity of a tapered fiber, *J. Opt. Soc. Am. A* **16**, 1992–1996 (1999).

18. L. M. Tong, R. R. Gattass, J. B. Ashcom, S. L. He, J. Y. Lou, M. Y. Shen, I. Maxwell, and E. Mazur, Subwavelength diameter silica wires for low-loss optical wave guiding, *Nature* **426**, 816–818 (2003).

19. L. M. Tong, L. L. Hu, J. J. Zhang, J. R. Qiu, Q. Yang, J. Y. Lou, Y. H. Shen, J. L. He, and Z. Z. Ye, Photonic nanowires directly drawn from bulk glasses, *Opt. Express* **14**, 82–87 (2006).

20. J. Jäckle and K. Kawasaki, Intrinsic roughness of glass surfaces, *J. Phys.: Condens. Matter* **7**, 4351–4358 (1995).

21. G. Brambilla, V. Finazzi, and D. J. Richardson, Ultra-low-loss optical fiber nanotapers, *Opt. Express* **12**, 2258–2263 (2004).

22. P. J. Roberts, F. Couny, H. Sabert, B. J. Mangan, D. P. Williams, L. Farr, M. W. Mason et al., Ultimate low loss of hollow-core photonic crystal fibers, *Opt. Express* **13**, 236–244 (2005).

23. P. J. Roberts, F. Couny, H. Sabert, B. J. Mangan, T. A. Birks, J. C. Knight, and P. St. J. Russell, Loss in solid-core photonic crystal fibers due to interface roughness scattering, *Opt. Express* **13**, 7779–7793 (2005).

24. G. Y. Zhai and L. M. Tong, Roughness-induced radiation losses in optical micro or nanofibers, *Opt. Express* **15**, 13805–13816 (2007).

25. L. M. Tong, J. Y. Lou, Z. Z. Ye, G. T. Svacha, and E. Mazur, Self-modulated taper drawing of silica nanowires, *Nanotechnology* **16**, 1445–1448 (2005).

26. E. G. Rawson, Analysis of scattering from fiber waveguides with irregular core surface, *Appl. Opt.* **13**, 2370–2377 (1974).

27. D. Marcuse, *Theory of Dielectric Optical Waveguides*, Academic Press, New York (1974).

28. S. Leon-Saval, T. Birks, W. Wadsworth, P. St. J. Russell, and M. Mason, Supercontinuum generation in submicron fibre waveguides, *Opt. Express* **12**, 2864–2869 (2004).

29. R. R. Gattass, G. T. Svacha, L. M. Tong, and E. Mazur, Supercontinuum generation in submicrometer diameter silica fibers, *Opt. Express* **14**, 9408–9414 (2006).

30. L. M. Tong, J. Y. Lou, R. R. Gattass, S. L. He, X. W. Chen, L. Liu, and E. Mazur, Assembly of silica nanowires on silica aerogels for microphotonics devices, *Nano Lett.* **5**, 259–262 (2005).

31. M. Sumetsky, Y. Dulashko, J. M. Fini, and A. Hale, Optical microfiber loop resonator, *Appl. Phys. Lett.* **86**, 161108 (2005).

32. Y. H. Li and L. M. Tong, Mach—Zehnder interferometers assembled with optical microfibers or nanofibers, *Opt. Lett.* **33**, 303–305 (2008).

33. Y. Xiao, C. Meng, P. Wang, P. Ye, H. K. Yu, S. S. Wang, F. X. Gu, L. Dai, and L. M. Tong, Single-nanowire single-mode laser, *Nano Lett.* **11**(3), 1122–1126 (2011).

34. F. L. Kien, J. Q. Liang, K. Hakuta, and V. I. Balykin, Field intensity distributions and polarization orientations in a vacuum-clad subwavelength-diameter optical fiber, *Opt. Commun.* **242**, 445–455 (2004).

35. S. D. Gedney, An anisotropic perfectly matched layer absorbing media for the truncation of FDTD lattices, *IEEE Trans. Antennas Propag.* **44**, 1630–1639 (1996).

36. D. Roundy, M. Ibanescu, P. Bermel, A. Farjadpour, J. D. Joannopoulos, and S. G. Johnson, The Meep FDTD package, http://ab-initio.mit.edu/meep/

37. H. K. Yu, S. S. Wang, J. Fu, M. Qiu, Y. H. Li, F. X. Gu, and L. M. Tong, Modeling bending losses of optical nanofibers or nanowires, *Appl. Opt.* **48**, 4365–4369 (2009).

38. V. Bondarenko and Y. Zhao, Needle beam: Beyond-diffraction-limit concentration of field and transmitted power in dielectric waveguide, *Appl. Phys. Lett.* **89**(14), 141103 (2006).

39. L. K. van Vugt, S. Rühle, and D. Vanmaekelbergh, Phase-correlated nondirectional laser emission from the end facets of a ZnO nanowire, *Nano Lett.* **6**(12), 2707–2711 (2006).

40. Y. Nakayama, P. J. Pauzauskie, A. Radenovic, R. M. Onorato, R. J. Saykally, J. Liphardt, and P. D. Yang, Tunable nanowire nonlinear optical probe, *Nature* **447**(7148), 1098–1101 (2007).

41. W. L. She, J. H. Yu, and R. H. Feng, Observation of a push force on the end face of a nanometer silica filament exerted by outgoing light, *Phys. Rev. Lett.* **101**(24), 243601 (2008).

42. S. S. Wang, J. Fu, M. Qiu, K. J. Huang, Z. Ma, and L. M. Tong, Modeling endface output patterns of optical micro/nanofibers, *Opt. Express* **16**, 8887–8895 (2008).

43. S. K. Mondal, S. Gangopadhyay, and S. Sarkar, Analysis of an upside-down taper lens end from a single-mode step-index fiber, *Appl. Opt.* **37**, 1006–1009 (2005).

44. Y. X. Mao, S. D. Chang, S. Sherif, and C. Flueraru, Graded-index fiber lens proposed for ultrasmall probes used in biomedical imaging, *Appl. Opt.* **46**, 5887–5894 (2007).

45. S. S. Wang, Z. F. Hu, H. K. Yu, W. Fang, M. Qiu, and L. M. Tong, Endface reflectivities of optical nanowires, *Opt. Express* **17**, 10881–10886 (2009).

46. Y. Chen, Z. Ma, Q. Yang, and L. M. Tong, Compact optical short-pass filters based on microfibers, *Opt. Lett.* **33**(21), 2565–2567 (2008).

47. K. J. Huang, S. Y. Yang, and L. M. Tong, Modeling of evanescent coupling between two parallel optical nanowires, *Appl. Opt.* **46**, 1429–1434 (2007).

48. R. G. Hunsperger, *Integrated Optics: Theory and Technology*, Springer-Verlag, Berlin, Germany (2002).

49. F. Bilodeau, K. O. Hill, D. C. Johnson, and S. Faucher, Compact, low-loss, fused biconical taper couplers: overcoupled operation and antisymmetric supermode cutoff, *Opt. Lett.* **12**, 634–636 (1987).

50. I. Brevik, Experiments in phenomenological electrodynamics and the electromagnetic energy-momentum tensor, *Phys. Rep.* **52**, 133 (1979).

51. R. N. C. Pfeifer, T. A. Nieminen, N. R. Heckenberg, and H. Rubinsztein-Dunlop, *Colloquium*: Momentum of an electromagnetic wave in dielectric media, *Rev. Mod. Phys.* **79**, 1197 (2007).

52. S. M. Barnett, Resolution of the Abraham-Minkowski dilemma, *Phys. Rev. Lett.* **104**, 070401 (2010).

53. M. Mansuripur, Comment on observation of a push force on the end face of a nanometer silica filament exerted by outgoing light, *Phys. Rev. Lett.* **103**, 019301 (2009).

54. M. Mansuripur and A. R. Zakharian, Theoretical analysis of the force on the end face of a nanofilament exerted by an outgoing light pulse, *Phys. Rev. A* **80**, 023823 (2009).

55. I. Brevik, Comment on observation of a push force on the end face of a nanometer silica filament exerted by outgoing light, *Phys. Rev. Lett.* **103**, 219301 (2009).

56. J. D. Jackson, *Classical Electrodynamics*, Wiley, New York (1999).

57. D. V. Thourhout and J. Roels, Optomechanical device actuation through the optical gradient force, *Nat. Photon.* **4**, 211 (2010).

58. H. K. Yu, W. Fang, F. X. Gu, M. Qiu, Z. Y. Yang, and L. M. Tong, Longitudinal Lorentz force on a subwavelength-diameter optical fiber, *Phys. Rev. A* **83**, 053830 (2011).

59. A. R. Zakharian, M. Mansuripur, and J. V. Moloney, Radiation pressure and the distribution of electromagnetic force in dielectric media, *Opt. Express* **13**, 2321 (2005).

60. E. Ozbay, Plasmonics: Merging photonics and electronics at nanoscale dimensions, *Science* **311**, 189–193 (2006).

61. X. Guo, M. Qiu, J. M. Bao, B. J. Wiley, Q. Yang, X. N. Zhang, Y. G. Ma, H. K. Yu, and L. M. Tong, Direct coupling of plasmonic and photonic nanowires for hybrid nanophotonic components and circuits, *Nano Lett.* **9**, 4515–4519 (2009).

62. S. Lal, S. Link, and N. J. Halas, Nano-optics from sensing to waveguiding, *Nat. Photon.* **1**, 641–648 (2007).

63. E. Hendry, T. Carpy, J. Johnston, M. Popland, R. V. Mikhaylovskiy, A. J. Lapthorn, S. M. Kelly, L. D. Barron, N. Gadegaard, and M. Kadodwala, Ultrasensitive detection and characterization of biomolecules using superchiral fields, *Nat. Nanotechnol.* **5**, 783–787 (2010).

64. A. V. Akimov, A. Mukherjee, C. L. Yu, D. E. Chang, A. S. Zibrov, P. R. Hemmer, H. Park, and M. D. Lukin, Generation of single optical plasmons in metallic nanowires coupled to quantum dots, *Nature* **450**, 402–406 (2007).

65. A. Cuche, O. Mollet, A. Drezet, and S. Huant, Deterministic quantum plasmonics, *Nano Lett.* **10**, 4566–4570 (2010).

66. A. Ridolfo, O. Di Stefano, N. Fina, R. Saija, and S. Savasta, Quantum plasmonics with quantum dot-metal nanoparticle molecules: influence of the Fano effect on photon statistics, *Phys. Rev. Lett.* **105**, 263601 (2010).

67. J. Zuloaga, E. Prodan, and P. Nordlander, Quantum plasmonics: Optical properties and tunability of metallic nanorods, *ACS Nano* **4**, 5269–5276 (2010).

68. C. A. Pfeiffer, E. N. Economou, and K. L. Ngai, Surface polaritons in a circularly cylindrical interface: Surface plasmons, *Phys. Rev. B* **10**, 3038–3051 (1974).

69. H. Khosravi, D. R. Tilley, and R. Loudon, Surface polaritons in cylindrical optical fibers, *J. Opt. Soc. Am. A* **8**, 112–122 (1991).

70. L. Novotny and C. Hafner, Light propagation in a cylindrical waveguide with a complex, metallic, dielectric function, *Phys. Rev. B* **50**, 4094–4106 (1994).

71. Y. P. Wang, Y. G. Ma, X. Guo, and L. M. Tong, Single-mode plasmonic waveguiding properties of metal nanowires with dielectric substrates, *Opt. Express*, 20, 19006–19015 (2012).

72. S. A. Maier, *Plasmonics: Fundamentals and Applications*, Springer, New York (2007).

73. R. F. Oulton, V. J. Sorger, D. A. Genov, D. F. P. Pile, and X. Zhang, A hybrid plasmonic waveguide for subwavelength confinement and long-range propagation, *Nat. Photon.* **2**, 496–500 (2008).

74. P. B. Johnson and R. W. Christy, Optical constants of the noble metals, *Phys. Rev. B* **6**, 4370–4379 (1972).

75. P. G. Etchegoin, E. C. Le Ru, and M. Meyer, An analytic model for the optical properties of gold, *J. Chem. Phys.* **125**, 164705 (2006).

76. J. Takahara, S. Yamagishi, H. Taki, A. Morimoto, and T. Kobayashi, Guiding of a one-dimensional optical beam with nanometer diameter, *Opt. Lett.* **22**, 475–477 (1997).

77. J. J. Mock, R. T. Hill, A. Degiron, S. Zauscher, A. Chilkoti, and D. R. Smith, Distance-dependent plasmon resonant coupling between a gold nanoparticle and gold film, *Nano Lett.* **8**, 2245–2252 (2008).

78. H. J. Chen, T. Ming, S. Zhang, Z. Jin, B. C. Yang, and J. F. Wang, Effect of the dielectric properties of substrates on the scattering patterns of gold nanorods, *ACS Nano* **5**, 4865–4877 (2011).

79. D. E. Chang, A. S. Sorensen, P. R. Hemmer, and M. D. Lukin, Strong coupling of single emitters to surface plasmons, *Phys. Rev. B* **76**, 035420 (2007).

6

Cavity Quantum Electrodynamics: Application to Quantum State Transfer through Nanophotonic Waveguidance

P.K. Choudhury and Md. Mijanur Rahman

CONTENTS

6.1 Introduction

Massless particle photon has been serving as the information carrier in a large number of classical communication systems in the form of electromagnetic waves. With the advancements in quantum optics, due to the progress in the design and development of high-quality optical cavities and improvements in the confinement of atoms, ions, and molecules using sophisticated optical traps and optical lattices, individual photons can now be generated, manipulated, and guided. These capabilities of controlling individual photons herald the emergence of quantum communication systems used in quantum optics [1–4]. Quantum communication is the art of communicating quantum states between spatially separated locations. A photon is generally used as the carrier for transferring quantum information (encoded into quantum states) from the transmit node to the receive node of the quantum communication system.

One of the key factors behind the modern revolution in informatics is the sustained progress in data processing and communication technology. During the last few decades, the capabilities of modern digital electronic devices have exponentially increased (according to the so-called Moore's law, doubled in every 2 years) [4–6]. For this pace of progress to be maintained, it is necessary to increasingly implement quantum devices as the replacement of classical ones.

The prospects of quantum networking technologies have brought quantum optics at the forefront of communication research. Investigators have been intensively studying the relevant phenomena/devices involved in quantum optics technologies such as cavity quantum electrodynamics (QED), optical lattice, optical trap, and photon source and detectors in order to enhance the quality of generation, manipulation, and guidance of individual photons [7–16].

6.1.1 Quantum Network

A quantum network consists of two major components, namely, quantum nodes and quantum channels (Figure 6.1). While a quantum node is equipped with devices to store and manipulate quantum information, a quantum channel is responsible for transferring quantum information from one node to another. In order to implement a quantum channel, it is required to identify an information carrier and a physical medium to guide the carrier from the transmit quantum node to the receive one. For quantum information carrier, photon is the only feasible candidate, and optical fibers (because of low loss) are proposed as suitable candidates for the physical medium for the transfer of photons. Photonic crystal fibers are also promising candidates to play the role of physical mediums [13,16].

The nodes in quantum network generally involve QED processes [7]. In the event of transmitting quantum information, in the transmit node, QED processes generate carrier photons, which carry information to the receive node. On the other hand, QED processes in the receive node absorb the carrier photons, mapping thereby the photonic states into the quantum states of the receive node.

6.1.2 Progresses in Quantum Optics

Although the origin of quantum optics dates back to the 1960s, the understanding of optical fields was greatly revolutionized in early twentieth century when Albert Einstein explained the experiment related to photoelectric effect through postulating discrete particle-like properties of light. Later developments in the theoretical and technological fields of quantum optics during the 1980s and 1990s enabled researchers to produce and manipulate light fields with properties dominated by quantum mechanical phenomena [1]. Quantum information science, which is closely related to quantum optics, was also established around the same time [2–4]. These two developments together led to

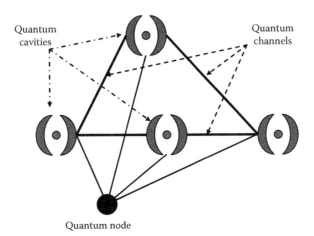

FIGURE 6.1
Schematic view of a simple quantum network.

rapid progress in quantum optics, particularly in the fields of quantum computing and quantum communications.

Several breakthrough advances have been made in quantum communication and quantum computing; quantum optics has been an attractive testing ground for these investigations [6,9,14,17]. One useful example of quantum communication protocol is quantum key distribution (QKD) systems, developed in early 1990, based on single-photon transmission [18–20]. These systems have now been implemented in *real-world* applications. Quantum teleportation, which was among the predictions of quantum optics, has applications in the implementation of quantum logic gates [5]. By engineering entanglements and measurements [21,22], it is now possible to realize quantum logic gates through the process of quantum teleportation. Quantum computing is also an active area of research, and a number of potential proposals for quantum computer have been reported in the literature [6,10,18,23]. While some of the proposed systems are based on superconductor, others utilize trapped ion, optical lattice, and nanodots.

Besides developments in quantum entanglement and quantum information, advancements have been made in many other areas of quantum optics, which have opened up new applications of quantum optics in diversified fields including sensing, imaging, and metrology.

6.2 Quantum Communication Systems

Quantum communication for transferring quantum state from one quantum node to another has been an attractive research field of quantum optics, which is evident from the substantial amount of research papers appearing in the literature. The systems described so far adopt various approaches to address the problem. This section is devoted to the description of present state of research in the relevant area, which essentially needs a review of existing important papers on quantum communications.

6.2.1 Proposed Quantum Communication Systems

In quantum communication systems, communications between distant quantum nodes are accomplished using photons as information carriers. A number of papers on closely related topics are reviewed, as organized in three categories—(1) carrier-mediated quantum state transfer, (2) entanglement-based quantum state transfer, and (3) miscellaneous quantum state transfer.

6.2.1.1 Carrier-Based Quantum Communication Systems

Quantum state transfer techniques in this category essentially require the information carrier (such as photon) to propagate from the transmit node to the receive node. A number of proposals within this category have been reported to address the problem of quantum communication. Among them, Gardiner [24] proposed a two-atom system wherein the first atom was driven by a coherent field and the second atom by the fluorescent light emitted from the first atom. These two atoms were placed in an electric field in spatially separated locations. The author pointed out that the system Hamiltonian must be formulated in such a way that the coupling could be allowed from atom 1 to atom 2, but not in the reverse direction. An equation of motion was derived to give the rate of change of field operator in terms of field and atomic operators and, also, coupling constants. The reflection invariance of the system was broken by making the atom–field coupling constant different for positive and negative wave numbers. Also, the author formulated the equation of motion for an arbitrary operator in the Hilbert space (of the two atoms) in terms of the atomic operators and coupling and decay constants, from which the quantum Langevin and master equations could be derived. A set of correlation functions was derived that gave conditional probability of counting a photon from the second atom at some finite time after counting a photon from the first atom. Finally, the equation of motion was numerically solved, and the intensity correlations between atoms were demonstrated.

Carmichael [25] formulated a quantum trajectory theory to deal with two spatially separated quantum systems, according to which the first quantum system A (the quantum source) emits a photon and the second quantum system B reacts to the emitted photon. A and B each consists of a cavity with leakage to allow input and output radiations. In order to eliminate the complexity of the correlation functions to describe the mediating field, the author formulated a stochastic wave function of the composite system $A \oplus B$. The spatial separation was eliminated by using Born–Markov approximation in Heisenberg picture to relate the field coupling of the first cavity to that of the second one. The author demonstrated the concept with a setup where the source consisted of a cavity and the destination consisted of an atom at the focus of the second cavity.

Cirac et al. [26] made a proposition of quantum communication between spatially separated atoms using photons as information carriers. In their approach, a quantum node consisted of an atom located inside a high-Q optical cavity. They modeled the atom as a three-level system where $|g\rangle$ and $|e\rangle$ were degenerate ground states and $|r\rangle$ was the excited state. Qubit was represented as a superposition of the $|g\rangle$ and the $|e\rangle$ states. These two states were coupled through Raman transition via the $|r\rangle$ state. A laser excites the atom from $|e\rangle$ to $|r\rangle$, which was followed by an atom–cavity coupling-mediated transition from $|r\rangle$ to $|g\rangle$ with a photon emitted into the cavity. The photon (leaked out of the cavity) reaches the receive node. The work was motivated by the idea that if the photon wave packet could be *time reversed* and sent back into the cavity, the original superposition state of the atom could be restored, provided the laser pulse was also *time reversed*. The authors estimated the laser pulse shape and numerically demonstrated the time evolution of the system with an effective Hamiltonian involving both transmit and receive subsystems.

Yao et al. [27] proposed that the time-symmetry requirement, as described by Cirac et al. [26], could be eliminated, allowing thereby independent control of photon transmission and reception processes. Instead of an atom used in a cavity (as in Ref. [26]), Yao et al. [27], however, used a nanodot coupled to the cavity. The nanodot was treated as a three-level system $|e\rangle$, $|t\rangle$, and $|g\rangle$ with $|e\rangle$ and $|g\rangle$ being the qubit states and $|t\rangle$ being the excited state for Raman transition. They designed the $|e\rangle$, $|t\rangle$, and $|g\rangle$ states, cavity mode, and laser frequency in such a way that the resonant condition $\omega_t = \omega_L + \omega_e = \omega_c + \omega_g$ was satisfied, ω_t, ω_e, and ω_g being the respective frequencies corresponding to the energies of $|t\rangle$, $|e\rangle$, and $|g\rangle$ states. Also, ω_L and ω_c are, respectively, the laser and the cavity mode frequencies. Thus, the laser could resonantly excite from $|e\rangle$ to $|t\rangle$ only, and the cavity mode allowed only $|t\rangle \rightarrow |g\rangle$ transition with the emission of photon having frequency ω_c. When the photon reached the receive node, the qubit state of the transmit node was restored into the receive node by the reverse process $|g\rangle \rightarrow |t\rangle \rightarrow |e\rangle$. By design, probabilities of off-resonant transitions were negligible. The authors derived the equation of motion for the resonant Raman interaction process within the Weisskopf–Wigner approximation and analytically demonstrated the laser pulse shape for a desired shape of the generated photon wave packet.

Nohama and Roversi [28] investigated quantum state transfer between atoms located inside and coupled to the corresponding cavities. According to their approach, the state transfer between the atoms would happen if one of the atoms is in a superposition state and the other subsystems are in their fundamental states. Under the assumptions that both atoms have the same frequency ω_a and that the coupling constants between the atoms and the corresponding cavities also have the same value g, the authors formulated a Hamiltonian for the system, which included terms corresponding to the cavity–atom coupling g as well as the coupling between the cavity modes.

While the proposed systems mostly used 2D quantum state to represent quantum information, Zhou et al. [29] represented the same using 3D quantum state. They devised a scheme to transfer this 3D quantum state between distant atoms. The atoms were trapped inside the corresponding cavities. As the physical medium, an optical fiber was connected to these two cavities. The fiber and the cavities were designed in such a way that the fiber mode was resonant with the cavity polarization modes. The authors performed an adiabatic passage along the dark states so that the fiber modes remained in the vacuum state and the population of the cavities in the excited states was negligible under appropriate conditions. Also, the transitions between atomic states were made detuned from the cavity modes so that the atomic spontaneous emission and the fiber decay could be effectively eliminated.

Photon-mediated quantum communication was used by Tang et al. [20] for QKD. In their approach, both quantum and classical communication channels were used—the former one was used for the synchronization purpose, whereas the latter one for QKD based on photon polarization states. In the transmit node, the quantum key consisting of pseudorandom bits was encoded into the photon polarization states. Bit "0" was encoded into a 45° polarization state and bit "1" into a vertical polarization state. A set of transmitted polarized photons formed a quantum packet. At the start of every quantum packet, a synchronization message was sent over the classical channel. At the receiving end, two avalanche photo diodes (APDs) were used for photon detection. Relevant circuitries were used to search for rising edges of the detected photon signals—the detected bits were parallelized by a serializer/deserializer and analyzed by an attached computer. Tang et al. [20] reported that the system was suitable for distributing quantum key over a distance of 1 km.

Li et al. [30] proposed an efficient scheme for the implementation of quantum information transfer via resonators cascaded serially in order to form a coupled resonator waveguide. A three-level atom was trapped in each of these resonators. The authors first demonstrated that the coupled system could be reduced to an effective configuration where Raman transitions between the first atom and the last one could be realized. This was possible because, in each of the resonators, transition from the ground state of the atom to the excited state is strongly detuned from the cavity mode, and this induced a long-range interaction between the atoms. Then they utilized the idea to implement quantum communications over short distances. In effect, the nonlocal interactions were combined with lasers to induce Raman transitions between the atoms trapped at the two ends of the resonator chain via the exchange of virtual cavity photons.

In order to achieve quantum communication efficiently, routing of photons carrying quantum information remains very important. Aoki et al. [31] proposed such a photon routing system. The proposed system consisted of a microtoroidal cavity and coupled cesium atom; the cavity is coupled to two inputs and two outputs. The system received photons from a coherent input and redirected to a separate output fiber. Near the atomic resonance frequency ω_a, the cavity supported two counterpropagating modes, both at the frequency ω_c. A tapered fiber was used to couple inputs and outputs to the internal cavity modes. Aoki et al. [31] argued that, when the rate of cavity decay was larger than any other rates in the system, the cavity modes could be adiabatically removed from the system dynamics, and the system could be described by the effective optical Bloch equations for a two-level atom.

6.2.1.2 Einstein-Podolsky-Rosen Entangled Quantum Communication Systems

In this category, entanglement is first generated between transmit and receive nodes, which is mostly mediated via photon polarization states. When the photon is generated, it remains in entanglement with the local quantum system. When the photon reaches the other end (or meets the photon generated from the other end), the two quantum nodes get entangled, and the exchange of information between quantum nodes becomes possible without explicit exchange of information carrier (such as photon).

Duan et al. [32] proposed a quantum communication system based on quantum entanglement of distant atom ensembles. They also addressed the problem associated with exponential fidelity decay of the entanglement-based quantum channel where the entanglement was generated between two atom ensembles. They solved the fidelity decay problem by adding entanglement purification at all of the entanglement, connection and application steps. They achieved robustness during entanglement generation by incorporating collective particle excitation rather than single-particle excitation in the atom ensemble. In the next step, the entanglement connection was achieved through simple linear optical operations and, thus, inherently robust from realistic imperfections. Finally, in the application step, the mixed state entanglement was purified automatically to the nearly perfect entanglement. Duan et al. [32] established connection between the two ensembles through a quantum swapping operation and extended the communication length by cascading a number of such connections. Once this cascaded entangled state covering a distance was generated, due to entanglement, a measurement on one end could affect the state on the other end. Thus, a communication could be performed end to end.

Boileau et al. [33] proposed a technique for entanglement-based communication that overcame the effect of noise and imperfections on the communication. In their scheme, quantum information was encoded in a pair of photons using a tag operation T_i, which corresponded to the time delay of the polarization mode $|i\rangle$. Through detailed calculations, the authors demonstrated that the effect of noise and imperfections could be overcome with this mechanism of entanglement-based communication.

Hong and Xiong [34] proposed a quantum entanglement-based communication system using linear optics and atom ensembles. They argued that, in the proposal put forward by Duan et al. [32], the probability of generating a single excitation in two ensembles is low to guarantee an acceptable quality of entanglement. In order to overcome the problem, Hong and Xiong [34] proposed a solution by generating the entanglement in every node in the cascaded system with on-demand single-photon source. Similar to the work in Ref. [32], nodes in the system of Ref. [34] consisted of a cloud of predefined number N_a of identical atoms having four-level structure with one excited state $|e\rangle$ and three split ground states $|g\rangle$, $|s\rangle$, and $|t\rangle$. A Raman transition through these atomic states generated photons, and nonlocal interactions between photons (generated from distant nodes) produced the expected entanglement. Once the entanglements were generated and nodes were cascaded, quantum communications could be performed through the entanglement-based channels.

A proposal for transferring quantum state from a two-level atomic system to a single photon was put forward by Matsukevich and Kuzmich [35]. The state transfer process involved three steps. In the first step, an entangled state was generated between a single photon and a single collective excitation distributed over many atoms in two distinct optically thick atomic samples (nodes). In the second step, a measurement was made on the photon in order to project the atomic ensembles into a desired state. The projection was conditioned on the choice of the basis and the outcome of the measurement. The authors argued that the resulting atomic state was a nearly maximally entangled state between the two distinct atomic ensembles. Finally, the entangled atomic state was converted into a single photon emitted into a well-defined mode. They expected that, if the second node was placed at a spatially separated location and if a joint detection of photons from the two nodes were made, quantum repeater protocol as well as distant teleportation of an atomic qubit might be realized.

Suchat et al. [36] proposed an approach to secure data transmission over a wireless link by using a single-photon entangled state. In their approach, the encrypted qubit (data) was generated in a biological tissue by the application of a simple optical system. When the optical input was applied on the tissue sample, interaction between the incident photon and the sample generated light emission/absorption relationship of the relevant physical parameter of the tissue. The fluorescence/luminescence of light was detected by a high-speed sensor. The authors proposed a quantum entanglement scheme in order to securely transmit these biological data to distant recipients, where the entanglement state consisted of Bell states involving the horizontal polarization state $|H\rangle$ and the vertical polarization state $|V\rangle$ of photons.

Pan et al. [37] proposed a feasible scheme for the entanglement purification of general mixed entangled states. In their entanglement purification scheme, quantum-controlled NOT operations were not required. Instead, they used simple linear optical elements, which could be manufactured using matured technology, and therefore, the perfection of such elements was very high, and the local operations (necessary for purification) could be accomplished with the required precision. In the proposed scheme, qubits were represented by the horizontally polarized photon state $|H\rangle$ (logic "0") and the vertically polarized photon state $|V\rangle$ (logic "1").

6.2.1.3 Miscellaneous Quantum Communication Systems

Many other systems related to quantum communication, in general, have been proposed that do not fit into the previously described two categories. For example, in order to achieve reliable quantum state transfer between nodes of a hypercube network, Strauch and Williams [38] proposed multilayered superconducting quantum circuits. The authors primarily focused on phase qubits coupled by capacitors. These capacitively coupled qubits were organized into a network, which consisted of nodes modeled as current-biased Josephson junctions, where each line requires a coupling capacitance. The interconnections of the network were implemented by superconducting wires in a crossover configuration using either insulating material or vacuum gaps between the layers. The authors argued that, for sufficiently large gaps (100 nm – 1 μm), the capacitance between the wires could be made negligible. They claimed that using tunable phase qubits, the network could be programmed to achieve high-fidelity quantum state transfer between the quantum nodes.

Moehring et al. [39] reported on an experiment to demonstrate the very existence of quantum entanglement that forms the basis for a class of quantum communication system. They prepared a probabilistic entangled state between an atom and a photon. For this entangled state, they performed the measurement of a Bell inequality violation and demonstrated that the violation occurred. The entanglement was probabilistic due to the reasons that (1) the photon collection optics involved small acceptance angle and transmission loss, (2) photon detectors had low quantum efficiency, and (3) there was a restriction on the excitation probability in order to suppress multiple excitations.

Volz et al. [40] reported the observation of atom–photon entanglement between the internal state of an atom trapped in an optical dipole trap and the polarization state of photon. The authors argued that the detection efficiency and the entanglement fidelity were high enough to allow the generation of entangled atoms at large distances.

Wang et al. [41] proposed a scheme to simulate the envisioned quantum network and implemented it using linear optical elements. The scheme was based on two Hong–Ou–Mandel interferometers [42], wherein β-barium borate (BBO), arranged in a Kwiat-type configuration [43], was pumped by ultraviolet laser beam. As a result, through a spontaneous parametric down-conversion (SPDC) process, the entangled photon pair was generated as entanglement source. The authors could prepare different kinds of entangled states and mixed states using simple optical elements and optical setups, such as a half-wave plate and a quarter-wave plate. Quantum state prepared in this manner was put into the quantum network and implemented in a simple fashion using linear optical devices. The scheme required two types of qubits—namely, control qubit and target qubit. The entangled photon pair was taken as the control qubit, whereas the target qubits could be realized using one of many choices such as polarization, spatial mode, and moment.

Bužek and Hillery [44] focused on the possibility of going beyond the no-cloning theory [45], which forbids the exact copying of a quantum state without changing the original state. This imposes a limit on the transfer of quantum states. The authors analyzed the quantum-copying machine used by Wootters and Zurek [45] and found that the machine suffered one major drawback—its operation depended on the state of the original input. Bužek and Hillery [44] analyzed the possibility of copying an arbitrary state of half-spin particle and argued that there existed a *universal quantum-copying machine* that approximately copied quantum mechanical states such that the quality of the output did not depend on the input.

6.2.2 Current Status of the Quantum Communication Systems

Investigators have put enormous efforts and proposed different techniques for the design and development of quantum communication systems. However, a reasonable solution to the problem leading to the development of commercially viable quantum communication systems is still lacking, which makes quantum communication as an active area of research and demands more efforts to come up with an effective solution. In particular, in many photon carrier-based quantum communication approaches [26,27], while transferring the quantum state from the transmit node to the receive one, the transmit node toggles its quantum state by itself. This remains undesirable, and it is important to find an appropriate solution to quantum state transfer in which the transmit node does not toggle its state while communicating with the receive node.

6.3 Cavity Quantum Electrodynamics for Quantum Communication

Cavity QED has been an important area of research relevant to quantum communication as it provides an ideal mechanism for controlled absorption, confinement, and emission of photons essential for many of the quantum communication systems [8,15,22]. Two important applications of cavity QED in quantum communication are spin–photon and photon–spin interfaces, which have similar constructions and working principles, although these are responsible for reciprocal functions.

Spin–photon interface, an important quantum interface for transferring stationary spin state to flying photon state, constitutes the foremost important step in transferring qubit between quantum nodes. A spin–photon interface typically consists of a quantum structure with spin and a coupled cavity. Orthogonal spin states for representing qubits can be realized using a number of different quantum structures including nanodots, electrons, and nuclei. So far as the reported research on spin–photon interface is concerned, Yao et al. [27] proposed an interface consisting of a nanodot coupled to an optical fiber through a microcavity. In this work, spin states of the nanodot were mapped into the photon state through a cavity-assisted Raman process. Vamivakas et al. [46] demonstrated spin-selective photon emission from a resonantly driven quantum dot. Yao et al. [47] also presented a method of control through a shaped optical pulse, which can convert a quantum dot spin state into a linear combination of 0 and 1 photon states, propagating along a waveguide, and vice versa.

In this section, the application of cavity QED for quantum communication system is touched upon. In particular, the description is made of a spin–photon interface (consisting of nanodot coupled to microcavity) that generates photon with flying qubit state corresponding to the spin state of the nanodot. Within the context, a spin–photon interface model is adopted based on Ref. [27] and a simulation platform is developed to simulate the model.

6.3.1 Spin–Photon Interface

The spin–photon interface is largely related to two fundamental concepts of physics, namely, Raman effect and cavity QED; the present section discusses these two in short. Also, two other important topics related to the development of spin–photon interface— namely, Rabi oscillation and nanodot spin state—are briefly discussed.

6.3.1.1 Raman Effect

Raman effect is observed during an inelastic scattering of a photon [48]. In the present context, Raman scattering from a nanodot is considered, and the process is illustrated in Figure 6.2. When a light source, typically a laser, is applied on a nanodot, a photon with

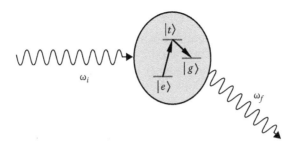

FIGURE 6.2
Raman scattering.

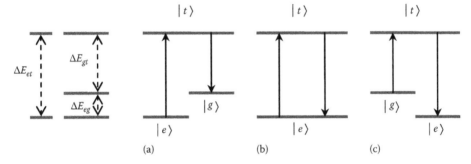

FIGURE 6.3
Three scattering types: (a) Stokes scattering with Raman effect, (b) Rayleigh scattering without Raman effect, and (c) anti-Stokes scattering with Raman effect.

energy ω_i is incident upon it, and a photon with energy ω_f emits from it due to scattering. The nanodot undergoes Raman effect if the energy ω_f of the emitted photon is different from that of the incident one (i.e., ω_i) (the case of inelastic scattering). The difference between ω_i and ω_f determines the type of Raman scattering. Within the context, $\omega_f < \omega_i$ gives Stokes scattering, as illustrated in Figure 6.3a. There is no Raman effect corresponding to the case $\omega_f > \omega_i$ (Figure 6.3b), yielding thereby Rayleigh (or elastic) scattering. The case of increased emitted photon energy, that is, $\omega_f > \omega_i$, provides anti-Stokes scattering, as shown in Figure 6.3c.

In quantum mechanical interpretation, during Raman scattering (Figure 6.3), when a photon is incident upon the nanodot, the nanodot itself absorbs the photon and jumps from the lower-energy state $|e\rangle$ to the excited state $|t\rangle$. Subsequently, the nanodot emits a photon of different frequencies (or energy) and decays to another lower state $|g\rangle$.

6.3.1.2 Cavity QED

Cavity QED can be defined as a theory of electromagnetic interactions between charged particles and electromagnetic fields in the presence of some macroscopic bodies like mirrors (metallic or dielectric), cavity walls, or waveguides [49]. The charge particle, typically free or bound electron in atom, is described by nonrelativistic quantum mechanics, and the electromagnetic field is quantized. Although cavity QED leads to many striking features, here the discussion is limited to the properties of vacuum, electromagnetic field quantization, and spontaneous emission.

6.3.1.2.1 *Properties of Vacuum in Cavity QED*

In cavity QED, the properties of vacuum are also greatly changed in the process of quantizing the electromagnetic field. The vacuum itself is no longer the state devoid of any fields. Instead, it is a state of the lowest energy level where *expectation values* of all fields are zero. But, the nonvanishing fluctuations of the fields lead to many observable effects such as spontaneous emission and Casimir force. These fluctuations are position dependent as they are largely affected by the presence of macroscopic object (cavity in the present case). The properties of vacuum, in this case, are different from those generally observed in the situation of free space. For example, the vacuum is no longer invariant under the Poincaré group transformations, because the homogeneity and isotropy properties of space-time are already destroyed by the presence of cavity. Furthermore, energy, momentum, and angular momentum of vacuum are no longer zero; they are dependent on the configuration of the cavity [49,50].

6.3.1.2.2 *Electromagnetic Field Quantization in Cavity QED*

In cavity QED, the electric/magnetic fields are operators and can be expressed as [49]

$$\hat{\vec{E}}(\vec{r},t) = \sum_i \left(\vec{E}_i(\vec{r},t)\,\hat{a}_i + \vec{E}_i^*(\vec{r},t)\,\hat{a}_i^\dagger \right) \tag{6.1a}$$

and

$$\hat{\vec{B}}(\vec{r},t) = \sum_i \left(\vec{B}_i(\vec{r},t)\,\hat{a}_i + \vec{B}_i^*(\vec{r},t)\,\hat{a}_i^\dagger \right) \tag{6.1b}$$

where the pairs of vectors $\vec{E}_i(\vec{r},t)$ and $\vec{B}_i(\vec{r},t)$ as well as $\vec{E}_i^*(\vec{r},t)$ and $\vec{B}_i^*(\vec{r},t)$ are *c*-number solutions to Maxwell equations. Because of the canonical commutation relations between the field operators, bosonic annihilation and creation operators $\hat{a}_i$ and $\hat{a}_i^\dagger$, respectively, obey the commutation relationships

$$\left[\hat{a}_i, \hat{a}_j^\dagger \right] = \delta_{ij} \tag{6.2}$$

and

$$\left[\hat{a}_i, \hat{a}_j \right] = 0 = \left[\hat{a}_i^\dagger, \hat{a}_j^\dagger \right] \tag{6.3}$$

If the cavity is at rest, the harmonic functions of time can be taken as mode functions, and thus, one can write

$$\vec{E}_i(\vec{r},t) = e^{-i\omega_i t}\vec{E}_i(\vec{r}) \tag{6.4a}$$

and

$$\vec{B}_i(\vec{r},t) = e^{-i\omega_i t}\vec{B}_i(\vec{r}) \tag{6.4b}$$

In this condition, the Hamiltonian operator of electromagnetic field can be expressed as

$$\hat{H} = \sum_j \omega_i \left(\hat{a}_i^\dagger \hat{a}_i + \frac{1}{2} \right) \tag{6.5}$$

6.3.1.2.3 Spontaneous Emission in Cavity QED

Spontaneous emission of radiation is induced by vacuum fluctuations [49,51] and leads to transitions between atoms or nanodots. As the cavity configuration significantly changes the vacuum fluctuations, it may greatly influence the spontaneous emission inside or near cavity. Modified and inhibited spontaneous emission of radiation has been experimentally verified [52–54].

Using Fermi's golden rule, one can estimate the spontaneous decay rate γ of an atomic excitation:

$$\gamma = 0.5 \left| \langle f | \hat{H} | i \rangle \right|^2 \rho(\omega_0) \tag{6.6}$$

with $\hat{H}$ as the system Hamiltonian, $|i\rangle$ and $|f\rangle$, respectively, as the initial and the final energy states of the atom; and $\rho(\omega_0)$ as the density of photon states for the transition frequency ω_0 between the atomic states. This density function depends on the boundary condition (i.e., the cavity configuration in the present case). In space, the function is given by

$$\rho_f(\omega) = \frac{\omega^2}{\pi^2 c^3} \tag{6.7}$$

Around the characteristic cavity frequency of a lossy cavity, the photon state density function $\rho_c(\omega)$ is given by the equation

$$\rho_c(\omega) = \frac{\Gamma}{\pi V} \frac{1}{\Gamma^2 + (\omega - \omega_c)^2} \tag{6.8}$$

with Γ and V, respectively, as the cavity loss rate and the cavity volume. Finally, the spontaneous decay rate in a cavity is given (in terms of spontaneous decay rate γ_f) by

$$\gamma_c = \frac{\rho_c(\omega_0)}{\rho_f(\omega_0)} \gamma_f \tag{6.9}$$

6.3.1.3 Rabi Oscillation

If a laser is applied on a two-level quantum system prepared in the ground state and if the laser mode frequency is resonant with the transition from the ground state to the excited state, the system absorbs a photon from the laser and makes a transition to the excited state. If the laser remains on while the system is in the excited state, a strange quantum mechanical activity happens—the (excited) system emits a photon into the laser mode with the same phase and frequency of the electromagnetic wave. Thus, with continuous application of laser of appropriate frequency on a two-level system, the system itself repeatedly absorbs and emits photons. This cyclic behavior of a two-level quantum system in the presence of an oscillatory driving field is called as Rabi oscillation. One cycle of absorption and emission is called as Rabi cycle, and the inverse of the duration of a Rabi cycle is termed as Rabi frequency of photon beam.

The time-dependent Hamiltonian of a two-level atomic system-driven oscillating electromagnetic field is given by [55]

$$
\hat{H} = \begin{bmatrix} \dfrac{1}{2}\omega_0 & \Omega_R e^{i\omega t/2} \\[2mm] \Omega_R e^{-i\omega t/2} & -\dfrac{1}{2}\omega_0 \end{bmatrix} \tag{6.10}
$$

In Equation 6.10, ω_0 is the atomic transition frequency and Ω_R is Rabi frequency. By solving the time-dependent Schrödinger equation with this Hamiltonian and assuming that the atom was initially in the ground state, the probability of finding the atom in the excited state can be found as a function of time as follows:

$$
P_e(t) = \left(\frac{\Omega_R}{\Omega_R'}\right)^2 \sin^2\left(\frac{\Omega_R'}{2}t\right) \tag{6.11}
$$

where Ω_R' is the effective Rabi frequency given by

$$
\Omega_R' = \sqrt{\Omega_R^2 + (\omega_0 - \omega)^2} = \sqrt{\Omega_R^2 + \delta^2} \tag{6.12}
$$

with δ as the detuning from the resonance frequency. The atom in the field oscillates between the two states at this effective Rabi frequency.

6.3.1.4 Spin State of a Nanodot

Following the proposal of *Loss–DiVincenzo quantum computer* in Ref. [23], the spin state of a nanodot is generally the spin state of an electron confined in the nanodot. As a fermion, electron has an intrinsic spin $s = 1/2$, which, in turn, gives the z component of angular momentum $S_z = m_s\hbar$ with $m_s = \pm 1/2$. Thus, two basis spin states for election may be obtained as

$$
|\uparrow\rangle \equiv \left|m_s = +\frac{1}{2}\right\rangle \tag{6.13a}
$$

and

$$
|\downarrow\rangle \equiv \left|m_s = -\frac{1}{2}\right\rangle \tag{6.13b}
$$

Typically, the ground-state electron is considered. If a static magnetic field is applied on the nanodot, the electronic ground state is split into two separate energy sublevels corresponding to the two spin states. If the lower and the higher sublevels are, respectively, denoted by $|g\rangle$ and $|e\rangle$, two basis states may be obtained for qubit realization using nanodot. A qubit can then be represented as a superposition of these two basis states as

$$|\psi\rangle = C_g|g\rangle + C_e|e\rangle \tag{6.14}$$

where C_g and C_e are scalars such that $|C_g|^2 + |C_e|^2 = 1$.

6.3.2 Interface Overview

The spin–photon interface consists of a nanodot and coupled cavity (Figure 6.4) along with other components of a point-to-point quantum channel—namely, a waveguide and a photon–spin interface to perform the reverse operation, that is, converting the flying photon state to the stationary spin state. The cavity is coupled to a fiber with a leakage constant k. As mentioned before, qubit is represented as a superposition of two degenerate ground states $|g\rangle$ and $|e\rangle$ split into two separate energy sublevels (upon application of a static magnetic field).

In order to transform the qubit state into photon state, a laser is applied on the nanodot, initiating thereby a Raman process. Following the work by Yao et al. [27], two higher-energy trion states $|t\rangle$ and $|t'\rangle$ can be defined. The cavity is designed in such a way that the cavity mode with frequency ω_c couples only with the transitions $|g\rangle \rightarrow |t\rangle$ and $|e\rangle \rightarrow |t'\rangle$. Similarly, the optical polarization of laser is set in such a way that the laser of frequency ω_L couples only with the transitions $|e\rangle \rightarrow |t\rangle$ and $|g\rangle \rightarrow |t'\rangle$ with complex Rabi frequency $\Omega(t)$. Raman process satisfies the resonance conditions

$$\omega_t = \omega_g + \omega_c \tag{6.15a}$$

and

$$\omega_t = \omega_e + \omega_L \tag{6.15b}$$

Here ω_t, ω_g, and ω_e are, respectively, frequencies corresponding to the energy levels of the states $|t\rangle$, $|g\rangle$, and $|e\rangle$. On the other hand, due to Zeeman splitting and selection rules, the other trion state $|t'\rangle$ off-resonantly couples with the cavity and laser modes.

6.3.3 Working Principle of Spin–Photon Interface

In spin–photon interface, the mapping of spin state to photon state is started with the application of a laser field of appropriate pulse shape and frequency on the nanodot, the

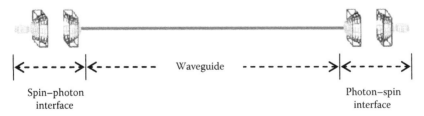

Spin–photon interface

Waveguide

Photon–spin interface

FIGURE 6.4
The spin–photon (and photon–spin) interface.

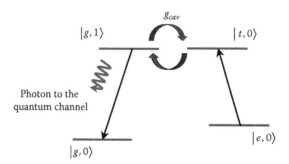

FIGURE 6.5
Photon generation through Raman transition.

fundamental controlling input depending on the pulse shape of the laser field. Figure 6.5 illustrates the state transitions during the generation of photon qubit through Raman process.

   As the laser field is incident upon the nanodot, a cavity-assisted Raman process raises the nanodot from the ground state $|e,0\rangle$ to the trion state $|t,0\rangle$ (with zero photon in the cavity) with a time-dependent complex Rabi frequency $\Omega(t)$. In Raman process, the trion state $|t,0\rangle$ is resonantly coupled to the cavity state $|g,1\rangle$ through coupling constant g_{cav}, where the state $|g,1\rangle$ is actually the state $|g\rangle$ with a photon in the cavity. Through this resonance coupling between the states $|t,0\rangle$ and $|g,1\rangle$, the nanodot emits a photon into the cavity resulting in a transition $|t,0\rangle \rightarrow |g,1\rangle$. Once the cavity mode is populated with a photon, the system is transformed from the state $|g,1\rangle$ to the ground spin state $|g,0\rangle$ through leakage of it into the associated quantum channel (optical fiber). The generated photon travels through the channel and reaches the receive node. This triggers a time-reversed sequence of actions in the receive node, and finally, the qubit state of the transmit node is restored into the receive node.

6.3.4 Mathematical Model

In order to develop a simulation platform for spin–photon interface, a suitable model involving spin states and transitions between them through Raman process must be devised. One may adopt the system model by Yao et al. [27], which concisely states the system behavior under the Weisskopf–Wigner approximation and provides a set of equations describing the evolution of system through Raman process. The overall quantum state of the system can be written as

$$|\Psi(t)\rangle = C_g|g,0\rangle|vac\rangle + C_e|\Psi^e(t)\rangle \tag{6.16a}$$

where

$$|\Psi^e(t)\rangle = \beta_e(t)|e,0\rangle|vac\rangle + \beta_t(t)|t,0\rangle|vac\rangle$$

$$+ \beta_c(t)|g,1\rangle|vac\rangle + \int_0^\infty d\omega\,\alpha_\omega(t)|g,0\rangle|\omega\rangle \tag{6.16b}$$

In Equations 6.16a and 6.16b, the notation $|vac\rangle$ stands for the vacuum state of electromagnetic mode and $|\omega\rangle$ denotes the one-photon Fock state of quantum channel mode with frequency ω. Further, it is noticed that the associated Hilbert space has two invariant subspaces, of which the subspace spanned by the basis $\{|e,0\rangle|vac\rangle, |t,0\rangle|vac\rangle, |g,1\rangle|vac\rangle, |g,0\rangle|\omega\rangle\}$ is of particular interest; the other subspace is spanned by the basis $\{|g,0\rangle|vac\rangle\}$. Also, β_e, β_t, and β_c are time-dependent Hilbert space parameters whose evolutions characterize the system behavior and must be tracked in order to estimate the output photon wave packet. The time evolution of these parameters during the resonant Raman interaction process, as given by Yao et al. [15], can be modeled (under the Weisskopf–Wigner approximation) as

$$\frac{d\beta_e}{dt} = \frac{-\Omega^*(t)\beta_t}{2} \tag{6.17a}$$

$$\frac{d\beta_t}{dt} = \frac{+\Omega(t)\beta_e}{2} + g_{cav}^*\beta_c \tag{6.17b}$$

$$\frac{d\beta_c}{dt} = \frac{-\gamma\beta_c}{2} - g_{cav}\beta_t - \sqrt{2\pi}\kappa^*\alpha_{in}(t) \tag{6.17c}$$

The incoming photon pulse $\alpha_{in}(t)$ at the spin–photon interface is assumed to be 0. In this condition, the outgoing photon pulse $\alpha_{out}(t)$ is given by

$$\alpha_{out}(t) = \frac{\gamma\beta_c}{\sqrt{2\pi}k^*} \tag{6.18}$$

where $\gamma = 2\pi|\kappa|^2$ corresponds to the cavity damping rate with κ as the loss rate due to the cavity–fiber coupling.

6.3.5 Simulation Platform

The simulation platform [56,57] consists of two major components, namely, control and evolution (Figure 6.6). The control unit is responsible for timing, configuration, and the overall control of simulation. The evolution unit updates Hilbert space parameters and the output photon wave packet at every step of simulation time provided by the control unit. The dimensionless time γt is used for simulation time. Since the system initially remains in the ground state $|e,0\rangle$, the simulation initialization is done as

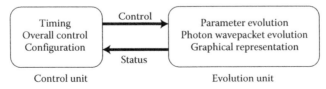

FIGURE 6.6
Overview of the simulation platform.

$$\beta_e(0) = 1.0 \tag{6.19a}$$

$$\beta_t(0) = 0.0 \tag{6.19b}$$

$$\beta_c(0) = 0.0 \tag{6.19c}$$

For the simulation purpose, the value of coupling constant κ is set to 5.64×10^{-3} eV and the nanodot–cavity coupling constant g_{cav} to 0.1×10^{-3} eV.

For simplification, the time evolution of quantum system may be approximated by piecewise linear functions, that is, the functions $\beta_e(t)$, $\beta_t(t)$, and $\beta_c(t)$ bare linear in the intervals between the successive points in time provided by the control unit. At every step of simulation time, the Hilbert space parameters $\beta_e(t)$, $\beta_t(t)$, and $\beta_c(t)$ are updated as

$$\beta_e(t + \Delta t) = \beta_e(t) + \frac{d\beta_e}{dt} \cdot \Delta t \tag{6.20a}$$

$$\beta_t(t + \Delta t) = \beta_t(t) + \frac{d\beta_t}{dt} \cdot \Delta t \tag{6.20b}$$

$$\beta_c(t + \Delta t) = \beta_c(t) + \frac{d\beta_c}{dt} \cdot \Delta t \tag{6.20c}$$

The time derivatives of $\beta_e(t)$, $\beta_t(t)$, and $\beta_c(t)$ are governed by the respective Equations 6.17a through 6.17c. Because of the aforesaid linear approximation, these derivatives are constant during the intervals between successive simulation points in time. Finally, the photon output $\alpha_{out}(t)$ is updated at every step of time, according to Equation 6.18.

6.3.6 Simulation Results

Simulations have been performed with different controlling laser pulse shapes, an example of which is illustrated in Figure 6.7 showing Rabi frequencies corresponding to the pulse. The application of laser pulse upon nanodot triggered a Raman transition, resulting in the evolution of β_e, β_t, and β_c. Figure 6.8 shows the evolution of β_e. As noticed, after the

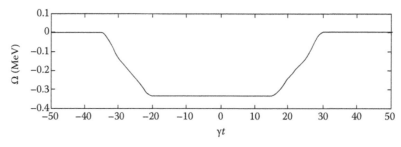

FIGURE 6.7
Pulse shape of laser input as a function of the dimensionless time γt.

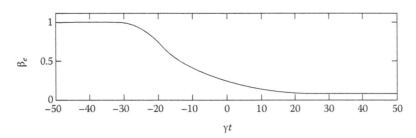

FIGURE 6.8
Time evolution of β_e for the input pulse in Figure 6.7.

start of laser pulse, the value of β_e started decreasing, which continued until it approached almost zero. The decrease of β_e from 1.0 to almost 0 can be interpreted as the transition of atom from the $|e,0\rangle|vac\rangle$ state to some other state that can be determined by examining the other two parameters β_t and β_c. Figure 6.9 shows the evolution of β_t. It can be observed that after the start of the input laser pulse, β_t decreased from 0 toward −0.5. That is, the squared magnitude $|\beta_t|^2$ increased from 0 to upward. Therefore, if Figures 6.8 and 6.9 are observed together, it can be understood that the occupation probability of the state $|e,0\rangle|vac\rangle$ decreased while that of the $|t,0\rangle|vac\rangle$ state increased. This shows that the system made a transition from the state $|e,0\rangle|vac\rangle$ to the state $|t,0\rangle|vac\rangle$.

Figure 6.10 shows that around the dimensionless time $\gamma t = -20$ (i.e., after applying the laser pulse), the value of β_c started increasing. $|\beta_c|^2$ represents the probability of occupation of the state $|g,1\rangle|vac\rangle$, and it increased with the increase in β_c. Thus, it is observed that

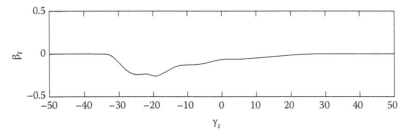

FIGURE 6.9
Time evolution of β_t for the input pulse in Figure 6.7.

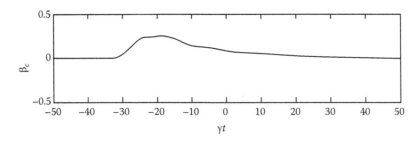

FIGURE 6.10
Time evolution of β_c for the input pulse in Figure 6.7.

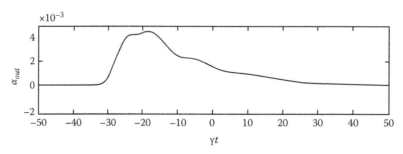

FIGURE 6.11
Dimensionless amplitude of the output wave packet for the input pulse in Figure 6.7.

the system would be in the $|g\rangle$ state with a photon in the cavity. Comparing Figures 6.10 and 6.11, one may notice that $|\beta_t|^2$ started increasing before $|\beta_c|^2$, and the increase in $|\beta_t|^2$ is followed by the increase in $|\beta_c|^2$. As such, if Figures 6.8 through 6.10 are seen together, it becomes explicit that with the application of laser, the system first made a transition from the state $|e,0\rangle|vac\rangle$ to the state $|t,0\rangle|vac\rangle$ and then another transition happened from the state $|t,0\rangle|vac\rangle$ to the state $|g,1\rangle|vac\rangle$. Thus, the system finally occupied the $|g\rangle$ state with the emission of a photon into the cavity.

After the dimensionless time $\gamma t = -20$, the laser pulse became flat at the minimum value. After this point of time, the value of β_t started increasing toward 0, that is, $|\beta_t|^2$ started decreasing. Afterward (around $\gamma t = 0$), these two parameters became zero. However, it was noticed before that $|\beta_e|^2$ also approached zero at this time. Therefore, the occupation probability of the remaining sate $|g,0\rangle|\omega\rangle$ must have increased significantly. This means that the emitted photon (into the cavity) leaked out of the cavity and propagated to the receive node; Figure 6.11 shows the output photon wave packet.

6.4 Framework for Quantum Communication

Quantum systems require in-depth theoretical analysis for both design and implementation phases. In particular, a comprehensive theoretical framework is essential to develop a quantum communication system. Several schemes have been proposed in the literature for transferring quantum states between spatially separated systems using mostly photon as the mediator. As stated before, Cirac et al. [26] proposed a well-accepted approach for transferring qubit from one quantum node to another, where each of the quantum nodes consisted of an atom inside a cavity, and time-symmetric photon wave packet acted as the information carrier. Yao et al. [27] enhanced the approach by removing the requirement of time symmetry and improving the controllability through the shape of the controlling laser pulse. However, as mentioned earlier, both of these schemes suffered from a common drawback—namely, while transferring the qubit state, the transmit node switched the state by itself. This motivates for the proposition of an approach in order to overcome the problem.

This section presents a theoretical framework with detailed mathematical formulation that would enable the implementation of the aforesaid approach for quantum communication. A comprehensive framework is developed mainly within the premise of quantum mechanics. In the development process, suitable quantum systems are first determined to represent stationary and flying qubit states. The next step is to design the transmit/receive nodes capable of quantum communication—namely, the mapping of the stationary qubit state (i.e., stored qubit state at the transmit node) to the flying qubit state (photon state), and vice versa, in addition to the capability of storing the qubit state at nodes. Also, a suitable physical medium is selected that would guide the flying qubit from the transmit to the receive node. The framework primarily includes the construction of relevant Hilbert space, the formulation of system Hamiltonian, the construction of density matrix, and specifying the system evolution.

6.4.1 System Overview

Figure 6.12 illustrates a quantum communication system comprised of transmit node, communication medium, and receive node. The transmit node consists of an atom trapped at the center of a two-mode optical cavity, whereas the receive node consists of an atom placed at the cavity center. The atoms at both nodes couple to the corresponding cavity modes through appropriate coupling constants. Both the nodes are connected through a fiber waveguide (through appropriate leakage constants). Qubit states at these nodes are represented by atomic energy levels.

Application of lasers in the transmit node initiates a Raman process by exciting the atom to a higher energy level. Once the atom completes transition to the higher energy level, lasers are turned off and the atom decays to a lower energy level, emitting thereby a photon into the cavity. Thus, the qubit state of the transmit node is mapped into the photon state (or the flying qubit state). Through cavity–fiber coupling, the emitted photon leaks out, travels through the fiber, and eventually reaches the receive node wherein a reverse process is initiated. The atom in the receive node absorbs the (incoming) photon and makes a transition to a higher energy level. When the atom completes the transition, lasers in the receive node are turned on. Application of lasers with this timing makes the atom to emit a photon into the appropriate laser mode and jump down to a lower energy level corresponding to the qubit state of the received photon. Thus, the qubit state of the transmit node is transferred to the receive node.

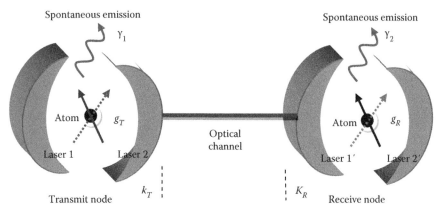

FIGURE 6.12
Overview of the system.

6.4.2 Physics of the System

Physics underlying the quantum processes at the transmit and the receive nodes includes architectural decisions such as qubit representation, cavity mode, and laser requirements for cavity QED design and specifications of the system working principles.

6.4.2.1 Transmit Node

The transmit node is responsible for realizing a stationary qubit and transmitting the qubit state into the waveguide on demand. While designing the transmit node, it is important to address the issues like realization of stationary qubit states, identification of flying qubit carrier, and selection of appropriate lasers. This section addresses these issues and presents a comprehensive design and mathematical formulation for the transmit node.

6.4.2.1.1 Qubit Realization

Qubit has been traditionally represented by two-level systems [26]. These two levels ($|1\rangle$ and $|2\rangle$) represent two basis states for the qubit, and the qubit state is represented as an arbitrary superposition of these basis states as

$$|\psi\rangle = C_1|1\rangle + C_2|2\rangle \tag{6.21}$$

where
    $|\psi\rangle$ is the qubit state
    C_1 and C_2 are two scalars such that

$$|C_1|^2 + |C_2|^2 = 1 \tag{6.22}$$

In some of the existing qubit representation systems, the aforesaid two levels are implemented by atomic energy levels [26,40], while in other systems, by two spin states of nanodot, split by the application of a static magnetic field [27,58].

In the investigative work, the hyperfine states of rubidium atom (^{87}Rb) are utilized to represent qubit states. In the transmit node, however, instead of using each hyperfine state as an individual qubit basis, the two subspaces S_0 and S_1 are defined in the Hilbert space of the hyperfine states of ^{87}Rb, and the logical basis of qubit into these subspaces is encoded following the principles given in Ref. [59]. This representation of qubit is adopted in order to overcome the drawback of the proposals in Refs. [26,27]—namely, when the transmit node (in the proposals) transmitted qubit state into the waveguide, the transmit node toggled its own qubit state. Although the qubit state can be represented by any superposition of the basis states, in the investigation addressed here, the qubit state is limited to two logical states "0" and "1" only. Thus, in the present system, the logic "0" is represented by the subspace S_0, whereas the logic "1" by the subspace S_1.

6.4.2.1.2 Rubidium Hyperfine Levels

Rubidium atom is an important ingredient in many quantum optics experiments [40,60]. A rubidium atom has the electronic configuration given as $1s^2 2s^2 p^6 3s^2 p^6 d^{10} 4s^2 p^6 5s^1$.

The coupling between the orbital angular momentum **L** of the outermost electron and its spin angular momentum **S** results in fine structure energy level splitting. The total angular momentum of electron is given by **J = L+S**, while the corresponding quantum number J lies in the range $|L - S| \leq J \leq |L+S|$. Again, the coupling between the total electronic angular momentum **J** and the total nuclear angular momentum **I** results in the hyperfine structure with the total atomic angular momentum **F = J+I**. The corresponding quantum number F lies in the range $|J - I| \leq F \leq |J+I|$.

Selected hyperfine energy levels of ^{87}Rb atom are shown in Figure 6.13 wherein the symbols are interpreted as follows.

The first number "5" is the principal quantum number, the superscript "2" is coming from 2S+1 (i.e., $S = 1/2$), the letter "S" refers to the value of L (e.g., S means $L = 0$, P means $L = 1$), and the subscript "1/2" is the value of J. As seen in Figure 6.13, the $5^2S_{1/2}$ energy level is split into two hyperfine levels $F = 1$ and $F = 2$. Similarly, the energy levels corresponding to $5^2P_{1/2}$ and $5^2P_{3/2}$ are split into two ($F = 1$ and $F = 2$) and four ($F = 0$, $F = 1$, $F = 2$, and $F = 3$) hyperfine levels, respectively.

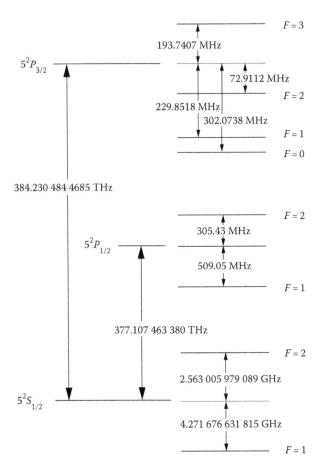

FIGURE 6.13
Selected hyperfine levels of rubidium ^{87}Rb atom. (From Steck, D.A., Rubidium-87 D line data (rev. 2.1.1), http://steck.us/alkalidata/rubidium87numbers.pdf, 2009.)

Further, each of these hyperfine energy levels contains $2F+1$ magnetic sublevels (m_F), where F is the hyperfine energy level. For example, the hyperfine level $F=2$ contains five magnetic sublevels with the magnetic quantum numbers $m_F = -2, -1, 0, 1, 2$. In the absence of a magnetic field, these sublevels are degenerate. However, the application of static magnetic field on atom removes the degeneracy (due to Zeeman effect), and the hyperfine energy level is split into $2F+1$ magnetic sublevels. For small magnetic fields, these energy levels split linearly and can be approximated by the linear equation [60]:

$$\Delta E_{|F,m_F\rangle} = \mu_B g_F m_F B_Z \qquad (6.23)$$

where
$\Delta E_{|F,m_F\rangle}$ is the energy shift due to Zeeman effect for the hyperfine level F
m_F is the magnetic quantum number
μ_B is the Bohr magneton
g_F is the hyperfine Landé g-factor
B_z is the z component of the magnetic field

The energy differences (i.e., transition frequencies) between various hyperfine levels of ^{87}Rb can be found from Figure 6.13. In addition, using Equation 6.23, one can calculate the energy differences between the magnetic sublevels due to the applied magnetic field. Now, as the necessary data corresponding to the hyperfine structure and the magnetic sublevels of ^{87}Rb atom are ready, one may proceed to identify the hyperfine states for representing qubit states.

6.4.2.1.3 ^{87}Rb Hyperfine Levels for Qubit Realization

The logic states "0" and "1" in the transmit node are represented using two subspaces S_0 and S_1 of the hyperfine state space of ^{87}Rb. The subspace S_0 is spanned by the hyperfine basis states:

$$\left.\begin{array}{l}
|\sigma_{00}\rangle \equiv (5^2 S_{1/2}, F=1, m_F=0) \\[6pt]
|\sigma_{01}\rangle \equiv (5^2 P_{1/2}, F=1, m_F=0) \\[6pt]
|\sigma_{02}\rangle \equiv (5^2 P_{1/2}, F=2, m_F=1)
\end{array}\right\} \qquad (6.24)$$

Similarly, the subspace S_1 is spanned by the basis states:

$$\left.\begin{array}{l}
|\sigma_{10}\rangle \equiv (5^2 S_{1/2}, F=1, m_F=1) \\[6pt]
|\sigma_{11}\rangle \equiv (5^2 P_{1/2}, F=1, m_F=1) \\[6pt]
|\sigma_{12}\rangle \equiv (5^2 P_{3/2}, F=2, m_F=0)
\end{array}\right\} \qquad (6.25)$$

The magnetic sublevels in these subspaces are split by the application of a static magnetic field. These two subspaces, each spanned by three hyperfine states, are pictorially illustrated in Figure 6.13. It can be observed that, in the subspace S_0, the state $|\sigma_{02}\rangle$ is at the highest energy level, while $|\sigma_{00}\rangle$ at the lowest. Also, $|\sigma_{01}\rangle$ is at the intermediate energy level. Thus, the state $|\sigma_{00}\rangle$ is designated as the ground state, $|\sigma_{02}\rangle$ as the excited state, and $|\sigma_{01}\rangle$ as an intermediate state of the subspace S_0. Similarly, $|\sigma_{10}\rangle$, $|\sigma_{11}\rangle$, and $|\sigma_{12}\rangle$, respectively, are the ground, excited, and intermediate states of the S_1 subspace. With these subspaces, the qubit representation in Equation 6.21 can be replaced by the following representations:

$$|\psi\rangle = C_0|\psi_0\rangle + C_1|\psi_1\rangle \tag{6.26a}$$

where

$$|\psi_0\rangle = C_{00}|\sigma_{00}\rangle + C_{01}|\sigma_{01}\rangle + C_{02}|\sigma_{02}\rangle \tag{6.26b}$$

$$|\psi_1\rangle = C_{10}|\sigma_{10}\rangle + C_{11}|\sigma_{11}\rangle + C_{12}|\sigma_{12}\rangle \tag{6.26c}$$

$$\sum_i C_{0i}^2 = 1 \quad \text{and} \quad \sum_i C_{1i}^2 = 1 \tag{6.26d}$$

In the earlier equations, $|\psi\rangle$ is the qubit state, which is a superposition of the states $|\psi_0\rangle$ and $|\psi_1\rangle$. The state $|\psi_0\rangle$, in turn, is a superposition of the hyperfine states of the subspace S_0, as defined by Equation 6.26b. Similarly, $|\psi_1\rangle$ is the superposition state in subspace S_1, as defined by Equation 6.26c. It is assumed that the probabilities of occupation of the other states of ^{87}Rb atom are negligible. With the limitation of the qubit values to logic "0" and logic "1" only, the atom is considered to be in the logic "0" state, if $C_0^2 \gg C_1^2$. Similarly, the condition $C_1^2 \gg C_0^2$ implies the atom to be considered in the logic "1" state.

6.4.2.1.4 Laser and Cavity Modes

In order to transform qubit state in the transmit node into photon state, ^{87}Rb atom needs to be excited from the ground state to the excited state within the subspace so that, while decaying back to the ground state, it emits a photon. Lasers are required to initiate these intra-subspace transitions. Two lasers, one for each subspace, could be used to raise the atom to the excited states through transitions $|\sigma_{00}\rangle \rightarrow |\sigma_{02}\rangle$ or $|\sigma_{10}\rangle \rightarrow |\sigma_{12}\rangle$ depending on the logic state of the transmit node, and then the atom could make transition to the corresponding ground state through the emission of a photon into the cavity mode. However, in that case, the laser frequency would be in resonance with the cavity mode frequency, and this would initiate undesirable complex Rabi oscillations with multiphoton states in the cavity. In order to avoid this, four lasers can be used in two steps. As seen in Figure 6.14, the laser $L_{01}^{(0)}$ is resonant to the transition from $|\sigma_{00}\rangle$ to $|\sigma_{01}\rangle$ and the laser $L_{12}^{(0)}$ to $|\sigma_{01}\rangle$ to $|\sigma_{02}\rangle$. Thus, the application of these two lasers initiates the transitions $|\sigma_{00}\rangle \rightarrow |\sigma_{01}\rangle$ and $|\sigma_{01}\rangle \rightarrow |\sigma_{02}\rangle$. Similarly, the laser $L_{01}^{(1)}$ initiates the transition $|\sigma_{10}\rangle \rightarrow |\sigma_{11}\rangle$ and the laser $L_{12}^{(1)}$, the transition $|\sigma_{11}\rangle \rightarrow |\sigma_{12}\rangle$. In Equations 6.24 and 6.25, it is seen that there are no changes in the magnetic quantum numbers m_F during the transitions $|\sigma_{00}\rangle \rightarrow |\sigma_{01}\rangle$ and $|\sigma_{10}\rangle \rightarrow |\sigma_{11}\rangle$. Therefore, π-polarized photons from lasers are absorbed during these transitions. On the other hand, during the transition $|\sigma_{01}\rangle \rightarrow |\sigma_{02}\rangle$, the value of m_F increases by 1, whereas during the transition $|\sigma_{11}\rangle \rightarrow |\sigma_{12}\rangle$, it decreases by 1. Therefore, right and left circularly polarized photons are absorbed

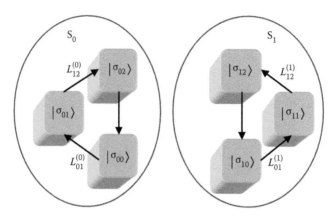

FIGURE 6.14
Subspaces of the Hilbert space of ^{87}Rb hyperfine states to represent logic "0" and logic "1."

during the transitions $|\sigma_{01}\rangle \rightarrow |\sigma_{02}\rangle$ and $|\sigma_{11}\rangle \rightarrow |\sigma_{12}\rangle$, respectively. Accordingly, these four lasers should consist of photons of appropriate frequencies and polarizations. Finally, during decay from the excited state $|\sigma_{02}\rangle$ to the ground state $|\sigma_{00}\rangle$, m_F decreases by 1, and during decay from $|\sigma_{12}\rangle$ to $|\sigma_{10}\rangle$, it increases by 1. Therefore, right and left circularly polarized photons emit during the transitions $|\sigma_{02}\rangle \rightarrow |\sigma_{00}\rangle$ and $|\sigma_{12}\rangle \rightarrow |\sigma_{10}\rangle$, respectively.

The decay transitions $|\sigma_{02}\rangle \rightarrow |\sigma_{00}\rangle$ and $|\sigma_{12}\rangle \rightarrow |\sigma_{10}\rangle$ are not initiated by lasers. In order to enhance these transitions and collect the emitted photons into the cavity, it is designed with two modes with one matching in frequency with the transition frequency of $|\sigma_{02}\rangle \rightarrow |\sigma_{00}\rangle$ and the other with that of $|\sigma_{12}\rangle \rightarrow |\sigma_{10}\rangle$.

6.4.2.1.5 Working Principle in the Transmit Node

At the start of quantum operations at the transmit node, that is, generation and transmission of photon wave packets, the transmit node is either in the logic "0" state or in the logic "1" state, and accordingly, ^{87}Rb atom is at the ground state $|\sigma_{00}\rangle$ or $|\sigma_{10}\rangle$, respectively. The fundamental working principle is to initiate a Raman process by raising the atom to the excited state $|\sigma_{02}\rangle$, if the initial state is $|\sigma_{00}\rangle$, or to the excited state $|\sigma_{12}\rangle$ otherwise and then allowing it to decay to the ground state $|\sigma_{00}\rangle$ or $|\sigma_{10}\rangle$, respectively, emitting thereby a photon of appropriate frequency and polarization into the cavity. In order to achieve this, the first two lasers $L_{01}^{(0)}$ and $L_{01}^{(1)}$ are *simultaneously* applied on the atom. If the current logic state is "0," the atom absorbs a photon from the laser $L_{01}^{(0)}$; otherwise, it absorbs a photon from the laser $L_{01}^{(1)}$. Depending on the frequency and polarization of the absorbed photon, the excited atom makes a transition to the intermediate state $|\sigma_{01}\rangle$ or $|\sigma_{11}\rangle$. At this stage, the lasers $L_{01}^{(0)}$ and $L_{01}^{(1)}$ must be turned off. Otherwise, due to Rabi oscillations, the atom will come back to the appropriate ground state depending on the initial logic state. Therefore, the lasers $L_{01}^{(0)}$ and $L_{01}^{(1)}$ are turned off, and the other two lasers $L_{12}^{(0)}$ and $L_{12}^{(1)}$ are *simultaneously* turned on. Now, ^{87}Rb atom absorbs the second photon from the laser $L_{12}^{(0)}$ or $L_{12}^{(1)}$, depending on the logic state, and undergoes the transition $|\sigma_{01}\rangle \rightarrow |\sigma_{02}\rangle$ or $|\sigma_{11}\rangle \rightarrow |\sigma_{12}\rangle$. Once the transition is complete, the lasers $L_{12}^{(0)}$ and $L_{12}^{(1)}$ are turned off. Now, due to spontaneous emission and coupling of atom to the appropriate cavity mode, the atom decays to $|\sigma_{00}\rangle$ or $|\sigma_{10}\rangle$ state through transitions $|\sigma_{02}\rangle \rightarrow |\sigma_{00}\rangle$ or $|\sigma_{12}\rangle \rightarrow |\sigma_{10}\rangle$, depending on the logic state. In the decay process, the atom emits a right circularly polarized photon in the former case and a left circularly polarized one in the latter.

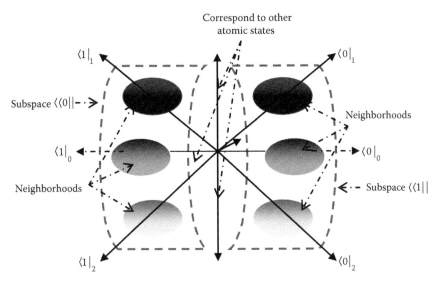

FIGURE 6.15
(See color insert.) Neighborhoods in the subspaces S_0 and S_1 of the Hilbert space.

The generated photon leaks out of cavity into a coupled optical fiber and propagates toward the receive node.

It is to be noted down that while communicating stationary qubit state from the transmit node to the receive one, the exact cyclic transitions $|\sigma_{00}\rangle \rightarrow |\sigma_{01}\rangle \rightarrow |\sigma_{02}\rangle \rightarrow |\sigma_{00}\rangle$ and $|\sigma_{10}\rangle \rightarrow |\sigma_{11}\rangle \rightarrow |\sigma_{12}\rangle \rightarrow |\sigma_{10}\rangle$ are forbidden [45,56]. Although these exact cyclic transitions are not allowed, transitions back to the neighborhoods of the ground states $|\sigma_{00}\rangle$ and $|\sigma_{10}\rangle$ are allowed. Thus, the subspaces S_0 and S_1 practically consist of three neighborhoods around the corresponding hyperfine states $|\sigma_{00}\rangle$, $|\sigma_{01}\rangle$, and $|\sigma_{02}\rangle$ (for S_0) and $|\sigma_{10}\rangle$, $|\sigma_{11}\rangle$, and $|\sigma_{12}\rangle$ (for S_1). These neighborhoods around the six hyperfine states are pictorially illustrated in Figure 6.15.

6.4.2.1.6 Framework Summary

Now the design of the transmit node can be concisely described. The node consists of a ^{87}Rb atom trapped at the center of a two-mode cavity. The two subspaces S_0 and S_1 constitute two logical basis states. The qubit is represented as a superposition of these two basis states. Mapping of the stationary qubit state into the photon state is achieved through a Raman process. Application of two lasers $L_{01}^{(0)}$ and $L_{01}^{(1)}$, followed by application of the other two lasers $L_{12}^{(0)}$ and $L_{12}^{(1)}$, raises the atom from the $|\sigma_{00}\rangle$ state to $|\sigma_{02}\rangle$ via the intermediate state $|\sigma_{01}\rangle$, all within the subspace S_0, if the atom was initially in the logic "0" state. Otherwise the atom is raised from the state $|\sigma_{10}\rangle$ to $|\sigma_{12}\rangle$ via the intermediate state $|\sigma_{11}\rangle$ within the subspace S_1. The two-mode cavity is designed in such a way that the cavity mode frequencies match the transition frequencies of the transitions $|\sigma_{02}\rangle \rightarrow |\sigma_{00}\rangle$ and $|\sigma_{12}\rangle \rightarrow |\sigma_{10}\rangle$. Thus, due to spontaneous emission and strong coupling of $|\sigma_{02}\rangle \rightarrow |\sigma_{00}\rangle$ and $|\sigma_{12}\rangle \rightarrow |\sigma_{10}\rangle$ transitions with the two cavity modes, the atom at the $|\sigma_{02}\rangle$ or $|\sigma_{12}\rangle$ state decays to the corresponding ground states and emits thereby a photon of appropriate frequency and polarization into the cavity. Photon in the cavity eventually leaks and goes to the coupled fiber.

During transforming the qubit state of the transmit node into the photon state, probabilities of occupation of the subspaces S_0 and S_1 do not significantly change, that is, classically speaking, the atom does not switch from one subspace to the other. Thus, the proposed approach overcomes the drawbacks of the existing proposals, as in Refs. [26,27]—namely, toggling the qubit state of the transmit node while transmitting the same.

6.4.2.2 Receive Node

The receive node is responsible for transforming the received photon state into the stationary qubit state and storing it in the node. As mentioned before, the receive node consists of a ^{87}Rb atom placed at the center of a two-mode cavity. The atom and the cavity modes are coupled through an appropriate coupling constant. While designing the receive node, it is important to address the issues like realization of the stationary qubit states and selection of the appropriate lasers. This section addresses such issues and presents a comprehensive design and mathematical formulation for the receive node.

6.4.2.2.1 Qubit Realization

In the receive node, instead of two separate hyperfine subspaces, two hyperfine states within a single subspace S are used to represent the basis qubit states. This is in order to allow the atom to switch between logic states, if required, upon reception of a photon from the coupled fiber waveguide. Although it is desirable to retain the qubit state in the transmit node while transmitting the qubit, it is necessary for the receive node to switch to the qubit state of the transmit node, if already not in that state. The subspace S is spanned by the following hyperfine states of ^{87}Rb atom [61,62]:

$$\left.\begin{aligned}
|\phi_0\rangle &\equiv (5^2 S_{1/2}, F=1, m_F=1) \\
|\phi_0'\rangle &\equiv (5^2 P_{3/2}, F=2, m_F=0) \\
|\phi_1\rangle &\equiv (5^2 S_{1/2}, F=1, m_F=0) \\
|\phi_1'\rangle &\equiv (5^2 P_{1/2}, F=2, m_F=1)
\end{aligned}\right\} \tag{6.27}$$

Similar to the transmit node, the magnetic sublevels in these hyperfine states are split by the application of a static magnetic field. Out of the four hyperfine states within the subspace S, as shown in Equation 6.27, two states $|\phi_0\rangle$ and $|\phi_1\rangle$ represent qubit basis states corresponding to logic "0" and "1," respectively. These two states are ground states split by the application of a static magnetic field. The other two states $|\phi_0'\rangle$ and $|\phi_1'\rangle$ function as the excited states for the quantum processes in the receive node. These four hyperfine states, along with their transitions, are pictorially represented in Figure 6.16.

It is seen from Equation 6.24 that in the subspace S_0 (representing logic "0"), the ground state is $|\sigma_{00}\rangle \equiv (5^2 S_{1/2}, F=1, m_F=0)$, and from Equation 6.27, the hyperfine state $|\phi_1\rangle$ (representing logic "1") is $|\phi_1\rangle = (5^2 S_{1/2}, F=1, m_F=0)$. They are the same hyperfine states with the same magnetic sublevels. Thus, the same hyperfine state represents logic "0" in the transmit node and logic "1" in the receive node. Similarly, upon comparing the definition of the state $|\sigma_{10}\rangle$ in Equation 6.25 and of $|\phi_0\rangle$ in Equation 6.27, we find that the same hyperfine state $(5^2 S_{1/2}, F=1, m_F=1)$ represents logic "1" in the transmit node and logic "0" in the receive node. This difference in the representation of logic states between the transmit and the receive nodes is required because of a fundamental difference between the quantum processes in the two nodes. The difference is that although the atom in the

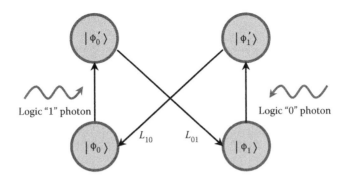

FIGURE 6.16
Relevant hyperfine states and interstate transitions in the receive node.

transmit node undergoes decay from the excited state to the ground state (while emitting a photon), the atom in the receive node is raised from the ground state to the excited state when it absorbs photon.

6.4.2.2.2 Laser and Cavity Modes

When a photon is received into the receive node cavity, the atom is raised to an excited state, if the selection rules for atomic transitions are met. It is known from quantum optics that, if a laser beam is applied on an excited atom, and if the laser frequency matches the transition frequency of an available decay path to a ground state, the atom emits a photon into the laser mode and decays to the ground state. Based on this principle, two lasers are needed in the receive node—one for causing transition from the state $|\phi_0'\rangle$ to the state $|\phi_1\rangle$ and the other for causing transition from $|\phi_1'\rangle$ to $|\phi_0\rangle$ state. These two lasers, denoted by L_{01} and L_{10}, are shown in Figure 6.16 along with the transitions $|\phi_0'\rangle \rightarrow |\phi_1\rangle$ and $|\phi_1'\rangle \rightarrow |\phi_0\rangle$, respectively. These transitions correspond to switching from logic "0" to logic "1," and vice versa. Naturally, the frequency and polarization of laser beams should match the transition frequency and the change in magnetic quantum number m_F in the corresponding atomic transitions.

The received photon into cavity represents either logic "0" or logic "1." There should be a cavity mode for each of these photon states so that the photon can populate the corresponding cavity mode and excite the atom to the corresponding excited state. This justifies the requirement of a two-mode cavity in the receive node. The mode corresponding to logic "0" (CM_R^0) supports frequency and polarization matching with the transition $|\phi_0\rangle \rightarrow |\phi_0'\rangle$, and corresponding to logic "1" (CM_R^1), that with the transition $|\phi_1\rangle \rightarrow |\phi_1'\rangle$.

6.4.2.2.3 Working Principle for Receive Node

Quantum process in the receive node is reverse of that which takes place in the transmit node. When a photon is received into the receive node cavity, based on the received flying qubit state and receive node stationary qubit state, one of the four possible scenarios would take place—(1) flying and stationary qubits represent logic "0," (2) flying qubit represents logic "0" and stationary qubit represents logic "1," (3) flying qubit represents logic "1" and stationary qubit represents logic "0," and (4) flying and stationary qubits represent logic "1."

In scenario 1, as the flying qubit represents logic "0," photon is generated due to the transition $|\sigma_{02}\rangle \rightarrow |\sigma_{00}\rangle$. In this representation, $|\sigma_{02}\rangle \equiv (5^2P_{1/2}, F=2, m_F=1)$ and $|\sigma_{00}\rangle \equiv (5^2S_{1/2}, F=1, m_F=0)$. Thus, frequency and polarization of photon match this transition.

From Equation 6.27, it is seen that this transition is equivalent to $|\phi_1'\rangle \rightarrow |\phi_1\rangle$, and thus, this photon can excite the atom only from $|\phi_1\rangle$ to $|\phi_1'\rangle$ state. As the cavity mode CM_R^1 supports this frequency and polarization, photon populates this cavity mode. However, the stationary qubit in the receive node represents logic "0," that is, the atom is in the $|\phi_0\rangle$ state. Now, due to selection rule of transition, the received photon cannot be absorbed by the atom. The photon rather decays over time due to the leakage of the cavity, and no change happens to the stationary qubit state of the receive node.

In scenario 2, the flying qubit state represents logic "0," implying thereby the photon matches the transition $|\sigma_{02}\rangle \rightarrow |\sigma_{00}\rangle$ in the transmit node. Comparing Equations 6.24 and 6.27, it is seen that this transition is equivalent to $|\phi_1'\rangle \rightarrow |\phi_1\rangle$ and, similar to scenario 1, the photon can excite the atom only from $|\phi_1\rangle$ to $|\phi_1'\rangle$ state. As the cavity mode CM_R^1 supports this photon frequency and polarization, photon populates this cavity mode. Now, because the stationary qubit state represents logic "1," the atom is in the $|\phi_1\rangle$ state, and it is excited to the state $|\phi_1'\rangle$ through absorbing photon. At this stage, if both the lasers L_{01} and L_{10} are turned on, the atom emits a photon into the L_{10} laser mode and decays to the ground state $|\phi_0\rangle$, which represents logic "0." Thus, the receive node switches its logic state from "1" to "0," meaning thereby the switching of the logic state of the transmit node.

In scenario 3, the flying qubit state represents logic "1." This implies that the received photon was generated through the transition $|\sigma_{12}\rangle \rightarrow |\sigma_{10}\rangle$. Comparing Equations 6.25 and 6.27, it is seen that this transition is equivalent to $|\phi_0'\rangle \rightarrow |\phi_0\rangle$, and thus, the received photon can excite the atom only from $|\phi_0\rangle$ to $|\phi_0'\rangle$ state. As the cavity mode CM_R^0 supports this mode, photon populates the mode. As the stationary qubit state represents logic "0," the atom is in the state $|\phi_0\rangle$ and is excited to the $|\phi_0'\rangle$ state through the absorption of photon. If both the lasers are turned on at this stage, the atom emits a photon into the L_{01} laser mode and decays to the ground state $|\phi_1\rangle$, which represents logic "1." Thus, the receive node, in this scenario, switches to the logic state of the transmit node.

In scenario 4, the flying qubit represents logic "1," and therefore, the photon was emitted during the transition from $|\sigma_{12}\rangle$ to $|\sigma_{10}\rangle$ state. Similar to scenario 3, this transition is equivalent to $|\phi_0'\rangle \rightarrow |\phi_0\rangle$, and photon can excite only from $|\phi_0\rangle$ to $|\phi_0'\rangle$ state. As the cavity mode CM_R^0 supports this mode, the photon populates the mode. However, as the stationary qubit state represents logic "1," the atom is in the state $|\phi_1\rangle$. Now, due to the restriction of the selection rule of transition, the received photon cannot be absorbed by the atom. The photon rather decays over time due to the leakage of the cavity, and no change happens to the stationary qubit state of the receive node.

From the earlier discussions on all the four possible scenarios, the operation strategy for the receive node can be determined. The strategy is to turn on the lasers for a predetermined amount of time after the reception of a photon. This delay in time is determined in such a way that the lasers turn on just after the atom is raised to the excited state, in case the receive node qubit state is different from that of the transmit node. As the receive node does not implement a photon detection device, the timing of turning on the lasers must be synchronized to the timings of the lasers in the transmit node. The timings of the lasers $L_{12}^{(0)}$ and $L_{12}^{(1)}$, and that of the lasers L_{01} and L_{10}, are set in such a way that the lasers L_{01} and L_{10} are turned on with the said delay constraints.

6.4.2.2.4 Framework Summary

The receive node consists of a ^{87}Rb atom trapped at the center of a two-mode optical cavity. A subspace S consisting of four hyperfine states in the Hilbert space of ^{87}Rb hyperfine states is identified. Out of these four hyperfine states, $|\phi_0\rangle$ and $|\phi_1\rangle$ are ground states split into different magnetic sublevels upon application of a static magnetic field. The other two states $|\phi_0'\rangle$ and $|\phi_1'\rangle$ are the excited states.

When a photon is received, if the receive node qubit state is the same as the qubit state of the transmit node, no change happens in the logic state, and photon decays due to the cavity leakage. On the other hand, if the receive node logic state is different from that of the transmit node, the atom absorbs photon and jumps to the excited state $|\phi_0'\rangle$ or $|\phi_1'\rangle$ depending on whether it was initially in the $|\phi_0\rangle$ or the $|\phi_1\rangle$ state. Exactly at this stage, the lasers L_{01} and L_{10} are simultaneously tuned on. This makes the atom to emit a photon into the L_{01} or L_{10} laser mode depending on whether it was in the $|\phi_0'\rangle$ or the $|\phi_1'\rangle$ excited state and switch to the ground state different from the initial ground state.

6.4.3 Mathematical Model

The mathematical model consists of density matrices to represent the system states and Hamiltonians as fundamental system operators to work on the density matrices to determine system evolutions. As the transmit and the receive nodes are independently controllable [27], separate mathematical models are proposed for these two nodes. Although the transmit and the receive nodes are given quantum mechanical treatment, the propagation of photon wave packet is treated classically.

6.4.3.1 Model Overview

In order to model the proposed system, the much popular Jaynes–Cummings model [63] is adopted. It provides a formulation for system Hamiltonian, which, along with a system state representation, describes the full system. This model is modified and extended to meet the requirements in order to deal with the hyperfine states, subspaces in the Hilbert space, and increased number of lasers in the system. In the proposed model, the quantum state of system is described by the density matrix formalism. The von Neumann equation [64] is adopted to determine the evolution of the density matrix for the initial closed-system analysis. Then the full-fledged open-system approach is adopted, and the master equation is used for the evolution of the density matrix in Lindblad form [65].

6.4.3.2 Jaynes–Cummings Model

In quantum optics, Jaynes–Cummings model [63] describes a two-level atomic system interacting with a quantized mode of an optical cavity. Modern usage of the model, generally, incorporates optical laser. In Jaynes–Cummings model, the Hamiltonian for a two-level atom is given by

$$\hat{H}_{JC} = \hat{H}_f + \hat{H}_a + \hat{H}_i \tag{6.28}$$

with $\hat{H}_f$ as the Hamiltonian for the radiation field, $\hat{H}_a$ as the Hamiltonian for the free atom, and $\hat{H}_i$ as the Jaynes–Cummings interaction Hamiltonian describing the interaction of atom with the radiation field.

6.4.3.2.1 Atom Hamiltonian

The Hamiltonian of a free atom is given by

$$\hat{H}_a = \sum_{i=1}^{n} E_i |E_i\rangle\langle E_i| \tag{6.29}$$

In Equation 6.29, the atom is considered to have n energy eigenstates $|E_i\rangle$; $i = 1, 2, \ldots, n$. However, in two-level approximation, which is applicable when the frequency of radiation field coincides with the relevant optical transition of atom, only two energy eigenstates are considered, and all the other eigenstates are neglected. Thus, in two-level approximation, the Hamiltonian of the atom is given by

$$\hat{H}_a = E_g |g\rangle\langle g| + E_e |e\rangle\langle e| \tag{6.30}$$

where
$|g\rangle$ and $|e\rangle$ are the ground and the excited states, respectively, of the two-level atom
E_g and E_e are, respectively, the energies of the ground and the excited levels

The energy difference between these two levels is

$$\Delta E = E_e - E_g = \omega_{eg}$$

If the zero point of the energy is taken at the midway between the $|g\rangle$ and the $|e\rangle$ states, the energy levels of the ground and the excited states are $-(1/2)\hbar\omega_{eg}$ and $(1/2)\hbar\omega_{eg}$, respectively. Thus, the atomic Hamiltonian is given by

$$\hat{H}_a = -\frac{1}{2}\omega_{eg}|g\rangle|g\rangle + \frac{1}{2}\hbar\omega_{eg}|e\rangle|e\rangle = \frac{1}{2}\hbar\omega_{eg}\left(|e\rangle|e\rangle - |g\rangle|g\rangle\right) \tag{6.31}$$

Equation 6.31 can be written in a further concise form as

$$\hat{H}_a = \frac{1}{2}\omega_{eg}\hat{\sigma}_z \tag{6.32}$$

where $\sigma_z = |e\rangle\langle e| - |g\rangle\langle g|$ is the atomic inversion operator.

6.4.3.2.2 Field Hamiltonian

The field Hamiltonian can be expressed in terms of annihilation and creation operators as

$$\hat{H}_f = \sum_k \sum_\lambda \omega_k \left(\hat{a}_{k\lambda}^\dagger \hat{a}_{k\lambda} + \frac{1}{2} \right) \tag{6.33}$$

where
ω_k is the angular frequency of field mode with wave vector $\vec{k}$
$\hat{a}_{k\lambda}$ and $\hat{a}_{k\lambda}^\dagger$ are the corresponding bosonic annihilation and creation operators, respectively

Also, the term $\sum_k \sum_\lambda (1/2)\hbar\omega_k$ corresponds to the zero-point energy, which can be neglected. Dropping the zero-point energy term and considering the simplest case of a single cavity mode, the field Hamiltonian can be written as

$$\hat{H}_f = \omega \hat{a}^\dagger \hat{a} \tag{6.34}$$

6.4.3.2.3 Interaction Hamiltonian

The interaction energy between atom and radiation field can be approximated by the electric-dipole interaction energy as

$$\hat{H}_i = -\hat{d}\cdot\hat{E} \tag{6.35}$$

where
  $\hat{E}$ is the electric field operator
  $\hat{d}$ is the electric-dipole moment of the atom, given by

$$\hat{d} = -e\hat{D} = -e\sum_n \hat{r}_n$$

summed over all electrons of the atom. In this approximation, it is assumed that the contributions from electric quadruple moment and magnetic dipole moment are much smaller than that of electric-dipole moment. Now, the operator $\hat{D}$ can be expressed by

$$\hat{D} = I\hat{D}I = \sum_i |i\rangle\langle i| \hat{D} \sum_j |j\rangle\langle j| \tag{6.36a}$$

where
  I is the identity matrix
  $|i\rangle$ and $|j\rangle$ are the atomic energy eigenstates

By reorganizing *bra*s and *ket*s in Equation 6.36a, it is found that

$$\hat{D} = \sum_{i,j} \vec{D}_{ij} |i\rangle\langle j| \tag{6.36b}$$

where

$$\vec{D}_{ij} = \langle i|\hat{D}|j\rangle$$

Now, if the electric field operator $\hat{E}$ is expressed as

$$\hat{E}(\vec{r}) = i\sum_k \sum_\lambda \vec{e}_{k\lambda} C_E \left(\hat{a}_{k\lambda} e^{i\vec{k}\cdot\vec{r}} - \hat{a}^\dagger e^{-i\vec{k}\cdot\vec{r}} \right) \tag{6.37}$$

the interaction Hamiltonian can be written as [66]

$$\hat{H}_I = \hat{d} \cdot \left\{ i \sum_k \sum_\lambda \vec{e}_{k\lambda} C_E \left(\hat{a}_{k\lambda} e^{i\vec{k}\cdot\vec{r}} - \hat{a}^\dagger_{k\lambda} e^{-i\vec{k}\cdot\vec{r}} \right) \right\}$$

$$= ie \sum_k \sum_\lambda \sum_{i,j} C_E \vec{e}_{k\lambda} \cdot D_{ij} \left(\hat{a}_{k\lambda} e^{i\vec{k}\cdot\vec{r}} - \hat{a}^\dagger_{k\lambda} e^{-i\vec{k}\cdot\vec{r}} \right) |i\rangle\langle j| \qquad (6.38)$$

where
$\vec{e}_{k\lambda}$ is a unit polarization vector
$\vec{k}$ is the wave vector
λ is the polarization state
$C_E = \sqrt{(\omega_k / 2\varepsilon_0 V)}$ with V as the cavity volume

Taking into account that, in a two-level system, the sum over i and j runs only over the eigenstates $|g\rangle$ and $|e\rangle$ and that $\vec{D}_{ge}$ and $\vec{D}_{eg}$ are equal vectors, the interaction Hamiltonian can be written as

$$\hat{H}_I = \sum_k \sum_\lambda \left(g_{k\lambda} \hat{a}_{k\lambda} + g^*_{k\lambda} \hat{a}^\dagger_{k\lambda} \right) \left(\hat{\sigma}_+ + \hat{\sigma}_- \right) \qquad (6.39)$$

In Equation 6.39, raising and lowering operators are $\hat{\sigma}_+ = |e\rangle\langle g|$ and $\hat{\sigma}_- = |g\rangle\langle e|$, respectively. Also, $g_{k\lambda}$ is the coupling constant between atom and radiation field, given by

$$g_{k\lambda} = ie \sqrt{\frac{\omega_k}{2\varepsilon_0 \hbar V}} \vec{e}_{k\lambda} \cdot \vec{D}_{ge} e^{i\vec{k}\cdot\vec{r}}$$

After performing the multiplication in Equation 6.39 and taking only the energy preserving terms (i.e., the terms $\hat{a}_{k\lambda}\hat{\sigma}_+$ and $\hat{a}^\dagger_{k\lambda}\hat{\sigma}_-$), the interaction Hamiltonian is obtained as

$$\hat{H}_I = \sum_k \sum_\lambda \left(g_{k\lambda} \hat{a}_{k\lambda} \hat{\sigma}_+ + g^*_{k\lambda} \hat{a}^\dagger_{k\lambda} \hat{\sigma}_- \right) \qquad (6.40)$$

Within the difference in complex phase of coupling constant $g_{k\lambda}$ alone, Equation 6.40 is equivalent to the equation in the following:

$$\hat{H}_I = -i \sum_k \sum_\lambda g_{k\lambda} \left(\hat{a}_{k\lambda} \hat{\sigma}_+ - a^*_{k\lambda} \hat{\sigma}_- \right) \qquad (6.41)$$

Finally, in the simplest case, where a single electromagnetic mode interacts with atom, the interaction Hamiltonian is given by

$$\hat{H}_I = -i\hbar g \left(\hat{a}\hat{\sigma}_+ - \hat{a}^\dagger\hat{\sigma}_- \right) \qquad (6.42)$$

6.4.3.3 Extensions to the Jaynes–Cummings Model

There are several differences between Jaynes–Cummings model and QED model considered in the investigation. Firstly, in Jaynes–Cummings model, a simple two-level system is considered, whereas the present case deals with multiple energy levels with hyperfine structure. In particular, the model for the transmit node is more complex as it involves two different subspaces, each with three hyperfine states. Secondly, in Jaynes–Cummings model, only one cavity mode is considered, whereas the present model incorporates two cavity modes—one corresponding to each logic state. Thirdly, in Jaynes–Cummings model, one laser is used at the most, whereas in the present model, four different lasers are used in the transmit node and two lasers in the receive node. Finally, the applied lasers are not resonant with the cavity modes; instead, contrary to Jaynes–Cummings model, the lasers are resonant with other atomic transitions. The original Jaynes–Cummings model is extended/modified to meet the new requirements, which include

1. Modification to atomic raising and lowering operators to deal with hyperfine states with magnetic sublevels
2. Incorporation of separate field Hamiltonian for each cavity mode
3. Incorporation of atom–laser interactions in the model instead of cavity–laser interactions, because the lasers are coupled with the atom
4. Incorporation of timing of four different lasers in the transmit node and two lasers in the receive node

6.4.3.3.1 Atomic Raising and Lowering Operators

For a two-level system with the ground state $|g\rangle$ and the excited state $|e\rangle$, the raising operator is defined as

$$\hat{\sigma}_+ = |e\rangle\langle g|$$

(6.43a)

An application of this operator on $|g\rangle$ gives $\hat{\sigma}_+|g\rangle = |e\rangle\langle g||g\rangle = |e\rangle$, which is the excited state. Similarly, the lowering operator is defined as

$$\hat{\sigma}_- = |g\rangle\langle e|$$

(6.43b)

An application of $\hat{\sigma}_-$ on the state $|e\rangle$ lowers the state to the ground state ($\hat{\sigma}_-|e\rangle = |g\rangle \langle e||e\rangle = |g\rangle$).

In the investigation, the operator $\hat{\sigma}_+$ is replaced with a new operator $\hat{S}_{\pm,0}$, which raises the atom from lower-energy state to a higher-energy state, and at the same time, it changes the magnetic sublevel (m_F) corresponding to magnetic sublevel splitting within the hyperfine states. The operator $\hat{S}_+$ increases the value of m_F while raising the atom and $\hat{S}_-$ decreases the magnetic sublevel. Finally, the operator $\hat{S}^*$ raises the atom to upper hyperfine state without changing the magnetic sublevel. Similarly, the operator $\hat{\sigma}_-$ is replaced with the new operator $\hat{S}_{\pm,0}$, which lowers the atom from a higher hyperfine state to a lower one and, at the same time, changes the magnetic quantum number. The operator $\hat{S}_+$ lowers the atom from higher to lower energy levels, and at the same time, it increases the magnetic quantum number. The operator $\hat{S}_-$ lowers the energy level and decreases the magnetic quantum number, while $\hat{S}$ lowers the energy level without changing the magnetic quantum number. In particular, the lowering operators are given by the following equations:

$$\hat{S}_+^{1/2} = \sum_{m_{F,g}} C(F_g, m_{F,g}, F_e, m_{F,g}+1) \left| F_g m_{F,g} \right\rangle \cdot \left\langle F_e m_{F,g}+1 \right| \tag{6.44a}$$

$$\hat{S}_-^{1/2} = \sum_{m_{F,g}} C(F_g, m_{F,g}, F_e, m_{F,g}-1) \left| F_g m_{F,g} \right\rangle \cdot \left\langle F_e m_{F,g}-1 \right| \tag{6.44b}$$

$$\hat{S}^{1/2} = \sum_{m_{F,g}} C(F_g, m_{F,g}, F_e, m_{F,g}) \left| F_g m_{F,g} \right\rangle \cdot \left\langle F_e m_{F,g} \right| \tag{6.44c}$$

where
  F_g and F_e are total atomic angular momentums at the ground and the excited states, respectively
  $m_{F,g}$ is the magnetic quantum number at the ground hyperfine state and indicates the magnetic sublevel
  The *ket* of the form $|F\ m\rangle$ represents the hyperfine state with total atomic angular momentum F and magnetic quantum number (specifying magnetic sublevel) m

In these equations, the expression $C(F_g, m_{F,g}, F_e, m_{F,g}+d)$ with $d = +1, -1, 0$ is the Clebsch–Gordan coefficient defined as

$$C(F_g, m_{F,g}, F_e, m_{F,g}+d)$$
$$= \left\langle F_g, m_{F,g}; \Delta F = F_e - F_g, \Delta m_F = d \middle| F_e, m_{F,e} = m_{F,g} + \Delta m_F \right\rangle \tag{6.45}$$

Clebsch–Gordan coefficients arise in angular momentum coupling within quantum mechanical framework. It gives the expansion coefficients for total angular momentum eigenstates in an uncoupled tensor product basis. The right-hand side of Equation 6.45 gives the expansion coefficient for the uncoupled state $|F_g, m_{F,g}\rangle|\Delta F, \Delta m_F\rangle$ when combining into the total angular state $|F_e, m_{F,e}\rangle$. Here $\Delta F (= F_e - F_g)$ and Δm_F, respectively, stand for the difference in total atomic angular momentum and magnetic quantum numbers at the excited and the ground hyperfine states.

6.4.3.3.2 Laser Modes

In the investigative work, laser modes are also given full quantum mechanical treatment. As such, separate Fock spaces are devised for individual lasers. Then creation and annihilation operators are formulated in the respective Fock spaces. In the transmit node, four annihilation operators $\hat{a}_{01}^{(0)}$, $\hat{a}_{12}^{(0)}$, $\hat{a}_{01}^{(1)}$, and $\hat{a}_{12}^{(1)}$ are defined for four laser modes $L_{01}^{(0)}$, $L_{12}^{(0)}$, $L_{01}^{(1)}$, and $L_{12}^{(1)}$, respectively, in four respective Fock spaces. In the receive node, two Fock spaces are created for two laser modes L_{01} and L_{10}, and the corresponding respective annihilation operators $\hat{a}_{01}$ and $\hat{a}_{10}$ are devised in those Fock spaces. In both the cases, creation operators are just the conjugate transpose of annihilation operators.

6.4.3.3.3 Atom–Laser Interaction

As lasers are treated quantum mechanically, their interaction with atom is also described using new atomic raising and lowering operators. If, for example, a laser mode L is resonant with an atomic transition $|\psi_1\rangle \rightarrow |\psi_2\rangle$ with associated raising and lowering operators $\hat{S}^*$ and $\hat{S}$, respectively, the interaction Hamiltonian of the laser with atom is given by

$$\hat{H}_{\text{int}} = -ig_L(\hat{a}_L\,\hat{S}^* - \hat{a}_L^*\,\hat{S}) \tag{6.46}$$

where $\hat{a}_L$ and $\hat{a}_L^*$ are, respectively, the annihilation and the creation operators for laser modes.

6.4.3.4 System Hamiltonian

In this section, system Hamiltonian is formulated for both the transmit and the receive nodes, and the overall Hamiltonian combines component Hamiltonians, such as cavity mode Hamiltonians, laser mode Hamiltonians, atomic Hamiltonians, and different interaction Hamiltonians. These individual Hamiltonians are formulated first and then combined to form the Hamiltonians for transmit and receive nodes.

6.4.3.4.1 Cavity Mode Hamiltonian

In the transmit node, the cavity incorporates two electromagnetic modes. Let the annihilation operators for these modes be $\hat{a}_{\tau av}^{(0)}$ and $\hat{a}_{\tau av}^{(1)}$. Therefore, the corresponding creation operators are $\hat{a}_{\tau av}^{(0)*}$ and $\hat{a}_{\tau av}^{(1)*}$, respectively. In the respective Fock space, the number operators are, therefore, $N^{(0)} = \hat{a}_{\tau av}^{(0)*}\,\hat{a}_{\tau av}^{(0)}$ and $N^{(1)} = \hat{a}_{\tau av}^{(1)*}\,\hat{a}_{\tau av}^{(1)}$. Thus, the total cavity mode Hamiltonian in the transmit node is

$$\hat{H}_T^{(cav)} = \omega_{cav}^{(0)}\,\hat{a}_{\tau av}^{(0)*}\,\hat{a}_{\tau av}^{(0)} + \omega_{cav}^{(1)}\,\hat{a}_{\tau av}^{(1)*}\,\hat{a}_{\tau av}^{(1)} \tag{6.47}$$

where $\omega_{cav}^{(0)}$ and $\omega_{cav}^{(1)}$ are the cavity mode frequencies. Similarly, in the receive node, the total cavity mode Hamiltonian can be given by

$$\hat{H}_R^{(cav)} = w_{cav}^{(0)}\,\hat{\alpha}_{cav}^{(0)*}\,\hat{\alpha}_{cav}^{(0)} + w_{cav}^{(1)}\,\hat{\alpha}_{cav}^{(1)*}\,\hat{\alpha}_{cav}^{(1)} \tag{6.48}$$

where
$w_{cav}^{(0)}$ and $w_{cav}^{(1)}$ are cavity mode frequencies
$\alpha_{cav}^{(0)}$ and $\alpha_{cav}^{(1)}$ are annihilation operators for the cavity modes
$\alpha_{cav}^{(0)*}$ and $\alpha_{cav}^{(1)*}$ are the corresponding creation operators

6.4.3.4.2 Laser Mode Hamiltonian

Similar to the cavity modes, laser modes are also associated with creation, annihilation, and number operators. As mentioned before, the transmit node incorporates four lasers $L_{01}^{(0)}$, $L_{01}^{(1)}$, $L_{12}^{(0)}$, and $L_{12}^{(1)}$ with the corresponding annihilation operators $\hat{a}_{01}^{(0)}$, $\hat{a}_{01}^{(1)}$, $\hat{a}_{12}^{(0)}$, and $\hat{a}_{12}^{(1)}$, respectively, with their respective creation operators $\hat{a}_{01}^{(0)*}$, $\hat{a}_{01}^{(1)*}$, $\hat{a}_{12}^{(0)*}$, and $\hat{a}_{12}^{(1)*}$. Thus, the laser mode Hamiltonian for the transmit node is

$$\hat{H}_T^{(Laser)} = \eta_{01}\left(\hbar\omega_{01}^{(0)}\,\hat{a}_{01}^{(0)*}\,\hat{a}_{01}^{(0)} + \hbar\omega_{01}^{(1)}\,\hat{a}_{01}^{(1)*}\,\hat{a}_{01}^{(1)}\right) + \eta_{12}\left(\hbar\omega_{12}^{(0)}\,\hat{a}_{12}^{(0)*}\,\hat{a}_{12}^{(0)} + \hbar\omega_{12}^{(1)}\,\hat{a}_{12}^{(1)*}\,\hat{a}_{12}^{(1)}\right) \qquad (6.49)$$

where
$\omega_{01}^{(0)}$, $\omega_{01}^{(1)}$, $\omega_{12}^{(0)}$, and $\omega_{12}^{(1)}$ are angular frequencies of the laser modes $L_{01}^{(0)}$, $L_{01}^{(1)}$, $L_{12}^{(0)}$, and $L_{12}^{(1)}$, respectively

The quantities η_{01} and η_{12} are timing factors for laser modes, defined as

$$\eta_{01} = \begin{cases} 1 & \text{when the lasers } L_{01}^{(0)} \text{ and } L_{01}^{(1)} \text{ are turned on} \\ 0 & \text{otherwise} \end{cases}$$

$$\eta_{12} = \begin{cases} 1 & \text{when the lasers } L_{12}^{(0)} \text{ and } L_{12}^{(1)} \text{ are turned on} \\ 0 & \text{otherwise} \end{cases}$$

In the receive node, two lasers L_{01} and L_{10} are used, and the laser mode Hamiltonian can be written as

$$\hat{H}_R^{(Laser)} = \eta\left(\hbar\omega_{10}\,\hat{a}_{10}^*\,\hat{a}_{10} + \omega_{01}\,\hat{a}_{01}^*\,\hat{a}_{01}\right) \qquad (6.50)$$

where
ω_{10}, $\hat{a}_{10}$, and $\hat{a}_{10}^*$ are, respectively, frequency, annihilation, and creation operators for the laser mode L_{10}
ω_{01}, $\hat{a}_{01}$, and $\hat{a}_{01}^*$ are those for the laser mode L_{01}

The quantity η is defined as

$$\eta = \begin{cases} 1 & \text{when the lasers } L_{01} \text{ and } L_{10} \text{ are turned on} \\ 0 & \text{otherwise} \end{cases}$$

6.4.3.4.3 Atomic Hamiltonian

The framework involves the hyperfine states of ^{87}Rb atom with magnetic sublevels. Therefore, the atomic Hamiltonian involves energy terms of the hyperfine states as well as that associated with Zeeman splitting. In the transmit node, six hyperfine states, grouped into two subspaces, are considered. If energies of the six hyperfine states ($|\sigma_{00}\rangle$, $|\sigma_{01}\rangle$, $|\sigma_{02}\rangle$, $|\sigma_{10}\rangle$, $|\sigma_{11}\rangle$, and $|\sigma_{12}\rangle$) are E_{si} (where $s = 0,1$ represents the subspace number and i represents the index within the subspace), the atomic Hamiltonian for the transmit node can be written as

$$\hat{H}_T^{(atom)} = \sum_{s,i}\left(E_{si} + g_{F,si}m_{F,si}\beta_T\right)\left| F_{si}, m_{F,si}\right\rangle\left\langle F_{si}, m_{F,si}\right| \qquad (6.51)$$

with $g_{F,si}$ as the Landé g-factor [60]. β_T is defined as $\beta_T = \mu_B B_T$ with μ_B as Bohr magneton and B_T as the applied magnetic field. Thus, the energy shift for Zeeman effect due to the applied external magnetic field is given by

$$\Delta E = g_F m_F \beta_T \tag{6.52}$$

In the receive node, four hyperfine states in a single subspace are involved. Similar to the transmit node, the atomic Hamiltonian for the receive node can be written as

$$\hat{H}_R^{(atom)} = \sum_i (E_i + g_{F,i} m_{F,i} \beta_R) |F_i, m_{F,i}\rangle\langle F_i, m_{F,i}| \tag{6.53}$$

where
E_i are the energies of the hyperfine state i
$\beta_R = \mu_B B_R$ with B_R as the applied magnetic field in the receive node

6.4.3.4.4 Cavity–Atom Interaction Hamiltonian
The cavity–atom interaction Hamiltonian is similar to the one given by Equations 6.41 and 6.42 except that the two cavity modes are involved, and conventional atomic raising and lowering operators are replaced by raising and lowering operators developed in Section 6.4.3.3.1. The Hamiltonian for the transmit node can be expressed as

$$\hat{H}_T^{(cav-atom)} = -i\hbar g_{T,0}\left(\hat{a}_{cav}^{(0)}\, \hat{S}_+^{(0)} - \hat{a}_{cav}^{(0)*}\, \hat{S}_+^{(0)} \right) - i\hbar g_{T,1}\left(\hat{a}_{cav}^{(1)}\, \hat{S}_-^{(1)} - \hat{a}_{cav}^{(1)*}\, \hat{S}_-^{(1)} \right) \tag{6.54}$$

where $g_{T,0}$ and $g_{T,1}$ are the atom–cavity coupling constants for the two cavity modes. Similarly, the atom–cavity interaction Hamiltonian can be written as

$$\hat{H}_T^{(cav-atom)} = -i\hbar g_{R,0}\left(\hat{a}_{cav}^{(0)}\, \hat{S}_-^{(0)} - \hat{a}_{cav}^{(0)*}\, \hat{S}_-^{(0)} \right) - i\hbar g_{R,1}\left(\hat{a}_{cav}^{(1)}\, \hat{S}_+^{(1)} - \hat{a}_{cav}^{(1)*}\, \hat{S}_+^{(1)} \right) \tag{6.55}$$

with $g_{R,0}$ and $g_{R,1}$ as the atom–field coupling constants.

6.4.3.4.5 Laser–Atom Interaction Hamiltonian
The laser–atom interaction Hamiltonian is obtained by repeating the Hamiltonian given in Equation 6.46 for each laser. For the transmit node, the Hamiltonian can be expressed as

$$\hat{H}_T^{(laser-atom)} = -i\hbar \eta_{01} g_{01}^{(0)}\left(\hat{a}_{01}^{(0)}\, \hat{S}_{01}^{(0)*} - \hat{a}_{01}^{(0)*}\, \hat{S}_{01}^{(0)} \right) - i\hbar \eta_{01} g_{01}^{(1)}\left(\hat{a}_{01}^{(1)}\, \hat{S}_{01}^{(1)*} - \hat{a}_{01}^{(1)*}\, \hat{S}_{01}^{(1)} \right)$$
$$- i\eta_{12} g_{12}^{(0)}\left(\hat{a}_{21}^{(0)}\, \hat{S}_{21+}^{(0)*} - \hat{a}_{12}^{(0)*}\, \hat{S}_{12+}^{(0)} \right) - i\eta_{12} g_{12}^{(1)}\left(\hat{a}_{12}^{(1)}\, \hat{S}_{12-}^{(1)*} - \hat{a}_{12}^{(1)*}\, \hat{S}_{12}^{(1)} \right) \tag{6.56}$$

The meanings of many of the terms in earlier equations are already explained before. The remaining terms have their meanings as follows:

$g_{01}^{(0)}$: Coupling constant between the atom and the laser mode $L_{01}^{(0)}$

$\hat{\sigma}_{01}^{(0)}$: Lowering operator between the states $|\sigma_{00}\rangle$ and $|\sigma_{01}\rangle$

$\hat{\sigma}_{01}^{(0)*}$: Raising operator between the states $|\sigma_{00}\rangle$ and $|\sigma_{01}\rangle$

$g_{01}^{(1)}$: Coupling constant between the atom and the laser mode $L_{01}^{(1)}$

$\hat{\sigma}_{01}^{(1)}$: Lowering operator between the states $|\sigma_{10}\rangle$ and $|\sigma_{11}\rangle$

$\hat{\sigma}_{01}^{(1)*}$: Raising operator between the states $|\sigma_{10}\rangle$ and $|\sigma_{11}\rangle$

$g_{12}^{(0)}$: The coupling constant between the atom and the laser mode $L_{12}^{(0)}$

$\hat{\sigma}_{12}^{(0)}$: Lowering operator between the states $|\sigma_{01}\rangle$ and $|\sigma_{02}\rangle$

$\hat{\sigma}_{12}^{(0)*}$: Raising operator between the states $|\sigma_{01}\rangle$ and $|\sigma_{02}\rangle$

$g_{12}^{(1)}$: The coupling constant between the atom and the laser mode $L_{12}^{(1)}$

$\hat{\sigma}_{12}^{(1)}$: Lowering operator between the states $|\sigma_{11}\rangle$ and $|\sigma_{12}\rangle$

$\hat{\sigma}_{12}^{(1)*}$: Raising operator between the states $|\sigma_{11}\rangle$ and $|\sigma_{12}\rangle$

Similarly, for the receive node, the laser–atom interaction Hamiltonian can be written as

$$\hat{H}_R^{(laser-atom)} = -i\hbar\eta g_{10}\left(\hat{a}_{10}\,\hat{\sigma}_{01} - \hat{a}_{10}^\dagger\,\hat{\sigma}_{10}\right) - i\eta g_{01}\left(\hat{a}_{01}\,\hat{\sigma}_{01} - \hat{a}_{01}^\dagger\,\hat{\sigma}_{01}\right) \tag{6.57}$$

The meanings of the terms in Equation 6.57 are as follows:

$$\eta = \begin{cases} 1 & \text{when both lasers in the receiving node are turned on} \\ 0 & \text{otherwise} \end{cases}$$

g_{10} : Coupling constant between the atom and the laser L_{10}

$\hat{a}_{10}$: Annihilation operator for the laser mode L_{10}

$\hat{a}_{10}^\dagger$: Creation operator for the laser mode L_{10}

g_{01} : Coupling constant between the atom and the laser L_{01}

$\hat{a}_{01}$: Annihilation operator for the laser mode L_{01}

$\hat{a}_{01}^\dagger$: Creation operator for the laser mode L_{10}

6.4.3.4.6 Total Hamiltonian

Now, combining all the Hamiltonians presented in the previous sections, the total Hamiltonian for the transmit node is obtained as

$$
\begin{aligned}
\hat{H}_T^{(total)} =\ & \omega_{cav}^{(0)} \hat{a}_{cav}^{(0)*} \hat{a}_{cav}^{(0)} + \omega_{cav}^{(1)} \hat{a}_{cav}^{(1)*} \hat{a}_{cav}^{(1)} \\
& + \eta_{01}\left(\omega_{01}^{(0)} \hat{a}_{01}^{(0)*} \hat{a}_{01}^{(0)} + \hbar\omega_{01}^{(1)} \hat{a}_{01}^{(1)*} \hat{a}_{01}^{(1)}\right) + \eta_{12}\left(\omega_{12}^{(0)} \hat{a}_{12}^{(0)*} \hat{a}_{12}^{(0)} + \hbar\omega_{12}^{(1)} \hat{a}_{12}^{(1)*} \hat{a}_{12}^{(1)}\right) \\
& + \sum_{s,i}\left(E_{si} + g_{F,si}m_{F,si}\beta_T\right)\left|F_{si}, m_{F,si}\right\rangle\left\langle F_{si}, m_{F,si}\right| \\
& - ig_{T,0}\left(\hat{a}_{cav}^{(0)} \hat{S}_+^{1/2} - \hat{a}_{cav}^{(0)*} \hat{S}_+^{1/2}\right) - ig_{T,1}\left(\hat{a}_{cav}^{(1)} \hat{S}_-^{1/2} - \hat{a}_{cav}^{(1)*} \hat{S}_-^{1/2}\right) \\
& - i\eta_{01}g_{01}^{(0)}\left(\hat{a}_{01}^{(0)} \hat{S}_{01}^{1/2(0)*} - \hat{a}_{01}^{(0)*} \hat{S}_{01}^{1/2(0)}\right) - i\eta_{01}g_{01}^{(1)}\left(\hat{a}_{01}^{(1)} \hat{S}_{01}^{1/2(1)*} - \hat{a}_{01}^{(1)*} \hat{S}_{01}^{1/2(1)}\right) \\
& - i\eta_{12}g_{12}^{(0)}\left(\hat{a}_{12}^{(0)} \hat{S}_{12+}^{1/2(0)*} - \hat{a}_{12}^{(0)*} \hat{S}_{12+}^{1/2(0)}\right) - i\eta_{12}g_{12}^{(1)}\left(\hat{a}_{12}^{(1)} \hat{S}_{12-}^{1/2(1)*} - \hat{a}_{12}^{(1)*} \hat{S}_{12-}^{1/2(1)}\right)
\end{aligned}
\tag{6.58}
$$

Similarly, the total Hamiltonian for the receive node can be presented as

$$
\begin{aligned}
\hat{H}_R^{(total)} =\ & w_{cav}^{(0)} \hat{a}_{cav}^{(0)*} \hat{a}_{cav}^{(0)} + \hbar w_{cav}^{(1)} \hat{a}_{cav}^{(1)*} \hat{a}_{cav}^{(1)} + \eta\left(\hbar\omega_{10} \hat{a}_{10}^* \hat{a}_{10} + \hbar\omega_{01} \hat{a}_{01}^* \hat{a}_{01}\right) \\
& + \sum_i\left(E_i + g_{F,i}m_{F,i}\beta_R\right)\left|F_i, m_{F,i}\right\rangle\left\langle F_i, m_{F,i}\right| \\
& - ig_{R,0}\left(\hat{a}_{cav}^{(0)} \hat{S}_+^{1/2} - \hat{a}_{cav}^{(0)*} \hat{S}_+^{1/2}\right) - ig_{R,1}\left(\hat{a}_{cav}^{(1)} \hat{S}_-^{1/2} - \hat{a}_{cav}^{(1)*} \hat{S}_-^{1/2}\right) \\
& - i\eta g_{10}\left(\hat{a}_{10} \hat{S}_{10}^{1/2} - \hat{a}_{10} \hat{S}_{10}^{1/2}\right) - i\eta g_{01}\left(\hat{a}_{01} \hat{S}_{01}^{1/2} - \hat{a}_{01} \hat{S}_{01}^{1/2}\right)
\end{aligned}
\tag{6.59}
$$

6.4.3.5 Hilbert Spaces

Two composite Hilbert spaces—one for the transmit node and the other for the receive node—are considered. In the transmit node, separate Hilbert spaces are constructed for the atom, each of the cavity modes and each of the laser modes. The composite Hilbert space is constructed by taking the tensor product of these individual Hilbert spaces and is defined as

$$
H_T = H_{atom} \otimes H_{cav}^{(0)} \otimes H_{cav}^{(1)} \otimes H_{L01}^{(0)} \otimes H_{L12}^{(0)} \otimes H_{L01}^{(1)} \otimes H_{L12}^{(1)}
\tag{6.60}
$$

where
H_{atom} is the Hilbert space of ^{87}Rb atom
$H_{cav}^{(0)}$ and $H_{cav}^{(1)}$ are the Hilbert spaces of the two cavity modes
$H_{L01}^{(0)}$, $H_{L12}^{(0)}$, $H_{L01}^{(1)}$, and $H_{L12}^{(1)}$ are the Hilbert spaces for the laser modes $L_{01}^{(0)}$, $L_{01}^{(1)}$, $L_{12}^{(0)}$, and $L_{12}^{(1)}$, respectively

Similarly, the Hilbert space for the receive node is defined as

$$
H_R = H_{atom} \otimes H_{cav}^{(0)} \otimes H_{cav}^{(1)} \otimes H_{L01} \otimes H_{L10}
\tag{6.61}
$$

where the component Hilbert spaces belong to ^{87}Rb atom, two cavity modes, and the laser modes L_{01} and L_{10}.

6.4.3.6 Density Matrices

Density matrix formalism is adopted to represent the system state. Because the component systems interact with each other, the tensor product of density matrices cannot be obtained to form the overall density matrix. However, it is assumed that initially at time $t=0$, the component systems are uncorrelated. Under this assumption, the initial overall density matrix is formed by taking the tensor product of the initial component density matrices. Thus, at time $t=0$, the overall density matrix is found for the transmit node as

$$\rho_T\big|_{t=0} = D_{atom}\big|_{t=0} \otimes D_{cav}^{(0)}\big|_{t=0} \otimes D_{cav}^{(1)}\big|_{t=0} \otimes D_{L01}^{(0)}\big|_{t=0} \otimes D_{L12}^{(0)}\big|_{t=0} \otimes D_{L01}^{(1)}\big|_{t=0} \otimes D_{L12}^{(1)}\big|_{t=0} \quad (6.62)$$

where

$D_{atom}\big|_{t=0}$ is the density matrix of the atom at $t=0$

$D_{cav}^{(0)}\big|_{t=0}$ and $D_{cav}^{(1)}\big|_{t=0}$ are the initial density matrices of the two cavity modes

$D_{L01}^{(0)}\big|_{t=0}$, $D_{L12}^{(0)}\big|_{t=0}$, $D_{L01}^{(1)}\big|_{t=0}$, and $D_{L12}^{(1)}\big|_{t=0}$ are, respectively, the density matrices of the laser modes $L_{01}^{(0)}$, $L_{12}^{(0)}$, $L_{01}^{(1)}$, and $L_{12}^{(1)}$

Once the initial density matrix is known, the state of density matrix can be found at other times using the system evolution equation, as discussed in the following section.

The initial density matrix for the receive node can also be formed as

$$\rho_R\big|_{t=0} = D_{atom}\big|_{t=0} \otimes D_{cav}^{(0)}\big|_{t=0} \otimes D_{cav}^{(1)}\big|_{t=0} \otimes D_{L01}\big|_{t=0} \otimes D_{L10}\big|_{t=0} \quad (6.63)$$

where $D_{L01}\big|_{t=0}$ and $D_{L01}\big|_{t=0}$ are the initial density matrices for the laser modes L_{01} and L_{10}, respectively.

6.4.4.7 System Evolution

An open-system approach is adopted for formulating the equation of motion wherein both transmit and receive nodes interact with the environment, generally called as the reservoir. In this stream, two composite systems $S_T \otimes \Re$ and $S_R \otimes \Re$ (S_T represents the transmit node, S_R the receive node, and $\Re$ the reservoir) may be given a thought, for which the equations of motion could be written using von Neumann equations [64]:

$$\frac{d}{dt}\rho_{S_T \otimes \Re} = \frac{1}{i\hbar}\left[H_{S_T \otimes \Re}, \rho_{S_T \otimes \Re}\right] \quad (6.64)$$

and

$$\frac{d}{dt}\rho_{S_R \otimes \Re} = \frac{1}{i\hbar}\left[H_{S_R \otimes \Re}, \rho_{S_R \otimes \Re}\right] \quad (6.65)$$

Here $\rho_{S_T \otimes \Re}$ and $\rho_{S_R \otimes \Re}$ are, respectively, the density matrices for the transmit and the receive composite subsystems. However, the implementation of these equations would require handling complex reservoir states. Master equations in the Lindblad form [65] are adopted in order to involve only S_T and S_R in the equations, avoiding thereby the reservoir. These new equations for system evolution are

$$\frac{d}{dt}\rho_T(t) = L_T\rho_T(t) \tag{6.66}$$

and

$$\frac{d}{dt}\rho_R(t) = L_R\rho_R(t) \tag{6.67}$$

where ρ_T and ρ_R are density matrices of the transmit and the receive nodes, respectively, and the linear operators L_T and L_R take the form

$$L\circ = \frac{1}{i\hbar}[\hat{H},\circ]+\sum_i\left(\hat{C}_i\circ\hat{C}_i^* - \frac{1}{2}\hat{C}_i^*\,\hat{C}_i\circ -\frac{1}{2}\circ\hat{C}_i^*\,\hat{C}_i \right) \tag{6.68}$$

where $\hat{C}_i$s are so-called jump operators, which represent interactions between the system and the reservoir. Here four such jump operators are defined for both the transmit and the receive nodes—two of these four jump operators correspond to photon leakage out of the cavity (due to cavity–fiber coupling), and the remaining two to spontaneous photon emission from the atom. Formally, jump operators in the transmit node are defined as

$$\hat{C}_{T0} = \sqrt{2\kappa_T}\,\hat{a}_{cav}^{(0)} \tag{6.69a}$$

$$\hat{C}_{T1} = \sqrt{2\kappa_T}\,\hat{a}_{cav}^{(1)} \tag{6.69b}$$

$$\hat{C}_{T+} = \sqrt{\gamma_T}\,\hat{S}_+ \tag{6.69c}$$

$$\hat{C}_{T-} = \sqrt{\gamma_T}\,\hat{S}_- \tag{6.69d}$$

Similarly, jump operators for the receive node are defined as

$$\hat{C}_{R0} = \sqrt{2\kappa_R}\,\hat{a}_{cav2}^{(0)} \tag{6.70a}$$

$$\hat{C}_{R1} = \sqrt{2\kappa_R}\,\hat{a}_{cav2}^{(1)} \tag{6.70b}$$

$$\hat{C}_{R+} = \sqrt{\gamma_R}\,\hat{S}_+ \tag{6.70c}$$

$$\hat{C}_{R-} = \sqrt{\gamma_R}\,\hat{S}_- \tag{6.70d}$$

In Equations 6.69 and 6.70, κ is the cavity mode decay rate and γ is the atomic decay rate. Equations 6.69a and 6.69b define cavity decay operators for the transmit mode, and

Equations 6.70a and 6.70b define those for the receive node. Equations 6.69c and 6.69d define spontaneous emission operators with the right and the left circularly polarized photonic states, respectively, for the transmit node, and Equations 6.70c and 6.70d define the same for the receive node.

6.5 Implementation Framework for Quantum Communication

This section presents implementation of the theoretical framework, as discussed in Section 6.4. A comprehensive implementation scheme is developed to concretize the abstract formulas. The implementation scope is limited to building a programmable model of the system that can be simulated to study the system behavior.

6.5.1 Some Preliminary Issues

As the activities in the quantum processes primarily include transitions between the hyperfine states of ^{87}Rb atom with the absorption and emission of photons, it is natural to think of a unit system that is convenient for working with physical quantities relevant to atomic physics and QED. Hartree atomic units (a.u.) [67] serve the purpose wherein the numerical values of six physical constants, namely, electronic mass m_e, elementary electronic charge e, Bohr radius a_0 of hydrogen atom, absolute value of the electric potential energy of hydrogen atom in its ground state E_h, reduced Planck constant $(h/2\pi)$, and Coulomb force constant $(1/4\pi\varepsilon_0)$, are tied to unity. These six fundamental units form the system of atomic units (Table 6.1).

Some of the derived atomic units are illustrated in Table 6.2.

6.5.1.1 Kronecker Product for Tensors

Kronecker product, denoted by $\otimes$, is adopted for matrix representation of tensor product. The output block matrix is a representation of tensor product of the two input matrices with respect to a standard set of basis. If $A = [a_{ij}]$ is an $m \times n$ matrix and $B = [b_{kl}]$ is a

TABLE 6.1

Fundamental Atomic Units

Quantity	Physical Constants	Symbol	SI Equivalent
Length	Bohr radius	a_0	$5.291772108(18) \times 10^{-11}$ m
Mass	Electron rest mass	m_e	$9.1093826(16) \times 10^{-31}$ kg
Charge	Elementary charge	e	$1.602176\,53(14) \times 10^{-19}$ C
Electrostatic force	Coulomb force constant	$(1/4\pi\varepsilon_0)$	8.9875516×10^9 C^{-2} N m^2
Angular momentum	Reduced Planck constant	$(h/2\pi)$	$1.05457168(18) \times 10^{-34}$ J s
Energy	Hartree energy	E_h	$4.35974417(75) \times 10^{-18}$ J

TABLE 6.2

Derived Atomic Units

Quantity	Expression	SI Equivalent
Time	$/E_h$	$2.418\ 884\ 326\ 505(16) \times 10^{-17}$ s
Velocity	$a_0 E_h/$	$2.187\ 691\ 2633(73) \times 10^6$ m s^{-1}
Force	E_h/a_0	$8.238\ 7225(14) \times 10^{-8}$ N
Temperature	E_h/k_B	$3.157\ 7464(55) \times 10^5$ K
Pressure	$E_h/(a_0)^3$	$2.942\ 1912(19) \times 10^{13}$ N m^{-2}

$p \times q$ matrix, the Kronecker product of these two matrices is defined by the $mp \times nq$ block matrix as

$$A \otimes B = \begin{bmatrix} a_{11}B & a_{12}B & \cdots & a_{1n}B \\ a_{21}B & a_{22}B & \cdots & a_{2n}B \\ \cdots & \cdots & \cdots & \cdots \\ a_{m1}B & a_{am2}B & \cdots & a_{mn}B \end{bmatrix} \tag{6.71}$$

Equation 6.71 can be further expanded for every element of the $A \otimes B$ matrix as

$$A \otimes B = \begin{bmatrix} a_{11}b_{11} & \cdots & a_{11}b_{1q} & \cdots & \cdots & \cdots & a_{1n}b_{11} & \cdots & a_{in}b_{1q} \\ \cdots & \cdots & \cdots & \cdots & \cdots & \cdots & \cdots & \cdots & \cdots \\ a_{11}b_{p1} & \cdots & a_{11}b_{pq} & \cdots & \cdots & \cdots & a_{1n}b_{p1} & \cdots & a_{in}b_{pq} \\ \cdots & \cdots & \cdots & \cdots & \cdots & \cdots & \cdots & \cdots & \cdots \\ \cdots & \cdots & \cdots & \cdots & \cdots & \cdots & \cdots & \cdots & \cdots \\ \cdots & \cdots & \cdots & \cdots & \cdots & \cdots & \cdots & \cdots & \cdots \\ a_{m1}b_{11} & \cdots & a_{m1}b_{1q} & \cdots & \cdots & \cdots & a_{mn}b_{11} & \cdots & a_{mn}b_{1q} \\ \cdots & \cdots & \cdots & \cdots & \cdots & \cdots & \cdots & \cdots & \cdots \\ a_{m1}b_{pq} & \cdots & a_{m1}b_{pq} & \cdots & \cdots & \cdots & a_{mn}b_{p1} & \cdots & a_{mn}b_{pq} \end{bmatrix} \tag{6.72}$$

6.5.1.2 Sparse Matrix Representation

In the mathematical subfield of numerical analysis, a sparse matrix is defined as a matrix populated primarily with zeros [68]. In order to efficiently store and manipulate these matrices using digital computer, it is important to use specialized algorithms and data structures that make use of sparse structure of the matrix.

The present investigation deals with large-size sparse matrices, with only a few non-zero elements. In particular, many of the matrices used in this investigation are diagonal matrices. For example, when the density matrices are formed from the statistical mixture of eigenstates, the density matrices contain only diagonal elements.

When storing and processing a sparse matrix, it is sufficient to store only the nonzero elements. For example, in order to store the following sparse matrix,

$$
A_{sparse} = \begin{bmatrix} 1 & 0 & 0 & 0 & 0 \\ 0 & 0 & 0 & 1 & 0 \\ 0 & 0 & 0 & 0 & 0 \\ 0 & 3 & 0 & 0 & 0 \\ 0 & 0 & 0 & 0 & 1 \end{bmatrix}
$$

it is sufficient to store only the dimensional information of the matrix and the nonzero elements as follows:

Number of rows $n_{rows} = 5$
Number of columns $n_{cols} = 5$
Nonzero elements $A(1,1) = 1$, $A(2,4) = 1$, $A(4,2) = 3$, and $A(5,5) = 1$

6.5.2 Transmit Node Implementation

This section presents a detailed implementation of the transmit node of the proposed quantum communication system. In the process, Hilbert space for the node is constructed, and system Hamiltonian and density matrix are devised. Also, the implementation level equation for the evolution of density matrices is formulated.

6.5.2.1 Hilbert Spaces

The overall Hilbert space of the transmit node is a tensor product of Hilbert spaces for atom, cavity modes, and laser modes. Therefore, Hilbert spaces are constructed first for atom, cavity, and laser modes, and finally the overall Hilbert space is constructed as a tensor product of the component Hilbert spaces.

6.5.2.1.1 Hilbert Space for Atom

In the transmit node, the Hilbert space for ^{87}Rb atom contains two subspaces, each spanned by three basis states. Thus, the Hilbert space is spanned by a total of six basis states as stated in Equations 6.24 and 6.25. Now, as the Hilbert space is a vector space, these basis states should be represented as a set of orthonormal basis vectors. These orthonormal bases are identified as

$$
\left.\begin{aligned}
|\sigma_{00}\rangle &\equiv [1 \quad 0 \quad 0 \quad 0 \quad 0 \quad 0]^T \\
|\sigma_{01}\rangle &\equiv [0 \quad 1 \quad 0 \quad 0 \quad 0 \quad 0]^T \\
|\sigma_{02}\rangle &\equiv [0 \quad 0 \quad 1 \quad 0 \quad 0 \quad 0]^T \\
|\sigma_{10}\rangle &\equiv [0 \quad 0 \quad 0 \quad 1 \quad 0 \quad 0]^T \\
|\sigma_{11}\rangle &\equiv [0 \quad 0 \quad 0 \quad 0 \quad 1 \quad 0]^T \\
|\sigma_{12}\rangle &\equiv [0 \quad 0 \quad 0 \quad 0 \quad 0 \quad 1]^T
\end{aligned}\right\} \tag{6.73}
$$

Now, any element in this 6D Hilbert space, corresponding to an arbitrary superposition of hyperfine states, can be expressed as

$$|\psi\rangle = \begin{bmatrix} c_1 & c_2 & c_3 & c_4 & c_5 & c_6 \end{bmatrix}^T \tag{6.74}$$

where c_i ($i = 1, 2, 3, 4, 5,$ and 6) represent components of the superposition in the ith dimension of the Hilbert space.

6.5.2.1.2 Hilbert Spaces for Cavity Modes

A separate Hilbert space is defined for each cavity mode. These Hilbert spaces are actually Fock spaces, which are useful when arbitrary number of particles occupy the same state. As photons are bosonic particles, an arbitrary number of photons can occupy a single state, and thus, Fock states are useful for representing electromagnetic modes. A useful and convenient basis for the Fock space is the occupancy number basis. If $|\psi_i\rangle$ represents a basis of the Hilbert space H, then a state in which n_0 particles are in $|\psi_0\rangle$, n_1 particles are in $|\psi_1\rangle$, and so on, can be represented with occupancy numbers as

$$|n_0, n_1, n_3, \ldots, n_N\rangle$$

If only one electromagnetic mode and an arbitrary number of photons in the cavity are considered, the Fock state can then be represented as $|n\rangle$, where n is the number of photons in the single mode. For different values of n, Fock states will be $|0\rangle, |1\rangle, |2\rangle, \ldots$. In order to make the cavity mode state manageable for computer programming, it is required to limit the maximum number of photons that can be present in a single cavity mode. Let the maximum number be N. Then the Fock states are

$$|0\rangle, |1\rangle, |2\rangle, \ldots, |N\rangle$$

where
$|0\rangle$ represents Fock state where there is no photon in the cavity
$|1\rangle$ represents Fock state where there is one photon, and so on

If these Fock states represent a complete set of bases, this basis set can be identified with an orthonormal vector basis set as

$$\left.\begin{aligned}
|0\rangle &\equiv \begin{bmatrix} 1 & 0 & 0 & 0 & 0 & \cdots & 0 \end{bmatrix}^T \\
|1\rangle &\equiv \begin{bmatrix} 0 & 1 & 0 & 0 & 0 & \cdots & 0 \end{bmatrix}^T \\
|2\rangle &\equiv \begin{bmatrix} 0 & 0 & 1 & 0 & 0 & \cdots & 0 \end{bmatrix}^T \\
|3\rangle &\equiv \begin{bmatrix} 0 & 0 & 0 & 1 & 0 & \cdots & 0 \end{bmatrix}^T \\
&\;\vdots \\
|N\rangle &\equiv \begin{bmatrix} 0 & 0 & 0 & 0 & 0 & \cdots & 1 \end{bmatrix}^T
\end{aligned}\right\} \tag{6.75}$$

Now, any arbitrary cavity mode state in this Fock space can be represented as a superposition of these basis vectors.

6.5.2.1.3 Hilbert Spaces for Lasers

Similar to the quantum states of cavity modes, the quantum states of laser modes are also denoted by the use of Fock space representation. If the number of photons in a laser mode is limited to N_L, the basis state vectors are then as follows:

$$\left.\begin{aligned}
|L0\rangle &\equiv [1 \quad 0 \quad 0 \quad 0 \quad 0 \quad \cdots \quad 0]^T \\
|L1\rangle &\equiv [0 \quad 1 \quad 0 \quad 0 \quad 0 \quad \cdots \quad 0]^T \\
|L2\rangle &\equiv [0 \quad 0 \quad 1 \quad 0 \quad 0 \quad \cdots \quad 0]^T \\
|L3\rangle &\equiv [0 \quad 0 \quad 0 \quad 1 \quad 0 \quad \cdots \quad 0]^T \\
&\vdots \\
|LN_L\rangle &\equiv [0 \quad 0 \quad 0 \quad 0 \quad 0 \quad \cdots \quad 1]^T
\end{aligned}\right\} \tag{6.76}$$

where
$\quad|L0\rangle$ is the Fock state representing no photon in the laser mode
$\quad|L1\rangle$ is the Fock state representing one photon in the laser mode, and so on

Any arbitrary quantum state of a laser mode can be represented by a superposition of these basis vectors.

6.5.2.1.4 Combined Hilbert Space

With the earlier definitions of Hilbert spaces for atom, cavity modes, and laser modes, the combined Hilbert space, which is the tensor product of seven individual Hilbert spaces (1 for the atom, 2 for the two cavity modes, and 4 for the laser modes), can be specified. As mentioned before, the atomic Hilbert space is 6D. Ideally, all the cavity and laser modes can accommodate infinite number of photons, and therefore, the corresponding Hilbert spaces are infinite dimensional. However, in order to keep the implementation of the system feasible, the Hilbert spaces of cavity and laser modes are truncated. If the maximum number of photons in the two cavity modes is N_0 and N_1 and if the maximum number of photons in the laser modes is N_{L0}, N_{L1}, N_{L2}, and N_{L3}, then considering the 6D atomic Hilbert space, the total number of dimensions in the combined Hilbert space will be

$$6 \times (N_0+1) \times (N_1+1) \times (N_{L0}+1) \times (N_{L1}+1) \times (N_{L2}+1) \times (N_{L3}+1)$$

6.5.2.2 System Hamiltonian

Once the Hilbert spaces for atom, cavity modes, and laser modes are defined, Hamiltonians for them can be formulated. Then the Hamiltonian for the transmit node can be obtained by adding the Hamiltonians of atom, cavity modes, and laser modes in combined Hilbert space.

6.5.2.2.1 Atomic Energy Levels

Formulation of atomic Hamiltonian needs the knowledge about energy levels. As laser and cavity modes are resonant to the atomic transition frequencies, their Hamiltonians also ultimately depend on the atomic energy levels. The energy levels of ^{87}Rb atom can be written as

$$E_{\sigma_{ij}} = E_{\sigma_{ij}}^{(F)} + E_{\sigma_{ij}}^{(m_F)} \tag{6.77}$$

where

$E_{\sigma_{ij}}^{(F)}$ is the hyperfine energy level for the state $|\sigma_{ij}\rangle$

$E_{\sigma_{ij}}^{(m_F)}$ is the magnetic sublevel energy due to Zeeman splitting

$E_{\sigma_{ij}}^{(F)}$ is evaluated using the formula

$E_{\sigma_{ij}}^{(f)} = \hbar\omega_{\sigma_{ij}}^{(F)} = 2\pi\hbar f_{\sigma_{ij}}^{(F)}$

with $f_{\sigma_{ij}}^{(F)}$ as the frequency corresponding to the energy level of the $|\sigma_{ij}\rangle$ state, as defined in Equations 6.24 and 6.25. Considering the $5^2S_{1/2}$ energy level of ^{87}Rb atom equals to zero, the relative values of $f_{\sigma_{ij}}^{(F)}$ are obtained from Figure 6.13 as

$$f_{\sigma_{00}}^{(F)} = -4.271676631815181 \times 10^3 \, \text{MHz}$$

$$f_{\sigma_{01}}^{(F)} = 377.107463380 \times 10^6 - 509.05 \, \text{MHz}$$

$$f_{\sigma_{02}}^{(F)} = 377.107463380 \times 10^6 + 305.43 \, \text{MHz}$$

$$f_{\sigma_{10}}^{(F)} = -4.271676631815181 \times 10^3 \, \text{MHz}$$

$$f_{\sigma_{11}}^{(F)} = 377.107463380 \times 10^6 - 509.05 \, \text{MHz}$$

$$f_{\sigma_{12}}^{(F)} = 384.2304844685 \times 10^6 - 72.9112 \, \text{MHz}$$

Correspondingly, the frequencies in atomic unit can be stated as

$$f_{\sigma_{00}}^{(F)} = 1.033269165259541 \times 10^{-7} \, \text{a.u.}$$

$$f_{\sigma_{01}}^{(F)} = 0.00912178101245 \, \text{a.u.}$$

$$f_{\sigma_{02}}^{(F)} = 0.00912180071378 \, \text{a.u.}$$

$$f_{\sigma_{10}}^{(F)} = 1.033269165259541 \times 10^{-7} \, \text{a.u.}$$

$$f_{\sigma_{11}}^{(F)} = 0.00912178101245 \, \text{a.u.}$$

$$f_{\sigma_{12}}^{(F)} = 0.00929408920283 \, \text{a.u.}$$

Also, the values of $E_{\sigma_{ij}}^{(m_F)}$ are obtained from Equation 6.52 as

$$E_{\sigma_{ij}}^{(m_F)} = \mu_B g_{F_{ij}} m_{F_{ij}} B_z \tag{6.78}$$

where
  μ_B (=0.5 a.u.) is Bohr magneton
  m_F is the magnetic quantum number representing magnetic sublevel within the hyperfine energy level
  B_z is the z component of the applied magnetic field. The direction of applied magnetic field is chosen as the quantization axis and set as
  $B_z = B = 100 \, \text{Ga} = 100 \times 10^{-4} \, \text{T} = 4.2553 \times 10^{-8} \, \text{a.u.}$

Also, $g_{F_{ij}}$ in Equation 6.78 is the hyperfine Landé g-factor whose value depends on all the angular momentum quantities F, I, J, L, and S, and can be calculated by using

$$g_F = g_J \frac{F(F+1) - I(I+1) + J(J+1)}{2F(F+1)} \tag{6.79}$$

where g_J is the fine structure Landé g-factor, which can be calculated by

$$g_J = 1 + \frac{J(J+1) + S(S+1) - L(L+1)}{2J(J+1)} \tag{6.80}$$

For the hyperfine energy levels considered in this investigation, the calculated hyperfine Landé g-factors are as follows:

$$g_{F_{00}} = g_F\left(|\sigma_{00}\rangle\right) = -\frac{1}{2}, \quad g_{F_{01}} = g_F\left(|\sigma_{01}\rangle\right) = -\frac{1}{6},$$

$$g_{F_{02}} = g_F\left(|\sigma_{02}\rangle\right) = \frac{1}{6}, \quad g_{F_{10}} = g_F\left(|\sigma_{10}\rangle\right) = -\frac{1}{2},$$

$$g_{F_{11}} = g_F\left(|\sigma_{11}\rangle\right) = -\frac{1}{6} \quad \text{and} \quad g_{F_{12}} = g_F\left(|\sigma_{12}\rangle\right) = \frac{2}{3}.$$

6.5.2.2.2 Atomic Hamiltonian
Now that the basis vectors for the Hilbert space of ^{87}Rb atom are defined and energies for the hyperfine states with magnetic sublevels are calculated, the Hamiltonian for atom can then be formulated as

$$
H_T^{(atom)} = E_{\sigma 00}
\begin{bmatrix}1\\0\\0\\0\\0\\0\end{bmatrix}
\begin{bmatrix}1\\0\\0\\0\\0\\0\end{bmatrix}^T
+ E_{\sigma 01}
\begin{bmatrix}0\\1\\0\\0\\0\\0\end{bmatrix}
\begin{bmatrix}0\\1\\0\\0\\0\\0\end{bmatrix}^T
+ E_{\sigma 0}
\begin{bmatrix}0\\0\\1\\0\\0\\0\end{bmatrix}
\begin{bmatrix}0\\0\\1\\0\\0\\0\end{bmatrix}^T
$$

$$
+ E_{\sigma 10}
\begin{bmatrix}0\\0\\0\\1\\0\\0\end{bmatrix}
\begin{bmatrix}0\\0\\0\\1\\0\\0\end{bmatrix}^T
E_{\sigma 11}
\begin{bmatrix}0\\0\\0\\0\\1\\0\end{bmatrix}
\begin{bmatrix}0\\0\\0\\0\\1\\0\end{bmatrix}^T
+ E_{\sigma 1}
\begin{bmatrix}0\\0\\0\\0\\0\\1\end{bmatrix}
\begin{bmatrix}0\\0\\0\\0\\0\\1\end{bmatrix}^T
\tag{6.81}
$$

Here the terms $E_{\sigma_{ij}}$ ($i = 0, 1$ and $j = 0, 1, 2$) are as given by Equation 6.77.

6.5.2.2.3 Cavity Mode Hamiltonian

In order to construct cavity mode Hamiltonian for the transmit node, as given by Equation 6.47, annihilation and creation operators are needed to be constructed first. For the basis vectors in the Hilbert space for the cavity mode, given by Equation 6.75, the annihilation operator is given by

$$
\hat{a}_{cav} =
\begin{bmatrix}
0 & 1 & 0 & 0 & \cdots & 0 \\
0 & 0 & \sqrt{2} & 0 & \cdots & 0 \\
0 & 0 & 0 & \sqrt{3} & \cdots & 0 \\
\cdots & \cdots & \cdots & \cdots & \cdots & \cdots \\
0 & 0 & 0 & 0 & \cdots & \sqrt{N} \\
0 & 0 & 0 & 0 & \cdots & 0
\end{bmatrix}
\tag{6.82}
$$

and the creation operator is given by the conjugate transpose of the annihilation operator, that is,

$$
\hat{a}_{cav}^{\dagger} =
\begin{bmatrix}
0 & 0 & 0 & \cdots & 0 & 0 \\
1 & 0 & 0 & \cdots & 0 & 0 \\
0 & \sqrt{2} & 0 & \cdots & 0 & 0 \\
0 & 0 & \sqrt{3} & \cdots & 0 & 0 \\
\cdots & \cdots & \cdots & \cdots & \cdots & \cdots \\
0 & 0 & 0 & \cdots & \sqrt{N} & 0
\end{bmatrix}
\tag{6.83}
$$

In the earlier two equations, N is the maximum number of photons allowed in the cavity mode. Specifically, if the cavity modes permit maximum N_0 and N_1 photons, the cavity mode Hamiltonians can be written as

$$\hat{H}_T^{(cav0)} = \omega_{cav}^{(0)}\, \hat{a}_{cav}^{(0)*}\, \hat{a}_{cav}^{(0)}$$

Performing the matrix multiplication, one gets

$$\hat{H}_T^{(cav0)} = \omega_{cav}^{(0)} \begin{bmatrix} 0 & 0 & 0 & 0 & \cdots & 0 \\ 0 & 1 & 0 & 0 & \cdots & 0 \\ 0 & 0 & 2 & 0 & \cdots & 0 \\ 0 & 0 & 0 & 3 & \cdots & 0 \\ \cdots & \cdots & \cdots & \cdots & \cdots & \cdots \\ 0 & 0 & 0 & 0 & \cdots & N_0 \end{bmatrix} \tag{6.84}$$

and similarly,

$$\hat{H}_T^{(cav1)} = \omega_{cav}^{(1)} \begin{bmatrix} 0 & 0 & 0 & 0 & \cdots & 0 \\ 0 & 1 & 0 & 0 & \cdots & 0 \\ 0 & 0 & 2 & 0 & \cdots & 0 \\ 0 & 0 & 0 & 3 & \cdots & 0 \\ \cdots & \cdots & \cdots & \cdots & \cdots & \cdots \\ 0 & 0 & 0 & 0 & \cdots & N_1 \end{bmatrix} \tag{6.85}$$

The terms $\omega_{cav}^{(0)}$ and $\omega_{cav}^{(1)}$ in Equations 6.84 and 6.85, respectively, can be obtained from the assumption that the cavity modes are resonant with the corresponding atomic transitions. Thus,

$$\left. \begin{aligned} \hbar\omega_{cav}^{(0)} &= E_{\sigma 02} - E_{\sigma 00} \\ \hbar\omega_{cav}^{(1)} &= E_{\sigma 12} - E_{\sigma 10} \end{aligned} \right\} \tag{6.86}$$

6.5.2.2.4 Laser Hamiltonian

Although many of the proposals treated lasers classically, in the present work, these are treated quantum mechanically. Also, all the four lasers, used in the transmit node, work in the similar principle. Therefore, instead of describing the construction of Hamiltonians of these lasers separately, the formulation of a single laser Hamiltonian is described with a general notation. In general terms, a laser Hamiltonian can be given by

$$\hat{H}_T^{(laser)} = \omega_{laser}\, \hat{a}_{laser}^*\, \hat{a}_{laser} \tag{6.87}$$

If the laser mode allows maximum N_{laser} photons, the annihilation operator $\hat{a}$ can then be written as

$$
\hat{a}_{laser} = \begin{bmatrix} 0 & 1 & 0 & 0 & \cdots & 0 \\ 0 & 0 & \sqrt{2} & 0 & \cdots & 0 \\ 0 & 0 & 0 & \sqrt{3} & \cdots & 0 \\ \cdots & \cdots & \cdots & \cdots & \cdots & \cdots \\ 0 & 0 & 0 & 0 & \cdots & \sqrt{N_{laser}} \\ 0 & 0 & 0 & 0 & \cdots & 0 \end{bmatrix} \tag{6.88}
$$

By taking the conjugate transport of the annihilation operator, the creation operator for the laser mode can be obtained as

$$
\hat{a}^*_{laser} = \begin{bmatrix} 0 & 0 & 0 & \cdots & 0 & 0 \\ 1 & 0 & 0 & \cdots & 0 & 0 \\ 0 & \sqrt{2} & 0 & \cdots & 0 & 0 \\ 0 & 0 & \sqrt{3} & \cdots & 0 & 0 \\ \cdots & \cdots & \cdots & \cdots & \cdots & \cdots \\ 0 & 0 & 0 & \cdots & \sqrt{N_{laser}} & 0 \end{bmatrix} \tag{6.89}
$$

Putting the values of $\hat{a}_{laser}$ and $\hat{a}^*_{laser}$ in Equation 6.88, one finally finds

$$
\hat{H}_T^{(laser)} = \omega_{laser} \begin{bmatrix} 0 & 0 & 0 & 0 & \cdots & 0 \\ 0 & 1 & 0 & 0 & \cdots & 0 \\ 0 & 0 & 2 & 0 & \cdots & 0 \\ 0 & 0 & 0 & 3 & \cdots & 0 \\ \cdots & \cdots & \cdots & \cdots & \cdots & \cdots \\ 0 & 0 & 0 & 0 & \cdots & N_{laser} \end{bmatrix} \tag{6.90}
$$

In the current investigation, frequencies of lasers are resonant with the transitions of atomic levels. Therefore, these frequencies can be determined as the difference in frequency between the relevant atomic energy levels. Also, the intensity of laser corresponds to the number of photons in the mode.

6.5.2.2.5 Atomic Raising and Lowering Operators

In the transmit node, following transitions take place during the quantum processes:

1. $|\sigma_{00}\rangle \rightarrow |\sigma_{01}\rangle$
2. $|\sigma_{01}\rangle \rightarrow |\sigma_{02}\rangle$
3. $|\sigma_{02}\rangle \rightarrow |\sigma_{00}\rangle$
4. $|\sigma_{10}\rangle \rightarrow |\sigma_{11}\rangle$
5. $|\sigma_{11}\rangle \rightarrow |\sigma_{12}\rangle$
6. $|\sigma_{12}\rangle \rightarrow |\sigma_{10}\rangle$

With each of the earlier transitions, an atomic raising operator and a lowering operator are involved. Again, three variations are associated with these operators due to the changes in magnetic quantum number during transitions. Out of these raising and lowering operators, 12 operators—namely, $\hat{S}_+^{1/2*}$, $\hat{S}_+$, $\hat{S}_-^{1/2}$, $\hat{S}_-$, $\hat{S}_{01}^{(40)*}$, $\hat{S}_{01}^{(40)}$, $\hat{S}_{01}^{(41)*}$, $\hat{S}_{01}^{(41)}$, $\hat{S}_{12+}^{(40)*}$, $\hat{S}_{12+}^{(40)}$, $\hat{S}_{12-}^{(41)*}$, and $\hat{S}_{12-}^{(41)}$—are directly relevant to the implementation, as is evident from Equation 6.56. These operators can be constructed following Equation 6.44. However, one important step in constructing these operators is to calculate the Clebsch–Gordan coefficients.

6.5.2.2.6 Extending Operators to the Combined Hilbert Space

So far the descriptions have been made on the work with operators within the respective Hilbert space. In order to combine these individual operators or to implement interactions between operators, these are needed to be extended, including Hamiltonians, to the total Hilbert space. After that the interaction Hamiltonians can be found, and finally, all the Hamiltonians can be added to find the total Hamiltonian.

First, the atomic Hamiltonian can be extended as

$$H_T^{(atom-ext)} = H_T^{(atom)} \otimes I_{N_0} \otimes I_{N_1} \otimes I_{N_{L0}} \otimes I_{N_{L1}} \otimes I_{N_{L2}} \otimes I_{N_{L3}} \tag{6.91}$$

where I_{N_0}, I_{N_1}, $I_{N_{L0}}$, $I_{N_{L1}}$, $I_{N_{L2}}$, and $I_{N_{L3}}$ are identity matrices of sizes N_0, N_1, N_{L0}, N_{L1}, N_{L2}, and N_{L2}, respectively. Similarly, the cavity mode Hamiltonians can be extended as

$$H_T^{(cav0-ext)} = I_6 \otimes H_T^{cav0} \otimes I_{N_1} \otimes I_{N_{L0}} \otimes I_{N_{L1}} \otimes I_{N_{L2}} \otimes I_{N_{L3}} \tag{6.92}$$

and

$$H_T^{(cav1-ext)} = I_6 \otimes I_{N_0} \otimes H_T^{(cav1)} \otimes I_{N_{L0}} \otimes I_{N_{L1}} \otimes I_{N_{L2}} \otimes I_{N_{L3}} \tag{6.93}$$

The laser mode Hamiltonians are also similarly extended as

$$H_T^{(laser0-ext)} = I_6 \otimes I_{N_0} \otimes I_{N_1} \otimes H_T^{(laser0)} \otimes I_{N_{L1}} \otimes I_{N_{L2}} \otimes I_{N_{L3}} \tag{6.94}$$

$$H_T^{(laser1-ext)} = I_6 \otimes I_{N_0} \otimes I_{N_1} \otimes I_{N_{L0}} \otimes H_T^{(laser1)} \otimes I_{N_{L2}} \otimes I_{N_{L3}} \tag{6.95}$$

$$H_T^{(laser2-ext)} = I_6 \otimes I_{N_0} \otimes I_{N_1} \otimes I_{N_{L0}} \otimes I_{N_{L1}} \otimes H_T^{(laser2)} \otimes I_{N_{L3}} \tag{6.96}$$

$$H_T^{(laser3-ext)} = I_6 \otimes I_{N_0} \otimes I_{N_1} \otimes I_{N_{L0}} \otimes I_{N_{L1}} \otimes I_{N_{L2}} \otimes H_T^{(laser3)} \tag{6.97}$$

Further, the raising and the lowering operators are needed to be extended as

$$S_{\pm,0}^{(ext)} = S_{\pm,0}^{1/2} \otimes I_{N_0} \otimes I_{N_1} \otimes I_{N_{L0}} \otimes I_{N_{L1}} \otimes I_{N_{L2}} \otimes I_{N_{L3}} \tag{6.98}$$

All the creation and annihilation operators for cavity and laser modes should also be extended (extensions of annihilation operators only are shown here). First, the cavity mode annihilation operators are extended as

$$\hat{a}_{cav}^{(0-ext)} = I_6 \otimes \hat{a}_{cav}^{(0)} \otimes I_{N_1} \otimes I_{N_{L0}} \otimes I_{N_{L1}} \otimes I_{N_{L2}} \otimes I_{N_{L3}} \tag{6.99}$$

$$\hat{a}_{cav}^{(1-ext)} = I_6 \otimes I_{N_0} \otimes \hat{a}_{cav}^{(1)} \otimes I_{N_{L0}} \otimes I_{N_{L1}} \otimes I_{N_{L2}} \otimes I_{N_{L3}} \tag{6.100}$$

Then, the laser mode annihilation operators are extended as

$$\hat{a}_{01}^{(0-ext)} = I_6 \otimes I_{N_0} \otimes I_{N_1} \otimes \hat{a}_{01}^{(0)} \otimes I_{N_{L1}} \otimes I_{N_{L2}} \otimes I_{N_{L3}} \tag{6.101}$$

$$\hat{a}_{12}^{(0-ext)} = I_6 \otimes I_{N_0} \otimes I_{N_1} \otimes I_{N_{L0}} \otimes \hat{a}_{12}^{(0)} \otimes I_{N_{L2}} \otimes I_{N_{L3}} \tag{6.102}$$

$$\hat{a}_{01}^{(1-ext)} = I_6 \otimes I_{N_0} \otimes I_{N_1} \otimes I_{N_{L0}} \otimes I_{N_{L1}} \otimes \hat{a}_{01}^{(1)} \otimes I_{N_{L3}} \tag{6.103}$$

$$\hat{a}_{12}^{(1-ext)} = I_6 \otimes I_{N_0} \otimes I_{N_1} \otimes I_{N_{L0}} \otimes I_{N_{L1}} \otimes I_{N_{L2}} \otimes \hat{a}_{12}^{(1)} \tag{6.104}$$

6.5.2.2.7 Cavity–Atom Interaction Hamiltonian

At the implementation level, the cavity–atom interaction Hamiltonian is expressed in terms of extended operators as

$$\hat{H}_T^{(cav-atom-ext)} = -ig_{T,0}\left(\hat{a}_{cav}^{(0-ext)}\, \hat{S}_+^{(ext)^*} - \hat{a}_{cav}^{(0-ext)^*}\, \hat{S}_+^{(ext)} \right)$$

$$-ig_{T,1}\left(\hat{a}_{cav}^{(1-ext)}\, \hat{S}_-^{(ext)^*} - \hat{a}_{cav}^{(1-ext)^*}\, \hat{S}_-^{(ext)} \right) \tag{6.105}$$

where $\hat{S}_\pm^{(ext)}$ and $\hat{S}_\pm^{(ext)^*}$ are atomic lowering and raising operators, respectively, corresponding to the cavity-assisted transitions. The suffix $\pm$ indicates the increase $(+)$ or the decrease $(-)$ in the magnetic quantum number.

6.5.2.2.8 Atom–Laser Interaction Hamiltonian

At the implementation level, the atom–laser interaction Hamiltonian is given in terms of extended operators as

$$\hat{H}_T^{(laser-atom-ext)} = -i\eta_{01}g_{01}^{(0)}\left(\hat{a}_{01}^{(0-ext)}\, \hat{S}_{01}^{(0-ext)^*} - \hat{a}_{01}^{(0-ext)^*}\, \hat{S}_{01}^{(0-ext)} \right)$$

$$-i\eta_{01}g_{01}^{(1)}\left(\hat{a}_{01}^{(1-ext)}\, \hat{S}_{01}^{(1-ext)^*} - \hat{a}_{01}^{(1-ext)^*}\, \hat{S}_{01}^{(1-ext)} \right) - i\eta_{12}g_{12}^{(0)}\left(\hat{a}_{12}^{(0-ext)}\, \hat{S}_{21+}^{(0-ext)^*} - \hat{a}_{12}^{(0-ext)^*}\, \hat{S}_{12+}^{(0-ext)} \right)$$

$$-i\eta_{12}g_{12}^{(1)}\left(\hat{a}_{12}^{(1-ext)}\, \hat{S}_{12-}^{(1-ext)^*} - \hat{a}_{12}^{(1-ext)^*}\, \hat{S}_{12-}^{(1-ext)} \right) \tag{6.106}$$

In this equation, $\hat{a}_{01}^{(0-ext)}$ and $\hat{a}_{01}^{(0-ext)^*}$ are, respectively, the extended annihilation and creation operators for the laser mode $L_{01}^{(0)}$, and $\hat{a}_{12}^{(0-ext)}$ and $\hat{a}_{12}^{(0-ext)^*}$ are similarly defined for the laser mode $L_{12}^{(0)}$. Also, $\hat{S}_{01}^{(0-ext)}$, $\hat{S}_{01}^{(0-ext)^*}$, $\hat{S}_{12+}^{(0-ext)}$, and $\hat{S}_{12+}^{(0-ext)^*}$ represent the extended atomic lowering and raising operators for the corresponding transitions. Noticeably, the operators $\hat{S}_{01}^{(0-ext)}$ and $\hat{S}_{01}^{(0-ext)^*}$ do not change the magnetic quantum number. However, the operators $\hat{S}_{12+}^{(0-ext)}$ and $\hat{S}_{12+}^{(0-ext)^*}$ decrease and increase, respectively, the magnetic quantum number by 1 during the respective lowering and raising of the atomic state. Finally, $\hat{S}_{12-}^{(1-ext)}$ and $\hat{S}_{12-}^{(1-ext)^*}$ are, respectively, the extended lowering and raising operators corresponding to the atomic transition $|\sigma_{11}\rangle \rightarrow |\sigma_{12}\rangle$, which increase and decrease the magnetic quantum number by 1 during the respective lowering and raising.

6.5.2.2.9 Total Hamiltonian

The total Hamiltonian can be formulated by adding all the extended component Hamiltonians as

$$\hat{H}_T = \hat{H}_T^{(atom-ext)} + \hat{H}_T^{(cav0-ext)} + \hat{H}_T^{(cav1-ext)} + \hat{H}_T^{(laser0-ext)} + \hat{H}_T^{(laser1-ext)} + \hat{H}_T^{(laser2-ext)}$$

$$+ \hat{H}_T^{(laser3-ext)} + \hat{H}_T^{(cav-atom-ext)} + \hat{H}_T^{(laser-atom-ext)} \tag{6.107}$$

where the extended laser mode Hamiltonians $\hat{H}_T^{(laser0-ext)}$, $\hat{H}_T^{(laser1-ext)}$, $\hat{H}_T^{(laser2-ext)}$, and $\hat{H}_T^{(laser3-ext)}$ correspond to the laser modes $L_{01}^{(0)}$, $L_{12}^{(0)}$, $L_{01}^{(1)}$, and $L_{12}^{(1)}$, respectively.

6.5.2.3 Density Matrices

The system state is represented by density matrices, defined for each of the components in the transmit node (i.e., for ^{87}Rb atom), cavity modes, and laser modes. When the components are interacting, it is not possible to obtain the overall density matrix simply by taking the tensor product of the component density matrices. However, it is assumed that, initially at time $t=0$, the components are noninteracting and the initial density matrix of the transmit node is given by the tensor product of initial component density matrices. In this section, the construction of initial density matrices is described.

6.5.2.3.1 Atomic Density Matrix

The atomic Hilbert space is spanned by the eigenstates $|\sigma_{ij}\rangle$ represented by orthonormal basis vectors, as stated in Equation 6.73. If the atom is initially in the $|\sigma_{00}\rangle$ eigenstate, probability of occupation of the $|\sigma_{00}\rangle$ state is 1, and that of all the other states is 0. In that case, the initial density matrix can be defined as

$$\rho_T^{(atom)}\Big|_{t=0} = 1.0\times|\sigma_{00}\rangle\langle\sigma_{00}| + 0.0\times|\sigma_{01}\rangle\langle\sigma_{01}| + 0.0\times|\sigma_{02}\rangle\langle\sigma_{02}| + 0.0\times|\sigma_{10}\rangle\langle\sigma_{10}|$$

$$+ 0.0\times|\sigma_{11}\rangle\langle\sigma_{11}| + 0.0\times|\sigma_{12}\rangle\langle\sigma_{12}|$$

Thus, in this case, the initial atomic density matrix is given by

$$\rho_T^{(atom)}\Big|_{t=0} = \begin{bmatrix} 1 \\ 0 \\ 0 \\ 0 \\ 0 \\ 0 \end{bmatrix} \begin{bmatrix} 1 \\ 0 \\ 0 \\ 0 \\ 0 \\ 0 \end{bmatrix}^T = \begin{bmatrix} 1 & 0 & 0 & 0 & 0 & 0 \\ 0 & 0 & 0 & 0 & 0 & 0 \\ 0 & 0 & 0 & 0 & 0 & 0 \\ 0 & 0 & 0 & 0 & 0 & 0 \\ 0 & 0 & 0 & 0 & 0 & 0 \\ 0 & 0 & 0 & 0 & 0 & 0 \end{bmatrix} \tag{6.108}$$

6.5.2.3.2 Cavity Mode Density Matrices

Initially, cavity modes are assumed to contain no photons. Thus, from the definition of cavity mode basis vectors in Equation 6.75, the initial density matrices for cavity modes are constructed. It can be shown that, for the first cavity mode, the initial density matrix will be given by

$$\rho_T^{(cav0)}\Big|_{t=0} = \begin{bmatrix} 1 & 0 & 0 & 0 & \cdots & 0 \\ 0 & 0 & 0 & 0 & \cdots & 0 \\ 0 & 0 & 0 & 0 & \cdots & 0 \\ 0 & 0 & 0 & 0 & \cdots & 0 \\ \cdots & \cdots & \cdots & \cdots & \cdots & \cdots \\ 0 & 0 & 0 & 0 & \cdots & 0 \end{bmatrix} \tag{6.109}$$

and for the second cavity mode, it will be

$$\rho_T^{(cav1)}\Big|_{t=0} = \begin{bmatrix} 1 & 0 & 0 & 0 & \cdots & 0 \\ 0 & 0 & 0 & 0 & \cdots & 0 \\ 0 & 0 & 0 & 0 & \cdots & 0 \\ 0 & 0 & 0 & 0 & \cdots & 0 \\ \cdots & \cdots & \cdots & \cdots & \cdots & \cdots \\ 0 & 0 & 0 & 0 & \cdots & 0 \end{bmatrix} \tag{6.110}$$

6.5.2.3.3 Laser Mode Density Matrices

In order to trigger quantum processes in the transmit node for generating photon state, the lasers $L_{01}^{(0)}$ and $L_{01}^{(1)}$ are simultaneously turned on, and afterward the remaining two lasers $L_{12}^{(0)}$ and $L_{12}^{(1)}$ are also turned on. Therefore, initially the two lasers $L_{01}^{(0)}$ and $L_{01}^{(1)}$ are prepared with N_0 photons each where $N_0 \gg 0$, and the other two lasers $L_{12}^{(0)}$ and $L_{12}^{(1)}$ with zero photon. Basis vectors of these laser modes in the corresponding Hilbert spaces were defined in Equation 6.76. Based on these basis vectors and the said initial conditions, the initial density matrices for the lasers $L_{01}^{(0)}$ and $L_{01}^{(1)}$ can finally be written as

$$\rho_T^{(L01-0)}\Big|_{t=0} = \begin{bmatrix} 0 & 0 & \cdots & 0 & 0 & \cdots & 0 \\ 0 & 0 & \cdots & 0 & 0 & \cdots & 0 \\ \cdots & \cdots & \cdots & \cdots & \cdots & \cdots & \cdots \\ 0 & 0 & \cdots & 1 & 0 & \cdots & 0 \\ 0 & 0 & \cdots & 0 & 0 & \cdots & 0 \\ \cdots & \cdots & \cdots & \cdots & \cdots & \cdots & \cdots \\ 0 & 0 & \cdots & 0 & 0 & \cdots & 0 \end{bmatrix} \tag{6.111}$$

with $\rho_T^{(L01-0)}(N_0, N_0)\big|_{t=0} = 1$, where $(N_0 - 1)$ is the number of photons in laser mode. Similarly,

$$
\rho_T^{(L01-1)}\Big|_{t=0} =
\begin{bmatrix}
0 & 0 & \cdots & 0 & 0 & \cdots & 0 \\
0 & 0 & \cdots & 0 & 0 & \cdots & 0 \\
\cdots & \cdots & \cdots & \cdots & \cdots & \cdots & \cdots \\
0 & 0 & \cdots & 1 & 0 & \cdots & 0 \\
0 & 0 & \cdots & 0 & 0 & \cdots & 0 \\
\cdots & \cdots & \cdots & \cdots & \cdots & \cdots & \cdots \\
0 & 0 & \cdots & 0 & 0 & \cdots & 0
\end{bmatrix}
\tag{6.112}
$$

As there are no photons in the other two laser modes $L_{12}^{(0)}$ and $L_{12}^{(1)}$, the corresponding density matrices are given by

$$
\rho_T^{(L12-0)}\Big|_{t=0} = \rho_T^{(L12-1)}\Big|_{t=0} =
\begin{bmatrix}
1 & 0 & 0 & 0 & \cdots & 0 \\
0 & 0 & 0 & 0 & \cdots & 0 \\
0 & 0 & 0 & 0 & \cdots & 0 \\
0 & 0 & 0 & 0 & \cdots & 0 \\
\cdots & \cdots & \cdots & \cdots & \cdots & \cdots \\
0 & 0 & 0 & 0 & \cdots & 0
\end{bmatrix}
\tag{6.113}
$$

6.5.2.3.4 Combined Density Matrix

As mentioned before, initially the components of system are not interacting with each other. Therefore, the total density matrix can be devised by taking the tensor product of component density matrices constructed before, as shown in Equations 6.108 through 6.113. Thus, the total density matrix will be of the form

$$
\rho_T\big|_{t=0} = \rho_T^{(atom)}\Big|_{t=0} \otimes \rho_T^{(cav0)}\Big|_{t=0} \otimes \rho_T^{(cav1)}\Big|_{t=0} \otimes \rho_T^{(L01-0)}\Big|_{t=0} \otimes \rho_T^{(L12-0)}\Big|_{t=0}
$$
$$
\otimes \rho_T^{(L01-1)}\Big|_{t=0} \otimes \rho_T^{(L12-1)}\Big|_{t=0}
\tag{6.114}
$$

the explicit form of which is not incorporated here.

6.5.2.4 System Evolution

This section presents implementation strategy for the evolution of system, that is, the evolution of the transmit node density matrix $\rho_T(t)$ with time. Given the initial density matrix by Equation 6.114, this evolution triggers transitions between the hyperfine states of ^{87}Rb atom causing emission of a photon with appropriate frequency and polarization. The system evolution is analyzed in two steps—in the first step, a closed-system analysis [56] is performed, and in the second step, a full-fledged open-system analysis is touched upon [61,62,69].

6.5.2.4.1 Closed-System Evolution

In the closed-system evolution, the transmit node is isolated from the environment. As such, the emitted photon from ^{87}Rb atom does not leak out of cavity. Instead, it stays in cavity, and the atom cyclically absorbs and emits it. Since no photon reaches the receive node, this node is excluded from the system during the closed-system analysis. In this mode, evolution of the transmit node density matrix ρ_T is given by the von Neumann equation [64] as follows:

$$\frac{d}{dt}\rho_T(t) = \frac{1}{i\hbar}\left[\hat{H}_T, \rho_T(t)\right] \tag{6.115}$$

where

$\rho_T(t)$ is the total density matrix of the transmit node, the initial value of which is given by Equation 6.114

$\hat{H}_T$ is the total Hamiltonian given by Equation 6.107

6.5.2.4.2 Open-System Evolution

In the open-system evolution, the system interacts with the environment, which triggers the emitted photon to leak out of cavity, propagate through the connected fiber, and reach the receive node. This makes one to implement an open-system evolution in the receive node, for which the master equation in the Lindblad form [65] is adopted, as given in Equation 6.66. However, the evolution starts with the initial density matrix as

$$\frac{d}{dt}\rho_T\Big|_{t=0} = L_T\, \rho_T\Big|_{t=0} \tag{6.116}$$

Afterward, the density matrix undergoes various changes as the atom, through quantum processes, makes transitions between energy levels, and the number of photons in laser and cavity modes fluctuate.

6.5.3 Receive Node Implementation

Implementation of the receive node is significantly different from that of the transmit node, although both the nodes consist of a ^{87}Rb atom placed at the center of an optical cavity. While six hyperfine states, grouped into two subspaces, are used to represent the qubit state in the transmit node, in the receive node, four hyperfine states are used in a single subspace. This allows the receive node to change the qubit state based on the received photon state. Also, two lasers are used in the receive node compared to four in the transmit node. However, in certain aspects, implementations of both nodes are exactly the same. In those aspects, implementation of the transmit node is referred to, and the main points are briefly mentioned.

6.5.3.1 Hilbert Spaces

The Hilbert space for the receive node is constructed wherein various states and operators are defined. The Hilbert spaces are first defined for atom, cavity modes, and laser modes, and then those are combined to obtain the overall Hilbert space for the receive node.

6.5.3.1.1 Hilbert Space for Atom

The Hilbert space for atom at the receive node is spanned by four hyperfine states (or basis states) given as

$$
\left.
\begin{aligned}
|\phi_0\rangle &\equiv (5^2S_{1/2}, F = 1, m_F = 1)\\
|\phi_0'\rangle &\equiv (5^2P_{3/2}, F = 2, m_F = 0)\\
|\phi_1\rangle &\equiv (5^2S_{1/2}, F = 1, m_F = 0)\\
|\phi_1'\rangle &\equiv (5^2P_{1/2}, F = 2, m_F = 1)
\end{aligned}
\right\}
$$

Basis vectors must be orthonormal for the Hilbert space, and therefore, the four basis states should be represented by four orthonormal basis vectors, which are identified as

$$
\left.
\begin{aligned}
|\phi_0\rangle &\equiv [1 \quad 0 \quad 0 \quad 0]^T\\
|\phi_0'\rangle &\equiv [0 \quad 1 \quad 0 \quad 0]^T\\
|\phi_1\rangle &\equiv [0 \quad 0 \quad 1 \quad 0]^T\\
|\phi_1'\rangle &\equiv [0 \quad 0 \quad 0 \quad 1]^T
\end{aligned}
\right\}
\tag{6.117}
$$

6.5.3.1.2 Hilbert Spaces for Cavity Modes

A Hilbert space is associated with each of the cavity modes. These Hilbert spaces are exactly similar to the cavity mode Hilbert spaces in the transmit node. The set of basis vectors for the Hilbert space (i.e., Fock space) of each cavity mode is

$$
\left.
\begin{aligned}
|0\rangle &\equiv [1 \quad 0 \quad 0 \quad 0 \quad 0 \quad \cdots \quad 0]^T\\
|1\rangle &\equiv [0 \quad 1 \quad 0 \quad 0 \quad 0 \quad \cdots \quad 0]^T\\
|2\rangle &\equiv [0 \quad 0 \quad 1 \quad 0 \quad 0 \quad \cdots \quad 0]^T\\
|3\rangle &\equiv [0 \quad 0 \quad 0 \quad 1 \quad 0 \quad \cdots \quad 0]^T\\
&\quad\vdots\\
|N\rangle &\equiv [0 \quad 0 \quad 0 \quad 0 \quad 0 \quad \cdots \quad 1]^T
\end{aligned}
\right\}
\tag{6.118}
$$

6.5.3.1.3 Hilbert Spaces for Lasers

Usage of lasers in the receive node is quite different from that in the transmit node. Similar to the case of cavity modes, Hilbert spaces for laser modes are also implemented as Fock spaces. If the maximum numbers of photons considered in the laser modes L_{01} and L_{10}

are N_{L01} and N_{L10}, respectively (in the receive node), the respective basis vectors for the corresponding Fock spaces are

$$
\left.\begin{aligned}
|0\rangle &\equiv [1 \quad 0 \quad 0 \quad 0 \quad 0 \quad \cdots \quad 0]^T \\
|1\rangle &\equiv [0 \quad 1 \quad 0 \quad 0 \quad 0 \quad \cdots \quad 0]^T \\
|2\rangle &\equiv [0 \quad 0 \quad 1 \quad 0 \quad 0 \quad \cdots \quad 0]^T \\
|3\rangle &\equiv [0 \quad 0 \quad 0 \quad 1 \quad 0 \quad \cdots \quad 0]^T \\
&\vdots \\
|N_{L01}\rangle &\equiv [0 \quad 0 \quad 0 \quad 0 \quad 0 \quad \cdots \quad 1]^T
\end{aligned}\right\} \tag{6.119}
$$

and

$$
\left.\begin{aligned}
|0\rangle &\equiv [1 \quad 0 \quad 0 \quad 0 \quad 0 \quad \cdots \quad 0]^T \\
|1\rangle &\equiv [0 \quad 1 \quad 0 \quad 0 \quad 0 \quad \cdots \quad 0]^T \\
|2\rangle &\equiv [0 \quad 0 \quad 1 \quad 0 \quad 0 \quad \cdots \quad 0]^T \\
|3\rangle &\equiv [0 \quad 0 \quad 0 \quad 1 \quad 0 \quad \cdots \quad 0]^T \\
&\vdots \\
|N_{L10}\rangle &\equiv [0 \quad 0 \quad 0 \quad 0 \quad 0 \quad \cdots \quad 1]^T
\end{aligned}\right\} \tag{6.120}
$$

where
$|0\rangle$ is the Fock state with no photon in the laser mode
$|1\rangle$ the Fock state with 1 photon, and so on

In the Fock space, any arbitrary superposition state can be represented as a vector of length $N_{L01} + 1$ as follows:

$$
|\psi_{01}\rangle = \begin{bmatrix} c_0 & c_1 & c_2 & \cdots & \cdots & c_{N_{L01}} \end{bmatrix}^T \tag{121}
$$

where c_i ($i = 0, 1, \ldots, N_{L01}$) represents projection of the state $|\psi_{01}\rangle$ onto the dimension i in the Fock space.

6.5.3.1.4 Combined Hilbert Space

A combined Hilbert space for the receive node can be defined as a tensor product of five individual Hilbert spaces (1 for atom, 2 for cavity modes, and 2 for laser modes). As mentioned before, the atomic Hilbert space is 4D. Ideally, all the cavity and laser modes can accommodate infinite number of photons, and thus, the corresponding Hilbert spaces are infinite dimensional. However, in order to keep the implementation of system feasible, Hilbert spaces of cavity and laser modes are truncated. If N_0 and N_1 are the maximum numbers of photons in the two cavity modes and N_{L01} and N_{L10} are the maximum numbers of photons in the laser modes, considering the 4D atomic Hilbert space, the total number of dimensions in the combined Hilbert space is $4 \times (N_0 + 1) \times (N_1 + 1) \times (N_{L01} + 1) \times (N_{L10} + 1)$.

6.5.3.2 System Hamiltonian

The total Hamiltonian for the receive node is obtained by first formulating the Hamiltonians for individual components, that is, atom, cavity modes, and laser modes, and then adding those. However, the individual Hamiltonians must be extended from the individual Hilbert spaces to the overall Hilbert space before adding them; building these Hamiltonians is the subject matter of this section.

6.5.3.2.1 Atomic Energy Levels

Similar to the transmit node, the atomic energy levels are needed to be known in order to formulate the atomic Hamiltonian. Furthermore, laser and cavity modes are resonant to the atomic transition frequencies, and thus, their Hamiltonians also ultimately depend on the atomic energy levels. The hyperfine energy levels of ^{87}Rb atom (relevant to the current investigation) can be written as

$$E_{\phi^{(i)}} = E_{\phi^{(i)}}^{(F)} + E_{\phi^{(i)}}^{(m_F)} \tag{6.121}$$

where

$E_{\phi^{(i)}}^{(F)}$ is the hyperfine energy level for the state $|\phi^{(i)}\rangle$

$E_{\phi^{(i)}}^{(m_F)}$ is the magnetic sublevel energy due to Zeeman splitting

Here the state $|\phi^{(i)}\rangle$ is a convenient notation for the atomic states, that is,

$$|\phi_0\rangle \to |\phi^{(0)}\rangle, |\phi_0'\rangle \to |\phi^{(1)}\rangle, |\phi_1\rangle \to |\phi^{(2)}\rangle, |\phi_1'\rangle \to |\phi^{(3)}\rangle$$

The first part of energy $E_{\phi^{(i)}}^{(F)}$ can be expressed as

$$E_{\phi^{(i)}}^{F} = \hbar\omega_{\phi^{(i)}}^{(F)} = 2\pi f_{\phi^{(i)}}^{(F)} \tag{6.122}$$

where $f_{\phi^{(i)}}^{(F)}$ is the frequency corresponding to the energy level of the $|\varphi^{(i)}\rangle$ state. Considering the $5^2S_{1/2}$ energy level of ^{87}Rb atom equals to zero, the relative values of $f_{\phi^{(i)}}^{(F)}$ may be obtained from Figure 6.13 as

$$f_{\varphi^{(0)}}^{(F)} = -4.271676631815181 \times 10^3 \, \text{MHz}$$

$$f_{\varphi^{(1)}}^{(F)} = 384.2304844685 \times 10^6 - 72.9112 \, \text{MHz}$$

$$f^{(F)}_{\varphi^{(2)}} = -4.271676631815181 \times 10^3 \, \text{MHz}$$

$$f^{(F)}_{\varphi^{(3)}} = 377.107463380 \times 10^6 + 305.43 \, \text{MHz}$$

Correspondingly, the frequencies in atomic unit can be stated as

$$f^{(F)}_{\varphi^{(0)}} = -1.033269165259541 \times 10^{-7} \, \text{a.u.}$$

$$f^{(F)}_{\varphi^{(1)}} = 0.00929408920283 \, \text{a.u.}$$

$$f^{(F)}_{\phi^{(2)}} = -1.033269165259541 \times 10^{-7} \, \text{a.u.}$$

$$f^{(F)}_{\varphi^{(3)}} = 0.00912180071378 \, \text{a.u.}$$

Also, the values of $E^{(m_F)}_{\phi^{(i)}}$ are obtained from Equation 6.23 as

$$E^{(m_F)}_{\phi^{(i)}} = \mu_B g_{F_i} m_{F_i} B_z \tag{6.123}$$

with the meanings of symbols as stated in Equation 6.78 corresponding to the transmit mode. Furthermore, the values of B_z and the hyperfine Landé g-factor g_{F_i} are also treated similarly as discussed before for the transmit node. The calculated hyperfine Landé g-factors are as follows:

$$g_{F_0} = g_F\left(\left|\phi^{(0)}\right\rangle\right) = -\frac{1}{2}, \quad g_{F_1} = g_F\left(\left|\phi^{(1)}\right\rangle\right) = \frac{1}{6},$$

$$g_{F_2} = g_F\left(\left|\phi^{(2)}\right\rangle\right) = -\frac{1}{2} \quad \text{and} \quad g_{F_3} = g_F\left(\left|\phi^{(3)}\right\rangle\right) = \frac{2}{3}$$

6.5.3.2.2 Atomic Hamiltonian

Using basis vectors for the Hilbert space of ^{87}Rb atom and energies for the atomic states associated with those basis vectors, the Hamiltonian for atom can be formulated as

$$H^{(atom)}_R = E_{\phi^{(0)}} \begin{bmatrix} 1 \\ 0 \\ 0 \\ 0 \end{bmatrix} \begin{bmatrix} 1 \\ 0 \\ 0 \\ 0 \end{bmatrix}^T + E_{\phi^{(1)}} \begin{bmatrix} 0 \\ 1 \\ 0 \\ 0 \end{bmatrix} \begin{bmatrix} 0 \\ 1 \\ 0 \\ 0 \end{bmatrix}^T + E_{\phi^{(2)}} \begin{bmatrix} 0 \\ 0 \\ 1 \\ 0 \end{bmatrix} \begin{bmatrix} 0 \\ 0 \\ 1 \\ 0 \end{bmatrix}^T + E_{\phi^{(3)}} \begin{bmatrix} 0 \\ 0 \\ 0 \\ 1 \end{bmatrix} \begin{bmatrix} 0 \\ 0 \\ 0 \\ 1 \end{bmatrix}^T \tag{6.124}$$

where the terms $E_{\phi^{(i)}}$ ($i = 0, 1, 2, 3$) are given by Equation 6.121.

6.5.3.2.3 *Cavity Hamiltonian*

With the previously stated definitions of cavity mode annihilation and creation operators, the Hamiltonian for cavity modes can be written as

$$\hat{H}_R^{(cav0)} = \hbar\omega_{cav}^{(0)} \, \hat{a}_{cav}^{(0)*} \, \hat{a}_{cav}^{(0)} = \omega_{cav}^{(0)} \begin{bmatrix} 0 & 0 & 0 & 0 & \cdots & 0 \\ 0 & 1 & 0 & 0 & \cdots & 0 \\ 0 & 0 & 2 & 0 & \cdots & 0 \\ 0 & 0 & 0 & 3 & \cdots & 0 \\ \cdots & \cdots & \cdots & \cdots & \cdots & \cdots \\ 0 & 0 & 0 & 0 & \cdots & N_0 \end{bmatrix} \tag{6.125}$$

and similarly,

$$\hat{H}_R^{(cav1)} = \omega_{cav}^{(1)} \begin{bmatrix} 0 & 0 & 0 & 0 & \cdots & 0 \\ 0 & 1 & 0 & 0 & \cdots & 0 \\ 0 & 0 & 2 & 0 & \cdots & 0 \\ 0 & 0 & 0 & 3 & \cdots & 0 \\ \cdots & \cdots & \cdots & \cdots & \cdots & \cdots \\ 0 & 0 & 0 & 0 & \cdots & N_1 \end{bmatrix} \tag{6.126}$$

Because cavity modes are resonant with corresponding atomic transitions, the terms $\omega_{cav}^{(0)}$ and $\omega_{cav}^{(1)}$ in Equations 6.125 and 6.126, respectively, can be obtained from the following relations:

$$\left.\begin{aligned} \hbar\omega_{cav}^{(0)} &= E_{\phi^{(1)}} - E_{\phi^{(0)}} \\ \hbar\omega_{cav}^{(1)} &= E_{\phi^{(3)}} - E_{\phi^{(2)}} \end{aligned}\right\} \tag{6.127}$$

6.5.3.2.4 *Laser Hamiltonian*

Using the laser mode creation and annihilation operators stated in Section 6.5.2.2.4, the Hamiltonian for the laser mode L_{01} can be written as

$$\hat{H}_R^{(L01)} = \omega_{L01} \begin{bmatrix} 0 & 0 & 0 & 0 & \cdots & 0 \\ 0 & 1 & 0 & 0 & \cdots & 0 \\ 0 & 0 & 2 & 0 & \cdots & 0 \\ 0 & 0 & 0 & 3 & \cdots & 0 \\ \cdots & \cdots & \cdots & \cdots & \cdots & \cdots \\ 0 & 0 & 0 & 0 & \cdots & N_{L01} \end{bmatrix} \tag{6.128}$$

and that for the laser mode L_{10} as

$$\hat{H}_R^{(L10)} = \omega_{L10} \begin{bmatrix} 0 & 0 & 0 & 0 & \cdots & 0 \\ 0 & 1 & 0 & 0 & \cdots & 0 \\ 0 & 0 & 2 & 0 & \cdots & 0 \\ 0 & 0 & 0 & 3 & \cdots & 0 \\ \cdots & \cdots & \cdots & \cdots & \cdots & \cdots \\ 0 & 0 & 0 & 0 & \cdots & N_{L10} \end{bmatrix} \tag{6.129}$$

As the lasers L_{01} and L_{10} are resonant with atomic transitions, the terms ω_{L01} and ω_{L10} can be determined as the difference in energy between the relevant atomic energy levels as follows:

$$\left. \begin{aligned} \hbar\omega_{L01} &= E_{\phi^{(1)}} - E_{\phi^{(2)}} \\ \hbar\omega_{L10} &= E_{\phi^{(3)}} - E_{\phi^{(1)}} \end{aligned} \right\} \tag{6.130}$$

6.5.3.2.5 Extending Operators to Combined Hilbert Space

In order to combine the individual operators (or to implement interactions between operators), the operators are needed to be extended to the total Hilbert space. Then the interaction Hamiltonians can be found, and finally, all the Hamiltonians can be added to obtain a total Hamiltonian. As mentioned before, there are five individual Hilbert spaces in the receive node. Taking all these Hilbert spaces into consideration, the atomic Hamiltonian can be extended as

$$\hat{H}_R^{(atom-ext)} = \hat{H}_R^{(atom)} \otimes I_{N_0} \otimes I_{N_1} \otimes I_{N_{L01}} \otimes I_{N_{L10}} \tag{6.131}$$

with I_{N_0}, I_{N_1}, $I_{N_{L01}}$, and $I_{N_{L10}}$ as the identity matrices of the sizes N_0, N_1, N_{L01}, and N_{L10}, respectively. Similarly, the cavity mode Hamiltonians are extended as

$$\hat{H}_R^{(cav0-ext)} = I_4 \otimes \hat{H}_R^{(cav0)} \otimes I_{N_1} \otimes I_{N_{L01}} \otimes I_{N_{L10}} \tag{6.132}$$

and

$$\hat{H}_R^{(cav1-ext)} = I_4 \otimes I_{N_0} \otimes \hat{H}_R^{(cav1)} \otimes I_{N_{L01}} \otimes I_{N_{L10}} \tag{6.133}$$

Here I_4 is the 4×4 identity matrix, the size of which being equal to that of the atomic Hamiltonian matrix. The laser mode Hamiltonians are also extended in the same way, that is,

$$\hat{H}_R^{(L01-ext)} = I_4 \otimes I_{N_0} \otimes I_{N_1} \otimes \hat{H}_R^{(L01)} \otimes I_{N_{L10}} \tag{6.134}$$

and

$$\hat{H}_R^{(L10-ext)} = I_4 \otimes I_{N_0} \otimes I_{N_1} \otimes I_{N_{L01}} \otimes \hat{H}_R^{(L10)} \tag{6.135}$$

Further, the raising and the lowering operators are extended as

$$S_{\pm,0}^{(ext)} = S_{\pm,0} \otimes I_{N_0} \otimes I_{N_1} \otimes I_{N_{L01}} \otimes I_{N_{L10}} \tag{6.136}$$

In the same way, creation and annihilation operators can be extended for cavity and laser modes. Because creation operators are just the conjugate transpose of annihilation operators, extensions of only annihilation operators are shown. The cavity mode annihilation operators are extended as

$$\hat{a}_{cav}^{(0-ext)} = I_4 \otimes \hat{a}_{cav}^{(0)} \otimes I_{N_1} \otimes I_{N_{L01}} \otimes I_{N_{L10}} \tag{6.137}$$

and

$$\hat{a}_{cav}^{(1-ext)} = I_4 \otimes I_{N_0} \otimes \hat{a}_{cav}^{(1)} \otimes I_{N_{L01}} \otimes I_{N_{L10}} \tag{6.138}$$

Similarly, annihilation operators for the laser modes L_{01} and L_{10} are extended as

$$\hat{a}_{01}^{(ext)} = I_6 \otimes I_{N_0} \otimes I_{N_1} \otimes \hat{a}_{01} \otimes I_{N_{L1}} \tag{6.139}$$

and

$$\hat{a}_{10}^{(ext)} = I_6 \otimes I_{N_0} \otimes I_{N_1} \otimes I_{N_{L0}} \otimes \hat{a}_{10} \tag{6.140}$$

respectively.

6.5.3.2.6 Cavity–Atom Interaction Hamiltonian

The cavity–atom interaction Hamiltonian in the receive node is expressed in terms of extended operators as

$$\hat{H}_R^{(cav-atom-ext)} = -i\hbar g_{R,0} \left(\hat{a}_{cav}^{(0-ext)} \hat{S}_-^{(ext)*} - \hat{a}_{cav}^{(0-ext)*} \hat{S}_-^{ext} \right)$$
$$- i g_{R,1} (\hat{a}_{cav}^{(1-ext)} \hat{S}_+^{(ext)*} - \hat{a}_{cav}^{(1-ext)*} \hat{S}_+^{(ext)}) \tag{6.141}$$

where $\hat{S}_\pm^{(ext)}$ and $\hat{S}_\pm^{(ext)*}$ are atomic lowering and raising operators corresponding to cavity-assisted transitions. Also, $g_{R,0}$ and $g_{R,1}$ are the atom–cavity mode coupling constants. The suffix $\pm$ indicates increase (+) or decrease (−) in magnetic quantum number while raising from the lower energy level to a higher energy level.

6.5.3.2.7 Atom–Laser Interaction Hamiltonian

At the receive node, the atom–laser interaction Hamiltonian is given in terms of extended operators as

$$\hat{H}_R^{(laser-atom-ext)} = -i\hbar g_{01} (\hat{a}_{01}^{(ext)} \hat{S}_{01}^{(ext)*} - \hat{a}_{01}^{(ext)*} \hat{S}_{01}^{(ext)})$$
$$- i\hbar g_{10} (\hat{a}_{10}^{(ext)} \hat{S}_{10}^{(ext)*} - \hat{a}_{10}^{(ext)*} \hat{S}_{10}^{(ext)}) \tag{6.142}$$

where $\hat{a}_{01}^{(ext)}$ and $\hat{a}_{01}^{(ext)*}$ are the extended annihilation and creation operators, respectively, for the laser mode L_{01}, and $\hat{a}_{10}^{(ext)}$ and $\hat{a}_{10}^{(ext)*}$ are those for the laser mode L_{10}. Also, $\hat{S}_{01}^{(ext)}$ and $\hat{S}_{01}^{(ext)*}$ are the extended lowering and raising operators, respectively, corresponding to the laser L_{01}-assisted transition, and $\hat{S}_{10}^{(ext)}$ and $\hat{S}_{10}^{(ext)*}$ are the same for the transitions corresponding to the laser L_{10}.

6.5.3.2.8 Total Hamiltonian

Once the extended Hamiltonians are derived for system components, the total Hamiltonian can be formulated by adding these extended component Hamiltonians to obtain

$$\mathcal{H}_R = \mathcal{H}_R^{(atom-ext)} + \mathcal{H}_R^{(cav0-ext)} + \mathcal{H}_R^{(cav1-ext)} + \mathcal{H}_R^{(L01-ext)} + \mathcal{H}_R^{(L10-ext)}$$

$$+ \mathcal{H}_R^{(cav-atom-ext)} + \mathcal{H}_R^{(laser-atom-ext)} \tag{6.143}$$

6.5.3.3 Density Matrices

In this section, the initial density matrix is developed for the receive node by taking the tensor product of the initial density matrices of rubidium atom, cavity modes, and laser modes. This construction is valid under the assumption that, initially at time $t=0$, the components (atom, cavity, and lasers) of the receive node are noninteracting.

6.5.3.3.1 Density Matrix of Atom

Using Equation 6.117, if the atom is initially in the $|\varphi_0\rangle$ eigenstate, the probability of occupation of this state is 1, and that of all the other states is 0. In that case, the initial density matrix of atom can be defined as

$$\rho_R^{(atom)}\Big|_{t=0} = 1.0 \times |\phi_0\rangle\langle\phi_0| + 0.0 \times |\phi_0'\rangle\langle\phi_0'| + 0.0 \times |\phi_1\rangle\langle\phi_1| + 0.0 \times |\phi_1'\rangle\langle\phi_1'|$$

Thus, the initial atomic density matrix is given by

$$\rho_R^{(atom)}\Big|_{t=0} = \begin{bmatrix}1\\0\\0\\0\end{bmatrix}\begin{bmatrix}1\\0\\0\\0\end{bmatrix}^T = \begin{bmatrix} 1 & 0 & 0 & 0 \\ 0 & 0 & 0 & 0 \\ 0 & 0 & 0 & 0 \\ 0 & 0 & 0 & 0 \end{bmatrix} \tag{6.144}$$

6.5.3.3.2 Cavity Mode Density Matrix

Initially, the cavity modes are assumed to contain no photons. Using this initial condition and basis vectors of Equation 6.118, the initial density matrix can be defined for the first cavity mode as

$$\rho_R^{(cav0)}\Big|_{t=0} = 1.0 \times |0\rangle\langle0| + 0.0 \times |1\rangle\langle1| + 0.0 \times |2\rangle\langle2| + \cdots + |N\rangle\langle N|$$

$$= \begin{bmatrix}1\\0\\0\\0\\ \cdots \\0\end{bmatrix}\begin{bmatrix}1\\0\\0\\0\\ \cdots \\0\end{bmatrix}^T = \begin{bmatrix} 1 & 0 & 0 & 0 & \cdots & 0 \\ 0 & 0 & 0 & 0 & \cdots & 0 \\ 0 & 0 & 0 & 0 & \cdots & 0 \\ 0 & 0 & 0 & 0 & \cdots & 0 \\ \cdots & \cdots & \cdots & \cdots & \cdots & \cdots \\ 0 & 0 & 0 & 0 & \cdots & 0 \end{bmatrix} \tag{6.145}$$

Similarly, the initial density matrix for the second cavity mode can be written as

$$
\rho_R^{(cav1)}\Big|_{t=0} = \begin{bmatrix}
1 & 0 & 0 & 0 & \cdots & 0 \\
0 & 0 & 0 & 0 & \cdots & 0 \\
0 & 0 & 0 & 0 & \cdots & 0 \\
0 & 0 & 0 & 0 & \cdots & 0 \\
\cdots & \cdots & \cdots & \cdots & \cdots & \cdots \\
0 & 0 & 0 & 0 & \cdots & 0
\end{bmatrix}
\tag{6.146}
$$

6.5.3.3.3 *Laser Density Matrix*

The basis vectors for laser modes are given in Section 6.5.3.1.3. Because, in the receive node, lasers are turned on only after a time interval in synchronization, initially laser modes are assumed to contain no photons. Therefore, the initial density matrices for the receive node laser modes L_{01} and L_{10} can finally be given as

$$
\rho_R^{(L01)}\Big|_{t=0} = \rho_R^{(L10)}\Big|_{t=0} = \begin{bmatrix}
1 & 0 & 0 & 0 & \cdots & 0 \\
0 & 0 & 0 & 0 & \cdots & 0 \\
0 & 0 & 0 & 0 & \cdots & 0 \\
0 & 0 & 0 & 0 & \cdots & 0 \\
\cdots & \cdots & \cdots & \cdots & \cdots & \cdots \\
0 & 0 & 0 & 0 & \cdots & 0
\end{bmatrix}
\tag{6.147}
$$

6.5.3.3.4 *Combined Density Matrix*

As initially atom, laser modes, and cavity modes are assumed to be noninteracting, the combined density matrix can be constructed for the receive node as

$$
\rho_R\big|_{t=0} = \rho_R^{(atom)}\Big|_{t=0} \otimes \rho_R^{(cav0)}\Big|_{t=0} \otimes \rho_R^{(cav1)}\Big|_{t=0} \otimes \rho_R^{(L01)}\Big|_{t=0} \otimes \rho_R^{(L10)}\Big|_{t=0}
\tag{6.148}
$$

the explicit form of which is not incorporated into the text.

6.5.3.4 **System Evolution**

Contrary to the case of transmit node, the system evolution is analyzed in one step—namely, the open-system evolution analysis. This is because in the receive node, quantum process starts only after the reception of a photon, which is possible only in the case of open system.

6.5.3.4.1 *Open-System Analysis*

For open-system evolution, the master equation in the Lindblad form [65] is adopted, as stated in Equation 6.67. The evolution starts with the initial density matrix as follows:

$$
\frac{d}{dt}\rho_R\Big|_{t=0} = L_T\,\rho_R\big|_{t=0}
\tag{6.149}
$$

Once the evolution starts, the density matrix undergoes various changes to reflect transition between hyperfine states and increase or decrease in the number of photons in cavity and laser modes.

6.6 Numerical Investigation

Complex systems are often studied by developing models supporting the underlying physics and using computationally intensive methods to acquire knowledge in respect of behavior of systems. An effective way to study a theoretically motivated model is to transform the original model equations into computable algorithms. In the present section, a simulation framework is first developed for the proposed system. In the process, Hilbert spaces and density matrices are customized so that the simulation can be performed within finite space (memory) and time. Then simulations in respect of the closed- and the open-system evolutions of the proposed system are performed. Finally, detailed analyses of simulation results are presented under different system configurations and initial conditions.

6.6.1 Simulation Framework

In this section, dynamics of the proposed system is explicitly presented as a differential equation. The equation of motion for the proposed system is given by the von Neumann equation [64] for the closed-system evolution and by the master equation in the Lindblad form [65] for the open-system evolution. The von Neumann equation for the closed-system evolution is as follows:

$$\frac{d}{dt}\rho_c(t) = \frac{1}{i\hbar}\left[\hat{H}, \rho_c(t)\right] \tag{6.150}$$

This differential equation is of the first order in t and gives the rate of change of density matrix in terms of commutator of the Hamiltonian $\hat{H}$ and the density matrix $\rho(t)$. Following the Euler integration formula [70], for sufficiently small step of time Δt, it can be written that

$$\rho_c(t+\Delta t) = \rho_c(t) + \Delta t \cdot \frac{1}{i\hbar}\left(\hat{H}\rho_c(t) - \rho_c(t)\hat{H}\right) \tag{6.151}$$

Similarly, for the open-system evolution, the master equation in the Lindblad form can be written as

$$\frac{d}{dt}\rho(t) = L\rho(t) \tag{6.152}$$

where the linear operator L is given by Equation 6.68. Again, following the Euler integration formula [70], for sufficiently small time step Δt, it can be finally written that

$$\rho(t+\Delta t) = \rho(t) + \Delta t \cdot L\rho(t)$$

which can be expanded as

$$\rho(t+\Delta t)=\rho(t)+\Delta t\left\{\frac{1}{i\hbar}\left(\hat{H}\rho(t)-\rho(t)\hat{H}\right)+\sum_i\left(\hat{C}_i\rho(t)\,\hat{C}_i^\dagger-\frac{1}{2}\hat{C}_i^\dagger\hat{C}_i\rho(t)-\frac{1}{2}\rho(t)\hat{C}_i^\dagger\hat{C}_i\right)\right\}$$

(6.153)

Equations 6.151 and 6.152 are recursive equations and, therefore, suitable for computer simulation.

6.6.2 Simulation and Results

A MATLAB® programming-based simulation platform is developed wherein the basic simulation process is the same for both the transmit and the receive nodes. Therefore, the basic simulation mechanism is described here in common terms. Simulation is performed in an iterative manner. At every iteration, two basic functions are performed—(1) timing control and (2) evolution update. The timing control consists of increasing the simulation time by Δt at every iteration, keeping track of simulation time and terminating the program when the total simulation time exceeds a predefined threshold; Δt is set at 200 a.u.

Evolution update consists of updating (at every iteration) the quantum state of system following Equation 6.151 or 6.153 depending on whether it is a closed-system or open-system simulation. As dictated by the equations, first, the rate of evolution ($d\rho/dt$) is calculated using Equation 6.150 or 6.152. This quantity is multiplied by the simulation time interval Δt, and then the product (which gives the change in density matrix ρ during the time interval Δt) is added to ρ. Thus, ρ, which contains the complete information of system state, evolves. This evolution process through the iteration cycles continues until the predefined time threshold is reached.

6.6.2.1 Closed-System Simulation

As mentioned earlier, behavior of the transmit node is simulated without any interaction with the environment. The basic simulation process is applied on the transmit node density matrix $\rho_T(t)$. The evolution of density matrix is given by Equation 6.151, and the relevant Hamiltonian $\hat{H}_T$ is given by Equation 6.68. Formally, the closed-system evolution equation can be written as

$$\rho_T(t+\Delta t)=\rho_T(t)+\Delta t\cdot\frac{1}{i\hbar}\left(\hat{H}_T\rho_T(t)-\rho_T(t)\hat{H}_T\right)$$

(6.154)

This evolution starts with an initial value of ρ_T, that is, $\rho_T|_{t=0}$, which is given by Equation 6.114. It is tested with two initial configurations. First, the simulation is tested with the transmit node in the logic state "0," and the results are shown in Figures 6.17 through 6.19. In these figures, probabilities of occupation of atomic eigenstates are shown as they evolve with time. As mentioned, closed-system simulation is naturally limited to the transmit node only.

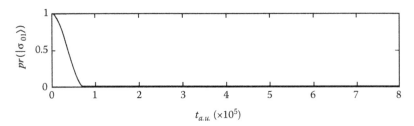

FIGURE 6.17
Probability of occupation of the state $|\sigma_{01}\rangle$.

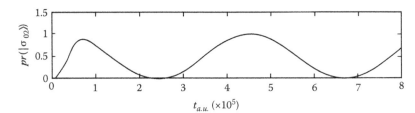

FIGURE 6.18
Probability of occupation of the state $|\sigma_{02}\rangle$.

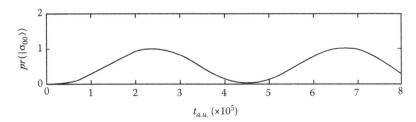

FIGURE 6.19
Probability of occupation of the state $|\sigma_{00}\rangle$.

Figure 6.17 shows the probability of occupation of the state $|\sigma_{01},C_00,C_10,L_23,L_33\rangle$ in the combined Hilbert space of the transmit node. Concentrating on the atomic state, it is considered to be the probability of occupation of the atomic state $|\sigma_{01}\rangle$, denoted by $pr(|\sigma_{01}\rangle)$. In the figure, it is seen that the probability of occupation, which was initially 1.0, decreased with time and reached 0.0 before time 1.0×10^5 a.u. This means that the atom left the state $|\sigma_{01}\rangle$ and entered some other state, which will be clear from the results presented in other figures. This transition was caused by the laser $L_{12}^{(0)}$ (i.e., L_2), which was turned on at $t = 0$. Figure 6.18 shows the probability of occupation of the state $|\sigma_{02},C_00,C_10,L_22,L_33\rangle$ in the combined Hilbert space. Concentrating on the atomic state, it can be said that Figure 6.18 shows the probability of occupation of the atomic state $|\sigma_{02}\rangle$, denoted by $pr(|\sigma_{02}\rangle)$.

At time $t = 0$, the probability $pr(|\sigma_{02}\rangle)$ was 0.0. With the passage of simulation time, it increased and reached almost 1.0 before time $t = 1.0 \times 10^5$ a.u. Now, analyzing Figures 6.17 and 6.18 together during the said interval, it is seen that while the probability of occupation of the atomic state $|\sigma_{01}\rangle$ decreased from 1.0 to 0.0, the probability of the atomic state $|\sigma_{02}\rangle$ increased from 0.0 to almost 1.0. That is, the rubidium atom made a transition from the state $|\sigma_{01}\rangle$ to $|\sigma_{02}\rangle$. Again, if the combined states $|\sigma_{01},C_00,C_10,L_23,L_33\rangle$ and

$|\sigma_{02},C_0 0,C_1 0,L_2 2,L_3 3\rangle$ are compared together, it is observed that the latter state contains one photon less than the former one. That is, during the transition, one photon was absorbed. Now, this transition can be concisely described as follows. When the lasers $L_{12}^{(0)}$ and $L_{12}^{(1)}$ were turned on, the atom in the state $|\sigma_{01}\rangle$ absorbed a photon from the laser $L_{12}^{(0)}$; after the photon absorption, being in the excited state, the atom jumped to the state $|\sigma_{02}\rangle$.

After approximately $t = 0.60 \times 10^5$ a.u., $pr(|\sigma_{02}\rangle)$ started to decrease, as is evident from Figure 6.18. During this time, as seen in Figure 6.19, the probability of occupation of the atomic state $|\sigma_{00}\rangle$, denoted by $pr(|\sigma_{00}\rangle)$, started to increase. Considering the total Hilbert space, Figure 6.19 shows the probability of occupation of the state $|\sigma_{00},C_0 1,C_1 0,L_2 2,L_3 3\rangle$. When Figures 6.18 and 6.19 are considered together, it can be observed that, after $t = 0.60 \times 10^5$ a.u., while $pr(|\sigma_{02}\rangle)$ decreased, $pr(|\sigma_{00}\rangle)$ increased. That is, the atom made a transition from the state $|\sigma_{02}\rangle$ to the state $|\sigma_{00}\rangle$. Also, from the corresponding states in the combined Hilbert space $|\sigma_{02},C_0 0,C_1 0,L_2 2,L_3 3\rangle$ and $|\sigma_{00},C_0 1,C_1 0,L_2 2,L_3 3\rangle$, it can be understood that a photon was generated during this time in the cavity mode C_0.

Summarizing the observations, it can be said that, after $t = 0.60 \times 10^5$ a.u., the atom emitted a photon into the cavity C_0 due to atom–cavity coupling and made a transition from the excited state $|\sigma_{02}\rangle$ to the ground state $|\sigma_{00}\rangle$. This transition completed at around $t = 2.3 \times 10^5$ a.u. when $pr(|\sigma_{02}\rangle)$ was reduced to 0, and $pr(|\sigma_{00}\rangle)$ became 1.0. Also, from the definitions of $|\sigma_{02}\rangle$ and $|\sigma_{00}\rangle$ in Equation 6.24, it is known that the $|\sigma_{02}\rangle$ state has higher magnetic quantum number than the state $|\sigma_{00}\rangle$. Therefore, the emitted photon is right circularly polarized. Thus, the logic "0" state of rubidium atom was converted into the photon state with right-circular polarization.

Interestingly, after time around $t = 2.3 \times 10^5$ a.u., $pr(|\sigma_{02}\rangle)$ started to increase while $pr(|\sigma_{00}\rangle)$ started to decrease. This transition was completed when $pr(|\sigma_{02}\rangle)$ got maximized to 1.0 and $pr(|\sigma_{00}\rangle)$ minimized to 0.0 at around $t = 4.5 \times 10^5$ a.u. That is, the atom reabsorbed the photon from the cavity and went back to the excited state. This completed one cycle of oscillation. As seen from Figures 6.18 and 6.19, this oscillation between the ground state $|\sigma_{00}\rangle$ and the excited state $|\sigma_{02}\rangle$ continues without damping. The reason for this undamped oscillation is that the performed simulation was a closed-system one. Therefore, the transmit node was isolated from the environment, and the emitted photon could not be dispatched out. As a result, the recycled photon was reabsorbed by the atom, and the excited atom jumped to the higher-energy state $|\sigma_{02}\rangle$; the cycle continued.

Similarly, the closed-system simulation was performed with the transmit node initialized in logic "1." The corresponding simulation results are shown in Figures 6.20 through 6.22. Contrary to the case of logic "0," in the case of logic "1," when the two lasers $L_{12}^{(0)}$ and $L_{12}^{(1)}$ were turned on with the atom in the $|\sigma_{11}\rangle$ state, the atom absorbed a photon from the laser $L_{12}^{(1)}$ and raised to the state $|\sigma_{12}\rangle$. This is evident from Figures 6.20 and 6.21. Again,

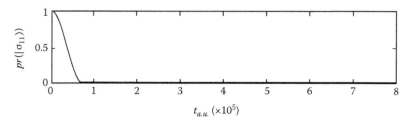

FIGURE 6.20
Probability of occupation of the state $|\sigma_{11}\rangle$.

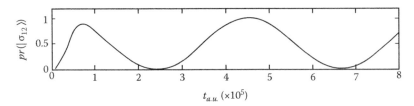

FIGURE 6.21
Probability of occupation of the state $|\sigma_{12}\rangle$.

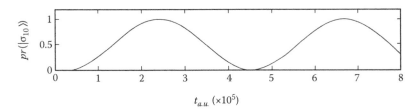

FIGURE 6.22
Probability of occupation of the state $\langle\sigma_{10}|$.

if Figures 6.21 and 6.22 are examined together, it can be observed that, after $t = 0.60 \times 10^5$ a.u., the probability of occupation of the state $|\sigma_{12}\rangle$ decreased, while that of the state $|\sigma_{10}\rangle$ increased. That is, the atom emitted a photon into the cavity and jumped down to the ground state $|\sigma_{10}\rangle$. As the magnetic quantum number in the $|\sigma_{10}\rangle$ state is more than that in the state $|\sigma_{12}\rangle$, the emitted photon is left circularly polarized. Thus, the logic "1" state of the atom was mapped into a photon state with left-circular polarization. Because the node is isolated from the environment, similar to the case of logic "0," the emitted photon is reabsorbed by the atom, and the excited atom made a transition to the state $|\sigma_{12}\rangle$. Thus, an oscillation between the ground state $|\sigma_{10}\rangle$ and the excited state $|\sigma_{12}\rangle$ started and continued the same without damping.

6.6.2.2 Open-System Simulation

In the open-system simulation, both the transmit and the receive nodes are simulated simultaneously. The evolution procedure is applied on the density matrices of both nodes. Evaluation equations are derived from Equation 6.153 as

$$\rho_T(t + \Delta t) = \rho_T(t) + \Delta t \times \left\{ \frac{1}{i\hbar} \left(\hat{H}_T \rho_T(t) - \rho_T(t) \hat{H}_T \right) \right.$$

$$\left. + \sum_i \left(\hat{C}_{Ti} \rho_T(t) \hat{C}_{Ti}^\dagger - \frac{1}{2} \hat{C}_{Ti}^\dagger \hat{C}_{Ti} \rho_T(t) - \frac{1}{2} \rho_T(t) \hat{C}_{Ti}^\dagger \hat{C}_{Ti} \right) \right\} \qquad (6.155)$$

$$\rho_R(t + \Delta t) = \rho_R(t) + \Delta t \times \left\{ \frac{1}{i\hbar} \left(\hat{H}_R \rho_R(t) - \rho_R(t) \hat{H}_R \right) \right.$$

$$\left. + \sum_i \left(\hat{C}_{Ri} \rho_R(t) \hat{C}_{Ri}^\dagger - \frac{1}{2} \hat{C}_{Ri}^\dagger \hat{C}_{Ri} \rho_R(t) - \frac{1}{2} \rho_R(t) \hat{C}_{Ri}^\dagger \hat{C}_{Ri} \right) \right\} \qquad (6.156)$$

Equations 6.155 and 6.156, respectively, represent evolution equations for the transmit and the receive nodes. The jump operators $\hat{C}_{Ti}(i=0,1,+,-)$ are given by Equation 6.69, and $\hat{C}_{Ri}(i=0,1,+,-)$ by Equation 6.69.

In the open-system evolution, photon generated in the transmit node propagates through the fiber, enters the receive node cavity, and triggers relevant quantum processes there. However, the simulation of photon propagation is not included here. As the proposed system is designed to transfer the quantum state of the transmit node to the receive one, the transmit/receive nodes are prepared in different logic states, and it is examined (through simulation) how the quantum state of the transmit node propagates to the receive one. Thus, two different configurations were tested, namely, (1) the transmit node was prepared in the logic state "0" and the receive node in the logic state "1," and (2) the transmit node was prepared in the logic state "1" and the receive node in the logic state "0." The simulation was performed with $\kappa = 0.6 \times 10^{-3}$ eV, $\gamma = 10^{-5}$ eV in the transmit node and $\kappa = 0.1 \times 10^{-3}$ eV, $\gamma = 3.0 \times 10^{-6}$ eV in the receive node.

Corresponding to the previously mentioned first initial configuration, the simulation results are shown in Figures 6.23 through 6.30. Figure 6.23 shows the probability of occupation of the state $|\sigma_{01}, C_0 0, C_1 0, L_2 3, L_3 3\rangle$ in the combined Hilbert space of the transmit

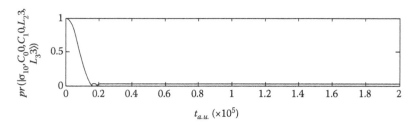

FIGURE 6.23
Probability of occupation of the state $|\sigma_{01}, C_0 0, C_1 0, L_2 3, L_3 3\rangle$.

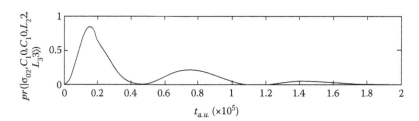

FIGURE 6.24
Probability of occupation of the state $|\sigma_{02}, C_1 0, C_1 0, L_2 2, L_3 3\rangle$.

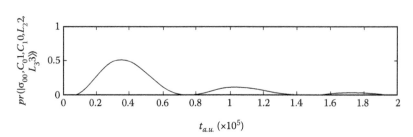

FIGURE 6.25
Probability of occupation of the state $|\sigma_{00}, C_0 1, C_1 0, L_2 2, L_3 3\rangle$.

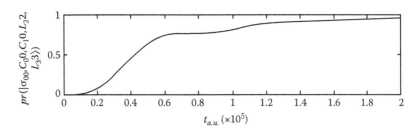

FIGURE 6.26
Probability of occupation of the state $|\sigma_{00},C_00,C_10,L_22,L_33\rangle$.

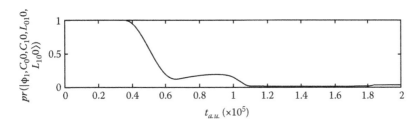

FIGURE 6.27
Probability of occupation of the state $|\varphi_1,C_00,C_10,L_{01}0,L_{10}0\rangle$.

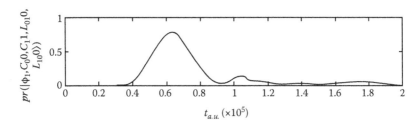

FIGURE 6.28
Probability of occupation of the state $|\varphi_1,C_00,C_11,L_{01}0,L_{10}0\rangle$.

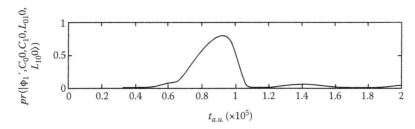

FIGURE 6.29
Probability of occupation of the state $|\phi_1',C_00,C_10,L_{01}0,L_{10}0\rangle$.

node, that is, the state $|\sigma_{01}\rangle$ in the atomic Hilbert space. As seen in the figure, probability decreased from 1.0, reached 0.0 at $t = 0.15 \times 10^5$ a.u., and stayed there throughout the simulation time. This implies that ^{87}Rb atom left the state $|\sigma_{01}\rangle$ and never came back to the state again.

Figure 6.24 depicts the probability of occupation of the state $|\sigma_{02},C_00,C_10,L_22,L_33\rangle$, that is, the state $|\sigma_{02}\rangle$. This probability increased starting from 0.0 and reached the maximum

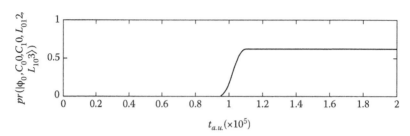

FIGURE 6.30
Probability of occupation of the state $|\varphi_0, C_0 0, C_1 0, L_{01} 2, L_{10} 3\rangle$.

value (around 0.85) at time $t = 0.15 \times 10^5$ a.u. Therefore, the atom entered the state $|\sigma_{02}\rangle$ during this time. Now, if the results in Figures 6.23 and 6.24 are considered together, the inference can be drawn that during this time interval, the atom made a transition from the state $|\sigma_{01}\rangle$ to the state $|\sigma_{02}\rangle$. Also, comparing the combined states $|\sigma_{01}, C_0 0, C_1 0, L_2 3, L_3 3\rangle$ and $|\sigma_{02}, C_0 0, C_1 0, L_2 2, L_3 3\rangle$, it can be seen that the atom absorbed a photon from the laser L_2, that is, $L_{12}^{(0)}$ during this transition. Beyond the time $t = 0.15 \times 10^5$ a.u., the probability for the state $|\sigma_{02}\rangle$ started to decrease and reached 0.0 at around $t = 0.45 \times 10^5$ a.u. This means, during the interval $t = 0.15 \times 10^5$ a.u. to $t = 0.45 \times 10^5$ a.u., the atom left the state $|\sigma_{02}\rangle$. Beyond $t = 0.45 \times 10^5$ a.u., the probability value had undergone damped oscillations and, finally, decayed to 0.

The probability of occupation of the state $|\sigma_{00}, C_0 1, C_1 0, L_2 2, L_3 3\rangle$, that is, the probability for the atomic state $|\sigma_{00}\rangle$ with a photon in the cavity, is shown in Figure 6.25. This probability is denoted by $pr(|\sigma_{00}, C_0 1, C_1 0, L_2 2, L_3 3\rangle)$ and is observed to be initially 0. It started increasing after $t = 0.10 \times 10^5$ a.u. and reached the maximum value (around 0.55) at $t = 0.35 \times 10^5$ a.u. Now, if the results in Figures 6.24 and 6.25 are compared during the interval $t = 0.15 \times 10^5$ a.u. to $t = 0.35 \times 10^5$ a.u., it can be concluded that the atom decayed from the excited state $|\sigma_{02}\rangle$ to the ground state $|\sigma_{00}\rangle$ through the emission of a photon into the cavity C_0. Also, comparing the states $|\sigma_{02}\rangle \equiv (5^2 P_{1/2}, F = 2, m_F = 1)$ and $|\sigma_{00}\rangle \equiv (5^2 S_{1/2}, F = 1, m_F = 0)$, the magnetic quantum number m_F is observed to be decreased by 1 during the transition. Thus, the emitted photon was a right circularly polarized photon. This generation of a right circularly polarized photon from the atom at the logic state "0" is the mapping of the logic state into the photon state. After $t = 0.35 \times 10^5$ a.u., $pr(|\sigma_{00}, C_0 1, C_1 0, L_2 2, L_3 3\rangle)$ decreased and reached 0 at $t = 0.75 \times 10^5$ a.u. Afterward, some oscillations were observed, which eventually damped to 0.

Figure 6.26 shows the probability of the state $|\sigma_{00}, C_0 0, C_1 0, L_2 2, L_3 3\rangle$, denoted by $pr(|\sigma_{00}, C_0 0, C_1 0, L_2 2, L_3 3\rangle)$. This probability followed the probability $pr(|\sigma_{00}, C_0 1, C_1 0, L_2 2, L_3 3\rangle)$, increased from 0, and continued to increase after $pr(|\sigma_{00}, C_0 1, C_1 0, L_2 2, L_3 3\rangle)$ started to decrease beyond $t = 0.35 \times 10^5$ a.u. The results in Figures 6.25 and 6.26 can be interpreted together as follows.

When the atom made a transition from the state $|\sigma_{02}\rangle$ to the state $|\sigma_{00}\rangle$, it emitted a photon in the cavity C_0. The emitted photon started to leak out of the cavity. This resulted in decrease in the probability $pr(|\sigma_{00}, C_0 1, C_1 0, L_2 2, L_3 3\rangle)$ and increase in the probability $pr(|\sigma_{00}, C_0 0, C_1 0, L_2 2, L_3 3\rangle)$. Finally, $pr(|\sigma_{00}, C_0 1, C_1 0, L_2 2, L_3 3\rangle)$ became 0 and $pr(|\sigma_{00}, C_0 0, C_1 0, L_2 2, L_3 3\rangle)$ approached unity, implying thereby the photon leaked out completely and the atom settled down to the ground state $|\sigma_{00}\rangle$.

Figures 6.27 through 6.30 depict the quantum processes that took place in the receive node after the emitted photon reached the receive node through fiber. Figure 6.28 shows the probability $pr(|\phi_1,C_00,C_10,L_{01}0,L_{10}0\rangle)$ of occupation of the state $|\phi_1,C_00,C_10,L_{01}0,L_{10}0\rangle$, that is, the probability of the state $|\phi_1\rangle$ that represents the logic state "1." After time $t = 0.35 \times 10^5$ a.u., the probability started decreasing, while the probability $pr(|\phi_1,C_00,C_11,L_{01}0,L_{10}0\rangle)$ of occupation of the state $|\phi_1,C_00,C_11,L_{01}0,L_{10}0\rangle$, as depicted in Figure 6.28, started increasing after $t = 0.35 \times 10^5$ a.u. If seen together, Figures 6.27 and 6.28 indicate that a photon was received in the cavity C_1 of the receive node after $t = 0.35 \times 10^5$ a.u. However, the probability of a photon leak out was seen after $t = 0.15 \times 10^5$ a.u. (Figure 6.26). Thus, the delay between $t = 0.15 \times 10^5$ a.u. and $t = 0.35 \times 10^5$ a.u., that is, 0.20×10^5 a.u., is the propagation delay in fiber.

As seen in Figure 6.28, the probability $pr(|\phi_1,C_00,C_11,L_{01}0,L_{10}0\rangle)$ started decreasing after $t = 0.65 \times 10^5$ a.u. when the probability $pr(|\phi_1',C_00,C_10,L_{01}0,L_{10}0\rangle)$ of occupation of the state $|\phi_1',C_00,C_10,L_{01}0,L_{10}0\rangle$, as shown in Figure 6.29, started increasing. Therefore, if seen together, Figures 6.28 and 6.29 indicate that the atom in the receive node absorbed the received photon from the cavity mode C_1 and, being excited, jumped to the higher-energy state $|\phi_1'\rangle$.

In Figure 6.29, it can be seen that the probability $pr(|\phi_1',C_00,C_10,L_{01}0,L_{10}0\rangle)$ reached its maximum value at $t = 0.90 \times 10^5$ a.u. At this time, lasers in the receive node were turned on. Due to the interaction of atom with laser, beyond this point in time, the probability $pr(|\phi_1',C_00,C_10,L_{01}0,L_{10}0\rangle)$ started decreasing and decayed to 0 with small oscillations. As expected, after 0.90×10^5 a.u., the probability $pr(|\phi_0,C_00,C_10,L_{01}2,L_{10}3\rangle)$ of occupation of the state $|\phi_0,C_00,C_10,L_{01}2,L_{10}3\rangle$, as depicted in Figure 6.30, started increasing. This probability reached the maximum value (≈ 0.7) at around $t = 1.1 \times 10^5$ a.u. and maintained the same for the remaining simulation time. Therefore, comparing the simulation results in Figures 6.29 and 6.30, it can be concluded that, after $t = 0.90 \times 10^5$ a.u., the atom emitted a photon into the laser mode L_{10} and decayed to the ground state $|\phi_0\rangle$, which represents the logic state "0" in the receive node. Thus, the logic state "0" of the transmit node was transferred to the receive node via an intermediate photon state.

Corresponding to the second initial configuration, where the transmit node is in the logic "1" state and the receive node in the logic "0," simulation results are shown in Figures 6.31 through 6.38. An analysis of these figures (similar to the one made for the first initial configuration) shows that, in this case, the logic state "1" was transferred from the transmit to the receive node via a photon state. Decrease in the probability of occupation of the state $|\sigma_{11},C_00,C_10,L_23,L_33\rangle$ (denoted by $pr(|\sigma_{11},C_00,C_10,L_23,L_33\rangle)$), as seen in Figure 6.31, and increase in the probability of occupation of the state $|\sigma_{12},C_00,C_10,L_23,L_32\rangle$ (denoted by $pr(|\sigma_{12},C_00,C_10,L_23,L_32\rangle)$), as seen in Figure 6.32, together imply that the atom in the transmit node made a transition from the state $|\sigma_{11}\rangle$ to the state $|\sigma_{12}\rangle$ with the absorption of a photon from laser L_3 (i.e., laser $L_{12}^{(1)}$).

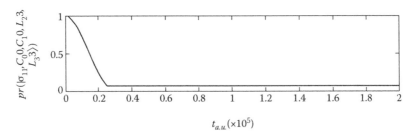

FIGURE 6.31
Probability of occupation of the state $|\sigma_{11},C_00,C_10,L_23,L_33\rangle$.

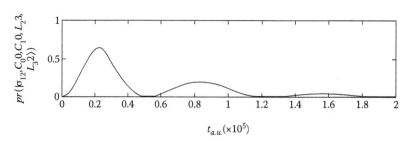

FIGURE 6.32
Probability of occupation of the state $|\sigma_{12}, C_00, C_10, L_23, L_32\rangle$.

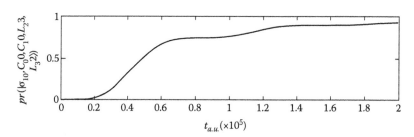

FIGURE 6.33
Probability of occupation of the state $|\sigma_{10}, C_00, C_11, L_23, L_32\rangle$.

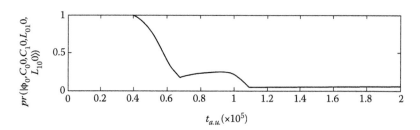

FIGURE 6.34
Probability of occupation of the state $|\sigma_{10}, C_00, C_10, L_23, L_32\rangle$.

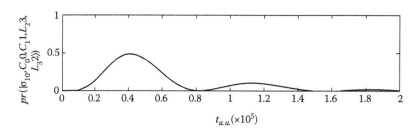

FIGURE 6.35
Probability of occupation of the state $|\phi_0, C_00, C_10, L_{01}0, L_{10}0\rangle$.

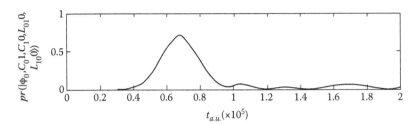

FIGURE 6.36
Probability of occupation of the state $|\phi_0,C_0 1,C_1 0,L_{01} 0,L_{10} 0\rangle$.

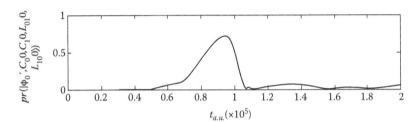

FIGURE 6.37
Probability of occupation of the state $|\phi_0',C_0 0,C_1 0,L_{01} 0,L_{10} 0\rangle$.

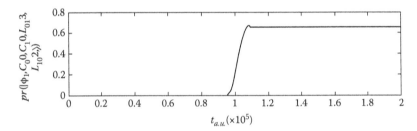

FIGURE 6.38
Probability of occupation of the state $|\phi_1,C_0 0,C_1 0,L_{01} 3,L_{10} 2\rangle$.

The probability $pr(|\sigma_{12},C_0 0,C_1 0,L_2 3,L_3 2\rangle)$ started to decrease after time $t=0.25\times 10^5$ a.u., as seen in Figure 6.32. On the other hand, the probability of occupation of the state $|\sigma_{10},C_0 0,C_1 1,L_2 3,L_3 2\rangle$, denoted by $pr(|\sigma_{10},C_0 0,C_1 1,L_2 3,L_3 2\rangle)$, increased at and after time $t=0.25\times 10^5$ a.u (Figure 6.33). Considering these two events together, it can be concluded that the atom emitted a photon in the cavity C_1 and decayed to the ground state $|\sigma_{10}\rangle$. As the magnetic quantum number increases during the transition from $|\sigma_{12}\rangle \equiv (5^2 P_{3/2}, F=2, m_F=0)$ to $|\sigma_{10}\rangle \equiv (5^2 S_{1/2}, F=1, m_F=1)$, the emitted photon was left circularly polarized. Thus, the logic state "1" of the atom was mapped on the left circularly polarized photon state.

As seen in Figure 6.33, the probability $pr(|\sigma_{10},C_0 0,C_1 1,L_2 3,L_3 2\rangle)$ also started decreasing after time $t=0.40\times 10^5$ a.u. and reached 0 at $t=0.80\times 10^5$ a.u. Afterward, the probability decayed to 0 with minor oscillations. Increase in the probability of occupation of the state $|\sigma_{10},C_0 0,C_1 0,L_2 3,L_3 2\rangle$ at and after $t=0.40\times 10^5$ a.u. (Figure 6.34) implied leakage of the emitted photon out of cavity and coupled into the fiber.

Figure 6.35 shows decrease in the probability of occupation of the state $|\phi_0, C_0 0, C_1 0, L_{01} 0, L_{10} 0\rangle$, whereas Figure 6.36 illustrates decrease in the probability of occupation of the state $|\phi_0, C_0 1, C_1 0, L_{01} 0, L_{10} 0\rangle$ after time $t = 0.40 \times 10^5$ a.u. These two events together imply the reception of a photon into cavity with atom in the $|\phi_0\rangle$ state.

It can be observed in Figure 6.37 that the probability of occupation of the state $|\phi_0', C_0 0, C_1 0, L_{01} 0, L_{10} 0\rangle$ increased after time $t = 0.80 \times 10^5$ a.u., which means that the atom in the receive node absorbed the received photon and jumped to the excited state $|\phi_0'\rangle$. Finally, decrease in the occupation probability for the state $|\phi_0', C_0 0, C_1 0, L_{01} 0, L_{10} 0\rangle$ and increase in the occupation probability for the state $|\phi_1, C_0 0, C_1 0, L_{01} 3, L_{10} 2\rangle$ (Figure 6.38), both around $t = 0.10 \times 10^5$ a.u., indicated that the atom emitted a photon into the laser mode L_{01} and made a transition to the state $|\phi_1\rangle$, which represents the logic "1" state in the receive node. Thus, the logic state "1" of the transmit node was transferred to the receive node via a left circularly polarized photon state.

6.7 Conclusion

The aforesaid descriptions would lead to the conclusion that an enhanced framework for the construction of nanophotonic waveguidance system is presented using an optical fiber as photon waveguide. Simulation platform is developed for the (previously reported) spin–photon interface. Studies are made of the system behavior involving nanophotonic processes taking place at the interface upon application of lasers of particular pulse shape.

In order to come up with an enhanced framework for nanophotonic waveguidance system, investigations are made of the hyperfine structures of ^{87}Rb atom that include associated magnetic sublevels split upon the application of static magnetic field (due to Zeeman effect). Proposals are made to represent qubit using selected hyperfine energy levels of ^{87}Rb atom followed by working out detailed nanophotonic processes involving application of a number of lasers. In the transmit node, a cavity-assisted Raman process generated photon with polarization and frequency corresponding to the logic state of the node; that is, the logic state of the node was mapped into the photon state. In the receive node, a reverse process absorbed photon and changed the logic state of the receive node to that of the transmit node. Hilbert spaces are constructed for the transmit and the receive nodes, and system Hamiltonians and density matrices are formulated. Finally, the equations of motion are derived (formulated as the evolution of density matrices) that would allow one to study the system dynamics through computer simulations under closed- and open-system configurations.

In the closed-system simulation, right and left circularly polarized photons with appropriate frequencies were generated based on the logic state of the transmit node. That is, the logic state of the transmit node was successfully mapped into the corresponding photon state. The open-system simulations revealed that the transmit node successfully mapped its logic state into the photon state. Photon leaked out of the cavity and reached the receive node. The analysis showed that, upon reception of photon, the receive node was able to successfully change its logic state to that of the transmit node. Thus, the closed-system and open-system simulation results confirmed the efficacy of the implemented approach.

Acknowledgment

One of the authors (PKC) is grateful to Profs. B.Y. Majlis and S. Shaari for continued support.

References

1. D. Bouwmeester, J.W. Pan, K. Mattle, M. Eibl, H. Weinfurter, and A. Zeilinger, Experimental quantum teleportation, *Nature* 390, 575–579, 1997.
2. Li Yongmin and Z. Kuanshou, Quantum optical implementation of quantum communication, *China Commun.* 68–73, 2005.
3. M. Fox, *Quantum Optics—An Introduction*, Oxford University Press, Oxford, U.K., 2006.
4. D.A. Steck, *Quantum and Atom Optics*, Available online: http://atomoptics.uoregon.edu/~dsteck/teaching/quantum-optics/quantum-optics-notes.pdf, 2009.
5. C. Monroe, D.N. Meekhof, B.E. King, W.M. Itano, and D.J. Wineland, Demonstration of a fundamental quantum logic gate, *Phys. Rev. Lett.* 75, 4714–4717, 1995.
6. J.I. Cirac, A.K. Ekert, S.F. Huelga, and C. Macchiavello, Distributed quantum computation over noisy channels, *Phys. Rev. A* 59, 4249–4254, 1999.
7. B. Yurke, Use of cavities in squeezed-state generation, *Phys. Rev. A* 29, 408–410, 1984.
8. G. Rempe, R.J. Thompson, R.J. Brecha, W.D. Lee, and H.J. Kimble, Optical bistability and photon statistics in cavity quantum electrodynamics, *Phys. Rev. Lett.* 67, 1727–1730, 1991.
9. M.C. Teich and B.E.A. Saleh, Squeezed and antibunched light, *Phys. Today* 43, 26–34, 1990.
10. J.I. Cirac and P. Zoller, Quantum computations with cold trapped ions, *Phys. Rev. Lett.* 74, 4091–4094, 1995.
11. X.D. Fan, P. Palinginis, S. Lacey, H.L. Wang, and M.C. Longeran, Coupling semiconductor nanocrystals to a fused-silica microsphere: A quantum-dot microcavity with extremely high Q factors, *Opt. Lett.* 25, 1600–1602, 2000.
12. R. Grimm, M. Weidemüller, and Y.B. Ovchinnikov, Optical dipole traps for neutral atoms. *Adv. Atom. Mol. Opt. Phys.* 42, 95–170, 2000.
13. A. Scherer, O. Painter, J. Vuckovic, M. Loncar, and T. Yoshie, Photonic crystals for confining, guiding, and emitting light, *IEEE Trans. Nanotechnol.* 1, 4–11, 2002.
14. P. Chen, C. Piermarocchi, L.J. Sham, D. Gammon, and D.G. Steel, Theory of quantum optical control of a single spin in a quantum dot, *Phys. Rev. B* 69, 075320.1–075320.8, 2004.
15. W. Choi, J. Lee, K. An, C. Fang-Yen, R.R. Dasari, and M.S. Feld, Observation of sub-Poisson photon statistics in the cavity-QED microlaser, *Phys. Rev. Lett.* 96, 093603.1–093603.4, 2006.
16. M.M. Rahman and P.K. Choudhury, On the investigation of field and power through photonic crystal fibers—A simulation approach, *Optik* 122, 963–969, 2011.
17. M.M. Rahman and P.K. Choudhury, A review of the state-of-the-art of particle physics with extra dimensions, *Asian J. Phys.* 17, 263–272, 2008.
18. C.H. Bennett and G. Brassard, Quantum cryptography: Public key distribution and coin tossing, *Proc. IEEE Int. Conf. Comput. Syst. Signal Process.* 175, 175–179, 1984.
19. C. Gobby, Z.L. Yuan, and A.J. Shields, Quantum key distribution over 122 km of standard telecom fiber, *Appl. Phys. Lett.* 84, 3762–3764, 2004.
20. X. Tang, L. Ma, A. Mink, A. Nakassis, B. Hershman, J. Bienfang, R.F. Boisvert, C. Clark, and C. Williams, High speed fiber-based quantum key distribution using polarization encoding, *Proc. SPIE* 5893, 58931A.1–58931A.9, 2005.
21. B.B. Blinov, D.L. Moehring, L.M. Duan, and C. Monroe, Observation of entanglement between a single trapped atom and a single photon. *Nature* 428, 153–157, 2004.

22. W. Yao, R.B. Liu, and L.J. Sham, Nanodot-cavity electrodynamics and photon entanglement, *Phys. Rev. Lett.* 92, 217402.1–217402.4, 2004.

23. D. Loss and D.P. DiVincenzo, Quantum computation with quantum dots, *Phys. Rev. A* 57, 120–126, 1998.

24. C.W. Gardiner, Driving a quantum system with the output field from another driven quantum system, *Phys. Rev. Lett.* 70, 2269–2272, 1993.

25. H.J. Carmichael, Quantum trajectory theory for cascaded open systems, *Phys. Rev. Lett.* 70, 2273–2276, 1993.

26. J.I. Cirac, P. Zoller, H.J. Kimble, and H. Mabuchi, Quantum state transfer and entanglement distribution among distant nodes in a quantum network, *Phys. Rev. Lett.* 78, 3221–3224, 1997.

27. W. Yao, R.B. Liu, and L.J. Sham, Theory of control of the spin-photon interface for quantum networks, *Phys. Rev. Lett.* 95, 030504.1–030504.4, 2005.

28. F.K. Nohama and J.A. Roversi, Quantum state transfer between atoms located in coupled optical cavities, *J. Mod. Opt.* 54, 1139–1149, 2007.

29. Y.J. Zhou, Y.M. Wang, L.M. Liang, and C.Z. Li, Quantum state transfer between distant nodes of a quantum network via adiabatic passage, *Phys. Rev. A* 79, 044304.1–044304.4, 2009.

30. P.B. Li, Y. Gu, O.H. Gong, and G.C. Guo, Quantum information transfer in a coupled resonator waveguide, *Phys. Rev. A* 79, 042339. 1–042339.4, 2009.

31. T. Aoki, A.S. Parkins, D.J. Alton, C.A. Regal, B. Dayan, E. Ostby, K.J. Vahala, and H.J. Kimble, Efficient routing of single photons by one atom and a microtoroidal cavity, *Phys. Rev. Lett.* 102, 083601.1–083601.5, 2009.

32. L.M. Duan, M. Lukin, J.I. Cirac, and P. Zoller, Long-distance quantum communication with atomic ensembles and linear optics, *Nature* 414, 413–418, 2001.

33. J.C. Boileau, R. Laflamme, M. Laforest, and C.R. Myers, Robust quantum communication using a polarization-entangled photon pair, *Phys. Rev. Lett.* 93, 220501.1–220501.4, 2004.

34. F.Y. Hong and S.J. Xiong, Long-distance quantum communication with polarization maximally entangled states, arXiv: 0809.3661v2, Available online: http://arxiv.org/PS_cache/arxiv/pdf/0809/0809.3661v2.pdf, 2009.

35. D.N. Matsukevich and A. Kuzmich, Quantum state transfer between matter and light, arXiv:quant-ph/0410092v1, Available online: http://arxiv.org/PS_cache/quant-ph/pdf/0410/0410092v1.pdf, 2004.

36. S. Suchat, P.P. Yupapin, and S. Chiangga, Communication and measurement, *Songklanakarin J. Sci. Technol.* 26, 84–91, 2004.

37. J.W. Pan, C. Simon, C. Brukner, and A. Zeilinger, Feasible entanglement purification for quantum communication, *Nature* 410, 1067–1070, 2004.

38. F.W. Strauch and C.J. Williams, Theoretical analysis of perfect quantum state transfer with superconducting qubits, *Phys. Rev. B* 78, 094516.1–094516.7, 2008.

39. D.L. Moehring, M.J. Madsen, B.B. Blinov, and C. Monroe, Experimental bell inequality violation with an atom and a photon, *Phys. Rev. Lett.* 93, 090410.1–090410.4, 2004.

40. J. Volz, M. Weber, W. Rosenfeld, D. Schlenk, J. Vrana, K. Saucke, C. Kurtsiefer, and H. Weinfurter, Observation of atom-photon entanglement, *Proc. Quantum Electron. Conf. 2005, EQEC '05 (European)*, 313–317, 2005.

41. Z.W. Wang, J. Li, Y.F. Huang, Y.S. Zhang, X.F. Ren, P. Zhang, and G.C. Guo, Linear optical implementation of a quantum network for quantum estimation, arXiv:quant-ph/0607189v3, Available: http://arxiv.org/PS_cache/quant-ph/pdf/0607/0607189v3.pdf, 2006.

42. C.K. Hong, Z.Y. Ou, and L. Mandel, Measurement of subpicosecond time intervals between two photons by interference, *Phys. Rev. Lett.* 59, 2044–2046, 1987.

43. A.G. White, D.F.V. James, P.H. Eberhard, and P.G. Kwiat, Nonmaximally entangled states: Production, characterization and utilization, *Phys. Rev. Lett.* 83, 3103–3107, 1999.

44. V. Bužek and M. Hillery, Quantum copying: Beyond the no-cloning theorem, *Phys. Rev. A* 54, 1844–1852, 1996.

45. W.K. Wootters and W.H. Zurek, A single quantum cannot be cloned, *Nature* 299, 802–803, 1982.

46. A.N. Vamivakas, Y. Zhao, C.Y. Lu, and M. Atatüre, Spin-resolved quantum-dot resonance fluorescence, *Nature Phys.* 5, 198–202, 2009.
47. W. Yao, R.B. Liu, and L.J. Sham, Optically manipulating spins in semiconductor quantum dots, *J. Appl. Phys.* 101, 081721.1–081721.5, 2007.
48. W. Demtröder, Experimental techniques in atomic and molecular physics, In *Atoms, Molecules and Photons: An Introduction to Atomic-, Molecular- and Quantum-Physics*, Springer Verlag, Würzburg, Germany, 2010, Chapter 11.
49. Z.B. Birula, Cavity QED. *Acta Physica Polonica B* 27, 2409–2420, 1996.
50. M.V. Cougo-Pinto, C. Farina, F.C. Santos, and A.C. Tort, QED vacuum between a conducting and a permeable plate, *J. Phys. A: Math. Gen.* 32, 4463–4474, 1999.
51. M.L. Terraciano, R. Olson, D.L. Freimund, L.A. Orozco, and P.R. Rice, Spectrum of spontaneous emission into the mode of a cavity QED system, arXiv:quant-ph/0601064v1 Available online: http://arxiv.org/PS_cache/quant-ph/pdf/0601/0601064v1.pdf, 2006.
52. P. Goy, J.M. Raimond, M. Gross, and S. Haroche, Observation of cavity-enhanced single-atom spontaneous emission, *Phys. Rev. Lett.* 50, 1903–1906.
53. R.G. Hulet, E.S. Hilfer, and D. Kleppner, Inhibited spontaneous emission by a Rydberg atom, *Phys. Rev. Lett.* 55, 2137–2140, 1985.
54. D.J. Heinzen, J.J. Childs, J.E. Thomas, and M.S. Feld, Enhanced and inhibited visible spontaneous emission by atoms in a confocal resonator, *Phys. Rev. Lett.* 58, 1320–1323, 1987.
55. H.J. Lewandowski, *Coherences and Correlations in an Ultracold Bose Gas*, Unpublished PhD thesis, Department of Physics, University of Colorado, 2002.
56. M.M. Rahman and P.K. Choudhury, Polarized photon generation for the transport of quantum states—A closed-system simulation approach, *Prog. Electromagn. Res. M* 8, 249–26, 2009.
57. M.M. Rahman and P.K. Choudhury, Nanophotonic waveguidance in quantum networks—A simulation approach for quantum state transfer. *Optik* 121, 1649–1653, 2010.
58. D. Klauser, W.A. Coish, and D. Loss, Nuclear spin state narrowing via gate–controlled Rabi oscillations in a double quantum dot, *Phys. Rev. B* 73, 205302.1–205302.13, 2006.
59. M. Grace, C. Brif, H. Rabitz, I. Walmsley, R. Kosut, and D. Lidar, Encoding a qubit into multilevel subspaces, *New J. Phys.* 8, 35–65, 2006.
60. D.A. Steck, Rubidium-87 D line data (rev. 2.1.1) Available online: http://steck.us/alkalidata/rubidium87numbers.pdf, 2009.
61. M.M. Rahman and P.K. Choudhury, Towards a novel simulation approach for the transport of atomic states through polarized photons, *Optik* 122, 84–88, 2011.
62. M.M. Rahman, and P.K. Choudhury, On the quantum link for transport of logic states, *Optik* 122, 660–665, 2011.
63. E.T. Jaynes and F.W. Cummings, Comparison of quantum and semiclassical radiation theories with application to the beam maser, *Proc. IEEE* 51, 89–109, 1963.
64. M.O. Scully and M.S. Zubairy, *Quantum Optics*, Cambridge University Press, Cambridge, U.K., 1997.
65. G. Lindblad, On the generators of quantum dynamical semigroups, *Commun. Math. Phys.* 48, 119–130, 1976.
66. R. Loudon, *The Quantum Theory of Light* (3rd edn.), Oxford University Press, New York, 1973.
67. J.D. Patterson and B.C. Bailey, *Solid-State Physics: Introduction to the Theory*, Springer, New York, 2007.
68. J. Stoer and R. Bulirsch, *Introduction to Numerical Analysis* (3rd edn.), SpringerVerlag, Berlin, New York, 2002.
69. M.M. Rahman and P. K. Choudhury, Cavity quantum electrodynamics for photon mediated transfer of quantum states, *J. Appl. Phys.* 109, 113110-1–113110-8, 2011.
70. P. Zarchan and H. Musoff, *Fundamentals of Kalman Filtering: A Practical Approach* (3rd edn.), American Institute of Aeronautics and Astronautics, Inc., Reston, VA, 2009.

7

Nanopatterned Photonics on Probe: Modeling, Simulations, and Applications for Near-Field Light Manipulation

Lingyun Wang and Xiaojing Zhang

CONTENTS

7.1 Introduction

Near-field scanning optical microscopy (NSOM/SNOM) is a microscopy technique for nanostructure investigation that breaks the far-field resolution limit by exploiting the properties of evanescent waves with lateral resolution of 20 nm and vertical resolution of 2–5 nm that have been demonstrated. However, the current widely used NSOM probe by metal-coated tapered optical fiber has intrinsic disadvantage of confining light in an inefficient manner due to its strong light propagation mode mismatch at the tip aperture. In this book chapter we propose a novel light-confining probe of the combination by the embedded metallic grating coupler and photonic crystal (PhC) slotted waveguide on

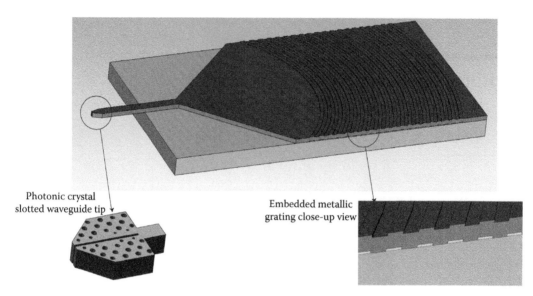

Photonic crystal
slotted waveguide tip

Embedded metallic
grating close-up view

FIGURE 7.1
(See color insert.) Schematic of the proposed NSOM tip composed of PhC tip and embedded metallic grating coupler as a whole system.

a single-chip design (Figure 7.1) with careful consideration of light propagation mode match that overcomes the disadvantages of the optical fiber-based probe. The grating coupler and the sub-wavelength light-focusing processing circuit are the most dominating basic design elements of the scanning optical imaging system. In this contribution, we describe the design and performance of embedded metal grating coupler and its application for PhC slotted waveguide probe tip embedded with Fabry–Perot (FP) nano-resonator for near-field light focusing at visible range. All the claims and findouts in this chapter are backed up by the extensive numerical modeling, for example, CAvity Modeling Framework (CAMFR) [1], MIT Photonic Bands (MPB) [2], and finite-difference time-domain (FDTD) [3] methods. This chapter is divided into two parts. The first part presents a unique compact light feeding mechanism by noble metal grating structure through numerical electromagnetic field modeling. A common design flow by finding the vertical resonance mode of a 2D grating by CAMFR is presented to derive the effective index of the grating waveguide composite structure for the 3D FDTD modeling of the compact focus grating structure. The far-field radiation pattern expressed in 3D directivity of the conventional etching compact focusing grating structure shows less free-space directivity enhancement than that of the embedded noble metal grating coupler. The second part investigates the nanoscale sub-wavelength light focusing and enhancement in near field through PhC probe tip embedded with slotted FP nano-resonator. The 2D field distribution of the tip near field is calculated to show the light focusing and enhancement due to the embedded nano-resonators. Thus, this new near-field light-focusing system is demonstrated to avoid the common drawbacks of the conventional scanning tip optical imaging system. Finally, possible biological applications, such as spatial differential field distribution of the fluorescence signal by embedded metal grating coupler and water perturbation in the PhC probe tip near field, are also discussed.

7.2 Design of Embedded Metal Focus Grating Coupler

In order to feed light efficiently into the PhC probe tip or other similar nanoscale light processing circuits, the light feeding method has to be carefully examined for such a nanometer-scale size probe tip to avoid unwanted light contamination and enhance signal-to-noise ratio for light detection or coupling. Conventional light-coupling technologies to sub-wavelength waveguide include bulky backplate prism coupling, which is not suitable for compact dense system integration [4], and optical fiber end-fire coupling with waveguide facet [5,6], which requires stringent alignment accuracy, especially at visible wavelength range due to the even smaller waveguide core size. The grating structure is defined by either metal liftoff on top of the waveguide layer [7] or etching the grating grooves partially into the higher-index waveguide layer [8,9]. As it is shown by simulation in this chapter, the etching depth is a very critical parameter for 3D far-field pattern control and coupling efficiency of the etching grating coupler. However, the etching approach suffers from inaccuracy in etching depth control due to nonstopping layer etching. Meanwhile, the metal liftoff approach excels for its relatively easy fabrication and low absorption loss of the thin metal layer thickness (usually less than 10th of the wavelength). In the fully etched grating case [9], which requires less on etching depth control, however, at least half of the light will unavoidably propagate into the substrate layer that strongly decreases the transmission efficiency into free space. In many grating designs, the high coupling efficiency with free space is the utmost design goal, and the light scattering into the substrate layer should be well managed to minimize the energy loss.

The concept of on-chip grating coupler design has been well studied at longer infrared (IR) communication wavelength due to the rapid development of telecommunication industry and the silicon-on-insulator (SOI) fabrication technologies during the last several decades. However, their counterpart at visible wavelength did not share equal attention and met even a greater challenge, because of much smaller feature size restriction on fabrication and opaque property of silicon at visible range. The FDTD modeling of metal in optical range also requires extensive dispersion modeling due to the negative real part of its permittivity property. Sub-wavelength optical detection, such as in NSOM [10–13], through nano-optical antennas [14], or silicon nitride PhC and nano-resonator on probe tip for near-field light localization [15,16], mandates efficient and compact light coupling from sub-wavelength light processing circuits.

A compact noble metal grating coupler design is studied for free-space directional light coupling from or into sub-wavelength silicon nitride optical waveguide working at HeNe laser wavelength of 632.8 nm. This study fills the research gap from device modeling perspective for noble metal gratings by placing the noble metal grating layer just above the substrate surface to block potential coupling loss into the substrate. The fabrication of the proposed noble metal grating coupler is fully compatible with silicon micromachining technology, and it offers a new concept of light coupling at 632.8 nm wavelength with high free-space directivity and coupling efficiency. Based on the theoretical first-order diffraction and effective refraction index of the waveguide, the elliptical patterned grating lines are demonstrated to further reduce the coupler size by enhanced focusing capability on the lateral plane through 3D FDTD simulation. Far-field radiation pattern, derived from the 3D FDTD model, indicates its high directivity that is suitable for angular far-field light detection.

7.2.1 Analysis of the Etched Dielectric Grating Coupler

Before we discuss the embedded metal grating design, it is useful to analyze the partially etched grating coupler and evaluate its performance. The partially etched grating coupler is usually fabricated by electron beam lithography (EBL) and dry etching method [17]. Other grating fabrication methods, such as polymer-based grating [18], deep-UV lithography [19], and achromatic interferometric lithography [20], have also been utilized. An out-of-plane grating coupler is adapted to efficiently butt-couple light from single-mode fiber into the ridged waveguide [17]. The grating coupler also achieves beam spot converting from the grating square into the narrow ridged waveguide. The eigenmode expansion technique can be used to greatly simplify the model to yield the reflection resonance peak in the 1D grating structures. In this research, we use CAMFR, an eigenmode expansion frequency-domain modeling tool, to obtain the optimal grating period and groove depth for the 1D grating in the 2D simulation [1]. Only transverse electric (TE) polarization (E-field parallel to the grating grooves) is considered in this study, since all the following PhC design is also based on TE-mode calculation. As illustrated in Figure 7.2a, the reflection rate R and transmission T rate can be calculated by finding the

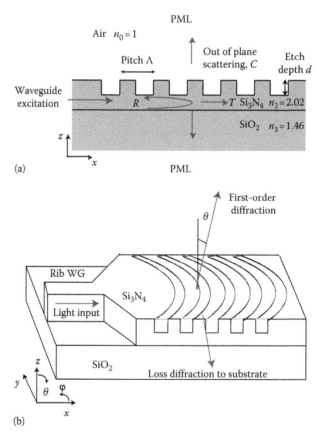

FIGURE 7.2
Design of etched silicon nitride grating coupler. (a) CAMFR calculation model on eigenmodes of the 1D partially etched grating on Si_3N_4 waveguide layer on top of the SiO_2 substrate; (b) 3D FDTD model with spherical axis as labeled.

eigenmode of each discretized slab in the vertical direction. The total out-of-plane light coupling can be expressed as $C = 1 - R - T$. The peak reflection corresponds to the grating coupler resonance condition that maximum out-of-plane coupling radiates in the vertical direction. In this grating structure, there is no special mechanism to prevent the light-coupling loss into the substrate. Silicon nitride of 250 nm thickness is chosen as the waveguide layer due to its transparent optical property at 632.8 nm wavelength. Silicon dioxide serves as the substrate layer. The CAMFR calculation domain is surrounded in upper and bottom boundaries with sufficient perfectly matched layer (PML) thickness in order to reduce the light reflection at the calculation boundaries.

The reflection and out-of-plane efficiency is shown in Figure 7.3 by sweeping the grating pitching size within the range from 300 to 400 nm and the etching depth from 70 to 120 nm in 10 nm interval with 50% grating duty cycle. At the reflection resonance pitch period Λ_p, the out-of-plane light-coupling curve also shows the minimum value due to strong reflection. We also need to increase the out-of-plane coupling rate in order to increase the free-space coupling rate. By finding out the pitch period that corresponds to the maximum out-of-plane coupling, the effective index of the waveguide grating composite structure can be calculated using effective index equation under vertical radiation condition:

$$n_{eff} = \frac{\lambda_0}{\Lambda_p} \tag{7.1}$$

It is found that 110 nm etching depth with 361 nm pitch period gives the maximum out-of-plane coupling rate with effective index of 1.7529 at the reflection resonance pitch period.

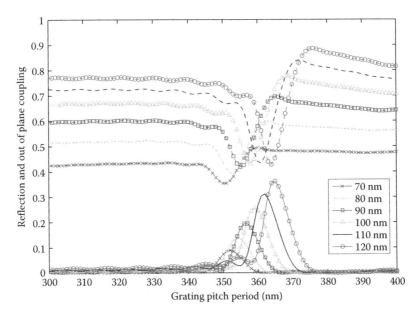

FIGURE 7.3
Reflection (bottom solid lines) and out-of-plane coupling (top dashed lines) rates as the function of grating pitch period for etching depth ranging from 70 to 120 nm.

For arbitrary out-of-plane coupling angle θ_0, the effective index can be estimated from the first-order grating diffraction equation [18]:

$$n_{eff} = n_0 \sin\theta_1 + \frac{\lambda_0}{\Lambda} \tag{7.2}$$

where
    n_0 is the refraction of top cover material (free space in the proposed grating coupler)
    Λ is the grating pitch period
    λ_0 is the wavelength in free space

CAMFR provides a convenient way to predict the effective index for maximum out-of-plane coupling under vertical radiation condition. However, for arbitrary 3D grating coupler design for radiation angles other than vertical direction, the light-coupling distribution has to be verified through 3D FDTD modeling. We choose the compact focusing grating coupler as 3D FDTD simulation example as illustrated in Figure 7.2b. For polarized plane-wave light source, the wave vector front is regulated by constructive interference according to the first-order Bragg diffraction theory. The focusing capability is realized by constructive interference of the wavefront phase between the wave coming from the dielectric waveguide and the scattered wave from the grating groove lines into the free space at the first-order diffraction angle. The 3D constructive interference on the center *xy* plane of the silicon nitride waveguide layer can be described by the phase-matching equation as follows:

$$q\lambda_0 = n_{eff} \cdot \sqrt{x^2 + y^2} - xn_0\sin\theta_0, \quad q = 1, 2, 3, \ldots \tag{7.3}$$

where q is the grating line numbers. In order to achieve focusing effects on the *xy* plane, the silicon nitride waveguide ending facet needs to be positioned at the focal point of the elliptical curves [18]. Detailed expansion of Equation 7.3 in elliptical curve form is also presented in Equation 7.5.

Such a 3D compact focusing grating coupler is modeled as in Figure 7.2b with azimuth angle φ and polar angle θ as indicated. A waveguide port with broadband excitation centered at 632.8 nm wavelength is used as excitation source, which is located at the single-mode silicon nitride ridge waveguide with TE-mode linear polarization. All the boundaries are set up as open space with PML to reduce the back-reflection calculation error. The size of the grating coupler is 10 μm in the *y* direction and 7 μm in the *x* direction, which would fit the approximate core size of single-mode optical fiber at 632.8 nm wavelength. The overall relative small size of the grating coupler also helps to greatly reduce FDTD calculation time. The grating coupler FDTD simulation problem can be treated as a typical antenna problem by converting the near-field light distribution into far-field radiation pattern [21]. The far-field directivity of the etching compact grating coupler is shown in Figure 7.4, in which both vertical out-of-plane coupling of the peak reflection resonance case and the 10°-tilted radiation of the same effective index are plotted. There are two radiation lobes for vertical radiation coupler and 10°-tilted coupler. Since there is no special treatment for blocking the radiation into substrate, almost half of the light energy is lost into the bottom. The main design requirement on the free-space grating coupler is to reduce the secondary radiation lobe into the substrate layer for enhanced free-space coupling. Based on the CAMFR calculation,

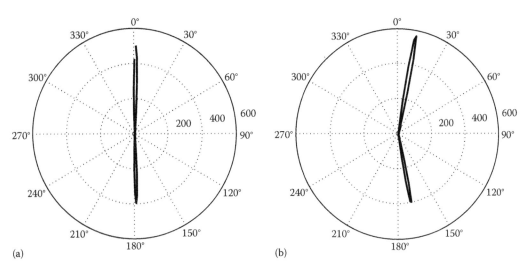

FIGURE 7.4
Far-field directivity plots of 3D focusing grating coupler fabricated by the etching method for polar angle θ on φ = 0° plane. The grating etching depth is 110 nm with effective grating index of 1.7529. (a) Coupler with peak reflection resonance (main-lobe directivity 504); (b) same effective grating coupler index but with tilted radiation angle of θ = 10° (main-lobe directivity 567).

under maximal vertical radiation case, the total out-of-plane light coupling is 45% and the free-space coupling rate is about slightly more than half of it as estimated from the far-field pattern, if we assume lossless antenna radiation. So the free-space coupling is quite low as compared to that of the embedded metal ones as shown in later sections. The main design requirement on the free-space grating coupler is to reduce the secondary radiation lobe into the substrate layer for enhanced free-space coupling. The reduction of the radiation into the substrate has been realized by introducing a distributed Bragg reflector in the substrate layer as that in the IR grating coupler design on SOI platform but with extra burden on fabrication process [17]. Also as shown in Figure 7.3, the grating etching depth is also very critical as it would also change coupling efficiency and beam pattern dramatically. Another novel way to reduce substrate coupling will be introduced next by placing the metal grating on top of the substrate.

7.2.2 Ag and Au Dispersion Optical Properties

The compatibilities with the EBL, stable chemical properties at ambient environment, and low light absorption make silver and gold excellent candidates for grating materials at visible wavelengths. For the noble metals, such as Ag and Au, the real part of the permittivity is negative (well <−10) throughout the visible range, and the imaginary part is weakly positive [22]. The metallic grating grooves serve as the diffraction element leading to the directional radiation due to their refraction index contrast with that of the waveguide layer material. Also due to the high negative permittivity, the surface plasmonic resonance waves exist on top of the metal surface [23]. Therefore, the propagation distance of the waves on the lateral grating plane is greatly restricted. The main-lobe size in the far-field radiation pattern is also reduced, which enhances the far-field directivity. The absorption loss for 632.8 nm wavelength for gold and silver is even smaller due to the smaller loss tangent, as compared to their IR counterparts. The relative permittivities of gold and silver are modeled as dispersive materials by the Drude–Lorentz (DL) model:

$$\varepsilon(\omega) = \varepsilon_\infty - \frac{\omega_D^2}{\omega + i\gamma_D\omega} + \sum_n \frac{\sigma_n \cdot \omega_n^2}{\omega_n^2 - \omega^2 - i\omega\gamma_n} \tag{7.4}$$

where
  ω_D is the plasma frequency
  γ_D is the damping term
  γ_n represents spectral width
  ω_n represents oscillator strength of the Lorentz oscillators
  σ_n stands for weighting factor [24–26]

The DL model with two Lorentzian terms for both metals is calculated by least-squares nonlinear curve-fitting method with experiment data in Ref. [22] for optical wavelength range from 500 to 1000 nm, which cover the interested wavelength of 632.8 nm as listed in Table 7.1 with data plotted in Figure 7.5. At the wavelength of 632.8 nm, the imaginary part of both metals, especially gold, shows minimal value as compared to other wavelengths.

TABLE 7.1

DL Parameters Optimized for Dispersive Dielectric Properties of Gold and Silver in 500 to 1000 nm Wavelength Range for FDTD Modeling

	ε_∞	$\omega_D/2\pi$ (THz)	$\gamma_D/2\pi$ (THz)	$\omega_1/2\pi$ (THz)	$\gamma_1/2\pi$ (THz)	σ_1	$\omega_2/2\pi$ (THz)	$\gamma_2/2\pi$ (THz)	σ_2
Au	5.2518	2089.7	16.07	564.1	103.23	0.1543	654.6	64.53	1.0827
Ag	2.8049	2207	2.28	700.9	10.03	0.1267	1186.7	1634.2	0.5128

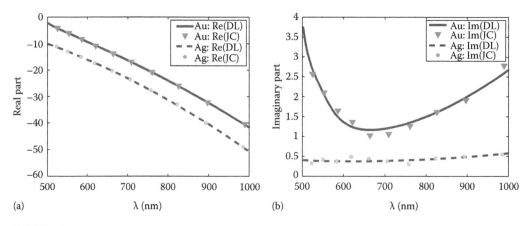

(a)   λ (nm)   (b)   λ (nm)

FIGURE 7.5
Relative permittivity as calculated by the DL model as compared with the experiment data in Ref. [22] by Johnson and Christy (JC) for both gold and silver for wavelength range from 500 to 1000 nm. (a) Real part of permittivity; (b) imaginary part of permittivity.

7.2.3 Compact Embedded Metallic Focus Grating Coupler Overview

As demonstrated in Ref. [7] for IR wavelength, a metal grating can be placed on top of the waveguide layer to form a grating coupler on SOI platform, which makes the design fully compatible with CMOS fabrication. However, the free-space coupling efficiency is greatly related to the buried oxide thickness, since maximum upward coupling only occurs when constructive interference forms between the wave reflection from the oxide substrate interface and the upward light propagation [7]. In order to avoid the oxide thickness control restriction, a bottom-placed metallic grating coupler is illustrated in a 2D vertical cross-sectional view in Figure 7.6a and a 3D overview of the 3D FDTD model in Figure 7.6b. A metal grating layer of 40 nm is placed on top of the silicon dioxide layer with duty cycle of 50%, which can be defined by EBL and fabricated by metal liftoff method. Only one EBL alignment is needed to align the silicon nitride waveguide pattern that is finalized by dry etching. Silicon nitride with thickness of 250 nm is chosen to form the single-mode (TE0, E-field polarized in the y direction) rib waveguide with cross-sectional width of 500 nm working at 632.8 nm for its relatively high index ($n \approx 2.02$) and near-zero absorption loss. Other transparent waveguide layer material with relatively high index like indium tin

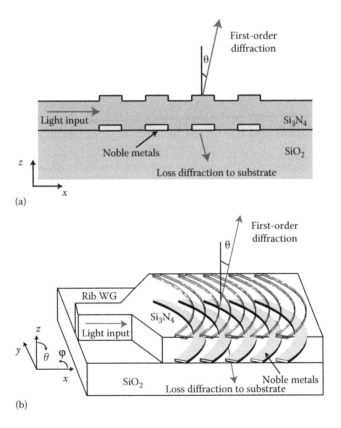

(a)

(b)

FIGURE 7.6
Noble metal grating on top of the silicon dioxide layer for interfacing far-field free space and silicon nitride dielectric rib waveguide (geometry not to scale). (a) Vertical cross-sectional view on the xz plane where the first-order diffraction angle is θ from the z-axis; (b) 3D illustration of the coupler including the metal focus grating ellipses and silicon nitride single-mode dielectric rib waveguide.

oxide (ITO) can also be used. The non-annealed ITO layer grown by sputtering deposition also provides similar index as silicon nitride and low transmission loss. The conducting material property of ITO also alleviates the common charging phenomena in the EBL pattern definition process. The ITO waveguide can also be patterned through the liftoff process by using low-power ITO sputtering deposition process, instead of dry etching. In this chapter, silicon nitride is selected as a modeling norm material due to its zero loss at visible range. Also, silicon nitride is compatible with silicon fabrication technology and can be deposited by either low-pressure chemical vapor deposition (LPCVD) or plasma-enhanced chemical vapor deposition (PECVD). The annealing temperature for silicon nitride can be achieved at around 800°C [27]. So gold and silver are compatible with silicon nitride deposition since their melting points are 1064°C and 962°C, respectively [28]. Also, the fused silica can be used as the SiO_2 substrate material for its low thermal expansion and annealing temperature above 1000°C [28]. The bottom placement of the metal layer under the waveguide layer also further enhances the free-space light-coupling efficiency by reducing the light energy diffracted with higher-order modes into the silicon dioxide substrate given by the penetrating properties of such noble metals. The top conformal groove pattern is also modeled to simulate the uniform conformal silicon nitride deposition. The top silicon nitride grooves do not serve as an active grating element due to their thin thickness.

7.2.4 2D FDTD Analysis of the Embedded Grating Coupler

A 2D FDTD model that requires less computing time, other than CAMFR (lossy metal is hard to model with stability problem) or computational-intensive and time-consuming 3D FDTD, has been implemented to estimate free-space coupling efficiency before extending it to the 3D model [29]. In this 2D FDTD model as illustrated in Figures 7.6a and 7.8a, a light source of TE_0 mode is launched from the silicon nitride rib waveguide ending at $x = 0$. For each pitch period, FDTD calculations are performed twice to separate the incident and reflected field. The first FDTD calculates field distribution of a simple silicon nitride waveguide without grating coupler perturbation to separate out the incident fields. The second FDTD calculates the field distribution of the grating coupler [3]. The free-space transmission efficiency, as shown in Figure 7.7, is calculated using the ratio between the flux of electromagnetic energy through the 5 μm line along the x direction at 4 μm above the grating coupler center and the incident flux energy at the waveguide input. The 5 μm line length in the x direction models a typical single-mode optical fiber at 632.8 nm wavelength, though the 2D FDTD model does not count the coupling loss into the optical fiber. A Gaussian broadband current signal with center wavelength at 632.8 nm is used as the excitation signal with metal-dispersive permittivity property described by the DL model. The flux of the electromagnetic energy is derived from the Fourier-transformed field into frequency domain [29]. In this 2D FDTD model, PML layers are surrounded by the calculation domain to avoid back-reflection error. Since only the total flux across the defined line in the x direction is involved in this calculation, the free-space transmission efficiency only quantifies the energy flux without taking account of the directional radiation information. The directional radiation pattern will be quantified through far-field directivity by the 3D FDTD model. The silver grating and the gold grating give peak free-space coupling efficiencies of 68% and 55%, respectively. The free-space coupling for both metal grating couplers is much higher than those of the etched grating coupler with the same waveguide layer material. The smaller loss tangent of silver explains the higher transmission rate than that of the gold one. The higher transmission rate for silver grating is also demonstrated by the higher free-space radiation directivity calculation in the following 3D FDTD simulation.

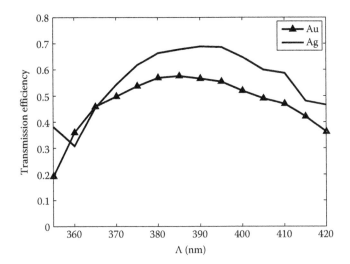

FIGURE 7.7
Free-space transmission efficiencies by sweeping grating pitch period from 355 to 420 nm for Au and Ag through the 2D FDTD model.

7.3 Far-Field Radiation Pattern of the Embedded Metal Focus Grating Coupler

To verify the field distribution and far-field radiation pattern, a 3D FDTD model is carried out in this research and one term, DL model, has been used to simplify the 3D model [21]. Based on the optical antenna radiation theory, the far-field directivity radiation pattern, combined with antenna radiation efficiency, can be used to quantify both the beam pattern and the transmission efficiency within a certain angle range. In this section, the optical antenna radiation theory is briefly reviewed. The 3D radiation patterns in terms of far-field directivity for both Au and Ag grating material are quantified by 3D FDTD simulation to reveal the focusing and enhanced directivity properties of such embedded metal focus grating coupler.

7.3.1 Optical Antenna Theory

The efficiency of a radio-frequency (RF) antenna describes how efficient the local oscillating driving current is transformed into a far-field radiation energy, P_{rad}, for radiation antenna or vice versa for receiving antenna. Since most of the RF antenna deals with lossless metal material, most of the RF antenna radiation efficiency loss is caused by strong impedance mismatch. However, in the optical range, metal may no longer be simplified as lossless material due to the dispersive dielectric properties as discussed in Section 7.2.2. The radiation loss is largely caused by the absorption or ohmic loss of the metal materials, P_{loss}, which cannot be ignored in the electric field calculation. If the impedance mismatch loss is simplified to be zero, the radiation efficiency of an optical antenna is defined as [30]

$$\varepsilon_{rad} = \frac{P_{rad}}{P_{rad} + P_{loss}} \tag{7.5}$$

The radiation efficiency only quantifies the total radiation efficiency but not the radiation in certain direction. The total radiation power, P_{rad}, can be defined as the integration for all directions of the normalized angular radiation density, $P_n(\theta,\varphi)$ [30]:

$$P_{rad} = \int_0^\pi \int_0^{2\pi} P_n(\theta, \varphi)\sin\theta d\varphi d\theta \qquad (7.6)$$

The average angular radiation intensity P_{av} of an equivalent isotropic radiator can be expressed as

$$P_{av} = \frac{P_{rad}}{4\pi} \qquad (7.7)$$

which calculates the angular radiation intensity with the same total radiation power. The directivity of the antenna is a dimensionless parameter that describes the focusing capability of radiation beam at a certain angle at far field. It is quoted in the John D. Kraus's classic antenna textbook [31], "The *directivity D* of an antenna is given by the ratio of the maximum radiation intensity (power per unit solid angle) to the average radiation intensity (averaged over a sphere)," which refers to the maximum directivity direction at the far field. For arbitrary directions, the angular directivity D is defined as

$$D = \frac{P_n(\theta, \varphi)}{P_{av}} = 4\pi \frac{P_n(\theta, \varphi)}{P_{rad}} \qquad (7.8)$$

The gain of the antenna combines both the antenna directivity and the antenna efficiency:

$$G = \varepsilon_{rad} D \qquad (7.9)$$

which is a parameter that relates to the antenna radiation efficiency in a certain direction. In this chapter, both the angular directivity D and the maximum directivity will be used to describe the far-field radiation pattern of the proposed embedded metallic grating coupler, since the radiation efficiency is almost a constant value due to the almost fixed amount of ohmic loss caused by metal material.

7.3.2 3D Directivity Analysis

The design of the grating coupler needs to achieve two aims, efficient light coupling and compact coupler size. Also the size of the grating coupler needs to match that of the free-space source or detector for maximal coupling efficiency. At the cross-sectional plane of $y=0$, the elliptical grating pitch period is a constant as defined in the 2D FDTD model. The effective index of the silicon nitride dielectric waveguide can be calculated from Equation 7.2. Following similar design ideology as the etched grating coupler for compact focusing on the xy center plane at the waveguide layer, Equation 7.3 can be further expanded in elliptical curve form as follows:

$$\frac{\left(x - \frac{q\lambda n_0 \sin\theta_1}{n_{eff}^2 - n_0^2 \sin^2\theta_1}\right)^2}{\left(\frac{q\lambda n_{eff}}{n_{eff}^2 - n_0^2 \sin^2\theta_1}\right)^2} + \frac{y^2}{\left(\frac{q\lambda}{(n_{eff}^2 - n_0^2 \sin^2\theta_1)^{0.5}}\right)^2} = 1 \tag{7.10}$$

to design the embedded metallic grating coupler. Equation 7.10 describes a set of ellipses that share a common focal point at $(0, 0)$ on the xy plane. By fixing the grating pitch period and the diffraction angle in free space, the focus grating curves can be exactly determined. The size of the grating coupler is determined by the starting grating number of q and can be adjusted for different free-space source size or detector size. To achieve focusing effects on the xy plane, the silicon nitride waveguide ending facet needs to be positioned at the focal point of the ellipse curves.

The size of the focusing grating coupler is set as 10 μm in the y direction with 20 grating grooves of 50% duty cycle, and the distance from the nearest grating line center to the waveguide is set as 7 μm to cover the optical fiber ending facet size. An E_y waveguide port current source with broad wavelength range with center wavelength of 632.8 nm is modeled as the excitation source for the 3D structure. The surrounding boundaries of the calculation domain are set as open absorbing boundary condition. And the E_y field distribution is plotted in Figure 7.8, based on the 3D FDTD simulation results of gold focusing grating coupler with center pitch period of $\Lambda = 397.5$ nm. The center pitch period in the 3D FDTD model is chosen by optimizing the directional radiation angle at $\theta_1 = 10°$. Figure 7.8a demonstrates the directional radiation along the 10°-tilt direction as designed. Shown in Figure 7.8b, the light exit from the waveguide facet has matched the wavefront that collides with the grating curvatures on the lateral center plane, which greatly reduces the impedance mismatch loss. The silver grating coupler has similar field distribution but with slightly higher field intensity in the free-space direction radiation, which will be demonstrated in the following far-field distribution pattern discussion.

To quantify the radiation difference, the 3D far-field radiation pattern is derived from near-field to far-field conversion for these two materials as shown in Figure 7.9. The effective refraction indexes of silver and gold grating coupler are 1.7711 (397.5 nm center pitch) and 1.7611 (400 nm center pitch), respectively, for optimized directional radiation for the first-order diffraction angle $\theta_1 = 10°$. The far-field radiation pattern shows strong directional angular radiation pattern oriented in $\theta_1 = 10°$ for first diffraction angle. As it can be seen, the pitch period for peak free-space coupling as calculated by the 2D FDTD model is not used for the 3D FDTD model due to the fact that the 2D FDTD does not take account of the directional radiation information. Thus, correspondingly the 2D free-space coupling efficiencies for the 3D designs are reduced to be 65% and 50% for silver and gold, respectively, as read from Figure 7.7. As expected, there is also a side lobe pointing toward the substrate layer that counts for the light-coupling loss but with much smaller magnitude as compared to the etched one. The peak free-space directivity is also greatly increased for both metallic grating designs. This phenomenon also echoes with the initial design discussion for bottom placement of the metallic grating. The radiation intensity difference is not apparent as shown in the directivity plot in Figure 7.9c due to the dBi unit used. However, both grating structures show a 4.2° 3 dB angular width for the main-lobe radiation pattern. On the other hand, the total radiation efficiencies for the Ag and Au grating coupler are 85% and 75%, respectively, which also explains the lower transmission

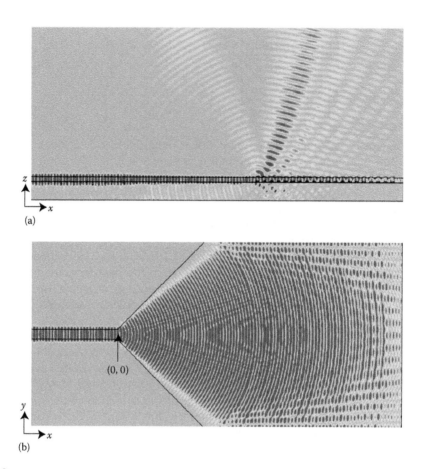

(a)

(b)

FIGURE 7.8

E_y field plots of an optimized gold focusing grating coupler excited by 632.8 nm TE_0 waveguide source on the center cross-sectional planes of the coupler. (a) Field plot on the xz center plane; (b) field plot on the xy plane along the silicon nitride slab center. The focusing point at (0, 0) on the xy plane is as indicated. Grating pitch period $\Lambda = 397.5$ nm, thickness of gold layer = 40 nm, and silicon nitride waveguide layer thickness = 250 nm.

efficiency for Au grating coupler design due to high loss tangent for Au material. From radiation antenna point of view, this kind of far radiation pattern corresponds to the super high gain or super-high directivity antenna, which is suitable for directional light coupling or detection with minimal sway light loss.

In order to test the design accuracy, a 3D FDTD simulation with changed design radiation angle $\theta_1 = 25°$ is implemented for Au grating coupler. Equation 7.3 is only valid by assuming equal effective index in the grating region and the waveguide region, and it is only fulfilled for shallow gratings less than 50 nm [18]. The metal grating grooves only have 40 nm thickness, which can be regarded as shallow grating. As it is observed from Equation 7.2, by finding out the effective index of the grating coupler at $\theta_1 = 10°$ and simplifying it as a constant for all radiation angles, the optimized grating pitch period for $\theta_1 = 25°$ peak radiation is only related to the designed radiation angle. By calculating the model using the updated grating pitch period, the 3D compact metallic focus grating ellipses can also be uniquely derived for new first-order diffraction angle. Thus, the effective index of the grating coupler is a very important parameter for grating coupler design to control main-lobe radiation angle. Figure 7.10 shows the radiation pattern for $\theta_1 = 25°$ with the

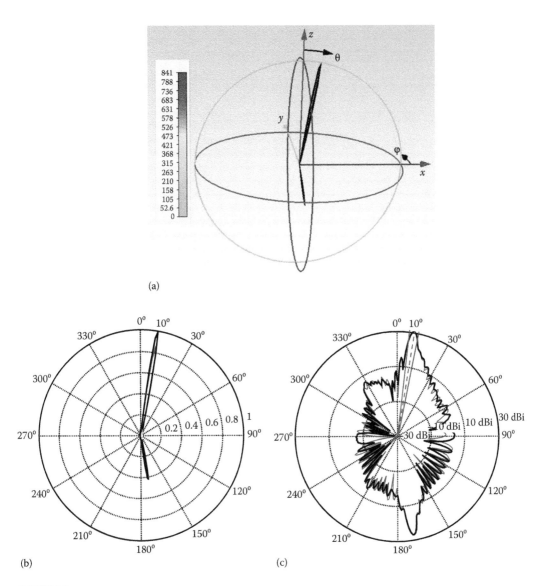

FIGURE 7.9
Far-field radiation pattern for incident waveguide source light at wavelength 632.8 nm. (a) 3D directivity pattern as functions of spherical angles for Ag grating coupler with peak directivity of 841; (b) normalized far-field power radiation pattern as function of θ on $\varphi=0°$ plane; (c) far-field directivity as function of θ on $\varphi=0°$ plane in unit of dBi. The dark black curve represents silver grating and the gray one is for gold grating.

exact same effective index estimation as $\theta_1 = 10°$ for Au grating coupler. The high first-order diffraction angle enables longer light propagation distance on top of the grating grooves, which explains higher material loss. The total radiation efficiency only reduces slightly to 73% for $\theta_1 = 25°$ case, but the free-space directivity is greatly enhanced with strongly suppressed side lobe toward the substrate. The side lobe for $\theta_1 = 25°$ toward the substrate is only half of the lower diffraction angle case. The side lobe is generated from the light leakage into the substrate and regulated by the metallic grating. The magnitude of the side lobe is strongly dependent on the penetration depth of the evanescent wave on top of the

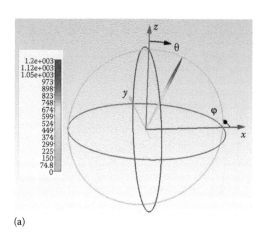

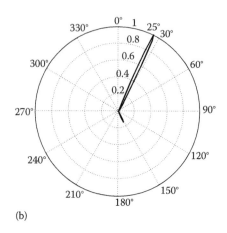

(a)   (b)

FIGURE 7.10
Far-field radiation pattern for first diffraction angle of 25° with same effective index estimation as Figure 7.10 for Au material. (a) 3D directivity pattern as functions of spherical angles for Au grating coupler with peak directivity of 1200; (b) normalized far-field power radiation pattern as function of θ on φ = 0° plane.

metal layer. The electric field of light reaches $1/e$ of its maximum value on the metal surface at the penetration depth that is defined as [32]

$$d(\theta_1) = \frac{\lambda}{2\pi\sqrt{n_m^2\sin^2\theta_1 - 1}} \tag{7.11}$$

where
  λ is the wavelength of the light
  n_m is the refraction index of metal
  θ_1 is the light incident angle

As shown in Equation 7.11, the penetration depth of light reduces for higher incident angle, which causes less light propagating through the metal layer and lowers the side lobe magnitude for the directivity far-field pattern. Even though the most grating coupler design including the one in this contribution chooses the near-vertical first-order diffraction angle to avoid the possible second reflection, the proposed embedded metallic grating design has unique advantage of better free-space transmission for higher radiation angle. In this research, $\theta_1 = 10°$ design would also be used for demonstrating fluorescence signal enhancement as it will be discussed in the last section. It is expected that silver grating design for $\theta_1 = 25°$ would have higher transmission rate for free space due to its lower material loss rate, though it has not been simulated.

7.4 Design of Slotted Photonic Crystal Nano-Resonator for Near-Field Light Focusing

In this section, we investigate the sub-wavelength near-field light-focusing effect on low in-plane refraction index meta-material. The embedded metallic grating coupler, as proposed in the last section, serves as the light-feeding mechanism and is fully compatible with the proposed light-focusing probe tip. Though the metallic material can be implemented in the relatively large area of the grating coupler and has relatively larger heat dissipating area, it is hard to be justified on probe tip design due to much stronger light focusing on nanometer-scale spot, which would cause strong heating damage. Thus, in this section, we propose an all-dielectric probe tip for near-field light focusing to avoid the possible heating damage. The probe is analyzed by FDTD method and it shows perturbation by water, and the substrate refraction index change of PhC has minimal impact on light confinement and throughput. Such a total dielectric probe tip design has great potential to complement the widely used metal-coated optical fiber-based light confinement probe. The sub-wavelength light source plays an important role in near-field scanning microscopy. By generating a light beam with spatial width far less than the wavelength, the imaging resolution reaches beyond the diffraction limit and instead is dominated by the light source size [33,34]. The low light throughput, due to mode mismatch, limits the applications of the optical fiber-based probe. To increase the light transmission rate by plasmonic effects, many kinds of metal-based light enhancement structures have been proposed, such as the nano-grating [35], superlens [36], bow tie [37,38], and dipole antenna [39] around the tip aperture. However, there are two main drawbacks with the designs aforementioned: light absorption due to metal material at the visible wavelength [40] and complicated manual assembly process. In this research, a total dielectric PhC-based near-field light confinement probe embedded with a center air slot and FP nano-resonators is analyzed numerically at the HeNe laser wavelength 632.8 nm. A typical low loss and low in-plane refraction index material is chosen to form the bulk material of PhC, such as silicon nitride with an index of 2.02 due to its compatibility with the planar silicon fabrication process.

The 2D PhC material is formed by arranging triangular air holes with radius $r = 0.3a$, where a is the lattice constant, through the substrate material, as shown in Figure 7.11a. The minimum light confinement realized by the 2D PhC waveguide is around a half wavelength through a single-line defect (W1) [15].

Other light-processing devices, such as the slot waveguide and nano-resonator, are needed to further confine light to the sub-wavelength level. The dielectric air slot waveguide provides a low loss transport mechanism for TE-mode light, where the electric field is linearly polarized in the y direction (Figure 7.11a) [41]. The light intensity peak concentrates in the air slot region with the full width at half-maximum (FWHM) beam size proportional to that of the air slot due to the continuity boundary condition of the TE displacement field [42]. A slot PhC waveguide serves as the base structure by placing an air slot with size of $0.2a$ along the ΓK direction of the triangular lattice in the center of the waveguide defect. By setting $D_t = 1.48a$, which defines the distance between the bulk PhC air hole and the air slot (Figure 7.11a), the PhC waveguide defect approximates a pseudo-W3 defect size. With the larger D_t, the new design relaxes the critical fabrication requirement, as compared to the W1 slot PhC waveguide design [15]. So the nano-resonator air holes later on can be placed around the center defect with ease without destructing the overall device features. The photonic band diagram, calculated by plane-wave expansion method using a supercell, as indicated at the right of Figure 7.11b, yields the guided mode, under which the optimal lattice size can be derived [2]. A guided mode in the gap center is chosen near the first Brillouin zone edge for the device design, because the magnetic field (Hz) distribution pattern is symmetric in the y direction for the TE mode, which excites symmetrically across the slot and provides light localization in the slot region. At the chosen guided mode, there is $\omega/(2\pi c) = 0.356$, where c is the speed of light in free space. For wavelength $\lambda_0 = 632.8$ nm, the lattice constant a is derived as $a = 0.356$, $\lambda_0 = 226$ nm, and the slot size $0.2a$ equals 45 nm,

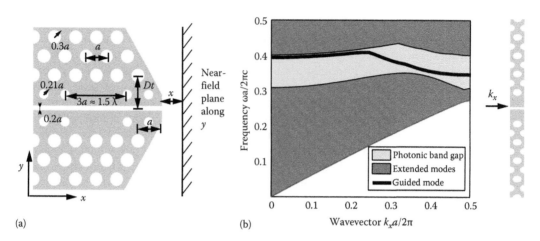

(a)   (b)

FIGURE 7.11
Design of a 2D slot FP PhC nano-resonator-based probe on triangular air hole lattice. (a) Probe design illustrated in terms of lattice constant a where $D_t = 1.48a$. White and gray color represents air and low refraction index material, respectively. (b) Projected band structure along the x direction of the slot PhC waveguide without nano-resonator formed by air hole radius of $0.3a$ (refraction index $n = 2.02$) for TE mode with the supercell lattice shown on the left.

which decides the minimum feature size of the overall device. The probe tip is shaped by removing the bulk PhC air holes along the ΓK direction, forming a 120° cone probe body. The probe body side cutting edges are aligned with the next line bulk PhC air hole edges. The final flat aperture surface is aligned in the y direction positioned $0.5a$ in the x direction away from the nearest mirror air holes. The probe is modeled by the FDTD method (Figure 7.12) with a grid size of 2 nm and a PML thickness of 1 μm surrounding the four boundaries of the calculation domain to reduce the reflection errors [29]. A TE light source with a Gaussian profile launches in a rectangular ridged waveguide with the same size of the PhC center defect. The FDTD calculation is converged after launching the source for time that allows the waves to propagate through the whole structure for at least 20 times and reaching a steady state before collecting the data at the monitoring planes. The performance of the probe is assessed in the near-field distance ranging from 0 to 30 nm with a step size of 2 nm. For straight slot PhC waveguide termination, there are strong reflections back to the probe body due to the PhC–air boundary, and the center peak inside the slot also splits into two strong side lobes (Figure 7.12c), which is not fit for near-field light confinement purposes.

In order to make a near-field light confinement tip, an FP resonator, formed by two smaller mirror air holes anchored in the bulk lattice position with a radius of $0.7r$ along the sides of the air slot (Figure 7.12a), is incorporated in the slotted PhC waveguide to enhance the light throughput and block the strong side lobes to make it a single peak probe at the near field. The cavity size of the FP resonator is around 1.5λ (λ is the wavelength in the substrate medium) by isolating mirroring holes with distance of $3a$ to form a constructive interference [43]. As the distance from the tip aperture increases, the center light intensity decreases, but the single peak is kept all the way to a 30 nm near-field distance. At a 10 nm near-field distance, the FWHM of the light beam is 87.8 nm, which is about $\lambda/7$ with peak intensity of 7.6%, twice that of the one without the resonator (all the light intensity is normalized to the light source intensity peak). In the vertical cross section, the light is confined in the waveguide layer through total internal reflection (TIR). Thus, in the near field, the light is confined in 3D with single one peak that is suitable for near-field light imaging purpose.

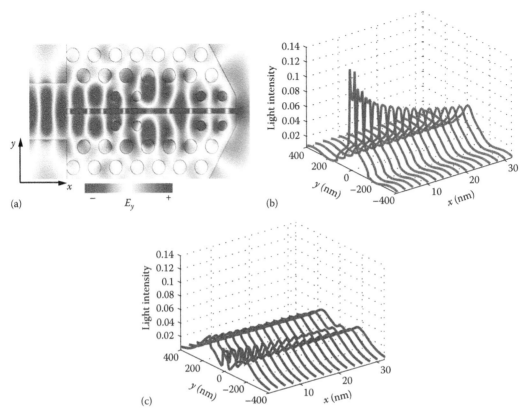

FIGURE 7.12
FDTD analysis of light confinement on probe without any perturbations. (a) Surface plot of electric field E_y distribution pattern of the slot PhC FP probe with bulk PhC index $n = 2.02$. (b) Light intensity distribution at near field of the probe, where x is the near-field distance from the tip aperture, ranging from 0 to 30 nm with 2 nm step. (c) Light distribution at near field of the probe without embedding nano-resonators.

7.5 Biological Applications

Biological applications of the proposed grating coupler and slotted PhC nano-resonator-based probe tip are analyzed using numerical modeling on possible Cy-5 fluorescence signal enhancement in biological imaging and biological water membrane impact of near-field light confinement, respectively.

7.5.1 Grating Coupler

7.5.1.1 Enhanced Detection of Cy-5 Fluorescence Radiation Signal

Cy-5 is a common cyanine-based fluorescent dye that can be excited by HeNe laser (632.8 nm) and has an emission max around 690 nm [44,45]. Also Cy-5, as water soluble, can be immobilized to the silicon nitride surface evenly through covalent bonding with DNA or antibody probe [46]. Thus, it is very useful to study the radiation pattern for fluorescence signal at wavelength of 690 nm. Based on the effective refraction index (Equation 7.2),

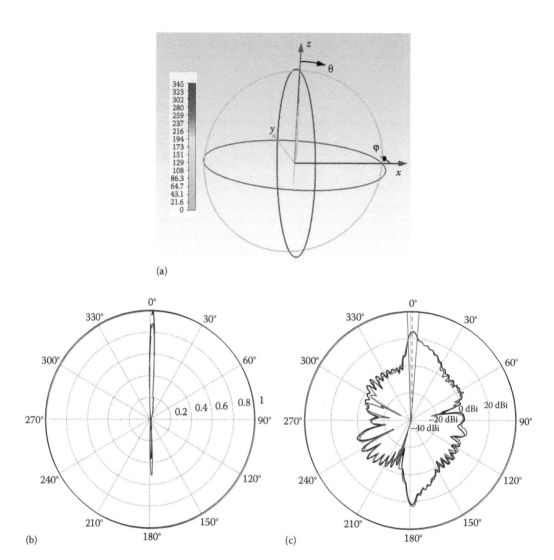

(a)

(b)

(c)

FIGURE 7.13
Far-field radiation pattern for fluorescence wavelength at 690 nm. (a) 3D directivity pattern as functions of spherical angles for Ag grating coupler with peak directivity of 345; (b) normalized far-field power radiation pattern as function of θ on $\varphi = 0$ plane; (c) far-field directivity as the function of θ on $\varphi = 0$ plane in unit of dBi. The dark black curve represents silver grating and the gray one is for gold grating.

the first-order diffraction angle relates to the light wavelength in a way that the increase of wavelength reduces the coupling angle. The peak radiation angle is predicted to be around $\theta_1 = 0°$ by Equation 7.2. Through 3D FDTD modeling of the exact same grating coupler structure as in Section 7.3, the radiation patterns are presented in Figure 7.13 for incident light wavelength at 690 nm for the TE mode. The peak directivities of both metallic grating couplers are pointing toward the vertical direction into free space, which would reduce the source light-coupling contamination at $\theta_1 = 10°$ and enhance the signal-to-noise ratio by utilizing a low numerical aperture objective lens. Similar to the case of the 632.8 nm wavelength, due to the material loss difference, the silver grating coupler still has slightly higher radiation intensity at the first-order diffraction angle.

7.5.2 Photonic Crystal Nano-Resonator

7.5.2.1 *Water Perturbation on Light Confinement*

Because the near-field light confinement is in the sub-wavelength level and very close to the tip aperture, it is necessary to study the light distribution changes due to the impacts from both the refraction index changes in the PhC substrate and the higher refraction index object in the near field. A flat water medium with refraction index of 1.33 is placed 10 nm away from the tip aperture (Figure 7.14a), because the normal near distance control for the near-field imaging system is most accurate in the sub-10 nm range [47]. This kind of configuration mimics the scenario of biological imaging applications, because most cells have high water content, with only a 4–5 nm lipid bilayer cell membrane [48]. Two monitoring planes at the near field of 5 and 15 nm along the y direction are chosen to assess the probe near-field focusing capabilities in terms of the FWHM and center peak intensity of the light beam as functions of the PhC substrate refraction index ranging from 1.97 to 2.07,

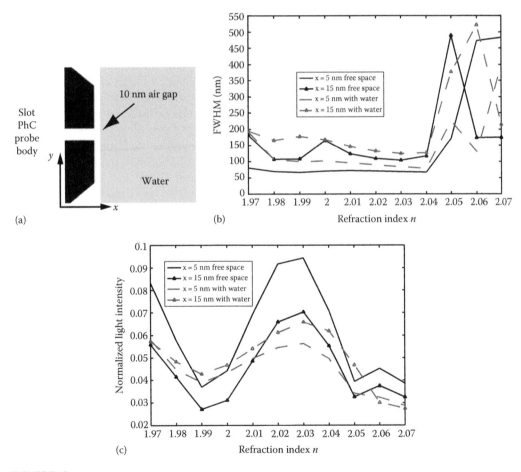

FIGURE 7.14
Perturbations on light confinement. (a) The illustration of water medium in the near field of the probe tip with 10 nm air gap between tip and water film. (b) FWHM plot of the light center peak and (c) normalized center peak intensity in 5 and 15 nm near-field distances with PhC bulk refraction index varying from 1.97 to 2.07 for both free-space case and water-film case in the near field.

which is about ±2.5% from the typical silicon nitride index ($n = 2.02$) (Figure 7.14b and c). Perturbation from the refraction index change of the bulk PhC causes beam splitting for a refraction index greater than 2.04. For an index ranging from 1.97 to 2.04, the FWHM is always less than 200 nm for all scenarios, and due to the impact from the water medium in the near field, the FWHM increases slightly about 30 nm in the center index range from 2.00 to 2.04. The "concave lens" effect, due to the near-field distance between the probe tip and the water film, causes the spreading of the wavefront that enlarges the beam size in the near field.

7.6 Conclusion

In conclusion, we investigate the nanopatterned photonic elements integrated on a single scanning probe for efficient light manipulations in the near field. Specifically, we propose a combination of embedded metallic grating coupler and PhC nano-resonator for near-field light confinement by using numerical modeling methods. A novel focus grating coupler design concept is proposed by using noble metals for directional light coupling at 632.8 nm wavelength for silicon nitride waveguide. Our modeling and simulations show that this focusing grating structure yields free-space coupling efficiency as high as 65% with minimal material loss. The noble metal grating is compatible with silicon fabrication technology through metal liftoff, EBL, and dry etching process. The far-field radiation pattern as derived from the 3D FDTD demonstrates the angular light-coupling capability for sub-wavelength light processing circuit. Also the fluorescence vertical radiation pattern can be further utilized to enhance fluorescence signal extraction for Cy-5 fluorescent dye. With the light coupled by the embedded metallic grating coupler, a near-field focusing effect of the slot FP PhC nano-resonator of a low in-plane index contrast and a pure dielectric material can provide sub-wavelength light confinement at the visible wavelength of 632.8 nm. The perturbation changes in the PhC substrate dielectric property cause beam splitting for refraction indexes higher than 2.04. Perturbation from the water medium causes slight beam spreading, but it is still under the sub-wavelength range. As an alternative application in biosensing, we show that the proposed embedded metallic focus grating coupler can also be used to realize a compact micrometer-scale fluorescence detector. Such compact devices can be used by biomedical researchers to image molecular interactions in vivo. Finally, the novel near-field light-focusing probe system, combining the embedded metallic grating coupler and PhC slotted waveguide, will likely provide the near-field imaging research community with an exciting tool using sub-wavelength light-focusing design and nano-manufacturing technology toward high imaging throughput at molecular resolution.

References

1. Bienstman, P. and R. Baets, Optical modelling of photonic crystals and VCSELs using eigenmode expansion and perfectly matched layers. *Optl Quantum Electron*, 2001. **33**(4): 327–341.
2. Johnson, S. and J. Joannopoulos, Block-iterative frequency-domain methods for Maxwell's equations in a planewave basis. *Opt Express*, 2001. **8**(3): 173–190.

3. Taflove, A. and S.C. Hagness, *Computational Electrodynamics: The Finite-Difference Time-Domain Method*, 3rd edn. Artech House Antennas and Propagation Library, Boston, MA: Artech House, 2005. p. xxii, 1006pp.

4. Ay, F. et al., Prism coupling technique investigation of elasto-optical properties of thin polymer films. *J Appl Phys*, 2004. **96**(12): 7147–7153.

5. Qian, W. et al., Design and analysis of optical coupling between silicon nanophotonic waveguide and standard single-mode fiber using an integrated asymmetric super-GRIN lens. Selected topics in quantum electronics. *IEEE J Quantum Electron*, 2011. **17**(3): 581–589.

6. Burns, W.K. and G.B. Hocker, End fire coupling between optical fibers and diffused channel waveguides. *Appl Opt*, 1977. **16**(8): 2048–2050.

7. Scheerlinck, S. et al., Efficient, broadband and compact metal grating couplers for silicon-on-insulator waveguides. *Opt Express*, 2007. **15**(15): 9625–9630.

8. Van Laere, F. et al., Compact and highly efficient grating couplers between optical fiber and nanophotonic waveguides. *J Lightwave Technol*, 2007. **25**(1): 151–156.

9. Doerr, C. et al., Wide bandwidth silicon nitride grating coupler. *IEEE Photon Technol Lett*, 2010. **22**(19): 1461–1463.

10. Hoshino, K. et al., Direct fabrication of nanoscale light emitting diode on silicon probe tip for scanning microscopy. *J Microelectromech Syst*, 2008. **17**(1): 4–10.

11. Hoshino, K. et al., Near-field scanning nanophotonic microscopy—Breaking the diffraction limit using integrated nano light-emitting probe tip. Selected topics in quantum electronics, *IEEE J Quantum Electron*, 2009. **15**(5): 1393–1399.

12. Hoshino, K. et al., Near-field scanning optical microscopy with monolithic silicon light emitting diode on probe tip. *Appl Phys Lett*, 2008. **92**(13): 131106–131106-3.

13. Wang, Y. et al., Plasmonic nanograting tip design for high power throughput near-field scanning aperture probe. *Opt Express*, 2010. **18**(13): 14004–14011.

14. Lee, Y. et al., Efficient apertureless scanning probes using patterned plasmonic surfaces. *Opt Express*, 2011. **19**(27): 25990–25999.

15. Wang, L. et al., Numerical simulation of photonic crystal based nano-resonators on scanning probe tip for enhanced light confinement, in *Scanning Microscopy*, T.P. Michael et al., Eds. Monterey, CA: SPIE, 2010. p. 77291M.

16. Wang, L. et al., Light focusing by slot Fabry-Perot photonic crystal nanoresonator on scanning tip. *Opt Lett*, 2011. **36**(10): 1917–1919.

17. Taillaert, D. et al., An out-of-plane grating coupler for efficient Butt-coupling between compact planar waveguides and single-mode fibers. *IEEE J Quantum Electron*, 2002. **38**(7): 949–955.

18. Waldhausl, R. et al., Efficient coupling into polymer waveguides by gratings. *Appl Opt*, 1997. **36**(36): 9383–9390.

19. Giuntoni, I. et al., Deep-UV technology for the fabrication of Bragg gratings on SOI rib waveguides. *Photon Technol Lett, IEEE*, 2009. **21**(24): 1894–1896.

20. Savas, T.A. et al. *Achromatic Interferometric Lithography for 100-nm-Period Gratings and Grids. J. Vac. Sci. Technol.*, 1995. **13**: p. 2732–2735.

21. *CST Microwave Studio*, 2010.

22. Johnson, P.B. and R.W. Christy, Optical constants of the noble metals. *Phys Rev B*, 1972. **6**(12): 4370–4379.

23. Schider, G. et al., Optical properties of Ag and Au nanowire gratings. *J Appl Phys*, 2001. **90**(8): 3825–3830.

24. Hao, F. and P. Nordlander, Efficient dielectric function for FDTD simulation of the optical properties of silver and gold nanoparticles. *Chem Phys Lett*, 2007. **446**(1–3): 115–118.

25. Vial, A. et al., Improved analytical fit of gold dispersion: Application to the modeling of extinction spectra with a finite-difference time-domain method. *Phys Rev B*, 2005. **71**(8): 085416.

26. Lee, T.W. and S. Gray, Subwavelength light bending by metal slit structures. *Opt Express*, 2005. **13**(24): 9652–9659.

27. Yoo, J. et al. Annealing optimization of silicon nitride film for solar cell application. *Thin Solid Films*, 2007. **515**(19): 7611–7614.

28. Lide, D.R., *CRC Handbook of Chemistry and Physics*, 86th edn., 2005 Boca Raton, FL: CRC Press.
29. Oskooi, A.F. et al., Meep: A flexible free-software package for electromagnetic simulations by the FDTD method. *Comp Phys Commun*, 2010. **181**(3): 687–702.
30. Bharadwaj, P. et al., Optical antennas. *Adv Opt Photon*, 2009. **1**(3): 438–483.
31. Kraus, J.D., Antennas. *McGraw-Hill Series in Electrical Engineering Radar and Antennas*, 2nd edn. New York: McGraw-Hill, 1988. p. xxv, 892pp.
32. Papathanassoglou, D.A. and B. Vohnsen, Direct visualization of evanescent optical waves. *Am J Phys*, 2003. **71**(7): 670–677.
33. Synge, E., A suggested method for extending microscopic resolution into the ultra-microscopic region. *Phil Mag*, 1928. **6**(356–362): 1.
34. Lewis, A. et al., Near-field optics: From subwavelength illumination to nanometric shadowing. *Nat Biotech*, 2003. **21**(11): 1378–1386.
35. Lezec, H.J. et al., Beaming light from a subwavelength aperture. *Science*, 2002. **297**(5582): 820–822.
36. Fang, N. et al., Sub–diffraction-limited optical imaging with a silver superlens. *Science*, 2005. **308**(5721): 534–537.
37. Kim, S. et al., High-harmonic generation by resonant plasmon field enhancement. *Nature*, 2008. **453**(7196): 757–760.
38. Ishihara, K. et al., Terahertz-wave near-field imaging with subwavelength resolution using surface-wave-assisted bow-tie aperture. *Appl Phys Lett*, 2006. **89**(20): 201120.
39. Taminiau, T.H. et al., Optical antennas direct single-molecule emission. *Nat Photon*, 2008. **2**(4): 234–237.
40. Gucciardi, P.G. et al., Thermal-expansion effects in near-field optical microscopy fiber probes induced by laser light absorption. *Appl Phys Lett*, 1999. **75**(21): 3408–3410.
41. Barrios, C.A. et al., Demonstration of slot-waveguide structures on silicon nitride/silicon oxide platform. *Opt Express*, 2007. **15**(11): 6846–6856.
42. Almeida, V.R. et al., Guiding and confining light in void nanostructure. *Opt Lett*, 2004. **29**(11): 1209–1211.
43. Kudryashov, A.V., H. Weber, and Society of Photo-optical Instrumentation Engineers, *Laser Resonators: Novel Design and Development*. Bellingham, WA: SPIE Optical Engineering Press, 1999. p. xiii, 301pp.
44. Block, I.D. et al., A detection instrument for enhanced-fluorescence and label-free imaging on photonic crystal surfaces. *Opt Express*, 2009. **17**(15): 13222–13235.
45. Mathias, P.C., H.Y. Wu, and B.T. Cunningham, Employing two distinct photonic crystal resonances to improve fluorescence enhancement. *Appl Phys Lett*, 2009. **95**(2): 21111.
46. Martiradonna, L. et al., Spectral tagging by integrated photonic crystal resonators for highly sensitive and parallel detection in biochips. *Appl Phys Lett*, 2010. **96**(11): 113702–113703.
47. Betzig, E., P.L. Finn, and J.S. Weiner, Combined shear force and near-field scanning optical microscopy. *Appl Phys Lett*, 1992. **60**(20): 2484–2486.
48. Kuchel, P.W., *Schaum's Outline of Theory and Problems of Biochemistry*. Schaum's outline series Schaum's outline series in science. New York: McGraw-Hill, 1988. 555pp.

8

Coupled Mode Theory and Its Applications on Computational Nanophotonics

Jianwei Mu and Yasha Yi

CONTENTS

Coupled-mode theory (CMT) has been widely utilized in the analysis of electromagnetic wave coupling and conversions due to its mathematical simplicity and physical intuitiveness. The early version of CMT was proposed by Pierce and Miller in the 1950s to study microwaves, and was mathematically formulated by Schelkunoff using mode expansion, and by Haus with variational principles. The CMT was later introduced to the investigation of optical waveguides by Marcuse [6], Snyder [7], and Kogelnik [8] in the early 1970s. Since then a series of formulations and applications in optical waveguide has been proposed and studied. The physical model of CMT is expanding the total field inside and optical waveguide in terms of the field of a reference waveguide structure, by applying the orthogonality condition, a set of ordinary differential equations are obtained. Generally, the CMT focused on the guided modes with the assumption that only a limited number of guided modes (usually one or two) close to phase matching play significant roles in the interaction of the modal fields. In situations of applications involving the radiation mode coupling, the application of CMT becomes cumbersome due to the continuous spectrum of radiation modes. One possible solution to circumvent the problem of radiation modes is to introduce leaky modes to approximate the radiation modes [9–20].

The leaky modes are, however, not orthogonal and normalizable in real domain. For this reason, it is difficult to deal with leaky mode formulations analytically and even more so numerically for practical applications.

Recently, a new computation model was introduced to the mode-matching method (MMM) in which the waveguide structure is enclosed by a perfectly matched layer (PML) terminated by a perfectly reflecting boundary (PRB) condition [21–23]. This seemingly paradoxical combination of PML and PRB leads to a somewhat unexpected yet remarkable result: it creates an open and reflectionless environment in a close and finite computation domain. A set of complex modes can be derived from this waveguide model that are well behaved in terms of orthogonality and normalization and can be readily solved by standard analytical and numerical techniques. By utilizing the complex modes as an orthogonal base functions to represent the radiation fields, the CMT can be applied as if all the modes are discrete and guided. The complex CMT was subsequently applied to simulation and analysis of slab, circular, and channel optical waveguide structures and shown to be highly accurate and versatile.

8.1 Modal Analysis with Perfectly Matching Layers

We make the following assumptions: (1) the medium in the waveguide structure is lossless, linear, and isotropic. The permittivity and the permeability of vacuum are denoted as ε_0 and μ_0, respectively. The permeability μ in the medium is equal to the free space value μ_0 throughout this chapter. (2) The time dependency is expressed as $\exp(j\omega t)$. The wave is propagating along z, and the z dependency is expressed as $\exp(-j\gamma z)$, which refers to the propagation in the positive z direction, or $\exp(-j\gamma z)$ in the negative z direction. ω and γ are the angular frequency and the propagation constant, respectively.

Considering a waveguide structure where the transverse index profile $n(x, y)$ is arbitrary and defined in the Cartesian coordinate system, if we enclose the optical waveguide by an anisotropic PML terminated by a PRB, the Maxwell's equations can be written as

$$\nabla \times E = -j\omega\mu_0 [\Lambda] H \tag{8.1}$$

$$\nabla \times H = j\omega\varepsilon_0 n^2 [\Lambda] E \tag{8.2}$$

The tensor $[\Lambda]$ accounts for the PML and is given by

$$[\Lambda] = \begin{bmatrix} S_y/S_x & 0 & 0 \\ 0 & S_x/S_y & 0 \\ 0 & 0 & S_x/S_y \end{bmatrix} \tag{8.3}$$

where S_x and S_y are called the coordinate stretching factors and given by

$$S_x = k_x - j\frac{\sigma_x}{\omega\varepsilon} \tag{8.4}$$

$$S_y = k_y - j\frac{\sigma_y}{\omega\varepsilon}$$ (8.5)

and κ_x (κ_y) and σ_x (σ_y) are the parameters to control the phase shift and absorption of the travelling waves in the PML along x (y). For non-PML media, we have $\kappa_x = \kappa_y = 1$, $\sigma_x = \sigma_y = 0$.

The vector wave equation for the electric field is derived from Equations 8.1 and 8.2

$$\nabla \times \left([\Lambda]^{-1} \nabla \times E \right) = n^2 k^2 [\Lambda] E$$ (8.6)

where $k = \omega\sqrt{\varepsilon_0\mu_0}$ is the wave number in free space. Further, we have the transverse wave equation as

$$\nabla_t \times \left(\frac{1}{s_x s L y} \nabla_t \times E_t \right) + \hat{z} \times \left([\Lambda]_t^{-1} \nabla_t \times (-j\gamma E_z \hat{z}) \right) - \gamma^2 \hat{z} \times \left([\Lambda]_t^{-1} \hat{z} \times E_t \right) = n^2 k^2 [\Lambda]_t E_t$$ (8.7)

where

$$[\Lambda]_t = \begin{pmatrix} S_y/S_x & 0 \\ 0 & S_x/S_y \end{pmatrix}$$ (8.8)

Using $\nabla \cdot (\varepsilon_0 n^2 [\Lambda] E) = 0$, we have

$$E_z = \frac{\left(\nabla_t \cdot \left(\varepsilon_0 n^2 [\Lambda]_t E_t \right) \right)}{j\gamma n^2 s_x s_y}$$ (8.9)

One can derive the vector wave equation for the transverse electric (TE) field,

$$\nabla_t \times \left(\frac{1}{s_x s_y} \nabla_t \times E_t \right) - \hat{z} \times \left\{ [\Lambda]_t^{-1} \nabla_t \times \left(\frac{\nabla_t \cdot \left(n^2 [\Lambda]_t E_t \right)}{n^2 s_x s_y} \hat{z} \right) \right\} - \gamma^2 \hat{z} \times \left\{ [\Lambda]_t^{-1} \hat{z} \times E_t \right\} = n^2 k^2 [\Lambda]_t E_t$$ (8.10)

Equation 8.10 can be written in the matrix form

$$\begin{bmatrix} P_{xx} & P_{xy} \\ P_{yx} & P_{yy} \end{bmatrix} \begin{bmatrix} E_x \\ E_y \end{bmatrix} = \gamma^2 \begin{bmatrix} E_x \\ E_y \end{bmatrix}$$ (8.11)

where the differential operators are defined as

$$P_{xx} E_x = \frac{\partial}{\partial x} \left[\frac{1}{n^2} \frac{1}{s_x} \frac{\partial}{\partial x} \left(n^2 \frac{1}{s_x} E_x \right) \right] + \frac{s_x}{s_y} \frac{\partial}{\partial y} \left[\frac{1}{s_y} \frac{\partial}{\partial y} \left(\frac{1}{s_x} E_x \right) \right] + n^2 k^2 E_x$$ (8.12)

$$P_{xy}E_y = \frac{s_y}{s_x}\frac{\partial}{\partial x}\left[\frac{1}{n^2}\frac{1}{s_y}\frac{\partial}{\partial y}\left(n^2\frac{1}{s_y}E_y\right)\right] - \frac{\partial}{\partial y}\left[\frac{1}{s_x}\frac{\partial}{\partial x}\left(\frac{1}{s_y}E_y\right)\right]$$ (8.13)

$$P_{yy}E_y = \frac{\partial}{\partial y}\left[\frac{1}{n^2}\frac{1}{s_y}\frac{\partial}{\partial y}\left(n^2\frac{1}{s_y}E_y\right)\right] + \frac{s_y}{s_x}\frac{\partial}{\partial x}\left(\frac{1}{s_x}\frac{\partial}{\partial x}\left(\frac{1}{s_y}E_y\right)\right) + n^2k^2E_y$$ (8.14)

$$P_{yx}E_x = \frac{s_x}{s_y}\frac{\partial}{\partial y}\left[\frac{1}{n^2}\frac{1}{s_x}\frac{\partial}{\partial x}\left(n^2\frac{1}{s_x}E_x\right)\right] - \frac{\partial}{\partial x}\left[\frac{1}{s_y}\frac{\partial}{\partial y}\left(\frac{1}{s_x}E_x\right)\right]$$ (8.15)

Equation 8.11 is a full vectorial equation. All the vectorial properties of the electromagnetic field are included. $P_{xx}\neq P_{yy}$ causes the polarization dependence, whereas $P_{xx}\neq 0$ and $P_{yy}\neq 0$ induce the polarization coupling between the two components E_x, E_y. If the coupling between the two polarizations is weak and negligible, by neglecting the cross terms P_{xy} and P_{yx}, the full vectorial (8.11) reduces to two decoupled equations:

$$P_{xx}E_x = \gamma^2 E_x$$ (8.16)

$$P_{yy}E_y = \gamma^2 E_y$$ (8.17)

There is only one transverse field component in Equations 8.16 and 8.17, and the eigenmodes associated with those equations are commonly named as quasi-transverse magnetic (TE) modes and quasi-TM modes. The equations are referred to as semivectorial equations. If the structures are weakly-guiding, even the polarization dependence may be neglected. Equation 8.11 reduces to

$$P_{xx} = P_{yy} = \frac{\partial^2}{\partial x^2} + \frac{\partial^2}{\partial y^2} + n^2k^2$$ (8.18)

Since the polarization difference is negligible, we call this scalar approximation. Equations 8.16 and 8.17 are then replaced by a single equation

$$P\Psi = \gamma^2\Psi$$ (8.19)

If we have the TE field components obtained from Equation 8.11, we can find the TM transverse field components by

$$H_x = -Y_0 N_{eff}\frac{s_x}{s_y}E_y + \frac{Y_0}{N_{eff}k^2}\frac{s_x}{s_y}\left\{\frac{\partial}{\partial y}\left[\frac{1}{n^2}\frac{1}{s_y}\frac{\partial}{\partial y}\left(n^2\frac{1}{s_y}E_y\right)\right] + \frac{\partial}{\partial y}\left[\frac{1}{n^2}\frac{1}{s_x}\frac{\partial}{\partial x}\left(n^2\frac{1}{s_x}E_x\right)\right]\right\}$$ (8.20)

$$H_y = +Y_0 N_{eff} \frac{s_y}{s_x} E_x - \frac{Y_0}{N_{eff} k^2} \frac{s_y}{s_x} \left\{ \frac{\partial}{\partial x} \left[\frac{1}{n^2} \frac{1}{s_x} \frac{\partial}{\partial x} \left(n^2 \frac{1}{s_x} E_x \right) \right] + \frac{\partial}{\partial x} \left[\frac{1}{n^2} \frac{1}{s_y} \frac{\partial}{\partial y} \left(n^2 \frac{1}{s_y} E_y \right) \right] \right\} \quad (8.21)$$

where $Y_0 = \sqrt{\varepsilon_0/\mu_0}$ is the admittance of light in vacuum and $N_{eff} = \gamma/k_0$ is the effective modal index.

For waveguides with the weakly polarization coupling

$$\frac{\partial}{\partial y} \left[\frac{1}{n^2} \frac{1}{s_x} \frac{\partial}{\partial x} \left(n^2 \frac{1}{s_x} E_x \right) \right] \approx \frac{\partial}{\partial x} \left[\frac{1}{n^2} \frac{1}{s_y} \frac{\partial}{\partial y} \left(n^2 \frac{1}{s_y} E_y \right) \right] \approx 0$$

the semivector solutions are given by

$$H_x = -Y_0 N_{eff} \frac{s_x}{s_y} E_y + \frac{Y_0}{N_{eff} k^2} \frac{s_x}{s_y} \frac{\partial}{\partial y} \left[\frac{1}{n^2} \frac{1}{s_y} \frac{\partial}{\partial y} \left(n^2 \frac{1}{s_y} E_y \right) \right] \quad (8.22)$$

$$H_y = +Y_0 N_{eff} \frac{s_y}{s_x} E_x - \frac{Y_0}{N_{eff} k^2} \frac{s_y}{s_x} \frac{\partial}{\partial x} \left[\frac{1}{n^2} \frac{1}{s_x} \frac{\partial}{\partial x} \left(n^2 \frac{1}{s_x} E_x \right) \right] \quad (8.23)$$

For very weakly guided waveguide structures with $\partial n^2/\partial x \approx \partial n^2/\partial y \approx 0$, we have the scalar approximation:

$$\tilde{H} = \pm N_{eff} Y_0 \tilde{E} \quad (8.24)$$

where
$\tilde{H}$ stands for (H_x, H_y)
$\tilde{E}$ stands for $((S_x/S_y)E_y, (S_y/S_x)E_x)$.

For general media, the time average power for each guided mode is real and finite and can be normalized as

$$\frac{1}{4} \iint\limits_{Entire\ region} \left(E_{tn} \times H_{tn}^* + E_{tn}^* \times H_{tn} \right) \cdot \hat{z}\, dA = 1 \quad (8.25)$$

where
t denotes transverse components
n denotes the mode number

The expression reduces to

$$\frac{1}{2} \iint\limits_{Entire\ region} \left(E_{tn} \times H_{tn} \right) \cdot \hat{z}\, dA = 1 \quad (8.26)$$

for lossless media. However, it is noted for general media

$$\frac{1}{2} \iint\limits_{Entire\ region} \left(E_{tn} \times H_{tn} \right) \cdot \hat{z}\, dA = N_n \tag{8.27}$$

where N_n may not be equal to unity and may even be complex!

Let m and n denote the modal indices of two distinct guided modes with corresponding propagation constants represented by β_m and β_n; the orthogonal relation is given by

$$\iint\limits_{Entire\ region} \left(E_{tn} \times H_{tm} \right) \cdot \hat{z}\, da = 0 \tag{8.28}$$

This orthogonality relation is valid even for active or lossy media. In particular, if the media are passive and lossless, the orthogonality relation maybe rewritten as

$$\iint\limits_{Entire\ region} \left(E^{*}_{tm} \times H_{tn} \right) \cdot \hat{z}\, da = 0 \tag{8.29}$$

In practice, we normally set the phase-shift parameter κ_x and κ_y to unity since they are significant only for dealing with evanescent fields. On the other hand, the profiles of the absorption coefficients σ_x and σ_y are critical for effectively reducing the reflections from the PRB. A commonly used expression for the absorption profile is

$$\sigma = \sigma_{max} \left(\frac{\rho}{T_{PML}} \right)^m \tag{8.30}$$

where
T_{PML} is the thickness of the PML layer
ρ is the distance measured from the starting position of the PML

A good measure for the effectiveness of the PML is the reflection coefficient defined by [61]

$$R_{PML} = \exp \left\{ -\frac{2\sigma_{max}}{n\sqrt{\varepsilon_0/\mu_0}} \int_0^{T_{PML}} \left(\frac{\rho}{T_{PML}} \right)^m d\rho \right\} \tag{8.31}$$

In terms of the PML reflection coefficient, the PML coefficient can be conveniently expressed as

$$S = k - j\frac{\lambda}{4\pi n T_{PML}} \left[(m+1)\ln\left(\frac{1}{R_{PML}} \right) \right]\left(\frac{\rho}{T_{PML}} \right)^m \tag{8.32}$$

It was shown that when $m = 2$, the *PML* seems to be the most effective.

8.2 Complex Coupled-Mode Theory Based on Normal Modes

8.2.1 Derivation of Complex Coupled-Mode Equations Based on Normal Modes

To derive the coupled-mode equations based on the complex modes described in Section 8.1, we assume that the permittivity function distribution along the waveguide with perturbations can be expressed as

$$\tilde{\varepsilon}(x,y,z) = \varepsilon(x,y) + \Delta\varepsilon(x,y,z) \tag{8.33}$$

where the index perturbation $\Delta\varepsilon$ is defined as the difference between the index profiles of the practical waveguides under investigation and the reference waveguides for which the complex modes are known. Maxwell equations for the perturbed waveguides are

$$\nabla \times E(x,y,z) = -j\omega\mu_0 [\Lambda] H(x,y,z) \tag{8.34}$$

$$\nabla \times H(x,y,z) = +j\omega\tilde{\varepsilon}(x,y,z)[\Lambda] E(x,y,z) \tag{8.35}$$

Suppose that the difference between the perturbed and the reference waveguides is sufficiently small so that we can expand the unknown transverse electromagnetic fields of the perturbed waveguides in terms of the transverse modal fields of the reference waveguides, that is,

$$E_t(x,y,z) = \sum_{n=1} \left[a_n(z) + b_n(z) \right] \mathbf{e}_{tn}(x,y) \tag{8.36}$$

$$H_t(x,y,z) = \sum_{n=1} \left[a_n(z) - b_n(z) \right] \mathbf{h}_{tn}(x,y) \tag{8.37}$$

The functions $a_n(z)$ and $b_n(z)$ are the mode amplitudes for the forward- and backward-propagating waves, respectively. The longitudinal fields can be expressed in terms of transverse components as

$$E_z(x,y,z) = \sum_{n=1} \left[a_n(z) - b_n(z) \right] \frac{\varepsilon}{\tilde{\varepsilon}} \mathbf{e}_{2n}(x,y) \tag{8.38}$$

$$H_z(x,y,z) = \sum_{n=1} \left[a_n(z) + b_n(z) \right] \mathbf{h}_{2n}(x,y) \tag{8.39}$$

The next step is to derive the coupled-mode equations governing the mode amplitudes. To do so, we simply substitute Equations 8.36 through 8.39 into 8.34 and 8.35. After mathematical manipulations, we derive the following coupled equations:

$$N_m \frac{da_m}{dz} + j\gamma_m a_m = -j\sum_{n=1} k_{mn} a_n - j\sum_{n=1} \chi_{mn} b_n \tag{8.40}$$

$$N_m \frac{db_m}{dz} - j\gamma_m b_m = +j\sum_{n=1} k_{mn} b_n + j\sum_{n=1} \chi_{mn} a_n \tag{8.41}$$

in which the coupling coefficients are given by

$$k_{mn} = \frac{\omega \varepsilon_o}{4} \int_A (\tilde{n}^2 - n^2) \left(\mathbf{e}_{tn} \cdot \mathbf{e}_{tm} - \frac{n^2}{\tilde{n}^2} e_{2n} \cdot e_{2m} \right) da \tag{8.42}$$

$$\chi_{mn} = \frac{\omega \varepsilon_o}{4} \int_A (\tilde{n}^2 - n^2) \left(\mathbf{e}_{tn} \cdot \mathbf{e}_{tm} + \frac{n^2}{\tilde{n}^2} e_{2n} \cdot e_{2m} \right) da \tag{8.43}$$

where the refractive indices are used to replace the permittivity functions, that is, $\varepsilon = n^2 \varepsilon_o$ and $\tilde{\varepsilon} = \tilde{n}^2 \varepsilon_o$. The coefficient N_m is defined as

$$N_m = \frac{1}{2} \iint (\mathbf{e}_{tm} \times \mathbf{h}_{tm}) \cdot \hat{z} \, da \tag{8.44}$$

Under the normalization condition (8.45), $N_1 = 1$ for the real core guided modes and $N_m \approx 1$ for the complex cladding leaky modes with relatively small mode losses. In general, however, N_m may not be equal to unity and can be complex.

The coupled-mode Equations 8.40 and 8.41 and the expressions for the coupling coefficients 8.42 and 8.43 are formally identical to those derived for guided modes in waveguides made of reciprocal media in the presence of loss and/or gain and hence can be solved by the same analytical and numerical techniques as previously done in literature. In the classical CMT, to deal with radiation fields based on the integral of the continuous radiation modes is possible but extremely cumbersome. On the other hand, however, both guided and radiation fields are considered in the new complex CMT in a unified fashion. Also, it is noted that the coupling coefficients for the co- and contra-coupling modes are symmetrical in the sense that

$$\kappa_{mn} = \kappa_{nm} \tag{8.45}$$

$$\chi_{mn} = \chi_{nm} \tag{8.46}$$

Further, we may rewrite the mode amplitudes by separating the slowly varying envelopes with the fast oscillating carriers so as to

$$a_n = A_n \exp(-j\gamma_n z) \tag{8.47}$$

$$b_n = B_n \exp(+j\gamma_n z) \tag{8.48}$$

On substitution of Equations 8.51 by Equation 8.55, we derive

$$N_m \frac{dA_m}{dz} = -j \sum_{n=1} k_{mn} A_n \exp\left[-j(\gamma_n - \gamma_m)z\right] - j \sum_{n=1} \chi_{mn} B_n \exp\left[+j(\gamma_n + \gamma_m)z\right] \quad (8.49)$$

$$N_m \frac{dB_m}{dz} = +j \sum_{n=1} k_{mn} B_n \exp\left[+j(\gamma_n - \gamma_m)z\right] + j \sum_{n=1} \chi_{mn} A_n \exp\left[-j(\gamma_n + \gamma_m)z\right] \quad (8.50)$$

For grating structures whose index perturbations are periodic along the waveguide axis, the coupling coefficients (Equations 8.42 and 8.43) are also periodic functions with the same period. We may expand the coupling coefficients in terms of the Fourier series as

$$k_{mn} = \sum_{l=-\infty}^{+\infty} D_{mn}^{(l)} \exp\left(jl\frac{2\pi}{\Lambda}z\right) \quad (8.51)$$

$$\chi_{mn} = \sum_{l=-\infty}^{+\infty} C_{mn}^{(l)} \exp\left(jl\frac{2\pi}{\Lambda}z\right) \quad (8.52)$$

Substitute Equations 8.51 and 8.52 into Equations 8.49 and 8.50 and we derive

$$N_m \frac{dA_m}{dz} = -j \sum_{n=1} A_n \sum_{l=-\infty}^{+\infty} D_{mn}^{(l)} \exp\left[-j\left(\gamma_n - \gamma_m - l\frac{2\pi}{\Lambda}\right)z\right]$$

$$-j \sum_{n=1} B_n \sum_{l=-\infty}^{+\infty} C_{mn}^{(l)} \exp\left[+j\left(\gamma_n + \gamma_m + l\frac{2\pi}{\Lambda}\right)z\right] \quad (8.53)$$

$$N_m \frac{dB_m}{dz} = +j \sum_{n=1} B_n \sum_{l=-\infty}^{+\infty} D_{mn}^{(l)} \exp\left[+j\left(\gamma_n - \gamma_m + l\frac{2\pi}{\Lambda}\right)z\right]$$

$$+j \sum_{n=1} A_n \sum_{l=-\infty}^{+\infty} C_{mn}^{(l)} \exp\left[-j\left(\gamma_n + \gamma_m - l\frac{2\pi}{\Lambda}\right)z\right] \quad (8.54)$$

It can be shown that the phase factors $\gamma_n - \gamma_m \pm l2\pi/\Lambda$ and $\gamma_n + \gamma_m \pm l2\pi/\Lambda$ in the exponential terms in Equations 8.53 and 8.54 are most critical in determining the strength of interactions between the different modes over distance. Only when these factors are close to zero, there will be appreciable power exchange between the coupled modes, a condition referred to as the phase-matching conditions. In fact, the role of grating is to facilitate the phase-matching between two propagation modes with different propagation constants by providing a grating space harmonic component related to the grating period and profile.

Note that the phase-matching conditions for the co- and contra-propagating modes are quite distinct and may not be readily realized by the same grating. In practice, we normally design the grating to assist coupling for either contra-propagating (e.g., Bragg gratings) or co-propagating (e.g., long-period gratings) waves only. Under this assumption, the coupled-mode equations are further reduced to two groups as described in the following.

For the contra-propagation waves, the coupled-mode equations reduce to

$$N_m \frac{dA_m}{dz} = -j \sum_{n=1} B_n \sum_{l=-\infty}^{+\infty} C_{mn}^{(l)} \exp\left[+j\left(\gamma_n + \gamma_m + l\frac{2\pi}{\Lambda}\right)z\right] \tag{8.55}$$

$$N_m \frac{dB_m}{dz} = +j \sum_{n=1} A_n \sum_{l=-\infty}^{+\infty} C_{mn}^{(l)} \exp\left[-j\left(\gamma_n + \gamma_m - l\frac{2\pi}{\Lambda}\right)z\right] \tag{8.56}$$

and for the co-directional propagation waves, we have

$$N_m \frac{dA_m}{dz} = -j \sum_{n=1} A_n \sum_{l=-\infty}^{+\infty} D_{mn}^{(l)} \exp\left[-j\left(\gamma_n - \gamma_m - l\frac{2\pi}{\Lambda}\right)z\right] \tag{8.57}$$

$$N_m \frac{dB_m}{dz} = +j \sum_{n=1} B_n \sum_{l=-\infty}^{+\infty} D_{mn}^{(l)} \exp\left[+j\left(\gamma_n - \gamma_m + l\frac{2\pi}{\Lambda}\right)z\right] \tag{8.58}$$

Further, for a given grating, the phase-matching condition is a function of wavelength through wavelength dependence of the mode propagation constants, that is, $\gamma_n(\lambda)$ and $\gamma_m(\lambda)$. It also depends on the mode index ($m,n = 1,2,\ldots$) and the order of space harmonics in the Fourier expansion. Any combination of these parameters that lead to a phase matching condition will likely yield a distinct resonant signature in the mode coupling as illustrated later in the transmission and reflection spectra in this chapter. By identifying these phase-matching conditions in the coupled-mode equations, we can greatly simplify the solutions and also gain great insight into the salient features underlying the interaction of the modes in presence of the gratings.

First of all, we may consider only the largest Fourier expansion coefficient, that is, $l = \pm 1$ or the first-order grating effect and ignore all these other high-order space harmonics. Investigation of the higher-order gratings can be performed in the similar fashion, but will not be pursued further in this work. The coupled-mode equations 8.55 through 8.58 are subsequently decoupled into two separate sets such that

$$N_m \frac{dA_m}{dz} = -j \sum_{n=1} B_n C_{mn}^{(-1)} \exp\left[+j\left(\gamma_n + \gamma_m - \frac{2\pi}{\Lambda}\right)z\right] \tag{8.59}$$

$$N_m \frac{dB_m}{dz} = +j \sum_{n=1} A_n C_{mn}^{(+1)} \exp\left[-j\left(\gamma_n + \gamma_m - \frac{2\pi}{\Lambda}\right)z\right] \tag{8.60}$$

for the contra-directional propagation waves and

$$N_m \frac{dA_m}{dz} = -j \sum_{n=1} A_n D_{mn}^{(-1)} \exp\left[-j\left(\gamma_n - \gamma_m + \frac{2\pi}{\Lambda}\right)z\right]$$

(8.61)

$$N_m \frac{dB_m}{dz} = +j \sum_{n=1} B_n D_{mn}^{(+1)} \exp\left[+j\left(\gamma_n - \gamma_m + \frac{2\pi}{\Lambda}\right)z\right]$$

(8.62)

for the codirectional propagation waves.

In many practical situations, the phase-matching conditions can only be realized at a distinct wavelength (i.e., λ_{mn}) for a given pair of modes (i.e., m and n). For the sake of simplicity, we suppose that the mth mode is the forward-propagating fundamental guided mode with largest real propagation constant, that is, $m = 1$ and $\gamma_m = \beta_1$ (real). In the proximity of the phase-matching wavelength $\lambda \cong \lambda_{1n}$, we may consider only the two modes that are close to the phase match (whenever it is possible!) so that Equations 8.59 through 8.62 are simplified to

$$N_1 \frac{dA_1}{dz} = -jC_{1n}^{(-1)}B_n \exp\left[+j\left(\gamma_n + \beta_1 - \frac{2\pi}{\Lambda}\right)z\right]$$

(8.63)

$$N_n \frac{dB_n}{dz} = +jC_{n1}^{(+1)}A_1 \exp\left[-j\left(\gamma_n + \beta_1 - \frac{2\pi}{\Lambda}\right)z\right]$$

(8.64)

for the contra-directional propagation modes, and

$$N_1 \frac{dA_1}{dz} = -jD_{1n}^{(-1)}A_n \exp\left[-j\left(\gamma_n - \beta_1 + \frac{2\pi}{\Lambda}\right)z\right]$$

(8.65)

$$N_n \frac{dA_n}{dz} = -jD_{n1}^{(+1)}A_1 \exp\left[+j\left(\gamma_n - \beta_1 + \frac{2\pi}{\Lambda}\right)z\right]$$

(8.66)

for the codirectional propagating modes. Note that $\gamma_n = \beta_n - j\alpha_n$ and we define the phase-detuning factors such that

$$\Delta\beta_n = \frac{1}{2}\left(\gamma_1 + \gamma_n - \frac{2\pi}{\Lambda}\right)$$

(8.67)

for the contra-directional propagation modes is

$$\Delta\beta_n = \frac{1}{2}\left(\gamma_1 - \gamma_n - \frac{2\pi}{\Lambda}\right)$$

(8.68)

for the co-directional propagation modes. The phase matching conditions happened as

$$\Re(\Delta\beta_n) = 0 \tag{8.69}$$

so that the grating period for the phase matching conditions for the contra- and co-directional modes are

$$\Lambda = \frac{2\pi}{\Re(\gamma_1 + \gamma_n)} \tag{8.70}$$

and

$$\Lambda = \frac{2\pi}{\Re(\gamma_1 - \gamma_n)} \tag{8.71}$$

We may recast Equations 8.65 and 8.66 into more revealing forms as follows:

$$N_1 \frac{dA_1}{dz} = -jC_{1n}^{(-1)} B_n \exp\left[+j(2\Delta\beta_n)z\right] \tag{8.72}$$

$$N_n \frac{dB_n}{dz} = +jC_{n1}^{(+1)} A_1 \exp\left[-j(2\Delta\beta_n)z\right] \tag{8.73}$$

for the contra-directional modes and

$$N_1 \frac{dA_1}{dz} = -jD_{1n}^{(-1)} A_n \exp\left[-j(2\Delta\beta_n)z\right] \tag{8.74}$$

$$N_n \frac{dA_n}{dz} = -jD_{n1}^{(+1)} A_1 \exp\left[+j(2\Delta\beta_n)z\right] \tag{8.75}$$

for the co-directional modes.

8.2.2 Solutions of the Complex Coupled-Mode Equations

Computationally, we may solve the full coupled mode equations 8.59 through 8.62 (referred to as the full CMT) or the reduced Equations 8.72 through 8.75 (referred to as the reduced CMT). The former can be readily carried out by a standard numerical algorithm such as the Runge–Kutta method, whereas the latter can be solved to yield simple analytical formulas.

Suppose that all the power is initially launched in the forward-propagating fundamental mode, that is, $a_1(0) = 1$. For the contra-directional modes, we assume that no power is associated with the backward-propagating modes at the other side of the grating, that is, $b_n(L) = 0$. The analytical solutions are

$$a_1(z) = a_1(0) \frac{\Delta\beta_n \sinh S(z-L) - jS \cosh S(z-L)}{-\Delta\beta_n \sinh SL - jS \cosh SL} e^{-j(\beta_1 - \Delta\beta_n)z} \tag{8.76}$$

$$b_n(z) = -a_1(0) \frac{C_{n1}^{(+1)}}{N_n} \frac{\sinh S(z-L)}{\Delta\beta_n \sinh SL + jS \cosh SL} e^{j(\gamma_n - \Delta\beta_n)z} \tag{8.77}$$

where

$$S = \sqrt{k_n^2 - (\Delta\beta_n)^2} \tag{8.78}$$

$$k_n = \sqrt{\frac{C_{1n}^{(-1)} C_{n1}^{(+1)}}{N_1 N_n}} \tag{8.79}$$

For the co-directional modes, we assume that no power is associated with the secondary mode at the starting point of the grating, that is, $a_n(0) = 0(n \neq 1)$. The solutions are

$$a_1(z) = a_1(0) \frac{j\Delta\beta \sin(Q_n z) + Q_n \cos(Q_n z)}{Q_n} e^{-j(\beta_1 - \Delta\beta_n)z} \tag{8.80}$$

$$a_n(z) = -je^{j\Delta\beta_n z} \frac{D_{n1}^{(+1)}}{N_n} \frac{\sin(Q_n z)}{Q_n} a_1(0) e^{-j(\gamma_n + \Delta\beta_n)z} \tag{8.81}$$

where

$$Q_n = \sqrt{\chi_n^2 + (\Delta\beta_n)^2} \tag{8.82}$$

$$\chi_n = \sqrt{\frac{D_{n1}^{(+1)} D_{1n}^{(-1)}}{N_n N_1}} \tag{8.83}$$

Note that the analytical solutions of the reduced coupled-mode equations are formally identical to those derived previously for real modes, except that the effective coupling coefficients κ_n (or γ_n) and the equivalent phase-detuning factor $\Delta\beta_n$ may become complex.

Once we obtain the mode amplitudes, we will be able to calculate the guided powers carried by each of the modes. In general, the power flow along the waveguide is given as

$$P(z) = \frac{1}{2} \Re \int_A \left[\mathbf{E}_t(x,y,z) \times \mathbf{H}_t^*(x,y,z) \right] \cdot \hat{z} \, da \tag{8.84}$$

By substitution of the field expansions in terms of the complex modes into Equation 8.84, we obtain

$$P(z) = \sum_{m=1} \sum_{n=1} M_{mn} \left[a_m a_n^* - b_m b_n^* \right] - \sum_{m=1} \sum_{n=1} N_{mn} \left[a_m a_n^* - b_m b_n^* \right] \tag{8.85}$$

where the first summation is for the power associated with the co-directional propagating modes in the forward and backward directions, whereas the second summation is related to the power of the contra-directional propagating modes. The cross-power coefficients are defined as

$$M_{mn} = \frac{1}{4} \iint_A \left(\mathbf{e}_{tm} \times \mathbf{h}_{tn}^* + \mathbf{e}_{tn}^* \times \mathbf{h}_{tm} \right) \cdot \hat{z} \, da \tag{8.86}$$

$$N_{mn} = \frac{1}{4} \iint_A \left(\mathbf{e}_{tm} \times \mathbf{h}_{tn}^* - \mathbf{e}_{tn}^* \times \mathbf{h}_{tm} \right) \cdot \hat{z} \, da \tag{8.87}$$

Normally, the cross power associated with the contra-directional modes is negligible in comparison with that of the codirectional modes, and hence the second summation in Equation 8.77 may be ignored so that

$$P(z) \cong \sum_{m=1} \sum_{n=1} M_{mn} \left[a_m a_n^* - b_m b_n^* \right] \tag{8.88}$$

For the complex modes that are significant in the mode-coupling process, the transverse fields are almost real, and hence, these modes are almost power orthogonal so that we may further approximate Equation 8.88 as

$$P(z) \cong \sum_{n=1} \left[|a_n|^2 - |b_n|^2 \right] \tag{8.89}$$

Assume that the total power is launched into the forward-propagating fundamental mode at the input. For the contra-directional mode coupling in Bragg grating, the total power at the input of the grating is therefore

$$P(0) = 1 - R(\lambda) \tag{8.90}$$

in which the total reflected power is expressed as

$$R(\lambda) \cong \sum_{m=1} \sum_{n=1} M_{mn} b_m(0) b_n^*(0) \tag{8.91}$$

and the mode indices m and n can not be equal to unity simultaneously. If we further neglect the cross power associated with the codirectional propagating modes, that is, $M_{mn} \cong 0$ and $M_{nn} \cong 1$, we have

$$R(\lambda) \cong \sum_{n=1} |b_m(0)|^2 \tag{8.92}$$

Similarly, assume that at the output of the grating, no backward-propagating modes exist. The total power at the output of the grating is

$$P(L) = |a_1(L)|^2 = T(\lambda) \tag{8.93}$$

The power conservation of the waveguide gratings calls for $P(0) = P(L)$, that is,

$$T(\lambda) = 1 - R(\lambda) \tag{8.94}$$

8.2.3 Complex Coupled Mode Equations Based on Local Modes

The coupled mode equations based on local modes are as follows:

$$\frac{d}{dz} a_m = -j\beta_m a_m - \sum k_{mn} a_n - \sum \chi_{mn} b_n \tag{8.95}$$

$$\frac{d}{dz} b_m = j\beta_m a_m - \sum k_{mn} b_n - \sum \chi_{mn} a_n \tag{8.96}$$

where

$$k_{mn} = \frac{1}{4N_m} \iint \left(\frac{\partial e_{tn}}{\partial z} \times h_{tm} + e_{tm} \times \frac{\partial h_{tn}}{\partial z} \right) \cdot \not{z} \, da \tag{8.97}$$

$$\chi_{mn} = \frac{1}{4N_m} \iint \left(\frac{\partial e_{tn}}{\partial z} \times h_{tm} - e_{tm} \times \frac{\partial h_{tn}}{\partial z} \right) \cdot \not{z} \, da \tag{8.98}$$

The coupling coefficients can be further simplified as the approaches used by Marcuse [33]:

$$k_{mn} = \frac{1}{4N_m} \frac{\omega\varepsilon_0}{\beta_n - \beta_m} \iint \left(\frac{\partial n^2}{\partial z} e_{tm} \cdot e_{tn} \right) \cdot \not{z} \, da \tag{8.99}$$

$$\chi_{mn} = \frac{1}{4N_m} \frac{\omega\varepsilon_0}{\beta_n + \beta_m} \iint \left(\frac{\partial n^2}{\partial z} e_{tm} \cdot e_{tn} \right) \cdot \not{z} \, da \tag{8.100}$$

8.2.4 Applications of Complex Coupled-Mode Equations in Gratings

To illustrate the salient features of the complex modes theory for grating structures, we use the waveguide structure and grating structure shown in Figure 8.1

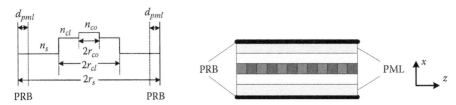

FIGURE 8.1
Slab waveguide and waveguide grating structure.

The step-index slab waveguide with refractive indices of the core, inner cladding, and outer cladding is denoted by n_{co}, n_{cl}, and n_s, respectively. The waveguide parameters are chosen such that the refractive indices of the core and the inner cladding are $n_{co} = 1.458$ and $n_{cl} = 1.450$, respectively. The half widths of the core and the inner cladding layers are $r_{co} = 2.5$ μm and $r_{cl} = 12.5$ μm.

8.2.4.1 Applications of Complex Coupled Mode Equations in Bragg Reflectors

For Bragg reflectors, we assume that $\Delta n_{grating} = 9 \times 10^{-4}$ and $L = 800$ μm and consider the following cases in our simulation.

8.2.4.1.1 Bragg Gratings with Lower Index Outer Cladding ($n_s < n_{cl}$)

By considering the phase matching condition, we readily identify three distinct wavelengths corresponding to the Bragg conditions between the forward-propagating fundamental mode and the first three backward-propagating modes, respectively. Note that under this situation, all modes are guided with real propagation constants and coupling coefficients so that the conventional CMT does apply. We subsequently calculated the coupling coefficients for these three pairs of mode coupling and show them in Figure 8.2a. By applying the analytical solutions of the reduced CMT around each of the phase-matching wavelengths, we obtained the transmission spectra as indicated by the solid, dotted, and dash lines in Figure 8.2a. It is observed that each of the phase-matching conditions produces a distinct dip in the transmission spectrum with magnitude proportional to the strength of the coupling coefficient $|\kappa_n|$. The entire transmission spectra may be obtained by first calculating the reflection spectra using the reduced CMT near each of the phase-matching points. The results are shown in Figure 8.2b as dashed lines. Also shown in the same figure are the results obtained by solving the full coupled-mode equations numerically with a total of three modes and also by applying the rigorous mode-matching method (MMM) with total of 40 modes. The results of the reduced CMT, the full CMT and the rigorous MMM are all in good agreement, indicating that the reduced CMT is sufficient.

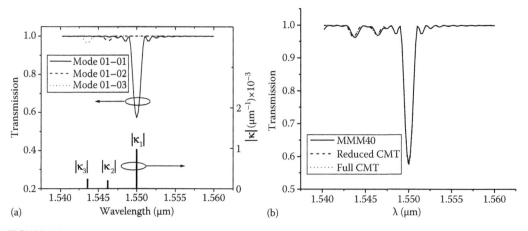

FIGURE 8.2
Transmission spectrum for case A with lower outer cladding index $n_s = 1.0$: (a) solutions from the reduced CMT involving only two phase-matching modes; (b) the transmission spectrum calculated by the reduced CMT (dash lines), the full CMT (dotted lines), and the rigorous MMM (solid lines).

8.2.4.1.2 Bragg Gratings with Equal Index Outer Cladding ($n_s = n_{cl}$)

For the infinite cladding gratings ($n_s = n_{cl}$), the real, guided cladding modes evolve to complex quasi-leaky modes as well as PML modes. The latter plays negligible roles in the interactions with the grating-assisted couplings between the forward-propagating guided mode and the backward-propagating complex modes due to their huge mode losses and small mode overlaps with the guided modes in the core. For the quasi-leaky modes, however, the spectral spacing between them is too small, and hence, the phase-matching wavelengths are hardly distinguishable relative to the spectral width of each transmission dip as illustrated in Figure 8.3a. Further, the coupling strengths $|\kappa_n|$ at these phase-matching modes are similar.

The results by solving the full CMT with consideration for coupling from 1 to 10 backward-propagating modes are illustrated in Figure 8.3a. A flat overall drop in the transmission spectrum is predicted by considering all the relevant modes. The accuracy of the complex CMT involving 10 modes is verified by from comparison with the results the rigorous MMM with a total of 60 modes as evident in Figure 8.3b. Shown on the same figure with dotted lines are results obtained using the reduced CMT considering only the phase-matched modes near the phase-matching wavelengths.

It is indeed surprising to see that the simple solutions of the reduced CMT yield remarkably accurate results even under the situation in which the phase-matching wavelengths are very close to each other. In comparison with the conventional CMT that has to resort to either continuous radiation modes or large number of box modes or tricky leaky modes, the complex CMT is much more straightforward in dealing with strong radiation fields in this case.

8.2.4.1.3 Bragg Gratings with Higher Index Outer Cladding ($n_s > n_{cl}$)

If the refractive index of the outer cladding is higher than that of the cladding, the waveguide structure becomes leaky in the sense that no guided modes exist for mode index lower than the cladding index. We identify a total of 7 phase-matching modes and calculate the transmission spectrum by solving the full CMT considering coupling from

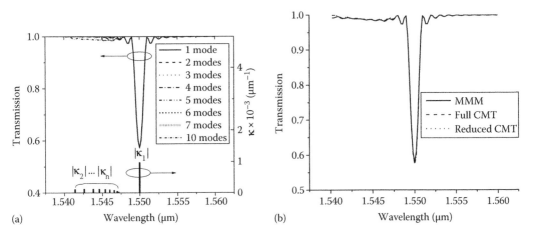

(a)   (b)

FIGURE 8.3
Transmission spectrum for case B with equal outer cladding index $n_s = 1.450$. (a) The phase-matching wavelengths, the corresponding coupling strengths, and the transmission spectrum predicted by the full CMT involving from 2 up to 11 modes; (b) the transmission spectrum calculated by the full CMT (dotted lines) and the rigorous MMM (solid lines).

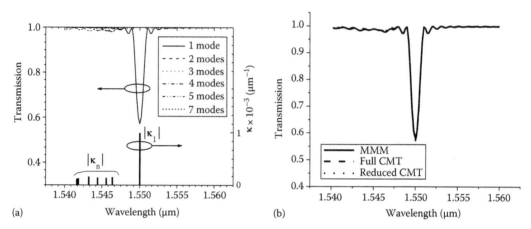

(a)   Wavelength (μm)   (b)   Wavelength (μm)

FIGURE 8.4

Transmission spectra with index of the outer cladding n_s slightly larger than the index of the inner cladding n_{cl} ($n_s = 1.455$). (a) A total of 7 phase-matching modes and (b) comparison between the full CMT and reduced CMT.

1 to 7 modes, respectively. In Figure 8.4b, it shows the comparison between the full CMT (total of 7 modes), the reduced CMT (2 modes near each phase-matching wavelengths), and the rigorous MMM (total of 60 modes), which are in excellent agreement with each other.

8.2.4.2 Applications of Complex Coupled Mode Equations in Transmission Gratings

We consider a long period grating with the following structure parameters: period $\Lambda = 300$ μm and length $L = 30$ mm; the grating is written in the core with the "dc" index change $\Delta n = 2 \times 10^{-4}$, core radius $r_{co} = 2.5$ μm, cladding radius $r_{cl} = 62.5$ μm, the refractive index of the core $n_{co} = 1.458$ and the refractive index of the inner cladding $n_{cl} = 1.45$. Long-period gratings with three different surrounding materials (n_s) will be analyzed as follows:

8.2.4.2.1 Long Period Gratings with Lower Index Outer Cladding ($n_s < n_{cl}$)

In this case ($n_s < n_{cl}$), the waveguide structure supports a number of guided cladding modes. As a result, the key parameters in the coupled mode equations, that is, the coupling coefficients, the normalizing factors, and the detuning factors, are all real. The complex CMT therefore is the same as the conventional CMT. Through identifying the phase-matching conditions, the transmission spectra can be easily obtained. We plot the transmission spectra for $n_s = 1.0$ and $n_s = 1.44$ in Figure 8.5. A blue shift of the transmission is observed as the increase of the refractive index of the surrounding material.

The power conservation is studied by assessing the power carried by the guided core mode and the cladding mode (which is the most closely phase-matched) when $n_s = 1.44$. The LPG is investigated at two wavelengths, that is, 1600 nm (the out of phase point) and 1645 nm (the in-phase point), as indicated in Figure 8.5. The simulation results in Figure 8.6 show that power exchange between the coupled modes becomes significant only at the phase-matched point. Also the power conservation along the propagation direction is observed.

8.2.4.2.2 Long Period Gratings with Higher Refractive Index Outer Cladding ($n_s > n_{cl}$)

If the surrounding material has a higher refractive index than that of the inner cladding, the cladding modes become complex cladding modes in which the field mainly is confined in

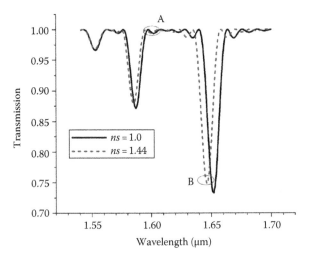

FIGURE 8.5
Transmission spectra of LPG with low refractive index outer cladding.

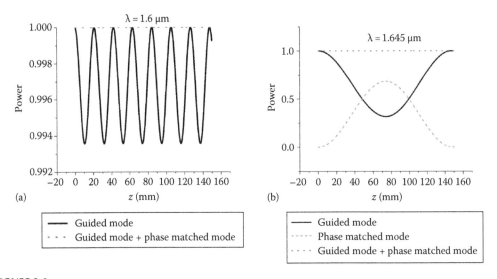

FIGURE 8.6
Power variation along propagation direction for LPG with lower refractive index of the outer cladding layer $(n_s = 1.44)$: (a) out-of-phase point; (b) in-phase point.

the inner cladding while oscillated in the surrounded materials. To disclose the impact of variations of the surrounding materials, we studied the LPG with two different refractive indices of the outer cladding medium. The detuning factors as functions of wavelengths in Figure 8.7 show that each dip in the transmission spectra corresponds to a specific phase-matching point. For this reason, the reduced coupled-mode equation is valid and can be used to predict the performance of the LPG. The transmission spectra shown in Figure 8.7b indicate that the coupling strength is enhanced with the increase of the refractive index for the outer cladding.

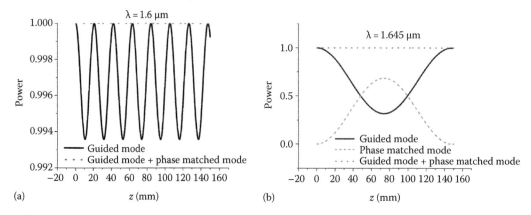

FIGURE 8.7
Characteristics of LPG with high refractive index surrounding media: (a) detuning factor as a function of wavelengths; (b) transmission spectra.

To further illustrate the salient features of the power exchange and attenuation in the presence of radiation, we examine power carried by the guided and the complex mode along the waveguide at two distinct wavelengths corresponding to the points A and C in Figure 8.7b. The results are shown in Figure 8.8a and b, respectively. In both cases, the guided mode is coupled with the phase-matched complex mode with significant power exchange as indicated by the oscillation of the power along the waveguide. On the other hand, the coupling strength at the point A is much stronger than that at the point C, while the leakage loss for the corresponding complex mode is much less (The effective indices of the complex modes at the points A and C are $1.45\,j\,10^{-7}$ and $1.45\,j\,10^{-6}$, respectively. The refractive index of the core mode is 1.451). Therefore, the power damping at the point A is smaller than that at the point C. On the other side, the total power carried by the guided mode and the complex mode decays monotonically due to the radiation loss.

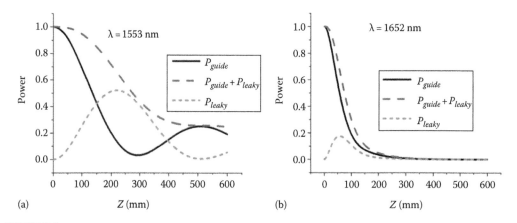

FIGURE 8.8
Power evolution of guided mode and complex mode as functions of propagation distance in LPG with $n_s = 1.60$ for phase-matched wavelengths: (a) wavelength equals to 1553 nm; (b) wavelength equals to 1652 nm.

8.2.4.2.3 *Long Period Gratings with Infinite Claddings* $(n_s = n_{cl})$

Now we consider the long period grating with the refractive index of the surrounding material equals to that of the cladding, for example, LPG with infinite cladding. This seemingly simple structure has the most challenging complexity comparing to other cases in regard to computation effort for two reasons: On one hand, the cladding modes are all continuous radiation modes, implying that a set of modes will satisfy the phase-matching condition; on the other hand, special attention has to be paid on choosing the PML parameters to obtain the accurate modal indices. Large computation window is preferred to reduce the mode spacing to yield a smooth transmission spectrum. In this study, the computation window is set to $r_s = 800$ μm; the PML reflection is set to 1e–50 with 5.5 μm thickness on both sides. The coupling length defined by $2\pi/(\beta_{co} - \beta_v)$ is plotted in Figure 8.9. It is shown that the many radiation modes are simultaneously close to the phase-matching condition.

Strictly speaking, the reduced coupled mode equations that utilize two phase-matched modes are not valid, as those modes close to resonant conditions will contribute to the power exchange and can not be neglected any more. On the other hand, by noting the fact that the interactions among the radiation modes are negligible, we may apply the approximated analytical solution as follows:

$$T = 1 - \sum_{v}^{N} \left(1 - |A_{co,v}|^2\right) \tag{8.101}$$

where $A_{co,v}$ is the amplitude of the core mode coupling to the vth radiation mode and is obtained through Equation 2.75. The convergence of the core-mode transmission has been investigated at wavelength equals to 1725 nm. The results in Figure 8.10a indicate that at least 30 modes are required to obtain a satisfactory precision. The transmission spectra in Figure 8.10b confirm that the analytical approximation agrees well with the numerical solutions from the full CMT equations.

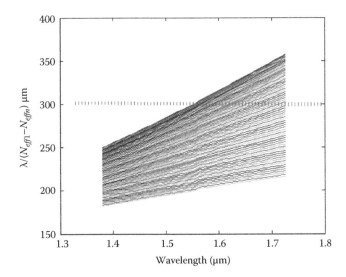

FIGURE 8.9
Coupling length for LPG with infinite cladding.

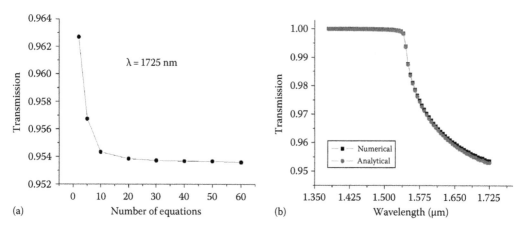

FIGURE 8.10

Transmission characteristics of LPG with infinite claddings. (a) Convergence of complex coupled equations and (b) transmission spectrum.

8.2.4.3 Applications of Complex Coupled Mode Equations in Waveguide Taper Structures

We consider a linear taper that has been investigated in the literature[111]. The structure is represented by

$$d(z) = d_{in} + z \tan \theta \qquad (8.102)$$

where θ is the taper angle.

The structure geometry is shown in Figure 8.11 and the physical parameters are the refractive index of substrate $n_s = 1.515$, the refractive index of the core $n_{co} = 1.517$, and the refractive index of the cladding $n_{cl} = 1.0$. The width of the input waveguide core is $d_{in} = 5$ μm and the width of the output waveguide core is $d_{out} = 10$ μm. The working wavelength is chosen to be $\lambda = 1.32$ μm. Though the ideal taper is continuous, we can use staircase approximation to evaluate the radiation loss. The number of steps (M) and grid size (Δz) along the propagation direction is determined by step height (Δd). The step height (Δd), on the other hand, is decided by the beam resolution.

For a linear taper shown earlier, the two refractive index profiles are the same everywhere except in the vicinity of the boundary between the cladding and the core. Moreover, for taper waveguide structure with small angles, the backward reflection is negligible. Based on the analysis, we may simplify the coupling coefficients for TE modes as

$$k_{mn}(z) = \frac{\omega \varepsilon_0}{4 N_m} \frac{n_{co}^2 - n_{cl}^2}{\beta_n - \beta_m} \tan \theta \cdot \left(e_{tm} \cdot e_{tn} \right)_{x=d} \qquad (8.103)$$

The computation parameters are as follows: the thickness of the substrate is $ds = 55$ μm, the computation window is $W = 70$ μm, the PML thickness is 2.5 μm on both sides, and the PML reflection is 1e−4. The power transmission coefficient obtained by solving the coupled Equations 2.84 and 2.85 will be referred to full CMT. For weakly coupled waveguide structure in which the mutual coupling among the higher-order modes is negligible, we

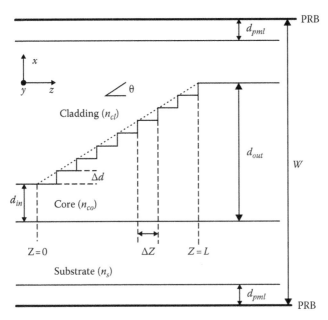

FIGURE 8.11
Geometry of linear taper waveguide structure.

may consider the coupling between the fundamental mode and the higher-order modes separately, for example, instead of using (2.84) and (2.85), the coupled equations can be expressed as

$$\frac{d}{dz}a_1 = -j\beta_1 a_1 - k_{1n}a_n \qquad (8.104)$$

$$\frac{d}{dz}a_n = -j\beta_n a_n - k_{n1}a_1 \qquad (8.105)$$

The transmission coefficient is approximated by

$$T = 1 - \sum_n |a_n|^2 \qquad (8.106)$$

The transmission obtained through (8.106) will be referred to reduced CMT. The power transmission of the studied linear taper as a function of taper angle is shown in Figure 8.12. Comparing to the benchmark obtained from MMM, the results from full CMT overestimate the transmission, on the other hand, the results of reduced CMT underestimate the power transmission. For small taper angle, full CMT and reduced CMT are good approximations. However, for large angle, full CMT is more accurate as the mutual coupling among higher modes becomes pronounced.

The convergence of the full CMT with the grid size has been investigated. The relative error of power transmission for different grid size (Δd) is has been shown in Figure 8.13.

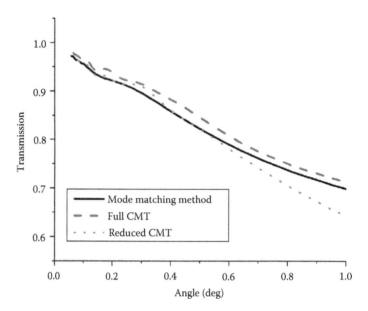

FIGURE 8.12
Power transmission of a linear taper.

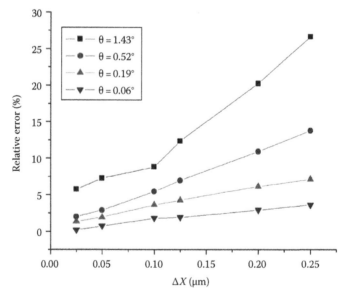

FIGURE 8.13
Convergence behavior of grid size (Δd).

It is observed that the relative error becomes smaller with the decreasing of mesh size for different taper angles. Further, the relative error decreases with the taper angle.

We also studied the convergence of the full CMT with respect to the number of modes being used. It is observed from Figure 8.14 that the relative error decreases with the number of modes. Similar to the convergence behavior of the mesh sizes, the accuracy is deteriorated with the increase of the taper angle.

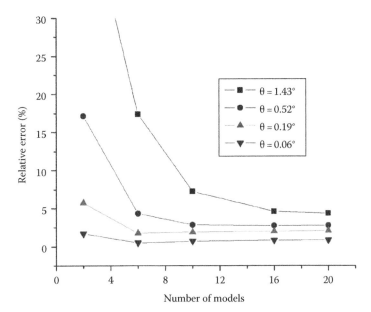

FIGURE 8.14
Convergence behavior of number of modes.

8.3 Summary

In this chapter, a CMT based on complex modes has been proposed and verified through examples of short-/long-period gratings. Further, we formulated the complex CMT based on local mode and applied it to study the transition loss of the tapered waveguide structure. An analytical solution of radiation loss for long-period grating has been derived. The effectiveness as well as the accuracy has been validated through the study of transmissive waveguide grating with different surrounding materials.

References

1. S. E. Miller, Integrated optics: An introduction, *Bell System Technical Journal*, 48(7), 2059–2069, 1969.
2. I. P. Kaminow, Optical integrated circuits: A personal perspective, *Journal of Lightwave Technology*, 26(9–12), 994–1004, May–June 2008.
3. D. Iazikov, C. Greiner, and T. W. Mossberg, Apodizable integrated filters for coarse WDM and FTTH-type applications, *Journal of Lightwave Technology*, 22(5), 1402–1407, May 2004.
4. M. Smit and Y. S. Oei, Photonic integrated circuits for advanced communication networks, *Journal of Optoelectronics*, 12(1), 25–30, January–March 1998.
5. J. J. G. M. van der Tol, Y. S. Oei, U. Khalique, R. Notzel, and M. K. Smit, InP-based photonic circuits: Comparison of monolithic integration techniques, *Progress in Quantum Electron*, 34(4), 135–172, July 2010.

6. Y. Sakamaki, K. Hattori, Y. Nasu, T. Hashimoto, Y. Hashizume, T. Mizuno, T. Goh, and H. Takahashi, One-chip integrated polarisation-multiplexed DQPSK demodulator using silica-based planar lightwave circuit technology, *Electronics Letters*, 46(16), 1152–1153, August 2010.

7. Y. Sakamaki, H. Yamazaki, T. Mizuno, T. Goh, Y. Nasu, T. Hashimoto, S. Kamei, K. Hattori, H. Takahashi, T. Kobayashi, and M. Ishikawa, Dual polarisation optical hybrid using silica-based planar lightwave circuit technology for digital coherent receiver, *Electronics Letters*, 46(1), 58–59, January 2010.

8. H. Fukuda, K. Yamada, T. Tsuchizawa, T. Watanabe, H. Shinojima, and S. I. Itabashi, Silicon photonic circuit with polarization diversity, *Optics Express*, 16(7), 4872–4880, March 2008.

9. W. Bogaerts, D. Taillaert, P. Dumon, D. Van Thourhout, and R. Baets, A polarization-diversity wavelength duplexer circuit in silicon-on-insulator photonic wires, *Optics Express*, 15(4), 1567–1578, February 2007.

10. R. Nagarajan, M. Kato, J. Pleumeekers, P. Evans, D. Lambert, A. Chen, V. Dominic et al., Single-chip 40-channel InP transmitter photonic integrated circuit capable of aggregate data rate of 1.6 Tbit/s, *Electronics Letters*, 42(13), 771–773, June 2006.

11. W. A. Zortman, D. C. Trotter, and M. R. Watts, Silicon photonics manufacturing, *Optics Express*, 18(23), 23598–23607, November 2010.

12. B. Jalali, S. Yegnanarayanan, T. Yoon, T. Yoshimoto, I. Rendina, and F. Coppinger, Advances in silicon-on-insulator optoelectronics, *IEEE Journal of Selected Topics in Quantum Electronics*, 4(6), 938–947, November–December 1998.

13. A. J. Lowery, Computer-aided photonics design, *IEEE Spectrum*, 34(4), 26–31, April 1997.

14. D. Gallagher, Photonic CAD matures, *IEEE LEOS Newsletter*, 8–14, February 2008.

15. R. Soref, The past, present, and future of silicon photonics, *IEEE Journal of Selected Topics in Quantum Electron*, 12(6), 1678–1687, November–December 2006.

16. C. Xu, CAD for photonic devices and circuit, in *Asia Optical Fiber Communication and Optoelectronic Exposition*, Shanghai, China, 2006.

17. A. Taflove and S. C. Hagness, *Computational Electrodynamics: The Finite-Difference Time-Domain Method*, Boston, MA: Artech House, 2005.

18. K. S. Yee, Numerical solution of initial boundary value problems involving Maxwell's equations in isotropic media, *IEEE Transactions of Antennas and Propagation*, 16(3), 302–307, 1966.

19. D. Marcuse, Coupled mode theory of round optical fibers, *Bell System Technical Journal*, 52(6), 817–842, 1973.

20. A. Yariv, Coupled-mode theory for guided-wave optics, *IEEE Journal of Quantum Electron*, 9(9), 919–933, 1973.

21. W. Snyder, Coupled-mode theory for optical fibers, *Journal of Optical Society of America*, 62(11), 1267–1277, 1972.

22. S. E. Miller, Coupled-wave theory and waveguide applications, *Bell System Technical Journal*, 33(3), 661–719, 1954.

23. M. D. Feit and J. A. Fleck, Light-propagation in graded-index optical fibers, *Applied Optics*, 17(24), 3990–3998, 1978.

24. J. Gerdes and R. Pregla, Beam-propagation algorithm based on the method of lines, *Journal of Optical Society of America B*, 8(2), 389–394, February 1991.

25. G. R. Hadley, Wide-angle beam propagation using pade approximant operators, *Optics Letters*, 17(20), 1426–1428, October 15, 1992.

26. H. Shigesawa and M. Tsuji, Mode propagation through a step discontinuity in dielectric planar wave-guide, *IEEE Transactions of Microwave Theory and Techniques*, 34(2), 205–212, February 1986.

27. T. E. Rozzi, Rigorous analysis of step discontinuity in a planar dielectric waveguide, *IEEE Transactions of. Microwave Theory and Techniques*, 26(10), 738–746, 1978.

28. H. A. Jamid and M. Z. M. Khan, 3-D full-vectorial analysis of strong optical waveguide discontinuities using Pade approximants, *IEEE Journal of Quantum Electron*, 43(3–4), 343–349, March–April 2007.

29. P. L. Ho and Y. Y. Lu, A bidirectional beam propagation method for periodic waveguides, *IEEE Photonics Technology Letters*, 14(3), 325–327, March 2002.

30. P. L. Ho and Y. Y. Lu, A stable bidirectional propagation method based on scattering operators, *IEEE Photonics Technology Letters*, 13(12), 1316–1318, December 2001.
31. J. Ctyroky, S. Helfert, R. Pregla, P. Bienstman, R. Baets, R. De Ridder, R. Stoffer et al., Bragg waveguide grating as a 1D photonic band gap structure: COST 268 modelling task, *Optical and Quantum Electronics*, 34(5), 455–470, May–June 2002.
32. K. Q. Le, Complex Pade approximant operators for wide-angle beam propagation, *Optics Communications*, 282(7), 1252–1254, April 1, 2009.
33. D. Marcuse, *Theory of Dielectric Optical Waveguides*. New York: Academic Press, 1974.
34. H. Kogelnik, Theory of dielectric waveguides, in *Integrated Optics*, T. Tamir, Ed. New York: Springer-Verlag, 1975.
35. A. Hardy and W. Streifer, Coupled mode theory of parallel wave-guides, *Journal of Lightwave Technology*, 3(5), 1135–1146, 1985.
36. H. A. Haus, W. P. Huang, S. Kawakami, and N. A. Whitaker, Coupled-mode theory of optical wave-guides, *Journal of Lightwave Technology*, 5(1), 16–23, January 1987.
37. H. A. Haus, W. P. Huang, and A. W. Snyder, Coupled-mode formulations, *Optics Letters*, 14(21), 1222–1224, November 1, 1989.
38. W. P. Huang and H. A. Haus, Self-consistent vector coupled-mode theory for tapered optical wave-guides, *Journal of Lightwave Technology*, 8(6), 922–926, June 1990.
39. H. A. Haus and W. P. Huang, Coupled-mode theory, *Proceedings of the IEEE*, 79(10), 1505–1518, October 1991.
40. S. L. Chuang, A coupled mode formulation by reciprocity and a variational principle, *Journal of Lightwave Technology*, 5(1), 5–15, January 1987.
41. A. W. Snyder and J. D. Love, *Optical Waveguide Theory*, New York: Chapman & Hall, 1983.
42. T. Erdogan, Cladding-mode resonances in short- and long-period fiber grating filters, *Journal of Optical Society America A*, 14(8), 1760–1773, August 1997.
43. T. Erdogan, Fiber grating spectra, *Journal of Lightwave Technology*, 15(8), 1277–1294, 1997.
44. R. Sammut and A. W. Snyder, Leaky modes on a dielectric waveguide—Orthogonality and excitation, *Applied Optics*, 15(4), 1040–1044, 1976.
45. S. L. Lee, Y. C. Chung, L. A. Coldren, and N. Dagli, On leaky mode approximations for modal expansion in multilayer open waveguides, *IEEE Journal of Quantum Electronics*, 31(10), 1790–1802, October 1995.
46. P. Bienstman and R. Baets, Optical modelling of photonic crystals and VCSELs using eigen-mode expansion and perfectly matched layers, *Optical Quantum Electronics*, 33(4–5), 327–341, April 2001.
47. P. Bienstman, H. Derudder, R. Baets, F. Olyslager, and D. De Zutter, Analysis of cylindrical waveguide discontinuities using vectorial eigenmodes and perfectly matched layers, *IEEE Transactions on Microwave Theory and Techniques*, 49(2), 349–354, February 2001.
48. H. Derudder, D. De Zutter, and F. Olyslager, Analysis of waveguide discontinuities using perfectly matched layers, *Electron Letters*, 34(22), 2138–2140, October 29, 1998.
49. H. Derudder, F. Olyslager, D. De Zutter, and S. Van den Berghe, Efficient mode-matching analysis of discontinuities in finite planar substrates using perfectly matched layers, *IEEE Transactions on Antennas and Propagation*, 49(2), 185–195, February 2001.
50. K. Jiang and W. P. Huang, Finite-difference-based mode-matching method for 3-D waveguide structures under semivectorial approximation, *Journal of Lightwave Technology*, 23(12), 4239–4248, December 2005.
51. Y. Shani, R. Alferness, T. Koch, U. Koren, M. Oron, B. I. Miller, and M. G. Young, Polarization rotation in asymmetric periodic loaded rib wave-guides, *Applied Physics Letters*, 59(11), 1278–1280, September 9, 1991.
52. C. Vassallo, 1993–1995 Optical mode solvers, *Optics in Quantum Electronics*, 29(2), 95–114, February 1997.
53. Y. P. Chiou, Y. C. Chiang, and H. C. Chang, Improved three-point formulas considering the interface conditions in the finite-difference analysis of step-index optical devices, *Journal of Lightwave Technology*, 18(2), 243–251, February 2000.

54. M. S. Stern, Semivectorial polarized finite-difference method for optical wave-guides with arbitrary index profiles, *IEE Proceedings Journal of Optoelectronics*, 135(1), 56–63, February 1988.

55. C. L. Xu, W. P. Huang, M. S. Stern, and S. K. Chaudhuri, Full-vectorial mode calculations by finite-difference method, *IEE Proceedings Journal of Optoelectronics*, 141(5), 281–286, October 1994.

56. E. Anemogiannis, E. N. Glytsis, and T. K. Gaylord, Determination of guided and leaky modes in lossless and lossy planar multilayer optical waveguides: Reflection pole method and wavevector density method, *Journal of Lightwave Technology*, 17(5), 929–941, May 1999.

57. J. Chilwell and I. Hodgkinson, Thin-films field-transfer matrix-theory of planar multilayer waveguides and reflection from prism-loaded waveguides, *Journal of Optical Society America A*, 1(7), 742–753, 1984.

58. J. Petracek and K. Singh, Determination of leaky modes in planar multilayer waveguides, *IEEE Photonics Technical Letters*, 14(6), 810–812, June 2002.

59. C. Vassallo, Improvement of finite-difference methods for step-index optical wave-guides, *IEE Proceedings Journal of Optoelectronics*, 139(2), 137–142, April 1992.

60. Y. C. Chiang, Y. P. Chiou, and H. C. Chang, Improved full-vectorial finite-difference mode solver for optical waveguides with step-index profiles, *Journal of Lightwave Technology*, 20(8), 1609–1618, August 2002.

61. W. P. Huang, C. L. Xu, W. Lui, and K. Yokoyama, The perfectly matched layer (PML) boundary condition for the beam propagation method, *IEEE Photonics Technical Letters*, 8(5), 649–651, May 1996.

62. M. Reed, P. Sewell, T. M. Benson, and P. C. Kendall, Efficient propagation algorithm for 3D optical waveguides, *IEE Proceedings-Optoelectronics*, 145(1), 53–58, February 1998.

63. E. Silberstein, P. Lalanne, J. P. Hugonin, and Q. Cao, Use of grating theories in integrated optics, *Journal of Optical Society America A*, 18(11), 2865–2875, November 2001.

64. T. Ando, T. Murata, H. Nakayama, J. Yamauchi, and H. Nakano, Analysis and measurement of polarization conversion in a periodically loaded dielectric waveguide, *IEEE Photonics Technical Letters*, 14(9), 1288–1290, September 2002.

65. K. Kawano, T. Kitoh, M. Kohtoku, T. Takeshita, and Y. Hasumi, 3-D semivectorial analysis to calculate facet reflectivities of semiconductor optical waveguides based on the bi-directional method of line BPM (MoL-BPM), *IEEE Photonics Technical Letters*, 10(1), 108–110, January 1998.

66. V. R. Almeida, Q. F. Xu, C. A. Barrios, and M. Lipson, Guiding and confining light in void nanostructure, *Optics Letters*, 29(11), 1209–1211, June 2004.

67. N. N. Feng, J. Michel, and L. C. Kimerling, Optical field concentration in low-index wave-guides, *IEEE Quantum Electronics*, 42(9–10), 885–890, September–October 2006.

68. M. Lipson, Guiding, modulating, and emitting light on silicon—Challenges and opportunities, *Journal of Lightwave Technology*, 23(12), 4222–4238, December 2005.

69. Q. F. Xu, V. R. Almeida, R. R. Panepucci, and M. Lipson, Experimental demonstration of guiding and confining light in nanometer-size low-refractive-index material, *Optics Letters*, 29(14), 1626–1628, July 15, 2004.

70. C. Vassallo, *Optical Waveguide Concepts*, Amsterdam, the Netherlands: Elsevier, 1991.

71. T. Baehr-Jones, M. Hochberg, G. X. Wang, R. Lawson, Y. Liao, P. A. Sullivan, L. Dalton, A. K. Y. Jen, and A. Scherer, Optical modulation and detection in slotted Silicon waveguides, *Optics Express*, 13(14), 5216–5226, July 11, 2005.

72. T. Fujisawa and M. Koshiba, Polarization-independent optical directional coupler based on slot waveguides, *Optics Letters*, 31(1), 56–58, January 1, 2006.

73. F. Dell'Olio and V. M. N. Passaro, Optical sensing by optimized silicon slot waveguides, *Optics Express*, 15(8), 4977–4993, April 16, 2007.

74. T. Fujisawa and M. Koshiba, Theoretical investigation of ultrasmall polarization-insensitive 1×2 multimode interference waveguides based on sandwiched structures, *IEEE Photonics Technical Letters*, 18(9–12), 1246–1248, 2006.

75. H. Zhang, J. W. Mu, and W. P. Huang, Improved bidirectional beam-propagation method by a fourth-order finite-difference scheme, *Journal of Lightwave Technology*, 25(9), 2807–2813, September 2007.

76. W. L. Barnes, A. Dereux, and T. W. Ebbesen, Surface plasmon subwavelength optics, *Nature*, 424(6950), 824–830, August 14, 2003.

77. L. Chen, J. Shakya, and M. Lipson, Subwavelength confinement in an integrated metal slot waveguide on silicon, *Optics Letters*, 31(14), 2133–2135, July 15, 2006.

78. G. Gay, O. Alloschery, B. V. De Lesegno, C. O'Dwyer, J. Weiner, and H. J. Lezec, The optical response of nanostructured surfaces and the composite diffracted evanescent wave model, *Nature Physics*, 2(4), 262–267, April 2006.

79. K. G. Lee and Q. H. Park, Coupling of surface plasmon polaritons and light in metallic nanoslits, *Physical Review Letters*, 95(10), 103902, September 2, 2005.

80. S. A. Maier, Plasmonics: The promise of highly integrated optical devices, *IEEE Journal of Selected Topics in Quantum Electronics*, 12(6), 1671–1677, November–December 2006.

81. E. Ozbay, Plasmonics: Merging photonics and electronics at nanoscale dimensions, *Science*, 311(5758), 189–193, January 13, 2006.

82. H. Raether, *Surface Plasmons*, New York: Springer-Verlag, 1988.

83. S. A. Darmanyan and A. V. Zayats, Light tunneling via resonant surface plasmon polariton states and the enhanced transmission of periodically nanostructured metal films: An analytical study, *Physical Review B*, 67(3), 035424-1–035424-7, January 15, 2003.

84. M. Kretschmann and A. A. Maradudin, Band structures of two-dimensional surface-plasmon polaritonic crystals, *Physical Review B*, 66(24), 245408-1–245408-7, December 15, 2002.

85. J. M. Steele, C. E. Moran, A. Lee, C. M. Aguirre, and N. J. Halas, Metallodielectric gratings with subwavelength slots: Optical properties, *Physical Review B*, 68(20), 205103-1–205103-7, November 2003.

86. A. Boltasseva, T. Nikolajsen, K. Leosson, K. Kjaer, M. S. Larsen, and S. I. Bozhevolnyi, Integrated optical components utilizing long-range surface plasmon polaritons, *Journal of Lightwave Technology*, 23(1), 413–422, January 2005.

87. Z. H. Han, E. Forsberg, and S. L. He, Surface plasmon Bragg gratings formed in metal-insulator-metal waveguides, *IEEE Photonics Technical Letters*, 19(2–4), 91–93, January–February 2007.

88. A. Hosseini and Y. Massoud, A low-loss metal-insulator-metal plasmonic bragg reflector, *Optics Express*, 14(23), 11318–11323, November 13, 2006.

89. S. Jette-Charbonneau, R. Charbonneau, N. Lahoud, G. A. Mattiussi, and P. Berini, Bragg gratings based on long-range surface plasmon-polariton waveguides: Comparison of theory and experiment, *IEEE Journal of Quantum Electronics*, 41(12), 1480–1491, December 2005.

90. T. Sondergaard, S. I. Bozhevolnyi, and A. Boltasseva, Theoretical analysis of ridge gratings for long-range surface plasmon polaritons, *Physical Review B*, 73(4), 045320-1–045320-7, January 2006.

91. V. M. Fitio and Y. V. Bobitski, High transmission of system dielectric grating thin film dielectric grating, in *Laser and Fiber Optical Network Modelling*, Yalta, Crimea, 2005.

92. J. Yamauchi, T. Yamazaki, K. Sumida, and H. Nakano, TE/TM wave splitters using surface plasmon polaritons, in *Integrated Photonics and Nanophotonics Research and Applications*, Salt Lake City, UT, 2007.

93. V. Bhatia and A. M. Vengsarkar, Optical fiber long-period grating sensors, *Optics Letters*, 21(9), 692–694, May 1, 1996.

94. H. J. Patrick, G. M. Williams, A. D. Kersey, J. R. Pedrazzani, and A. M. Vengsarkar, Hybrid fiber Bragg grating/long period fiber grating sensor for strain/temperature discrimination, *IEEE Photonics Technical Letters*, 8(9), 1223–1225, September 1996.

95. A. M. Vengsarkar, P. J. Lemaire, J. B. Judkins, V. Bhatia, T. Erdogan, and J. E. Sipe, Long-period fiber gratings as band-rejection filters, *Journal of Lightwave Technology*, 14(1), 58–65, January 1996.

96. O. Duhem, J. F. Henninot, M. Warenghem, and M. Douay, Demonstration of long-period-grating efficient couplings with an external medium of a refractive index higher than that of silica, *Applied Optics*, 37(31), 7223–7228, November 1, 1998.

97. Y. Jeong, B. Lee, J. Nilsson, and D. J. Richardson, A quasi-mode interpretation of radiation modes in long-period fiber gratings, *IEEE Journal of Quantum Electronics*, 39(9), 1135–1142, September 2003.

98. Y. Koyamada, Numerical analysis of core-mode to radiation-mode coupling in long-period fiber gratings, *IEEE Photonics Technical Letters*, 13(4), 308–310, April 2001.

99. H. J. Patrick, A. D. Kersey, and F. Bucholtz, Analysis of the response of long period fiber gratings to external index of refraction, *Journal of Lightwave Technology*, 16(9), 1606–1612, September 1998.

100. D. B. Stegall and T. Erdogan, Leaky cladding mode propagation in long-period fiber grating devices, *IEEE Photonics Technology Letters*, 11(3), 343–345, March 1999.

101. H. Yanagawa, T. Shimizu, S. Nakamura, and I. Ohyama, Index-and-dimensional taper and its application to photonic devices, *Journal of Lightwave Technology*, 10(5), 587–592, May 1992.

102. O. Mitomi, K. Kasaya, and H. Miyazawa, Design of a single-mode tapered wave-guide for low-loss chip-to-fiber coupling, *IEEE Journal of Quantum Electronics*, 30(8), 1787–1793, August 1994.

103. J. D. Love, W. M. Henry, W. J. Stewart, R. J. Black, S. Lacroix, and F. Gonthier, Tapered single-mode fibers and devices. 1. Adiabaticity criteria, *IEE Proceedings Journal of Optoelectronics*, 138(5), 343–354, October 1991.

104. J. Haes, J. Willems, and R. Bates, Design of adiabatic tapers for high contrast step index waveguides, in *Proceedings of SPIE*, 2212, 685–693, 1994.

105. D. Marcuse, Radiation losses of tapered dielectric slab waveguides, *Bell System Technical Journal*, 49, 273–290, 1970.

106. A. R. Nelson, Coupling optical waveguides by tapers, *Applied Optics*, 14(12), 3012–3015, 1975.

107. J. A. Fleck, J. R. Morris, and M. D. Feit, Time-dependent propagation of high-energy laser-beams through atmosphere, *Applied Physics*, 10(2), 129–160, 1976.

108. T. Nakamura and N. Suzuki, Spot-size converted laser diodes based on mode interference, *IEEE Photonics Technical Letters*, 12(2), 143–145, February 2000.

109. C. T. Lee, M. L. Wu, L. G. Sheu, P. L. Fan, and J. M. Hsu, Design and analysis of completely adiabatic tapered waveguides by conformal mapping, *Journal of Lightwave Technology*, 15(2), 403–410, February 1997.

110. P. G. Suchoski and V. Ramaswamy, Exact numerical technique for the analysis of step discontinuities and tapers in optical dielectric wave-guides, *Journal of Optical Society America A*, 3(2), 194–203, February 1986.

111. P. G. Suchoski and R. V. Ramaswamy, Design of single-mode step-tapered Waveguide sections, *IEEE Journal of Quantum Electronics*, 23(2), 205–211, February 1987.

112. A. W. Snyder, Coupling of modes in a tapered dielectric cylinder, *IEEE Transactions on Microwave Theory and Techniques*, 18(7), 383–392, 1970.

113. A. F. Milton and W. K. Burns, Mode-coupling in tapered optical-waveguide structures and electro-optic switches, *IEEE Transactions on Circuits and Systems*, 26(12), 1020–1028, 1979.

114. A. F. Milton and W. K. Burns, Mode-coupling in optical-waveguide horns, *IEEE Journal of Quantum Electronics*, 13(10), 828–835, 1977.

9

Multilayer Coupled Nanoplasmonic Structures and Related Computational Techniques

Mina Ray and Mahua Bera

CONTENTS

9.1 Introduction

Surface plasmons (SPs) are collective oscillations of conduction electrons at the surface of a thin metallic film adjacent to a dielectric layer, resulting in an SP wave (SPW) that propagates along the interface of two materials with their real part of permittivity having opposite signs. SPs are generally excited through a coupling of an evanescent field generated by incident p-polarized light in the attenuated total reflection (ATR) coupler mode. At a certain incident angle greater than the critical angle, if the phase-matching condition is satisfied, that is, if the wave vector of the SP matches with that of the incident light, then the energy of the incident light almost gets totally transferred to an SP mode and produces a sharp resonance dip in the reflectivity spectra. Two types of configuration based on prism coupling have been generally used to obtain SP resonance (SPR), namely, Kretschmann–Raether configuration [1] and Otto configuration [2]. SPR technique permits the real-time characterization that allows precise measurement of changes in the refractive index or thickness of the sensing medium adjacent to the metal [3,4]. Classical mechanics and Drude model [5] are used widely for the analysis of the behavior of these plasmons and their associated applications [6].

SPR sensor technology has been used widely for more than two decades [7]. It has also been commercialized for both the chemical and biological sensor applications. In view of the increasing need for detection and analysis of chemical and biochemical substances in many important areas including medicine, environmental monitoring, biotechnology, and drug and food monitoring, SPR sensor technology holds a significant potential. An intermediate dielectric layer may be introduced between the glass prism and the thin metallic film to have a better precision in chemical sensor [8]. Different types of geometrical configurations of metal–dielectric interfaces have also been investigated for sensing application of SPR [9,10]. Nonplanar structure plays an important role in SPR sensing within hollow conical devices in which it is easier to insert a tapered sensing element rather than a planar one [11]. Another numerical approach such as admittance loci method also helps in the detection of chemical and biological samples along with the effect of substrate dependency [12,13]. Different nanoplasmonic structures have been proposed by many researchers with a motivation to utilize the same with improved performance for certain nanophotonic applications [14–21]. Moreover, many SPR-based sensors with improved measurement precision along with several inherent advantages have also been reported [22–29].

Gold and silver are ideal metal films in the visible wavelength range, but gold is more suitable because it is less reactive for any chemical reaction. However, silver offers extra advantage of getting sharper resonance spectra than gold, but it is very much oxidation-prone. This calls for some optimization procedure to be followed regarding the choice of different materials as well as their individual thicknesses in the composite nanoplasmonic structure.

Recently, the concept of coupled plasmonics has been explored by many researchers [30–33] claiming enhanced measurement precision as a result of coupling of waveguide mode and SPR mode. Conventional SPR along with long-range SPR (LRSPR) [34–36], coupled plasmon waveguide resonance (CPWR) [37], and coupled waveguide SPR (CWSPR) [38] are the four prominent nanoplasmonic thin-film structures to be considered in our analysis. Multiple resonance peaks obtained in various CWSPR structures have been found to increase the sensitivity of certain SPR-based sensors and for some other nanophotonic applications [39,40]. Field analysis also helps to better understand the coupling between the SPR mode and waveguide resonance mode [41]. The basic purpose is to explore special types of mono- and bimetallic waveguide coupled nanoplasmonic structures, which consist of both the silver and the gold layer simultaneously along with a waveguide layer. The related computational aspects for the detailed analysis of such nanoplasmonic structures are also provided for the benefits of the budding researchers. Moreover, interesting modal analysis is also performed showing energy distribution in different thin-film layers for each of the structures.

The main focus of this chapter will be on the computational analysis of coupling effect of the SPR mode in thin metallic film and waveguide mode in dielectric layers in different types of multilayer thin-film coupled nanoplasmonic structures in order to find an optimized structure for better sensitivity and detection accuracy depending upon the application concerned. Characterization of dielectric material within a composite SPR-producing structure will also be discussed as a part of the application point of view [42].

9.2 Mathematical Formulations

9.2.1 Physics of SPR

SPs are free charge oscillations that occur at the interface between a metallic medium having a frequency-dependent complex dielectric constant $\varepsilon_1(\omega)$ and a dielectric medium having a real dielectric constant $\varepsilon_2(\omega)$. $Z=0$ represents the interface between the two media. SPW on metallic thin film or any arbitrary nanoplasmonic structures can be analyzed using the electromagnetic theories based on Maxwell equations [43,44]. Homogeneous solution of Maxwell equation can be obtained by solving the wave equation represented by

$$\nabla \times \nabla \times E(\mathbf{r}, \omega) - \frac{\omega^2}{c^2} \varepsilon(\mathbf{r}, \omega) = 0 \qquad (9.1)$$

with $\varepsilon(\mathbf{r},\omega) = \varepsilon_1(\omega)$ if $z<0$ and $\varepsilon(\mathbf{r},\omega) = \varepsilon_2(\omega)$ if $z>0$. The localization at the interface is characterized by the electromagnetic fields that exponentially decay with increasing distance from the interface into both half-spaces. Here we will only consider the transverse magnetic (TM) modes in both half-spaces because there is no solution that supports a bounded mode in case of transverse electric (TE) distribution.

In case of TM modes,

$$E_i = \begin{pmatrix} E_{j,x} \\ 0 \\ E_{j,z} \end{pmatrix} e^{ik_x x - i\omega t} e^{ik_{j,z} z}, \quad j = 1, 2 \qquad (9.2)$$

Applying the boundary conditions, we get the dispersion relation, that is, a relation between the wave vector along the propagation direction and the angular frequency ω:

$$k_x^2 = \frac{\varepsilon_1 \varepsilon_2}{\varepsilon_1 + \varepsilon_2} k^2 = \frac{\varepsilon_1 \varepsilon_2}{\varepsilon_1 + \varepsilon_2} \frac{\omega^2}{c^2} \qquad (9.3)$$

and the normal component of the wave vector is given by

$$k_{j,z}^2 = \frac{\varepsilon_j^2}{\varepsilon_1 + \varepsilon_2} k^2, \quad j = 1, 2 \qquad (9.4)$$

In order to obtain the conditions that must be fulfilled for the existence of an interface mode, for simplicity, we assume that the imaginary parts of the complex dielectric functions are small compared to the real parts so that they may be neglected. For the interface waves to be of the propagating nature, one requires a real k_x^2. So, this can be fulfilled only if the sum and product of the dielectric functions are either both positive or both negative. Moreover, in order to obtain a bound solution, we require that the normal components of the wave vector are purely imaginary in both media giving rise to exponentially decaying solutions. This can only be achieved if the sum in the denominator is negative. Hence, we can conclude that the conditions for an interface mode to exist are given by

$$\varepsilon_1(\omega).\varepsilon_2(\omega) < 0 \tag{9.5}$$

$$\varepsilon_1(\omega) + \varepsilon_2(\omega) < 0 \tag{9.6}$$

which means that one of the dielectric permittivity must be negative with an absolute value greater than of the other. Noble metals like silver, gold, and aluminum have a large but negative real permittivity along with a small imaginary part. As, per the free electron model, for metals,

$$\varepsilon_1 = \varepsilon_0 \left(1 - \frac{\omega_p^2}{\omega^2 + i\omega v} \right) \tag{9.7}$$

where
  v is the collision frequency
  ω_p is the plasma frequency given by

$$\omega_p = \sqrt{\frac{Ne^2}{\varepsilon_0 m_e}} \tag{9.8}$$

where
  N is the free electron concentration
  e and m_e are the charge and mass of electron, respectively

All the guided modes described by the wave vector k_x in Equation 9.3 are due to SPs. Now, if we consider the complex nature of the metal's dielectric function,

$$\varepsilon_1 = \varepsilon_1' + i\varepsilon_1'' \tag{9.9}$$

Again, assuming that the adjacent medium is a good dielectric with negligible losses, that is, ε_2 is assumed to be real, we can represent the complex wave number as

$$k_x = k_x' + ik_x'' \tag{9.10}$$

The real part k_x' determines the surface plasmon polariton (SPP) wavelength, and the imaginary part k_x'' represents the damping of the SPP along its propagation direction, that is, along the metal–dielectric interface:

$$k_x' \approx \sqrt{\frac{\varepsilon_1' \varepsilon_2}{\varepsilon_1' + \varepsilon_2}} \frac{\omega}{c} \tag{9.11}$$

$$k_x'' \approx \sqrt{\frac{\varepsilon_1' \varepsilon_2}{\varepsilon_1' + \varepsilon_2}} \frac{\varepsilon_1'' \varepsilon_2}{2\varepsilon_1' \left(\varepsilon_1' + \varepsilon_2 \right)} \frac{\omega}{c} \tag{9.12}$$

Therefore, the SPP wavelength,

$$\lambda_{SPP} \approx \frac{2\pi}{k_x'} \approx \sqrt{\frac{\varepsilon_1' \varepsilon_2}{\varepsilon_1' + \varepsilon_2}} \lambda \tag{9.13}$$

where λ is the wavelength of the excitation light in vacuum.

The attenuation of the SPW can be represented by associated propagation length L, which is defined as the distance in the direction of propagation at which the energy of the SP decreases by a factor of $1/e$ and represented by

$$L = \frac{1}{2k_x''} \tag{9.14}$$

The decay length of the SPP electric fields along the perpendicular direction of the metal–dielectric interface can be obtained from Equation 9.4 limiting to first order in $|\varepsilon_1''|/|\varepsilon_1'|$ and using Equation 9.9 with

$$k_{1,z} = \frac{\omega}{c} \sqrt{\frac{\varepsilon_1'^2}{\varepsilon_1' + \varepsilon_2}} \left(1 + i \frac{\varepsilon_1''}{2\varepsilon_1'} \right) \tag{9.15}$$

$$k_{2,z} = \frac{\omega}{c} \sqrt{\frac{\varepsilon_2'^2}{\varepsilon_1' + \varepsilon_2}} \left[1 - i \frac{\varepsilon_1''}{2 \left(\varepsilon_1' + \varepsilon_2 \right)} \right] \tag{9.16}$$

The most common approach for the excitation of SPs is based on prism coupling method. In order to understand the coupling phenomenon of incident light and SPs first of all, we have to consider the total internal reflection (TIR) phenomenon that occurs when light propagates from a higher refractive index medium to a lower index medium. The light is incident at the boundary at an angle greater than a particular angle, called critical angle with respect to the normal to the surface. So, if a beam of light strikes the interface at an angle greater than the critical angle, then no light will transmit through the second medium and all the light will reflect back to the first medium. An important side effect of TIR is the propagation of an evanescent wave across the boundary surface. Although the entire energy is reflected back, there is a power flowing in the second medium. Thus, evanescent field decays exponentially along the transverse direction of the surface and propagates along the boundary of the surface. Detailed investigation of the penetration of evanescent wave into the second medium has been reported a long back by E. E. Hall [45]. Now, when the medium is absorbing, then the TIR will be converted to ATR. Another phenomenon related to TIR can also be taken into account by the introduction of the third medium of refractive index close to the first one. Now, the second medium is a thin film with a thickness of the order of the wavelength of light. After the occurrence of TIR, the evanescent field will tunnel through the second medium into the third medium. This phenomenon is frustrated TIR (FTIR) [46]. Thin-film technology [47] has been advanced both for metal and dielectric materials through the applications of FTIR and ATR techniques. However, the study of surface waves greatly improved after the well-known pioneering work of Otto and Kretschmann. The Fresnel relations [48] characterize the reflected and transmitted waves in terms of incident wave amplitude, phase, and polarization for both p- and s-polarization.

Here, the theoretical simulation study is mainly based on excitation of SP by ATR coupler method in Kretschmann geometry. In this configuration, a high refractive index prism with permittivity ε_{pr} is interfaced with a metal—dielectric waveguide consisting of a thin metal film with a complex permittivity ε_1 (having a large but negative real part and small imaginary part) and a semi-infinite dielectric with a real permittivity ε_2. When light wave propagating in the prism is incident on the metal film in ATR condition, then light is totally internally reflected and evanescent field decays exponentially in the direction perpendicular to the prism–metal interface. If the metal film is thin enough (less than 100 nm for light in visible and near-infrared range), then the evanescent field will penetrate through the metal film and couple with the SPs at the outer boundary of the metal film. When the wave vector of the SPs matches with that of the incident light, then the energy of the incident light will be totally transferred to the SPs producing a zero reflectivity dip.

Therefore, the resonance condition of the light in the prism (*pr*) with the SP (*sp*) at metal–dielectric interface (Kretschmann–Raether configuration) is given by

$$K_x^{pr} = K_x^{sp} \tag{9.17}$$

$$\sqrt{\varepsilon_{pr}} \frac{\omega}{c} \sin\theta_{pr} = \frac{\omega}{c} \sqrt{\frac{\varepsilon_1 \varepsilon_2}{\varepsilon_1 + \varepsilon_2}} \tag{9.18}$$

where ε_1 and ε_2 are the dielectric permittivities of the metal and dielectric layers, respectively.

9.2.2 Generalized Mathematical Formulation (N-Layer Model) Based on Characteristic Transfer Matrix Method

The radiative properties like transmittance and reflectance of multilayer thin films can be expressed by either the field tracing method, the resultant wave method, or the transfer matrix method. The former two are rather cumbersome especially for multilayers, whereas the transfer matrix method can be easily extended to multilayers. Thus, we have used the transfer matrix method to investigate the optical characteristics of multilayers with metal film thickness comparable to the wavelength of incident light [49].

Generalized formula for the characteristic matrix for the *k*th layer is given by [50]

$$M_k = \begin{pmatrix} \cos\beta_k & -i\sin\beta_k/q_k \\ -iq_k\sin\beta_k & \cos\beta_k \end{pmatrix} \tag{9.19}$$

where

$$q_k = \sqrt{\left(\frac{\mu_k}{\varepsilon_k}\right)} \cos\theta \tag{9.20}$$

For TM mode,

$$q_k = \sqrt{\left(\frac{1}{\varepsilon_k}\right)} \cos\theta_k = \sqrt{\left(\varepsilon_k - n_1^2 \sin^2\theta_1\right)/\varepsilon_k} \tag{9.21}$$

For TE mode,

$$q_k = \sqrt{\varepsilon_k} \cos\theta_k = \sqrt{\left(\varepsilon_k - n_1^2 \sin^2\theta_1\right)} \tag{9.22}$$

The phase factor is given by,

$$\beta_k = \left(\frac{2\pi}{\lambda}\right) n_k \cos\theta_k \left(z_k - z_{k-1}\right)$$

$$= \left(\frac{2\pi}{\lambda}\right)\left(z_k - z_{k-1}\right)\sqrt{\left(\varepsilon_k - n_1^2 \sin^2\theta_1\right)} \tag{9.23}$$

The reflection coefficient for multilayer nanoplasmonic structure is given by

$$r = \frac{\left(\left(M_{11} + M_{12}q_N\right)q_1 - \left(M_{21} + M_{22}q_N\right)\right)}{\left(\left(M_{11} + M_{12}q_N\right)q_1 + \left(M_{21} + M_{22}q_N\right)\right)} \tag{9.24}$$

where

$$M_{ij} = \left(\prod_{k=2}^{N-1} M_k\right)_{ij}, \quad i,j = 1,2 \tag{9.25}$$

where
  i, j are row and column indices
  k is the number of layers

So, the reflectance of a multilayer structure is given by

$$R = |r|^2 \tag{9.26}$$

Reflection coefficient of any multilayer structure can also be written as

$$r = R^{1/2}e^{i\phi_r} \tag{9.27}$$

where

$$\phi_r = \arg(r) \tag{9.28}$$

The transmission coefficient for magnetic field,

$$t_H = \frac{2q_1}{\left(M_{11} + M_{12}q_N\right)q_1 + \left(M_{21} + M_{22}q_N\right)} \tag{9.29}$$

and for electric field,

$$t_E = \frac{\mu_N n_1}{\mu_1 n_N} t_H \tag{9.30}$$

So, the transmittance can be written as

$$T = \frac{\mu_N \operatorname{Re}\left(n_N \cos\theta_N / n_N^2\right)}{\mu_1 n_1 \cos\theta_1 / n_1^2} |t_H|^2 \tag{9.31}$$

$$\phi_t = \arg\left(t_E\right) \tag{9.32}$$

9.3 Computational Procedure for Reflectivity Calculation

Based on characteristic transfer matrix (CTM) method described in the previous section, we can calculate the reflectivity of any multilayer structure and the corresponding computational procedure is explained in the self-explanatory flowchart describing the basic algorithm followed as shown in Figure 9.1.

9.4 Classification of Different Nanoplasmonic Structures

In Figure 9.2, different types of nanoplasmonic structures have been demonstrated. First of all conventional SPR structure based on Kretschmann configuration consists of a thin metal layer deposited on a high refractive index glass prism, and a dielectric layer is in contact with the plasmon-generating thin metallic film. SPR occurs at the metal—dielectric interface. Change of the thickness or refractive index of the dielectric medium can be precisely detected by the SPR measurement technique. Different modified nanoplasmonic structures have also been analyzed in order to increase the measurement accuracy and for other nanophotonic applications. In CPWR structure, a waveguide layer is deposited on a metal layer of conventional Kretschmann configuration to give a chemical and mechanical protection to the plasmon-generating thin metallic film, and the sensing medium is in contact with the waveguide layer. This type of structure provides a coupling of SPR mode and waveguide resonance mode. Another plasmonic structure, LRSPR, consists of same types of layer combination as in CPWR structure, but the position of the metal layer and waveguide layer has been swapped. Two types of metallo-dielectric combinations such as monometallic and bimetallic structures are also taken into account. Silver is very much oxidation-prone than gold, but silver always gives sharper resonance spectra than gold. Simultaneous use of silver and gold removes the chemical instability problem and also increases the measurement accuracy. These types of bimetallic structure consist of dielectric–metal–metal (DMM) block. Another coupled nanoplasmonic structure, CWSPR, also provides coupling of SPR mode and waveguide resonance mode. It consists of two metal layers between which a waveguide layer is sandwiched, that is, the metallo-dielectric block for this case consists of metal–dielectric–metal (MDM) and the whole structure is deposited on a high refractive index glass prism layer by layer and the sensing medium is in contact with the lower metal film. Here also for the same purpose, two types of metallo-dielectric combinations have been analyzed: one is homo-bimetallic ($M_u = M_l$) and the other is hetero-bimetallic ($M_u \neq M_l$) structure (subscripts u and l denote upper and lower layers).

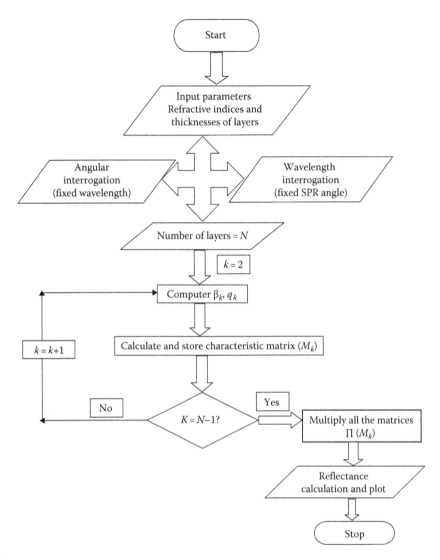

FIGURE 9.1
Algorithm flowchart for the reflectivity calculations using CTM method.

Analysis of the metallo-dielectric block has also been carried out with variations in the types of materials as well as in the thicknesses of the layers used.

9.5 Simulation of Resonance Curves for Different Nanoplasmonic Structures

9.5.1 2D Resonance Curves

Figure 9.3 demonstrates resonance curves for the different nanoplasmonic structures in angular and wavelength interrogation for both p-(TM) and s-(TE) polarization of the

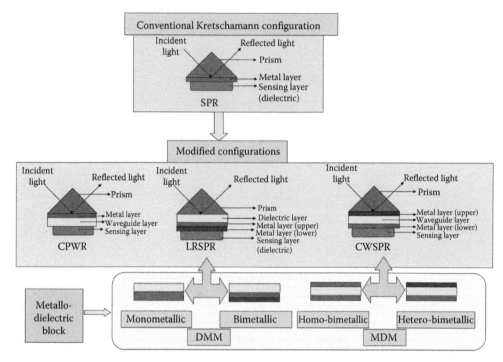

FIGURE 9.2
(See color insert.) Classification of different nanoplasmonic structures.

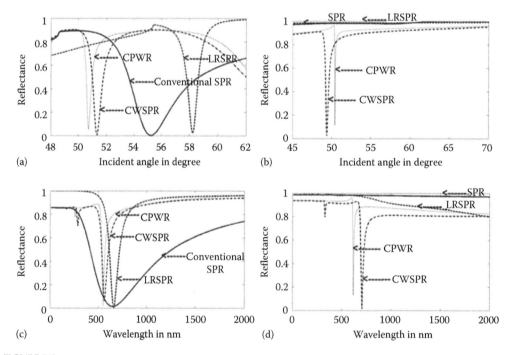

FIGURE 9.3
Reflectivity spectra of different nanoplasmonic structures for (a) TM mode in angular interrogation, (b) TE mode in angular interrogation, (c) TM mode in wavelength interrogation, and (d) TE mode in wavelength interrogation.

incident light. These plots help to compare the resonance phenomenon for those resonant structures. CPWR and CWSPR give resonance spectra for both p- and s-polarization. But conventional SPR and LRSPR do not produce resonance for s-polarization. The aforementioned two types of coupled plasmonic structure produce waveguide resonance along with the SPR. Though SPR phenomenon occurs only for p-polarization, waveguide resonance occurs for both p- and s-polarization. Hence, these nanoplasmonic structures that result in coupling of SPR mode and waveguide resonance mode support resonance for both TM and TE modes.

9.5.2 3D Resonance Curves

The 3D plots depict simultaneous interrogation of more than one optical parameter that gives clear understanding of the resonance phenomenon. Reflectance contour plot can be obtained for those optical parameters with specified reflectance values in the range 10%–100% with steps of 5%.

9.5.2.1 Analysis for Conventional SPR Structure

The 3D plot and reflectance contour plot with some specific reflectance value gives simultaneous angular and wavelength interrogation for conventional SPR structure as shown in Figure 9.4. If a line is drawn parallel to the horizontal axis of the reflectance contour plot with a fixed value of wavelength (which should be the working wavelength), we will get angular interrogation reflectance spectra. Similarly, if a line is drawn parallel to the vertical axis with a fixed value of angle (which is the SPR angle for a particular SPR configuration), we will get wavelength interrogation reflectance spectra. However, it is to be noted that such type of simultaneous interrogation plots will depend on the choice of working wavelength as evident from Figure 9.4a and b for two wavelengths of incident light, 633 and 850 nm, respectively. It is seen from these plots that the higher the wavelength, the sharper will be the resonance.

9.5.2.2 Analysis for Modified Coupled Nanoplasmonic Resonant Structure

The 3D resonance curves and reflectance contour plots have been demonstrated in Figure 9.5 along with individual metallo-dielectric block concerned with different nanoplasmonic structures. In reflectance contour plots of CPWR structure, two different zones can be designated as SPRZ for SPR zone and DLRZ for waveguide resonance zone due to the coupling of plasmon resonance and waveguide resonance.

In the same manner, 3D resonance curves and reflectance contour plots have also been simulated for modified multilayer DMM nanoplasmonic resonant structure. Both mono- and bimetallic structures have different effects with respect to one another. In monometallic structure, only one resonance dip and single contour zone will occur for the 50 nm gold layer. But in bimetallic configuration we can see that silver and gold produce separate resonance dip and also separate resonance zone. In 3D plot, the resonance effect due to the gold layer can be designated as SPRD-Au and for silver layer it is SPRD-Ag. Similarly, in reflectance contour plot, the zone for gold layer is SPRZ-Au and for silver layer SPRZ-Ag.

Same simulation for MDM plasmonic resonant structure produces a different effect than the DMM structure. Here the coupling takes place between SPR mode in thin plasmon-generating metallic film and waveguide resonance mode in sandwiched dielectric layer.

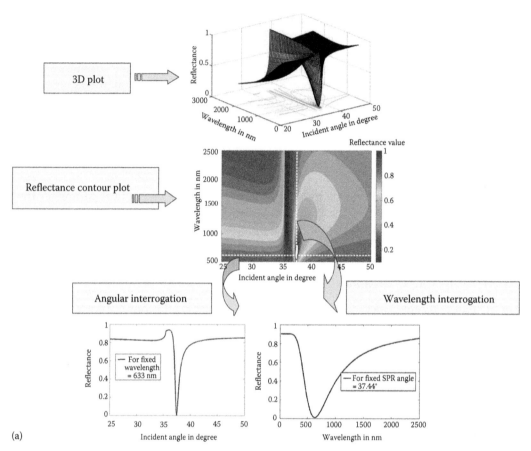

FIGURE 9.4
(See color insert.) 3D resonance curve with default contour plot and reflectance contour plot with specified reflectance value for conventional SPR structure with angular and wavelength interrogation in 2D representation for (a) 633 and (b) 850 nm.

The homo-MDM structure consists of silica dielectric layer sandwiched between two gold layers, and for hetero-MDM, upper layer is silver instead of gold. As shown in Figure 9.5, the effect of plasmon mode has been designated as SPRD in 3D plot and SPRZ in contour plot. Similarly, the effect of waveguide layer has been designated by dielectric layer resonance dip (DLRD) in 3D plot and DLRZ in contour plot. Moreover, it can be seen that DLRZ remains almost unaltered in both homo- and hetero-bimetallic MDM structures due to the fixed thickness of dielectric layer of 1000 nm. But from the contour plots, we can see that SPRZ is narrower in hetero-bimetallic structure than homo-bimetallic structure because of the presence of upper silver metal layer.

Multiple resonance dips are observed for certain wavelengths as a result of coupling of the upper metal layer with dielectric waveguide as evident from the magnified view shown in Figure 9.6a and b for both homo- and hetero-MDM structures, respectively. Blue regions in the contour represent the minimum reflectance value (DLRD) corresponding to different wavelengths.

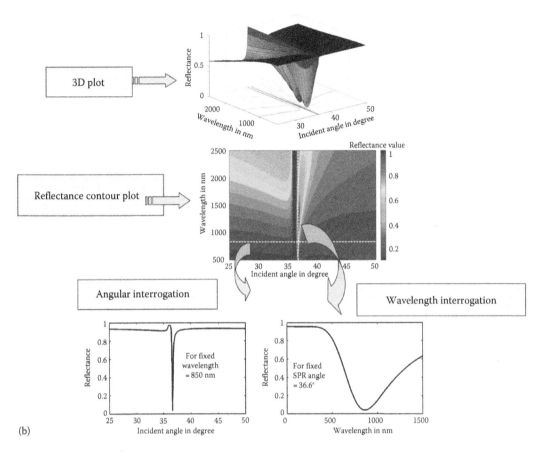

(b)

FIGURE 9.4 (continued)
(See color insert.) 3D resonance curve with default contour plot and reflectance contour plot with specified reflectance value for conventional SPR structure with angular and wavelength interrogation in 2D representation for (a) 633 and (b) 850 nm.

9.6 Metal Thickness Optimization

The reflectivity $R_{1/2/3}$ may be given by Fresnel equations of the prism/metal/air layer system. The total reflection of the three-layer model for p-polarization is given by

$$R_{123} = \left| r_{123}^{p} \right|^{2}$$

$$= \left| \frac{\left(r_{12}^{p} + r_{23}^{p} \exp\left(2ik_{z2}d_{2}\right) \right)}{\left(1 + r_{123}^{p} r_{23}^{p} \exp\left(2ik_{z2}d_{2}\right) \right)} \right|^{2} \tag{9.33}$$

where, from Fresnel equation, we can have reflection coefficient for TE and TM modes.

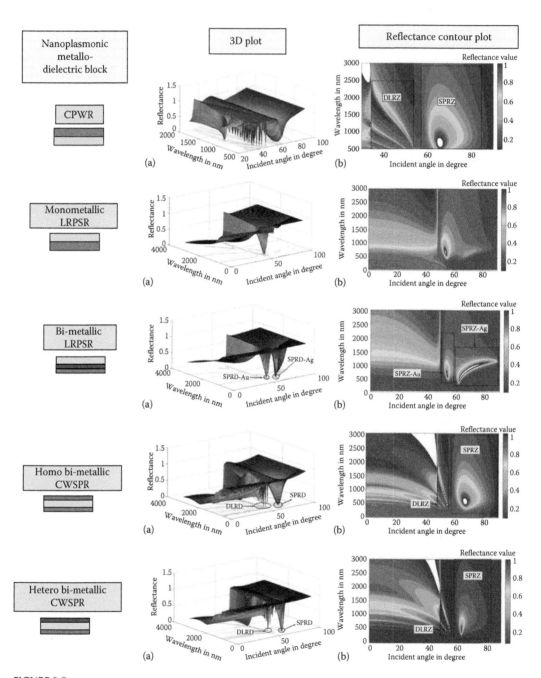

FIGURE 9.5
(See color insert.) (a) 3D plot and (b) reflectance contour plot for different modified nanoplasmonic structures with their specific metallo-dielectric block.

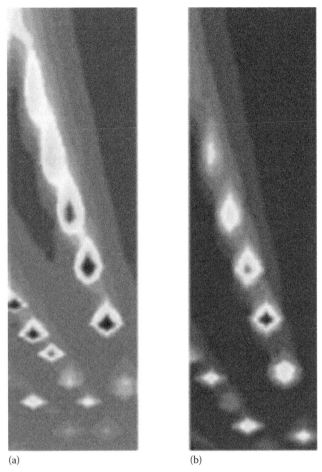

FIGURE 9.6
Magnified view of DLRZ for (a) homo-MDM structure and (b) hetero-MDM structure shown in Figure 9.5.

The reflection coefficient for TM mode is given as

$$r_{ik}^{p} = \frac{\left(n_k \cos\theta_i - n_i \cos\theta_k\right)}{\left(n_k \cos\theta_i + n_i \cos\theta_k\right)} \tag{9.34}$$

and for TE mode is given as

$$r_{ik}^{s} = \frac{\left(n_i \cos\theta_i - n_k \cos\theta_k\right)}{\left(n_i \cos\theta_i + n_k \cos\theta_k\right)} \tag{9.35}$$

$$\text{phase factor} = k_{z2}d_2 = k_0 d_2 \cos\theta_2$$

$$= \left(\frac{2\pi}{\lambda}\right) d_2 \left(\sqrt{\left(\varepsilon_2 - n_1^2 \sin^2\theta_1\right)}\right) \tag{9.36}$$

Similarly, we can have the expression of reflectivity for four-layer system as a function of layer thickness:

$$R_{1234} = \frac{r_{12}^p + r_{23}^p \exp(i2k_{z2}d_2) + r_{34}^p \exp(i2k_{z2}d_2)\exp(i2k_{z3}d_3) + r_{12}^p r_{23}^p r_{34}^p \exp(i2k_{z3}d_3)}{1 + r_{12}^p r_{23}^p \exp(i2k_{z2}d_2) + r_{23}^p r_{34}^p \exp(i2k_{z3}d_3) + r_{12}^p r_{34}^p \exp(i2k_{z2}d_2)\exp(i2k_{z3}d_3)} \quad (9.37)$$

Based on the preceding mathematical background, we can calculate the reflectivity and study the effect of variation in the layer thicknesses on the final resonance curve with the ultimate aim of achieving zero reflectance for some particular angle or wavelength depending upon whether we are considering angular or wavelength interrogation. Figure 9.7 shows the simple optimization procedure where we have considered

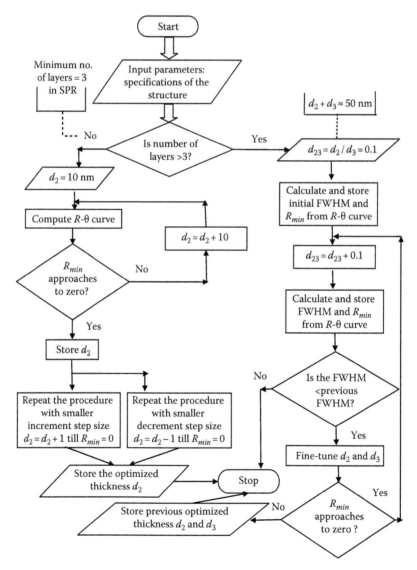

FIGURE 9.7
Algorithm for the metallic thickness optimization in a nanoplasmonic structure in the form of a flowchart.

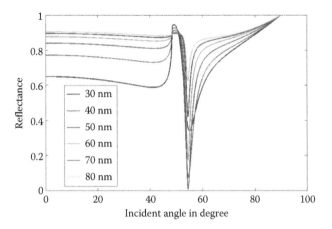

FIGURE 9.8
Metal thickness optimization of conventional SPR.

conventional three-layer SPR structure and also four-layer structure using bimetallic films instead of a single metal layer. This optimization procedure can also be applied to any multilayer nanoplasmonic structure having mono- or bimetallic nanofilms. In order to utilize the SPR phenomenon efficiently for the sensing purpose, it is quite necessary to theoretically determine the optimum thicknesses of the different metal films to be used. Using the principle of conservation of energy, if the reflected intensity reaches its lowest value, it implies that the maximum amount of energy from the incident light has been coupled to SPs. Therefore, the optimum thickness will be the thickness, which gives minimum reflectivity on one hand and maximum evanescent field enhancement on the other [51]. Figure 9.8 shows the resonance curves for metal (gold) thickness varying from 30 to 80 nm in conventional SPR structure, and we can conclude from this that zero-reflectance condition can only be achieved for 50 nm thickness. It is to be noted that this optimum thickness also varies from metal to metal.

Figure 9.9a shows 3D plot depicting reflectance with variation in thickness of silver and gold in bimetallic configuration where best performance of silver and its protection by gold

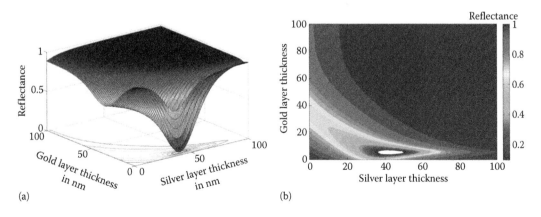

FIGURE 9.9
Metal thickness optimization of conventional SPR with bimetallic nanofilm. (a) 3D plot of reflectivity with variations of gold and silver layer thickness and (b) contour plot with the same variations.

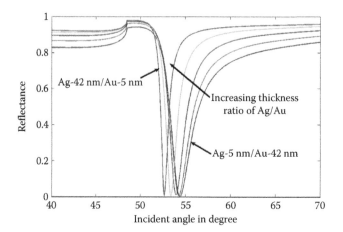

FIGURE 9.10
Metal thickness optimization of conventional SPR with bimetallic nanofilm for various combinations of silver–gold thickness.

both are simultaneously utilized. The contour plot for the same configuration is shown in Figure 9.9b. The resonance curves for such a bimetallic structure for various combinations of silver and gold thicknesses shown in Figure 9.10 indicate the increase in sharpness with increasing silver/gold thickness ratio, and one can use the optimization routine to arrive at the best possible combination of the thicknesses.

From our earlier simulations, we have seen that 3D variations of reflectivity with the thickness of both the metal layers always help to choose the proper working thickness in a bimetallic plasmonic resonant structure. Generally the greater the thickness of gold compared to that of silver, the broader will be the resonance contour. For our DMM structure, several combinations of Ag-Au thickness can be taken under consideration based on the location of the low reflectance contour as shown in the Figure 9.11, and as a result, SPRZ-Au and SPRZ-Ag will change in the reflectance contour of the variations of R-λ-θ plot. In case of MDM structure, we can see from Figure 9.12 that the contour

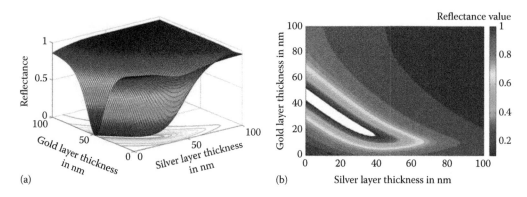

(a)   (b)

FIGURE 9.11
(a) 3D plot and (b) reflectance contour plot with same variations as in Figure 9.9 for bimetallic DMM plasmonic resonant structure.

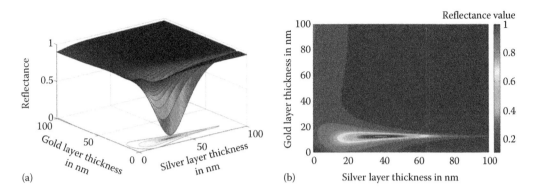

(a)   (b)

FIGURE 9.12
(a) 3D plot and (b) reflectance contour plot with a variation of silver and gold layer thickness for hetero-bimetallic MDM plasmonic resonant structure.

is almost parallel to the horizontal axis, so the thickness of the gold layer in order to achieve resonance is almost fixed and thickness of the silver layer can be chosen to approach zero-reflectance condition.

9.7 Applications in Nanophotonics

9.7.1 Thickness Monitoring of Dielectric Layer in a Multilayer Composite Plasmonic Resonant Structure

In this subsection, we are dealing with the CPWR structure where a waveguide or dielectric layer is deposited on the metallic film of a Kretschmann configuration. We have theoretically simulated different resonance curves in order to study the effect of change in material as well as the thickness of the waveguide layer for CPWR structure. Figure 9.13a depicts four resonance curves for magnesium fluoride (r.i. = 1.38) as a waveguide layer with thicknesses of 350, 700, 1050, and 1400 nm, respectively. It is seen that the number of resonance dips increases with increase in thickness of the waveguide layer and has repetitive nature. The same phenomena also occur for other dielectric materials such as silica (r.i = 1.4571) and titanium oxide (r.i. = 2.847). Progressive nature of multiple resonance of waveguide mode with unaltered SP mode has already been demonstrated with media clip in our previous work [42]. In these types of structures, the dielectric layer on a metal film can support both the TE and TM modes. But the lowest-order TM or p-polarized mode with broader resonance at higher incident angle is designated as the SP mode, because only this particular mode exists when the dielectric layer thickness tends to zero. If the metal layer is excited with TE or s-polarized light, then slightly different types of reflectivity curves can be found. The TIR phenomena and the guided modes remain unaltered, but the SP mode will not be excited.

The progressive multiple resonance peaks have been correlated with the change of dielectric film thickness, and as shown in Figure 9.13b, this is valid for the fractional dip also. When the resonance curve reaches almost zero-reflectance value, we call it one

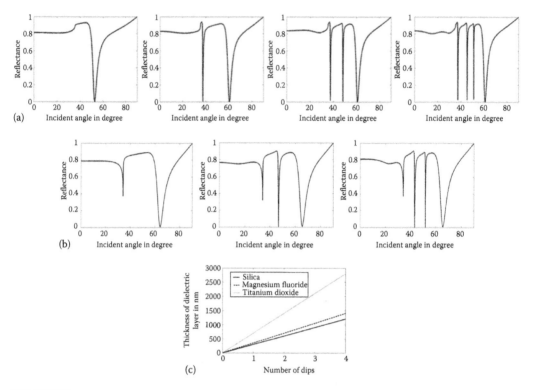

FIGURE 9.13
(a) Progressive multiple resonance curves in CPWR configuration with magnesium fluoride as a dielectric layer of different thicknesses, 350, 700, 1050, and 1400 nm. (b) The fractional dips as 1.5, 2.5, and 3.5 for the thickness of 450, 750, and 1050 nm of silica waveguide layer and (c) dielectric layer thickness versus number of resonance dips for three different dielectric materials.

complete dip, but when the lowest value acquired corresponds to higher reflectance value, it is termed as a fractional dip. The characteristic curves of number of dips versus thickness are linear with differing slope for different materials as depicted in Figure 9.13c. Gradual appearance of multiple resonance dips has unique characteristics having different repetition intervals (Table 9.1), which may find application in future research related to characterization of the dielectric material in a composite nanostructure.

TABLE 9.1

Parametric Overview of Multilayer Metallo-Dielectric Plasmonic Resonant Structure for Hetero-Bimetallic Nanofilms

Thickness of Dielectric Layer in nm			
Silica	Magnesium Fluoride	Titanium Dioxide	Number of Resonance Dips
300	350	700	1
600	700	1400	2
900	1050	2100	3
1200	1400	2800	4

9.7.2 Sensor Technology

9.7.2.1 Related Sensitivity Issues

For a given frequency of the light source and the dielectric constant of metal film, one can determine the dielectric constant of the sensing layer adjacent to the metal layer by knowing the value of the resonance angle (θ_{res}) or resonance wavelength (λ_{res}). The resonance angle and wavelength can be determined by using angular and wavelength interrogation method, respectively. The resonance angle or wavelength is very sensitive to variation in the refractive index (or dielectric constant) of the sensing layer as shown in Figure 9.14. Increase in refractive index of the dielectric sensing layer increases the resonance angle [52].

Sensitivity and detection accuracy or signal-to-noise ratio (SNR) are the two important parameters to be considered while judging the performance of any nanoplasmonic sensor. Sensitivity of a nanoplasmonic sensor depends on the shift of the resonance angle or wavelength as a result of change in refractive index of the sensing layer. Higher value of sensitivity corresponds to larger shift. Figure 9.14 shows a general plot of reflectance as a function of angle or wavelength of the incident light for a small change in sensing layer refractive index, say from n to $n+\delta n$. We denote the consequent shift in the resonance curve by $\delta\theta_{res}$ or $\delta\lambda_{res}$ for angular or wavelength interrogation, respectively. The sensitivity of an SPR sensor with angular interrogation is defined as

$$S_n = \frac{\delta\theta_{res}}{\delta n} \tag{9.38}$$

The detection accuracy of reflectivity curves further depends on the width of the SPR curve. The narrower the resonance curve, the higher is the detection accuracy. Therefore, if $\delta\theta_{0.5}$ is the angular width of the SPR response curve corresponding to 50% reflectance, the detection accuracy of the sensor can be assumed to be inversely proportional to $\delta\theta_{0.5}$. The *SNR* of the SPR sensor with angular interrogation is, thus, defined as

$$SNR = \frac{\delta\theta_{res}}{\delta\theta_{0.5}} \tag{9.39}$$

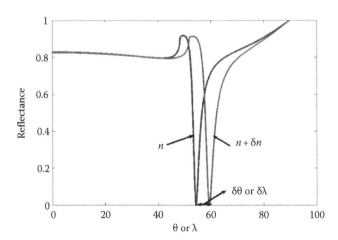

FIGURE 9.14
Reflectivity curve in angular or wavelength interrogation.

As sensitivity and detection accuracy are equally important, a figure of merit (FOM) can be defined, which includes both the parameters of sensing [53–55]. In angular interrogation, it can be called as angular FOM (AFOM), which can be defined as the ratio of sensitivity and the full width half maximum (FWHM) of resonance curve in angular interrogation mode for some specific wavelength:

$$AFOM = \frac{S_n}{FWHM} \qquad (9.40)$$

Table 9.2 and the Figure 9.15 represent the nature of AFOM for our different nanoplasmonic structures with the change of wavelength. From this plot, we can conclude that the hetero-bimetallic MDM structure is the best among the others.

TABLE 9.2

AFOM for Different Nanoplasmonic Structures

Nanoplasmonic Structure	AFOM at Different Wavelengths					
	450 nm	514 nm	633 nm	700 nm	750 nm	800 nm
Conventional SPR	0	2.131	18.082	36.115	48.388	62.234
Monometallic DMM	0	2.031	7.821	26.629	33.096	42.0146
Bimetallic DMM	4.279	5.24	29.907	65.789	88.822	93.321
Homo-bimetallic MDM	0	0.76	7.708	20.363	29.732	37.701
Hetero-bimetallic MDM with 100 nm silica layer	3.332	7.304	28.256	34.36	71.957	91.668
Homo-bimetallic MDM with 1000 nm silica layer	13.848	34.395	84.072	117.481	283.331	827.162
Hetero-bimetallic MDM with 1000 nm silica layer	90.094	208.182	917.1507	1174.812	1495.356	1811.88

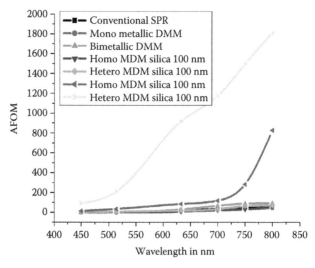

FIGURE 9.15
Comparison of AFOM of different nanoplasmonic structures.

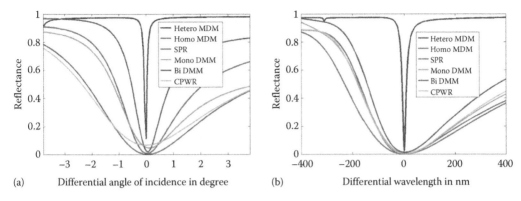

(a)   Differential angle of incidence in degree   (b)   Differential wavelength in nm

FIGURE 9.16
(a) Differential angular reflectivity curve and (b) differential wavelength reflectivity curve for different nanoplasmonic structures.

This analysis can also be done for wavelength interrogation also. In that case, the parameter will be such that

$$S_n = \frac{\delta\lambda_{res}}{\delta n} \quad \text{and} \quad SNR = \frac{\delta\lambda_{res}}{\delta\lambda_{0.5}} \tag{9.41}$$

and finally the FOM can be called as spectral FOM (SFOM) for different angles close to the SPR angle:

$$SFOM = \frac{S_n}{FWHM} \tag{9.42}$$

Material change, temperature change, and concentration change can cause a change in refractive index and that can be detected by the angular or wavelength shift of resonance curve. So, the material signature can be possible using the SPR phenomenon.

To investigate the influence of the different nanoplasmonic structure on the shape of the reflectivity curve, we have simulated those curves centered at zero as shown in Figure 9.16. The principle of computing this kind of curve is to plot the reflectivity with the variations of the difference of angle of incidence and the SPR angle. The same procedure has been applied for differential wavelength plot also. These types of simulations always help to understand clearly the difference in reflectance spectra due to the structural modification. From these comparisons of differential angular and wavelength interrogation mode, we can also conclude that the hetero-bimetallic MDM structure gives the narrowest resonance curve, hence producing greater detection accuracy [56,57].

9.8 Modal Analysis

In order to deal with the theory of light propagation in various nanoplasmonic structures described, we have used the beam-propagation method (BPM) (RSoft BeamPROP 8.1). It is a step-by-step method of simulating the propagation of light through any waveguiding medium so as to obtain the transverse field at any point along the propagation direction. The starting point to develop BPM algorithms is the Fresnel equation for paraxial propagation. In this method, an optical circuit parameter such as refractive index distribution and the type of input field parameters are required to be defined. Input optical field must be initially launched at a plane such as $z = 0$. Using the BPM analysis, we can have the output field either at the end of the structure or at any other value of z.

Of the two general approaches, BPM based on finite differences (FD-BPM) has superior performance than that based on fast Fourier transform (FFT-BPM) method for simulations related to integrated optical elements. The BPM can solve the paraxial form of the scalar wave equation as well as the vectorial wave equation (VBPM). Another method called semi-vectorial BPM (SVBPM) is based on an intermediate approximation that ignores the coupling terms between the transversal components and the fields. A semi-vectorial calculation is performed in which the minor component of the field is neglected (set to zero), and polarization is partially taken into account by employing the proper field equations for the major component. This option can discriminate between quasi-TE and quasi-TM fields and can be used to determine differences in field shapes and propagation constants. Modal analysis based on BPM allows the calculation of modes supported by a planar waveguide as well as the distribution of energy transfer within the composite structure. In our analysis, we have used the SVBPM for simplistic computational aspects as the minor components are absent. First of all we have defined the structure by their refractive index distribution as shown in Figure 9.17. Then input field is specified. The structure has its refractive index difference along the Y-axis and Z is the propagation direction along which refractive index of a particular layer of a structure remains the same. In Figure 9.18, the electric field distributions in TM mode using SVBPM of different nanoplasmonic structures have been demonstrated.

9.9 Concluding Remarks

This chapter summarizes some important aspects of coupling of plasmon modes with the waveguide modes arising out from different coupled plasmonic nanostructures along with brief comparison with conventional structure wherever necessary. Some simple computational techniques as described provide a lot of information to carry out the simulation to compare different structures and optimize different layer thicknesses in order to meet certain application-oriented performance criteria. Though the present simulations do not consider the effect of dispersion, we have further extended our analysis [58] incorporating the same. Such type of simulation studies have been recently carried out by us [59,60] and we feel that they are necessary before the actual implementation as a predesign procedures depending upon the applications concerned.

Acknowledgments

The authors would like to acknowledge Professor A. K. Datta for providing the valuable information regarding this book. The author M. Bera is grateful to the Council of Scientific and Industrial Research (CSIR), India, for providing the Senior Research Fellowship.

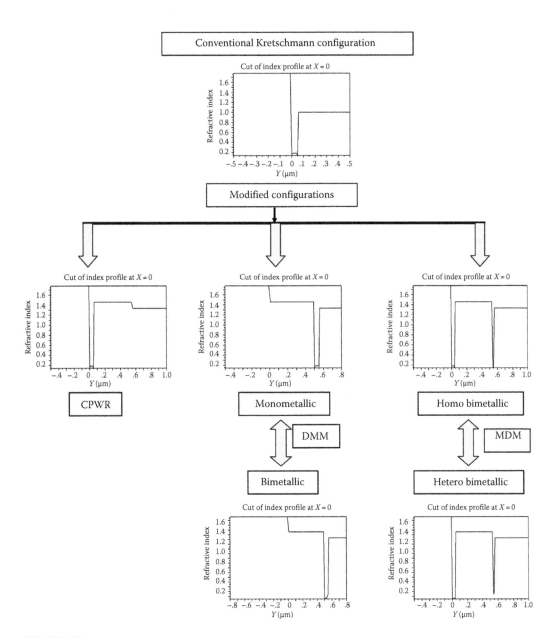

FIGURE 9.17
Refractive index profile of different nanoplasmonic structures.

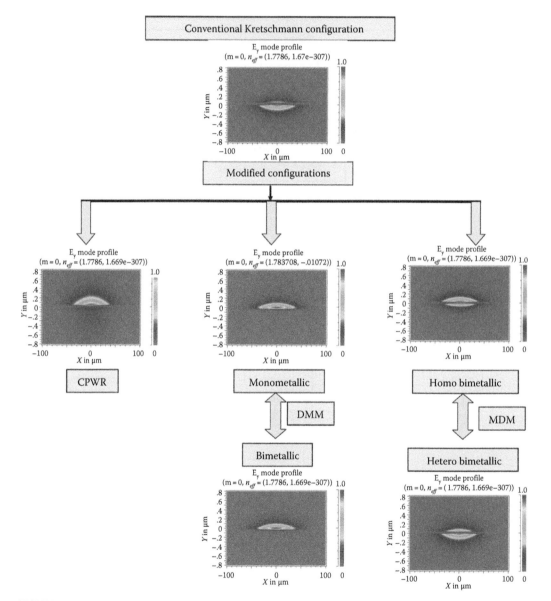

FIGURE 9.18
Modal profile of the electric field distribution in TM mode of different nanoplasmonic structures using SVBPM.

References

1. E. Kretschmann and H. Raether, Radiative decay of non radiative surface plasmons excited by light, *Zeitschrift Fur Naturforschung Part A: Astrophysik Physik Und Physikalische Chemie A* **23**, 2135–2136 (1968).
2. A. Otto, Excitation of surface plasma waves in silver by the method of frustrated total reflection, *Zeitschrift für Physik*, **216**, 398–410 (1968).
3. M. Yamamoto, Surface plasmon resonance (SPR) theory: Tutorial, *Rev Polarograms*, **48**, 209–237 (2002).
4. S. R. Karlsen, K. S. Johnston, R. C. Jorgenson, and S. S.Yee, Simultaneous determination of refractive index and absorbance spectra of chemical samples using surface plasmon resonance, *Sensors and Actuators B*, **24–25**, 747–749 (1995).
5. D. Paul, Zur Elektronentheorie der metallic, *Annalen der Physik*, **306**(3), 566–613 (1900).
6. H. Raether, Surface plasma oscillations and their applications, in *Physics of Thin Films*, G. Hass, M. Francombe, and R. Hoffman, eds. (Academic, New York, 1977) Vol. 9, Chapter 3.
7. J. Homola, S. S. Yee, and G. Gauglitz, Surface plasmon resonance sensor, *Review, Sensors and Actuators B*, **54**, 3–15 (1999).
8. M. Bera and M. Ray, Precise detection and signature of biological/chemical samples based on surface plasmon resonance (SPR), *Journal of Optics*, **38**(4), 232–248 (2009).
9. K. Bramhachari, S. Ghosh, and M. Ray, Experimental observation of surface plasmon resonance using various geometrical configurations of metal-dielectric interface, Paper ID: CP-11, presented at *International Symposium on Advances in Nanomaterials (ANM2010)*, Central Glass & Ceramic Research Institute, Kolkata, India, December 6–7, 2010.
10. S. Ghosh, K. Brahmachari, and M. Ray, Experimental investigation of surface plasmon resonance using a chemically deposited silver film on a tapered cylindrical glass rod, Paper ID: FO-4, presented at *International Conference on Specialty Glass & Optical Fiber: Materials, Technology & Devices (ICGF- 2011)*, Central Glass & Ceramic Research Institute, Kolkata, India, August 4–6, 2011.
11. S. Ghosh, K. Brahmachari, and M. Ray, Experimental investigation of surface plasmon resonance using tapered cylindrical light guides with metal-dielectric interface, *Journal of Sensor Technology*, **2**(1), 48–54 (2012).
12. K. Bramhachari, S. Ghosh, and M. Ray, Application of admittance loci method in surface plasmon resonance technology for sensing of different chemical and biological samples, Paper ID: GP-10, presented at *International Conference on Specialty Glass & Optical Fiber: Materials, Technology & Devices (ICGF- 2011)*, Central Glass & Ceramic Research Institute, Kolkata, India, August 4–6, 2011.
13. K. Brahmachari, S. Ghosh, and M. Ray, Substrate dependence of surface plasmon resonance sensor with a multilayer structure using admittance loci method, Trends in optics and photonics-II, *Proceedings of International Conference on Trends in Optics and Photonics*, Kolkata, India, pp. 402–407, December 7–9, 2011.
14. Y. Bian, Z. Zheng, X. Zhao, J. Zhu, and T. Zhou, Symmetric hybrid surface plasmon polariton waveguides for 3D photonic integration, *Optics Express*, **17**(23), 21320–21325 (2009).
15. P. B. Catrysse, G. Veronis, H. Shin, J.-T. Shen, and S. H. Fan, Guided modes supported by plasmonic films with a periodic arrangement of subwavelength slits, *Applied Physics Letters*, **88**, 031101 1–031101 3 (2006).
16. R. Wan, F. Liu, and Y. Huang, Ultra thin layer sensing based on hybrid coupler with short-range surface plasmon polariton and dielectric waveguide, *Optics Letters*, **35**(2), 244–246 (2010).
17. P. B. Catrysse and S. H. Fan, Understanding the dispersion of coaxial plasmonic structures through connection with the planar metal-insulator metal geometry, *Applied Physics Letters*, **94**, 231111 (2009).
18. R. Wan, F. Liu, Y. Huang, S. Hu, B. Fan, Y. Miura, D. Ohnishi, Y. Li, H. Li, and Y. Xia, Excitation of short range surface plasmon polariton mode based on integrated hybrid coupler, *Applied Physics Letters*, **97**, 141105 (2010).

19. A. Degiron, C. Dellagiacoma, J. G. McIlhargey, G. Shvets, and O. J. F. Martin, Simulations of hybrid long range plasmon modes with application to 90° bends, *Optics Letters*, **32**(16), 2354–2356 (2007).

20. S.-H. Kim, K.-S. Ock, J. H. Im, J.-H. Kim, K.-N. Koh, and S. W. Kang, Photoinduced refractive index change of self-assembled spiroxazaine monolayer based on surface plasmon resonance, *Dyes and Pigments*, **46**, 55–62 (2000).

21. S. Collin, F. Pardo, and J.-L. Pelouard, Waveguiding in nanoscale metallic apertures, *Optics Express*, **15**(7), 4310–4320 (2007).

22. E. K. Akowuah, T. Gorman, S. Haxha, and J. V. Oliver, Dual channel planar waveguide surface plasmon resonance biosensor for an aqueous environment, *Optics Express*, **18**(23), 24412–24422 (2010).

23. J. Shibayama, Three-dimensional numerical investigation of an improved surface plasmon resonance waveguide sensor, *IEEE Photonics Technology Letters*, **22**(9), 643–645 (2010).

24. M. Skorobogatiy and A. V. Kabashin, Photon crystal waveguide-based surface plasmon resonance biosensor, *Applied Physics Letters*, **89**, 143518 (2006).

25. X. Yin and L. Hesselink, Goos-Hanchen shift surface plasmon resonance sensor, *Applied Physics Letters*, **89**, 261108 (2006).

26. K. Takagi, H. Sasaki, A. Seki, and K. Watanabe, Surface plasmon resonances of a curved hetero-core optical fiber sensor, *Sensors and Actuators A: Physical*, **161**, 1–5 (2010).

27. W.-C. Kuo, C. Chou, and H.-T. Wu, Optical heterodyne surface plasmon resonance biosensor, *Optics Letters*, **28**(15), 1329–1331 (2003).

28. K. Johansen, H. Arwin, I. Lundstrom, and B. Leidberg, Imaging surface plasmon resonance sensor based on multiple wavelengths: Sensitivity considerations, *Review of Scientific Instruments*, **71**(9), 3530–3538 (2000).

29. K. Matsubara, S. Kawata, and S. Minami, Multilayer system for a high-precision surface plasmon resonance sensor, *Optics Letters*, **15**(1), 75–77 (1990).

30. D. M. Harnandez, J. Villatoro, D. Talavera, and D. L. Moreno, Optical—Fiber sensor with multiple resonance peaks, *Applied Optics*, **43**(6), 1216–1220 (2004).

31. K. Choi, H. Kim, Y. Lim, S. Kim, and B. Lee, Analytic design and visualization of multiple surface plasmon resonance excitation using angular spectrum decomposition for a Gaussian input beam, *Optics Express*, **13**(22), 8866–8874 (2005).

32. F. C. Chein and S. J. Chen, A sensitivity comparison of optical biosensors based on four different surface plasmon resonance modes, *Biosensors and Bioelectronics*, **20**, 633–642 (2004).

33. K.-S. Lee, J. M. Son, D.-Y. Jeong, T. S. Lee, and W. M. Kim, Resolution enhancement in surface plasmon resonance sensor based on waveguide coupled mode by combining a bimetallic approach, *Sensors*, **10**, 11390–11399 (2010).

34. D. Sarid, Long range surface plasma waves on very thin metal films, *Physical Review Letters*, **47**, 1927–1930 (1981), with Erratum in *Physical Review Letters*, **48**, 446 (1982).

35. J. C. Quail, J. G. Rako, and H. J. Simon, Long range surface plasmon modes in silver and aluminium films, *Optics Letters*, **8**(7), 377–379 (1983).

36. R. Slavik and J. Homola, Simultaneous excitation of long and short range surface plasmons in an asymmetric structure, *Optics Communications*, **259**, 507–512 (2006).

37. Z. Salamon, H. A. Macleod, and G. Tollin, Coupled plasmon waveguide resonator: A new spectroscopic tool for probing proteolipid film structure and properties, *Biophysical Journal*, **73**, 2791–2797 (1997).

38. F. C. Chien and S. J. Chen, Direct determination of the refractive index and thickness of a biolayer based on coupled waveguide—Surface plasmon resonance mode, *Optics Letters*, **31**(2), 187–189 (2006).

39. M. Bera and M. Ray, Multi-layer thin film modeling for observation of coupled waveguide-surface plasmon resonance, Paper ID: CP-10, presented at *International Symposium on Advances in Nanomaterials (ANM2010)*, Central Glass & Ceramic Research Institute, Kolkata, India, December 6–7, 2010.

40. M. Bera and M. Ray, 3D Simulation for simultaneous angular and wavelength interrogation with reflectance contour plot in bimetallic coupled plasmonic structure, Paper ID: GO-2, presented at *International Conference on Speciality Glass & Optical Fiber: Materials, Technology & Devices (ICGF- 2011)*, Central Glass & Ceramic Research Institute, Kolkata, India, August 4–6, 2011.

41. M. Bera and M. Ray, Field analysis of multilayer coupled plasmonic resonant structures using bimetallic nanofilms, Trends in optics and photonics-II, *Proceedings of International Conference on Trends in Optics and Photonics*, pp. 396–401, Kolkata, India, 7–9 December, 2011.

42. M. Bera and M. Ray, Coupled plasmonic assisted progressive multiple resonance for dielectric material characterization, *Optical Engineering*, **50**(10), 103801-1–103801-8, doi: 10.1117/1.3640823 (2011).

43. W. H. Weber and G. W. Ford, Optical electric field enhancement at a metal surface arising From surface-plasmon excitation, *Optics Letters*, **6**(3), 122–124 (1981).

44. M. P. Nezhad, K. Tetz, and Y. Fainman, Gain assisted propagation of surface plasmon polaritons on planar metallic waveguides, *Optics Express*, **12**(17), 4072–4079 (2004).

45. E. E. Hall, The penetration of totally reflected light into the rarer medium, *Physical Review* **15**, 73 (1902).

46. S. Zhu, A. W. Yu, D. Hawley, and R. Roy, Frustrated total internal reflection: A demonstration and review, *American Journal of Physics* **54**(7), 601–607 (1986).

47. E. N. Economou, Surface plasmons in thin films, *Physical Review*, **182**(2), 539–554 (1969).

48. M. Born and E. Wolf, *Principles of Optics*, Pergamon Press, New York (1989).

49. H. R. Gwon and S. H. Lee, Spectral and angular response of surface plasmon resonance based on the Kretschmann prism configuration, *Materials Transaction*, **51**(6), 1050–1055 (2010).

50. F. Abeles, Recherches sur la propagation des ondes electromagnetiques sinusoidales dans les milieux stratifies, Application aux couches minces, *Annales De Physique (Paris)*, **5**, 596–640 (1950).

51. B. H. Ong, X. Yuan, S. C. Tjin, J. Zhang, and H. M. Ng, Optimised film thickness for maximum evanescent field enhancement of a bimetallic film surface plasmon resonance biosensor, *Sensors and Actuators B*, **114**, 1028–1034 (2006).

52. J. Homola, I. Koudela, and S. S. Yee, Surface plasmon resonance sensors based on diffraction gratings and prism couplers: Sensitivity comparison, *Sensors and Actuators B*, **54**, 16–24 (1999).

53. A. Shalabney and I. Abdulhalim, Figure-of-merit enhancement of surface plasmon resonance sensors in the spectral interrogation, *Optics Letters*, **37**(7), 1175–1177 (2012).

54. B. D. Gupta and A. K. Sharma, Sensitivity evaluation of a multi-layered surface plasmon resonance based fiber optic sensor: A theoretical study, *Sensors and Actuators B*, **107**, 40–46 (2005).

55. A. K. Sharma, R. Jha, and B. D. Gupta, Fiber-sensor based on surface plasmon resonance: A comprehensive review, *IEEE Sensors Journal*, **7**(8), 1118–1129 (2007).

56. P. Lecaruyer, E. Maillart, M. Canva, and J. Rolland, Generalization of the Rourd method to an absorbing thin-film stack and application to surface plasmon resonance, *Applied Optics*, **45**(33), 8419–8423 (2006).

57. P. Lecaruyer, M. Canva, and J. Rolland, Metallic film optimization in a surface plasmon resonance biosensor, *Applied Optics*, **46**(12), 2361–2369 (2007).

58. M. Bera and M. Ray, Parametric analysis of multi-layer metallo-dielectric coupled plasmonic resonant structures using homo and hetero-bimetallic nanofilms, *Optics Communications* (in press), doi: 10.1016/j.opt com.2012.12.044

59. M. Bera and M. Ray, Role of waveguide resonance in coupled plasmonic structures using bimetallic nanofilms, *Optical Engineering*, **51**(10), 103801-1-10, doi: 10.1117/1OE.51.10.103801 (2012).

60. M. Bera and M. Ray, Long-range and short-range surface plasmon resonance in coupled plasmonic structure using bi-metallic nanofilms, *Proceedings of 5th International conference on Computers and Devices for Communications (CODEC 2012)*, Kolkata, India, December 17–19, 2012.

10

Advanced Techniques in Medical Computational Nanophotonics and Nanoplasmonics

Viroj Wiwanitkit

CONTENTS

Summary

Nanophotonics is a specific applied nanoscience that specifically deals with the behavior of light or optics on the nanometer scale. Applied in medicine, the nanophotonics can be useful in more aspects. Of many approaches, it is mainly set in ophthalmology, radiology, and laboratory medicine. Also, nanophotonics knowledge is the basic advanced concepts for explanation of nanostructure and nanophenomenon in medicine. Similar to any new nanoscience, the computational approach becomes an important part of nanophotonics.

Medical computational nanophotonics become one of the newest useful knowledge in medicine. In this chapter, we will briefly summarize and discuss on advanced new issues in medical computational nanophotonics. Furthermore, in this chapter, we introduce medical nanoplasmonics, which is an important specific part of medical nanophotonics.

10.1 Introduction

At present, the nanoscience is the new science that can be useful in several aspects. Many latest advents based on nanoscience can be seen. Nanoscience is the science of "small." Nanoscience is an actual bridging science that links between physical and biological sciences. Nanoscience helps provide the new way for solving of many previously difficult questions. Nanoscience is an example of convergence of sciences, which is the present trend of the novel science. Integration is the concept of nanoscience, and this is accepted as an actual multidisciplinary approach. Nanoscience is based on both physical and biological concern's principle. As "the integrative science for future," nanoscience can be applied in many basic physical and biological sciences. Many new branches of nanoscience are available at present. Good examples are nanobiology, nanophysics, nanochemistry, nanopharmacology, nanoengineering, nanomedicine, etc. Several advantages of those applied nanoscience are confirmed.

Focusing on a specific branch of nanoscience, nanomedicine is the specific nanoscience that deals with medical aspects. Nanomedicine becomes the novel thing in medicine. Diagnosis, treatment, and prevention of disease can be performed based on the new nanomedicine. Nanomedicine can provide fast highly accurate reliable medical services. At present, nanomedicine is already implemented and used in medicine. Several applications of nanomedicine for diagnostic and therapeutic purposes can be seen in many modern medical institutes around the world. Focusing on those applications, either in vivo, in vitro, or in silico usage can be expected [1].

Generally, in vivo and in vitro usages are known in medicine. These two usages are generally used in medical science. However, the new usage "in silico approach" might be a new thing that can be the myth for the nonfamilial users. This is the third usage in medicine for the medical scientist. In silico usage is an actual computational-based approach. In silico technique makes use of a simulating or imaginary medical approach. This means no real thing. One might question whether this is acceptable. It should be noted that in silico technique is acceptable and can be very useful and can be used as a standard new approach in medicine. The advantage of using in silico approach is based on the application of nanoinformatics technique. With systematic computation informatic approach, the solution of advanced medical question can be fast and accurately derived. Shortening the turnaround time, decreasing the real experiment cost, and controlling of confounding interference can be expected based on in silico approach.

Basically, computational simulation technique is the core concept of in silico nanoinformatics technology to help solve the medical question. The well-known in silico technique at present is the "omics" science [2]. There are several new "omics" sciences at present, and it can be said that it is presently the "omics" era [2]. Application of computational technology in nanomedicine helps deal with many medical problems. The problems on a very small object, such as molecular particle, can be solved with use of in silico nanoinformatics technique. This imaginary approach can give many data for the modern

TABLE 10.1

Usage of In Silico Nanoinformatics in Nanomedicine

Applications	Details
1. Clarification	A computational approach for answering the nanomedical question. The application of a computational tool can help clarify a process or phenomenon in medicine that cannot be simply done by standard in vivo or in vitro approaches. The clarification of the nanostructure of a medical molecule (such as enzyme, antibody, and viral particle) can be easily done with the help of in silico nanoinformatics.
2. Prediction	A computer-based approach can also be used for predicting nanomedical process or phenomenon. Computational simulation can be done aiming at predictions of the changes. The predictions of structural change and the consequent phenomenon after nanolevel interaction are good examples (such as interaction between molecule and its receptor ligand, antibody and antigen reaction, etc.)

Source: Haddish-Berhane, N. et al., *Int. J. Nanomed.*, 2(3), 315, 2007; Saliner, A.G. et al., *IDrugs*, 11(10), 728, 2008.

medicine and is the core strength of the present medical society. Hence, in silico nanoin-formatics techniques can help both in solving the nanomedical problem and predicting the nanomedical phenomenon (Table 10.1) [3,4].

At present, there are many new computational techniques that can be applied in nano-medicine research. To select the good technique is the art. Of several approaches, the author hereby discusses on the computational nanophotonics and nanoplasmonics, the two newest modalities of computational nanomedicine. Briefly, nanophotonics is a specific applied nanoscience that specifically deals with the behavior of light or optics on the nanometer scale. Applied in medicine, the nanophotonics can be useful in several aspects including ophthalmology, radiology, and laboratory medicine. A specific part of nanopho-tonics, nanoplasmonics can also be applied with computational informatics technology. In this chapter, we hereby will discuss on how computational nanophotonics and nanoplasmonics can be useful in nanomedicine. Also, a summary on important reports on application of computational nanophotonics and nanoplasmonics in nanomedicine is given in this chapter.

10.2 Overview of Nanophotonics and Nanoplasmonics and Its Application in Medicine

Nanophotonics is a new branch of nanoscience. By term nanophotonics is the study of photon or light at the nanoscale. Sometimes nanophotonics can be simply called as nano-optics.

This nanoscience is the advanced optical science. The science focuses on the optical engineering dealing with optics as well as interaction between light or photon and substances at nanolevel. Indeed, when one talks about light, the wavelength of the light should be mentioned. Generally, the wavelength is measured in nanometer; hence, it is not surprising that the study on light is an actual study in nanoscience [5]. In general, nano-photonics focus interest on both basic and applied issues. Focusing on basic issues, the fundamentals and principles of light and its interaction with particles at nanolevel are deeply studied. Focusing on applied issue, the devices that deals with optical nanoengineering is the main focus. Such newly developed devices are the examples of applied nanophotonics.

Focusing in medicine, nanophotonics can be used and applied in optical issue in medicine. Of several approaches, it is mainly set in ophthalmology, radiology, and laboratory medicine. Light is the basic thing for "seeing"; hence, the main focus is on visualization. This also includes the attempt to visualize at extraordinary scale in medicine. Such visualization that is widely used in medicine is that of small objects. This is the story of microscopy, a specific branch of laboratory medicine. The specific science of clinical microscopy has been set for many years and has become a core part in modern medical science. The nanophotonics techniques can be used in clinical microscopy [6]. Many new microscopic tools can work at nanoscale. The good examples include electronic electron microscopy, near-field scanning optical microscopy, photo-assisted scanning tunneling microscopy, and scanning probe microscopy. The attempt to develop better and better microscope for medical use at present can be done mainly by the knowledge on nanophotonics. Also, the nanophotonics can be applied as a diagnostic tool in laboratory medicine. The general principle of detection that depends of visualization can be seen in some specific laboratory tests especially for those with flow cytometry principle. The detection of particles or cells in flow cytometry system is usually based on detection of light scattering pulse. Hence, the application of nanophotonics in laboratory medicine in this issue can be expected. The new generation of flow cytometry–based tools as microfluidics and nanofluidics is a good example [7].

Focusing on radiology, the ray that can be classified as another form of photon can be studied via nanophotonics. X-ray spectroscopy is a good example of application in radiology [5,8]. The study of crystallography of medical molecules (such as new drugs or enzymes) can be based on the advanced nanophotonics technology. Focusing on normal visualization, nanophotonics can be the new principle in ophthalmology. Since ophthalmology is the study of eye and visualization, it is no doubt that nanophotonics can be applied [5]. Finally, adding to simple medical technology, nanophotonics can also help the medical communication. Since the optics is the new face of present communication. The nanophotonics can also be the new hope for communication. It can provide low-power, high-speed, interference-free devices [9]. Electro-optical and all-optical switches on a chip are good examples.

Next, the specific part of nanophotonics, nanoplasmonics will be further specifically discussed. By definition, nanophotonics is the specific study on multitude of nano-optical phenomena that is caused by resonant surface plasmons (SPs) localized in nanosystems [10,11]. This has to be based on advanced physics principle. Generally, the scenario in nanoplasmonics can be induced either by an external laser source exciting SPs in a nanosystem or by a stimulated emission of radiation (spaser). Similar to general nanophotonics, several new devices of nanoplasmonics have been developed for a few years. These tools can be useful for advanced scientific studies including medical studies. In medicine, the application in laboratory medicine can be seen. The use in clinical microscopy is the main application. Some new microscopes based on nanophotonic devices such as near-field scanning microscopy with chemical resolution and SP microscopy are introduced. With the use of nanoplasmonics tool, the detection of nanobiochemical molecule can be possible [12]. Furthermore, the application on the flow cytometry–based microfluidics and nanofluidics as nanochip is also available [13,14].

As mentioned earlier, the use of nanophotonics and nanoplasmonics in medicine can be either simple or advanced usage. The simple usage is the use of basic nanophotonics and nanoplasmonics tool for studying of real objects or phenomena. For more advanced usage, nanophotonics and nanoplasmonics can be used for studying of the imaginary objects or phenomena. This is an actual story of simulating or computational informatics technology, which will be further discussed in the next section. At this point, the reader might still not be able to imagine the exact application of nanophotonics and nanoplasmonics in medicine. To help the reader better understand the actual existing applications, some interesting reports on medical nanophotonics and nanoplasmonics applications will be further discussed.

10.2.1 Application in Clinical Microscopy

It is no doubt that both nanophotonics and nanoplasmonics are useful for clinical microscopy. This is the real advent in laboratory medicine. The application of nanophotonics and nanoplasmonics in clinical microscopy is now well established as a tool for advanced research, and the number of publications in which it is used is increasing rapidly.

The lists of important new interesting publications in this area are hereby presented:

- Fujikawa et al. [12] successfully used low-energy electron microscope spectromicroscopy for assessment of SPs, localized on micro- and nanoscale epitaxial Ag islands.
- Wells et al. [15] studied the use of silicon nanopillars for field-enhanced surface spectroscopy. Wells et al. reported that the enhancement factors derived from analysis of the acquired fluorescence microscopy images showed better enhancements from using silicon nanopillar structures than those typically achieved using plasmonic surface-enhanced fluorescence structures without the limitations of the metal-based substrates [15].
- Selmke et al. [16] proposed for further application of nanolens on development of single-molecule absorption microscopy based on photothermal single-molecule detection principle.
- Frimmer et al. [17] reported "an experimental technique to map and exploit the local density of optical states of arbitrary planar nanophotonic structures." This new method for scanning emitter lifetime imaging microscopy is based on "positioning a spontaneous emitter attached to a scanning probe deterministically and reversibly with respect to its photonic environment while measuring its lifetime [17]."
- Yarrow et al. [18] studied the sub-wavelength infrared imaging of lipids using combining optics and scanning probe microscopy. The nanostructure of lipids can be revealed within this nanophotonics study.

10.2.2 Application in Medical Instrumentation

In additional to the application in basic clinical microscopy, the nanophotonics and nanoplasmonics can also be applied in medical instrumentation. The new tools, especially for new chips and nanofluidics analyzers, can be developed based on nanophotonics and nanoplasmonics. These new devices can be useful in medical science.

The lists of important new interesting publications in this area are hereby presented:

- Chen et al. [13] reported a new real-time template-assisted manipulation of nanoparticles in a multilayer nanofluidic chip. An applied voltage to manipulate nanoparticles in a multilayer nanofluidic chip architecture is the core concept for development of nanochip in this work [13].

- Lovsky et al. [19] presented a new atomic-force-controlled capillary electrophoretic nanoprinting of proteins technique. Combination between combined capillary electrophoresis and atomic force microscope is the core for development of this new protein chip.
- Snoswell et al. [20] described a new technique of electrically induced colloidal clusters for generating shear mixing and visualizing flow in microchannels for possible development of new microfluids analyzer.
- Du et al. [21] reported "a simple and effective nanofabrication method for the pattern transfer of metallic nanostructures over a large surface area on a glass substrate." This technique of transferring can be helpful for further usage in nanofluidics analyzer development [21].
- Fuez et al. [22] described the new technique for "improving the limit of detection of nanoscale sensors by directed binding to high-sensitivity areas." This technique is based on using nanoplasmonics to increase protein-binding ability under physiological conditions in a label-free manner [22].
- Eftekhari et al. [23] introduced a combination between nanofluidics and nanoplasmonics for SP resonance (SPR). Eftekhari et al. concluded that "The flow-through format enables rapid transport of reactants to the active surface inside the nanoholes, with potential for significantly improved time of analysis and biomarker yield through nanohole sieving [23]."
- Rodríguez-Cantó et al. [24] reported experimentally the use of an array of gold nanodisks on functionalized silicon for chemosensing purposes. The developed metallic nanostructures could provided a very strong plasmonic resonance resulting in highly sensitive sensing [24].

10.2.3 Application in Clinical Oncology

The aforementioned applications of nanophotonics and nanoplasmonics are usually relating to the diagnostic purpose. The application of nanophotonics and nanoplasmonics for help diagnose in clinical oncology is also possible. A good example is the high-resolution-based fluorescence approach [25]. However, the nanophotonics and nanoinformatics can be useful for the therapeutic purpose as well. The application in cancer therapy can be a good example. This can add up to the basic phototherapy in clinical oncology.

The lists of important new interesting publications in both diagnostic and therapeutic issues in clinical oncology area are hereby presented:

- Lukianova-Hleb et al. [26] described a new plasmonic nanobubble for optically guided selective elimination of the target cancerous tissue with micrometer precision.
- Lukianova-Hleb et al. [27] reported on tunable plasmonic nanoprobes for theranostics of prostate cancer.
- Nguyen et al. [28] reported on the biological response to gold nanoshell-enabled photothermal therapy. In this work, Nguyen et al. observed that pro-inflammatory cytokines could not be sufficiently induced by the used technique [28].

TABLE 10.2

Application of Computational Nanophotonics and Nanoplasmonics in Medicine

Applications	Details
1. Structural approach	This application is corresponding to the structural question. Since the nanophotonics and nanoplasmonics can be applied in diagnostic purpose especially for visualization. The study on the structure via the computational approach is possible.
2. Functional approach	This application is corresponding to the functional problem. It is no doubt that there are many photophenomena in medicine and these phenomena cannot be assessed by the naked eye or simple technique. The use of nanophotonics and nanoplasmonics can be useful in simulating in this area.

10.3 Computational Nanophotonics and Nanoplasmonics

Basically light or photon relating phenomenon is the actual nanoscale phenomenon as already mentioned. It is no doubt that the phenomenon in this scale cannot be assessed by the naked eye or simple micro-assessment. Due to the limitation of the classical technique, the application of the nanotechnique has to be set. Several tools including the nanophotonics and nanoplasmonics tools can be helpful for solving the problem. The technique for assessing the nanolevel medical phenomena, either structural or functional aspects (Table 10.2), is required. In silico technique is a new acceptable solution [3,4]. In informatics technology, a specific nanoscience called "nanoinformatics" is available and this help approach nanoscale phenomena in nanomedicine [29]. Computational approach is required in this approaching.

Focusing on the concepts of nanophotonics and nanoplasmonics at the present computer era, the computational modeling and designing of focused objects is possible, and this becomes the new issue of the analysis. Applied as a bridge between physics and biology, the connection between engineering and medicine can be done, and the hybrid medical engineering using nanophotonics and nanoplasmonics is useful. For sure, there are many advantages of computational nanophotonics and nanoplasmonics in medicine. Simply the application of computer technique can be either for clarification or prediction as already noted. More details on examples of important reports on this topic can be seen in the next section.

10.3.1 Summary of Important Reports on Application of Computational Nanophotonics and Nanoplasmonics in Nanomedicine

As noted earlier, computational nanophotonics and nanoplasmonics have many advantages in nanomedicine. Here, the author will summarize important reports on application of computational nanophotonics and nanoplasmonics in nanomedicine (Table 10.3) [13,30–32].

1. Application for clarification

   A good example of application of nanophotonics and nanoplasmonics is the usage in nanomaterials. Clarification of new nanomaterial structure is possible [6]. The applied computational technology in advanced microscope is the best example.

TABLE 10.3

Some Important Reports on Computational Nanophotonics and Nanoplasmonics in Nanomedicine

Authors	Details
Chen et al. [13]	Chen et al. introduced a new real-time template-assisted manipulation of nanoparticles in a multilayer nanofluidic chip. In this work, Chen et al. used computational simulation of the nanoparticles' motion in the nanofluidic chip to help validate the new technique.
Yeng et al. [30]	Yeng et al. reported an approach for high-temperature nanophotonics to help new nanomaterial, high-purity tungsten in a precisely fabricated photonic crystal slab geometry design.
Khraiche et al. [31]	Khraiche et al. presented a new ultrahigh-sensitivity photodetector technology for nanoscale control over photodetector topography with the potential to reproduce the visual acuity of the retina that can support the computational design of retinal prosthesis.
Taylor et al. [32]	Taylor et al. studied on using cucurbit[n]uril "glue" for assembling subnanometer plasmonic junctions for SERS within gold nanoparticle [32]. Computational simulation was used in this work for prediction [32].

2. Application for prediction

A good example of application of computational nanophotonics and nanoplasmonics is the new nanomaterial design. Prediction of action of nanoparticles under different conditions is also possible.

As a summary, it can be seen that nanophotonics and nanoplamonics are applicable for various section of nanomedicine:

- In diagnosis (nanodiagnosis)

Nanophotonics and nanoplasmonics in nanodiagnosis is the main researching area at present. The aim is usually the visualization as already mentioned.

- In treatment (nanotherapy)

The use of nanophotonics and nanoplasmonics in nanotherapy is possible. The use in cancer therapy in clinical oncology as already mentioned is a good example.

10.4 Computational Nanophotonics and Nanoplamonics Tools for Management of Biomedical Data

Based on computational technology, a short time is required for manipulation of biomedical data. To manipulate these biomedical data, computational tools are required to help fasten the process. A selection of proper tool that fit for a specific work is the important key for success. It is necessary that practitioners have to understand the presently available tools. Here, the author will briefly summarize some important available nanophotonics and nanoplasmonics tools. These computational tools can be useful for the work in nanomedicine.

10.4.1 Usefulness of Computational Nanophotonics and Nanoplamonics in Nanomedicine

Basically, nanophotonics and nanoplasmonics are the sciences that deal with light. This is the real story of nanoscale phenomena. Using computational nanophotonics and nanoplasmonics to solve a problem or to simulate a case is possible and can be an effective solution to biomedical research questions relating to photon and optical issue.

Focusing on the specific program for help computational nanophotonics and nanoplasmonics analysis, several programs can be selected. MATLAB® is a good basic example that can be effectively used. It helps perform computational analysis with very few noises. Structural clarification and prediction of phenomena can be done. The ways that MATLAB can be used to support the nanophotonics and nanoplasmonics problems will be further discussed.

10.4.1.1 Creating of Graphical Model

MATLAB can help generation of graphical model for referencing in nanophotonics and nanoplasmonics. MATLAB can help create a 3D model as either contour, mesh, or surface plots. For example, this is a case of using MATLAB to create a model of a nanolaser irradiated blood cell in nanofluidics analyzers. The MATLAB code in a MATLAB Command Window can be

```
≫ [x, y, z] = peaks;
≫ c = contour (x, y, z, 125);
≫ clabel(c);
≫ title ('2-D Countour plot of nanolaser irradiated blood cell with
clable');
```

The result of this example of 2D graphical model in nanophotonics created by MATLAB is shown in Figure 10.1.

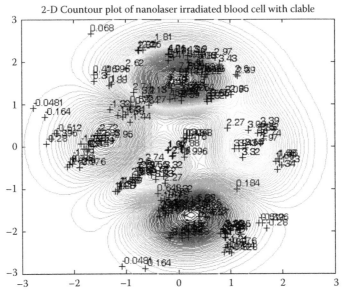

FIGURE 10.1
Example of a 2D graphical model in nanophotonics created by MATLAB®.

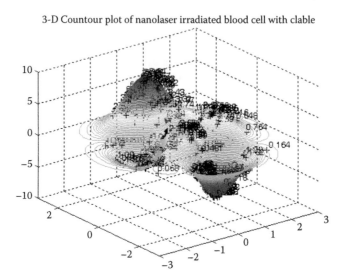

FIGURE 10.2
Example of a 3D graphical model in nanophotonics created by MATLAB.

Figure 10.2 shows an example of a 3D graphical model in nanophotonics created by the following MATLAB code:

```
≫ [x, y, z] = peaks;
≫ c = contour3 (x, y, z, 125);
≫ clabel (c);
≫ title ('3-D Countour plot of nanolaser irradiated blood cell with
clabel');
```

10.4.1.2 Solving the Problem of Differential and Integral Equation

Differentiation and integration are common mathematical manipulation in nanophotonics and nanoplasmonics. Solving these mathematical problems via MATLAB is possible. For example, this is a case of using MATLAB for solving the equation of size decreasing of cancer cells after response to nanophototherapy. An example code is hereby shown:

```
≫ p = - 2; delta = 0.02; y(1) = 1.5;
≫ k = 0;
≫ for X = [delta: delta: delta: 0.5]
k = k + 1;
y (k + 1) = y(k) + p * y(k) * delta;
end
≫ x = [0: delta: 0.5];
≫ y = 3 * exp (-4 * x);
≫ y_true = 1.5 * exp (-2 * x);
≫ plot (x, y, '*', x, y_true, '∧');
≫ legend ('predicted size', 'observed size');
```

Figure 10.3 shows the mathematical equation problem solved by the aforementioned MATLAB code.

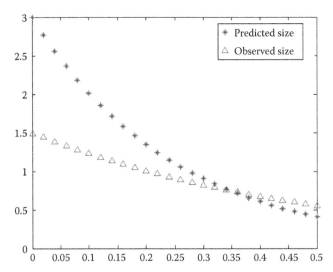

FIGURE 10.3
Example of a mathematical equation problem solved by MATLAB.

10.4.2 Important Computational Nanophotonics and Nanoplasmonics Tool for Management of Biomedical Data

There are several computational nanophotonics and nanoplasmonics tools for management of biomedical data. Some important tools that are specific will be hereby discussed:

1. MCML [33]

   Monte Carlo model of steady-state light transport in multilayered tissues (MCML) is an advanced program that is coded in ANSI standard [33]. This is for modeling light transportation in thick tissues [33]. Simulating data for separated grid element in the radial and angular directions can be performed based on MCML [33].

2. SARFIA [34]

   SARFIA is a newly developed tool that is designed for dynamic imaging experiments [34]. Detection of structures by SARFIA is based on Laplace operator, clustering, similar functional responses classification, and a database for storage and exchange [34].

3. MICAD [35,36]

   Molecular Imaging and Contrast Agent Database (MICAD) is an online tool [35,36]. It is a source on molecular imaging and contrast agents with confirmed in vivo data published in several international peer-reviewed scientific journals [35,36]. It can be a good database in nanophotonics [35,36].

4. CLRP [37]

   Carleton Laboratory for Radiotherapy Physics (CLRP) brachytherapy database is an online tool [37]. It is a specific tool on dosimetry of photon in photobrachytherapy [37]. It can be applied in radiology for cancer treatment [37]. This database tool is accessible via http://www.physics.carleton.ca/clrp/seed_database [37].

5. OME [38,39]

Open Microscopy Environment (OME) is a specific tool which is developed for analysis of optical microscope image data [38]. It can be a basic tool in biological image informatics and nano-optics [38,39].

6. KinetXBase [40]

KinetXBase is a new computational database for analyzing complex cAMP binding data and highly site-specific cAMP analogues [40]. The data are partially from collected SPR information [40].

7. ETMICRO [41]

Electron Transport code for MICROdosimetry (ETMICRO) is a tool for tracing the electrons in liquid water from collected from optical data set [41].

10.5 Common Applications of Computational Nanophotonics and Nanoplasmonics in General Nanomedicine Practice

1. Situation 1: Using computational nanophotonics and nanoplasmonics to determine flow pattern in flow cytometry

   This is an example on using computational nanophotonics and nanoplasmonics in nanomedicine. As already mentioned, the changes in light properties are the main determinant in functioning of a flow cytometry system. With continuous determination of such change, flow cytometry becomes a medical device in medical laboratory that is useful for tracking of small particles, especially for blood and cancerous cells. Tracing of change of cells during analysis can be done based on the principle of computational nanophotonics and nanoplasmonics. A good example is graphical modeling of the cells that pass irradiation by laser of the flow cytometry as shown in previous section. The finite element analysis principle can be successfully used in tracking of such changes [42]. Indeed, the study on the structural aspect in flow cytometry analytical process can give the information for further successful development of lab-on-a-chip [43]. Finally, the standardization and calibration are the main issues in biophotonics of cytometry that must not be forgotten [44].

2. Situation 2: Using computational nanophotonics and nanoplasmonics for finding the structure of hemoglobin

   This is another example on using computational nanophotonics and nanoplasmonics on medical structure. Basically, hemoglobin is a small nanomolecule that cannot be visualized by simple microscopy technique. Infrared optics with an atomic force microscope is required for getting infrared absorption imaging of hemoglobin [45]. This is very expensive and still has some limitations in study in metalloprotein of some rare abnormal hemoglobins [45]. However, with use of the computational approach, the hemoglobin imaging can be modeled in silico. This can be helpful in studying of molecular image of an extremely rare hemoglobin disorder such as hemoglobin Pakse [46].

10.5.1 Examples of Nanomedicine Researches Based on Computational Nanophotonics and Nanoplasmonics Application

Example 10.1: A study to determine the structure of hemoglobin South Florida

Basically, the biophotonics can be useful in studying hemoglobin. Basically, hemoglobin is an important metalloprotein within red blood cell. This metalloprotein has its main role in oxygen transportation process. The two main kinds of hemoglobin defects are thalassemia and hemoglobinopathy. The understanding on the pathogenesis of each abnormal hemoglobin defect is important in medicine and can be useful for development of new diagnostic and therapeutic approaches.

Of several hemoglobin defects, there are some rare hemoglobin defects that are less mentioned. Due to the difficulty in classical examinations, requirement of complex analyzers, and rarity of the hemoglobin sample, the computational nanophotonics can be the solution for this case. Here, the author focuses an interest on an uncommon hemoglobinopathy, namely, hemoglobin South Florida. The computational structural analysis of this hemoglobinopathy is hereby shown. Pathophysiologically, codon 1 GTG–>ATG is the main genetic underlying factor that results in hemoglobin South Florida [47–49]. This hemoglobin has no abnormal elongation but a single point changing. It can lead to confusion with hemoglobin A1C in electrophoresis and can be usually forgotten [49]. The abnormal changing part within the beta globin of hemoglobin South Florida molecule is believed to be the causal factor for the abnormal clinical presentation. Interaction with other hemoglobinopathies such as hemoglobin E might result in several thalassemic manifestations [50]. Although the primary structure of hemoglobin South Florida is well known, the specific knowledge on its secondary and tertiary structure is not well mentioned. In this work, a standard bioinformatic approach is used for assessing the secondary and tertiary structures based on the available information on amino acid sequence of this hemoglobin. A standard computer-based study for protein structure modeling is used in the present analysis.

The author starts the bioinformatics analysis by database searching and simulating for the abnormal primary amino acid sequence. Then the secondary structure was predicted using NNPREDICT server [51] followed by tertiary structure modeling by CPH models 2.0 server [52]. The calculated secondary and tertiary structures are shown in Figure 10.4.

Based on the prediction, the abnormal helix occurs at the starting point of the chain, which is according to the underlying abnormal primary structure within the hemoglobin South Florida. This abnormality is believed to be the underlying etiology of the detected abnormality. Indeed, retention of initiator methionine and N-alpha-acetylation might be due to this abnormal structure that leads to further abnormal biological processing [48].

Example 10.2: A generation of graphical model for referencing in pulse oximetry

A classical well-known diagnostic tool in medicine that is based on biophotonics principle is the pulse oximetry [53,54]. Pulse oximetry is a specific tool for rapid measurement of oxygenation. The measurement is via the skin translucent principle. The sensor of this kind is called optode in laboratory medicine. Pulse oximetry is presently widely used

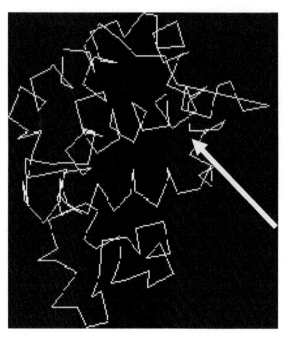

FIGURE 10.4
Calculated secondary structures of hemoglobin South Florida.

in general practice, especially in the emergency room [53]. It is useful for monitoring of hypoxia that can be seen in several conditions especially for lung disease, heart disease, and shock.

Here, the author performed an online simulation on the pulse oximetry test on the infantile skin situation. The standard MCML-type tool is used for all simulating study in this work. The tool is simply available online and supported by OTAGO research group (http://biophotonics.otago.ac.nz). The tool allows parameter adjustment to fit the simulating condition. The adjustment condition in this work is presented in Table 10.4. The primary conditioning is source diameter = 50, detector diameter = 100, separation = 200, and detector numerical aperture = 15. According to the simulation, the standard distribution of photon can be seen. The epidermis gets the most intensified photon with the lower and lower intensity varying by depth under the external skin of the infantile (Table 10.4).

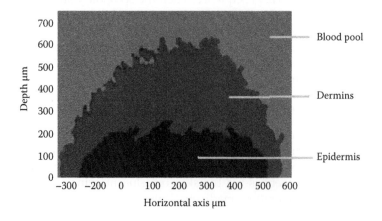

TABLE 10.4

Parameter Adjustment in the Simulation Study on Pulse Oximetry in Infantile Scenario

Layers	Scattering Coefficient	Absorption Coefficient	Refractive Index	Layer Thickness
Epidermis	45	0.1	1.3	50
Dermis	30	0.1	1.4	125
Blood pool	25	0.1	1.2	25

Example 10.3: A study to analyze the effect of photon stimulation of the nano-membrane of retina in a case with dengue fever

Basically, the effect of photon stimulation can be seen in various ways in the human body. Photon or light is exposed to all external parts of our body. In visualization, the most important part for every human being is the retina. Photon has to reach retina and stimulate and further cause the neurological signal for completeness of visualization process [55,56]. While it is a useful process, it is not completely understood at present.

Based on the advanced nanotechnology, the modeling of the change of retina membrane corresponding to the photon stimulation can be possible. The change of pattern in case with dengue fever is hereby focused. The simple computational finite element analysis approach can be used. A simple computer program, Excel, is hereby used. The membrane is primarily assumed as a very small and flat structure. The simple model of 2D uniform thickness structure is hereby used. The result can be seen in Table 10.5. It can be seen that there is an irregular change in photon flux on the nano-membrane of the retina when there is dengue fever.

Indeed, the dengue-induced visual complication is observed and it is still a myth [57–59]. Based on a recent study, the diffuse retinal thickening could be observed, and this might correlate to the finding of irregular flux change on retina membrane when there is high fever [59]. Additional recent electrophysiological findings on retina also support this fact [58]. The heterogeneous pattern of electroretinograms can be seen in dengue retinopathy [58]. This is also concordant with the finding on heterogeneous predicted flux in the simulation.

TABLE 10.5

Model of Photon Flux Change in Retinal Membrane due to Dengue Fever

A. At normal physiological stage

	Result at normal stage			
	1	1	1	
1	1	1	1	1
1	1	1	1	1
1	1	1	1	1
	1	1	1	

B. With dengue fever (assuming at 38°C)

	Result at pathological stage			
	1.0354	1.0354	1.0354	
1	1.0152	1.0187	1.0152	1
1	1.0067	1.0089	1.0067	1
1	1.0025	1.0035	1.0025	1
	1	1	1	

TABLE 10.6

Wavelength and Insertional Loss

Wavelength (nm)	Insertional Loss (dB)
400	−45.7
440	−44.2
480	−43.12
520	−42.99
560	−42.71

Example 10.4: A study on different wavelength application on nanosensor diagnostic property

In the present day, the nanosensor is new wave of diagnostic tools for nanodiagnosis. The development of nanosensor based on novel nanomaterial helps provide new generation of diagnostic tool. A good example is the surface acoustic wave (SAW) nanosensor. The main determinant of this kind of nanosensor is pH or hydrogen concentration. The finalized determined concentration at electrode varied on wavelength and frequency of the sensor. It can be written as "concentration = k X wavelength × frequency."

An example of a relationship between change wavelength and finalized insertional loss is presented in Table 10.6. With increasing the wavelength concentration, the insertion loss increases and this means reduction of the finalized concentration detected by nanosensors. This is an important concern for selection of proper wavelength for implementing in new nanosensors in medicine at present [60].

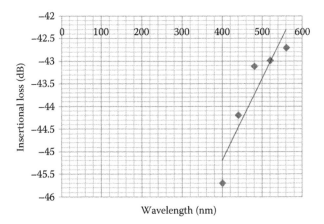

Wavelength (nm)

Example 10.5: A study on circular dichroism spectra of quantum dot–antibody conjugates

The quantum dot is the new technology applying nanophotonics for the diagnostic purpose in medicine. As a nanodiagnostic tool, it can provide simple and fast diagnosis. The development of new quantum dot molecule is the subject of present research in advanced laboratory medicine. Basic principles such as conjugation can be used for modification of the quantum dot. The use of quantum dot and antibody conjugation can be a good alternative approach. In developmental process, circular dichroism (CD) spectra analysis becomes an important step for validation of the new diagnostic tool [61]. The basic approach is the classical spectrometry. However, this is sometimes difficult and requires a complicated analyzer. The use of computational nanophotonics and nanoplasmonics approach can be the solution.

TABLE 10.7

Results from CD Spectra Analysis
of Quantum Dot–Antibody
Conjugation

Random: > 0.58
Square distance: 32.20
Max error: 0.080

Wavelength	Original	Computed
200	1.65	0.66
201	−0.63	−1.43
202	−2.64	−3.14
203	−4.22	−4.64
204	−5.38	−6.03
205	−6.44	−7.10

Here, the author uses an online analysis tool K2d to help predict the CD of quantum dot–antibody conjugation. The starting condition is the original quantum dot and the predicted condition is the conjugated molecule. The primary control is on pH and temperature. The given values of secondary structure are alpha helix = 0.30 and beta strand = 0.12. Table 10.7 shows the results from CD spectra analysis of quantum dot–antibody conjugation. The results can show favorable predicted CD spectra showing that the secondary structure remained well preserved after the conjugation.

10.6 Conclusion

Computational technology is applicable in nanomedicine. With the use of computational approach, several manipulations can be done aiming at clarifying and answering many nanomedical problems. Of several computational approaches in nanomedicine, the usage of computational nanophotonics and nanoplasmonics are of interest. These computational approaches are new and can be applied in medicine. Computational nanophotonics and nanoplasmonics can help medical scientists answer the complex questions in nanophotonics and nanoplasmonics. Both structural and functional approaches can be performed. Imaginary clarification and prediction via applied simulations can be derived from using computational nanophotonics and nanoplasmonics techniques. In addition, there are many available computational nanophotonics and nanoplasmonics tools for management of biomedical data at present. These new computational nanophotonics and nanoplasmonics tools can be very useful for many studies in nanomedicine.

References

1. Gehlenborg N, O'Donoghue SI, Baliga NS, Goesmann A, Hibbs MA, Kitano H, Kohlbacher O et al. Visualization of omics data for systems biology. *Nat Methods*. 2010 March;7(3 Suppl):S56–S68.

2. Haarala R, Porkka K. The odd omes and omics. *Duodecim.* 2002;118(11):1193–1195.

3. Haddish-Berhane N, Rickus JL, Haghighi K. The role of multiscale computational approaches for rational design of conventional and nanoparticle oral drug delivery systems. *Int J Nanomedicine.* 2007;2(3):315–331.

4. Saliner AG, Poater A, Worth AP. Toward in silico approaches for investigating the activity of nanoparticles in therapeutic development. *IDrugs.* 2008 October;11(10):728–732.

5. Behari J. Principles of nanoscience: an overview. *Indian J Exp Biol.* 2010 October;48(10):1008–1019.

6. Colliex C. From electron energy-loss spectroscopy to multi-dimensional and multi-signal electron microscopy. *J Electron Microsc (Tokyo).* 2011;60(Suppl 1):S161–S171.

7. Wu J, Gu M. Microfluidic sensing: State of the art fabrication and detection techniques. *J Biomed Opt.* 2011 August;16(8):080901.

8. Benson O. Assembly of hybrid photonic architectures from nanophotonic constituents. *Nature.* 2011 December;480(7376):193–199.

9. Brimont A, Thomson DJ, Sanchis P, Herrera J, Gardes FY, Fedeli JM, Reed GT, Martí J. High speed silicon electro-optical modulators enhanced via slow light propagation. *Opt Express.* 2011 October;19(21):20876–20885.

10. Duan H, Fernández-Domínguez AI, Bosman M, Maier SA, Yang JK. Nanoplasmonics: Classical down to the nanometer scale. *Nano Lett.* 2012 February;12(3):1683–1689.

11. Stockman MI. Nanoplasmonics: Past, present, and glimpse into future. *Opt Express.* 2011 October;19(22):22029–22106.

12. Fujikawa Y, Sakurai T, Tromp RM. Surface plasmon microscopy using an energy-filtered low energy electron microscope. *Phys Rev Lett.* 2008 March;100(12):126803.

13. Chen HM, Pang L, Gordon MS, Fainman Y. Real-time template-assisted manipulation of nanoparticles in a multilayer nanofluidic chip. *Small.* 2011 October;7(19):2750–2757.

14. Sannomiya T, Vörös J. Single plasmonic nanoparticles for biosensing. *Trends Biotechnol.* 2011 July;29(7):343–351.

15. Wells SM, Merkulov IA, Kravchenko II, Lavrik NV, Sepaniak MJ. Silicon nanopillars for field enhanced surface spectroscopy. *ACS Nano.* 2012 March;6(4):2948–2959.

16. Selmke M, Braun M, Cichos F. Photothermal single-particle microscopy: Detection of a nanolens. *ACS Nano.* 2012 February;6:2741–2749.

17. Frimmer M, Chen Y, Koenderink AF. Scanning emitter lifetime imaging microscopy for spontaneous emission control. *Phys Rev Lett.* 2011 September;107(12):123602.

18. Yarrow F, Kennedy E, Salaun F, Rice JH. Sub-wavelength infrared imaging of lipids. *Biomed Opt Express.* 2010 December;2(1):37–43.

19. Lovsky Y, Lewis A, Sukenik C, Grushka E. Atomic-force-controlled capillary electrophoretic nanoprinting of proteins. *Anal Bioanal Chem.* 2010 January;396(1):133–138.

20. Snoswell DR, Creaton P, Finlayson CE, Vincent B. Electrically induced colloidal clusters for generating shear mixing and visualizing flow in microchannels. *Langmuir.* 2011 November;27(21):12815–12821.

21. Du K, Wathuthanthri I, Mao W, Xu W, Choi CH. Large-area pattern transfer of metallic nanostructures on glass substrates via interference lithography. *Nanotechnology.* 2011 July;22(28):285306.

22. Feuz L, Jönsson P, Jonsson MP, Höök F. Improving the limit of detection of nanoscale sensors by directed binding to high-sensitivity areas. *ACS Nano.* 2010 April;4(4):2167–2177.

23. Eftekhari F, Escobedo C, Ferreira J, Duan X, Girotto EM, Brolo AG, Gordon R, Sinton D. Nanoholes as nanochannels: Flow-through plasmonic sensing. *Anal Chem.* 2009 June;81(11):4308–4311.

24. Rodríguez-Cantó PJ, Martínez-Marco M, Rodríguez-Fortuño FJ, Tomás-Navarro B, Ortuño R, Peransí-Llopis S, Martínez A. Demonstration of near infrared gas sensing using gold nanodisks on functionalized silicon. *Opt Express.* 2011 April;19(8):7664–7672.

25. Garcia-Parajo MF. The role of nanophotonics in regenerative medicine. *Methods Mol Biol.* 2012;811:267–284.

26. Lukianova-Hleb EY, Koneva II, Oginsky AO, La Francesca S, Lapotko DO. Selective and self-guided micro-ablation of tissue with plasmonic nanobubbles. *J Surg Res.* 2011 March;166(1):e3–e13.

27. Lukianova-Hleb EY, Oginsky AO, Samaniego AP, Shenefelt DL, Wagner DS, Hafner JH, Farach-Carson MC, Lapotko DO. Tunable plasmonic nanoprobes for theranostics of prostate cancer. *Theranostics*. 2011 January;1:3–17.

28. Nguyen HT, Tran KK, Sun B, Shen H. Activation of inflammasomes by tumor cell death mediated by gold nanoshells. *Biomaterials*. 2012 March;33(7):2197–2205.

29. De La Iglesia D, Chiesa S, Kern J, Maojo V, Martin-Sanchez F, Potamias G, Moustakis V, Mitchell JA. Nanoinformatics: New challenges for biomedical informatics at the nano level. *Stud Health Technol Inform*. 2009;150:987–991.

30. Yeng YX, Ghebrebrhan M, Bermel P, Chan WR, Joannopoulos JD, Soljacic M, Celanovic I. Enabling high-temperature nanophotonics for energy applications. *Proc Natl Acad Sci USA*. 2012 February;109(7):2280–2285.

31. Khraiche ML, Lo Y, Wang D, Cauwenberghs G, Freeman W, Silva GA. Ultra-high photosensitivity silicon nanophotonics for retinal prosthesis: Electrical characteristics. *Conf Proc IEEE Eng Med Biol Soc*. 2011 August;2011:2933–2936.

32. Taylor RW, Lee TC, Scherman OA, Esteban R, Aizpurua J, Huang FM, Baumberg JJ, Mahajan S. Precise subnanometer plasmonic junctions for SERS within gold nanoparticle assemblies using cucurbit[n]uril glue. *ACS Nano*. 2011 May;5(5):3878–3887.

33. Wang L, Jacques SL, Zheng L. MCML—Monte Carlo modeling of light transport in multi-layered tissues. *Comput Methods Programs Biomed*. 1995 July;47(2):131–146.

34. Dorostkar MM, Dreosti E, Odermatt B, Lagnado L. Computational processing of optical measurements of neuronal and synaptic activity in networks. *J Neurosci Methods*. 2010 April;188(1):141–150.

35. Chopra A, Shan L, Eckelman WC, Leung K, Latterner M, Bryant SH, Menkens A. Molecular imaging and contrast agent database(MICAD): Evolution and progress. *Mol Imaging Biol*. 2012 February;14(1):4–13.

36. *Molecular Imaging and Contrast Agent Database(MICAD)[Internet]*. Bethesda (MD): National Center for Biotechnology Information (US). 2004–2013.

37. Taylor RE, Rogers DW. EGSnrc Monte Carlo calculated dosimetry parameters for 192Ir and 169Yb brachytherapy sources. *Med Phys*. 2008 November;35(11):4933–4944.

38. Swedlow JR, Goldberg I, Brauner E, Sorger PK. Informatics and quantitative analysis in biological imaging. *Science*. 2003 April;300(5616):100–102.

39. Goldberg IG, Allan C, Burel JM, Creager D, Falconi A, Hochheiser H, Johnston J, Mellen J, Sorger PK, Swedlow JR. The open microscopy environment (OME) data model and XML file: Open tools for informatics and quantitative analysis in biological imaging. *Genome Biol*. 2005;6(5):R47.

40. Schweinsberg S, Moll D, Burghardt NC, Hahnefeld C, Schwede F, Zimmermann B, Drewianka S et al. Systematic interpretation of cyclic nucleotide binding studies using KinetXBase. *Proteomics*. 2008 March;8(6):1212–1220.

41. Kim EH. Electron track simulation using ETMICRO. *Radiat Prot Dosimetry*. 2006;122(1–4):53–55.

42. Huang L, Maerkl SJ, Martin OJ. Integration of plasmonic trapping in a microfluidic environment. *Opt Express*. 2009 April;17(8):6018–6024.

43. Reddy JN. *Introduction to the Finite Element Method*, 2nd edn. Boston, MA: McGraw-Hill Science, 1993.

44. Mittag A, Tárnok A. Basics of standardization and calibration in cytometry—A review. *J Biophoton*. 2009 September;2(8–9):470–481.

45. Kennedy E, Yarrow F, Rice JH. Nanoscale spectroscopy and imaging of hemoglobin. *J Biophotonics*. 2011 September;4(9):588–591.

46. Wiwanitkit V. Tertiary structural analysis of the elongated part of an abnormal hemoglobin, hemoglobin Pakse. *Int J Nanomed*. 2006;1(1):105–107.

47. Malone JI, Shah SC, Barness LA. Hemoglobin South Florida: A genetic variant with laboratory recognition of only 20% of its product. *Am J Med Genet Suppl*. 1987;3:227–231.

48. Boissel JP, Kasper TJ, Shah SC, Malone JI, Bunn HF. Amino-terminal processing of proteins: Hemoglobin South Florida, a variant with retention of initiator methionine and N alpha-acetylation. *Proc Natl Acad Sci USA*. 1985 December;82(24):8448–8452.

49. Shah SC, Malone JI, Boissel JP, Kasper TJ. Hemoglobin South Florida. New variant with normal electrophoretic pattern mistaken for glycosylated hemoglobin. *Diabetes*. 1986 October;35(10):1073–1076.
50. Tan JA, Tan KL, Omar KZ, Chan LL, Wee YC, George E. Interaction of Hb South Florida (codon 1; GTG— >ATG) and HbE, with beta-thalassemia (IVS1-1; G— >A): Expression of different clinical phenotypes. *Eur J Pediatr*. 2009 September;168(9):1049–1054.
51. Kneller DG, Cohen FE, Langridge R. Improvements in protein secondary structure prediction by an enhanced neural network. *J Mol Biol*. 1990;214:171–182.
52. Lund O, Nielsen M, Lundegaard C, Worning P. CPH models 2.0: X3M a computer program to extract 3D models. Abstract at the CASP5 conference A102, 2002, Asiloman, (CA).
53. Freeman J. Oxygen measurements in shock. *Int Anesthesiol Clin*. 1966 Spring;4(1):221–243.
54. Lübbers DW. Oxygen electrodes and optodes and their application in vivo. *Adv Exp Med Biol*. 1996;388:13–34.
55. Sung CH, Chuang JZ. The cell biology of vision. *J Cell Biol*. 2010 September;190(6):953–963.
56. Palczewski K. Chemistry and biology of vision. *J Biol Chem*. 2012 January;287(3):1612–1619.
57. Tan SY, Kumar G, Surrun SK, Ong YY. Dengue maculopathy: A case report. *Travel Med Infect Dis*. 2007 January;5(1):62–63.
58. Chia A, Luu CD, Mathur R, Cheng B, Chee SP. Electrophysiological findings in patients with dengue-related maculopathy. *Arch Ophthalmol*. 2006 October;124(10):1421–1426.
59. Teoh SC, Chee CK, Laude A, Goh KY, Barkham T, Ang BS. Eye institute dengue-related ophthalmic complications workgroup. Optical coherence tomography patterns as predictors of visual outcome in dengue-related maculopathy. *Retina*. 2010 March;30(3):390–398.
60. Hu Y, Jenkins RM, Gardes FY, Finlayson ED, Mashanovich GZ, Reed GT. Wavelength division (de)multiplexing based on dispersive self-imaging. *Opt Lett*. 2011 December;36(23):4488–4490. doi: 10.1364/OL.36.004488.
61. Hua XF, Liu TC, Cao YC, Liu B, Wang HQ, Wang JH, Huang ZL, Zhao YD. Characterization of the coupling of quantum dots and immunoglobulin antibodies. *Anal Bioanal Chem*. 2006 November;386(6):1665–1671.

11

Computational Modeling Aspects of Light Propagation in Biological Tissue for Medical Applications

Chintha Chamalie Handapangoda and Malin Premaratne

CONTENTS

11.1 Introduction

The application of optical methods for biomedical applications and clinical therapeutics is emerging as a new technological paradigm (Wilson et al. 2005). Optical techniques for tissue diagnosis that are currently being developed offer significant advantages over standard biopsy and cytology techniques, in terms of both patient care and medical costs (Mourant et al. 1998). A good understanding of the relationship between the biochemical and morphological structure of cells and light scattering will assist the development of diagnostic applications in nanophotonics (Dunn and Kortum, 1996). Many light-based and spectroscopic techniques are already being practiced in medical and other health-care fields.

There has been a rapid increase in the use of ultraviolet (UV), visible, and infrared (IR) radiation in both diagnostic and therapeutic medicine, and this has created a need to understand how this radiation propagates in tissue (Patterson et al. 1991). Such knowledge is necessary for the optimum development of therapeutic techniques and for the quantitative analysis of diagnostic measurements (Patterson et al. 1991). For example, the local tissue temperature is of prime importance in laser surgery and depends, in turn, on the spatial distribution of the incident radiation. This variable is also of central importance in the photodynamic therapy of cancer where the local biological effect is directly related to the light fluence (Patterson et al. 1991). Diagnostic methods, which use fluorescent, scattered, or transmitted light to measure parameters such as drug concentration and blood oxygenation, also require detailed information about the propagation of the excitation and observed light in the nanoscale regime (Patterson et al. 1991).

Optical techniques such as near-infrared (NIR) spectroscopy and mid-IR spectroscopy (which use absorption of light to determine the concentration of substances), Raman spectroscopy (which uses Raman scattering of laser light by target molecules), and photoacoustic spectroscopy (which uses laser excitation of fluids to generate an acoustic response and a spectrum as the laser is tuned) are preferable to invasive techniques, such as the finger-stick test, for detecting and determining the concentration of substances in blood (Waynant and Chenault, 1998).

Light-aided sensing of substances has been receiving tremendous interest recently (Kumar et al. 1996). These techniques of sensing substances in tissue or blood require foreign nanostructures to be embedded/implanted in tissue in order to condition optical signals. For safety reasons, the incident intensity used in these optical detection techniques should not be more than $0.1 \, \text{J/cm}^2$ per laser light pulse, and if continuous exposure is used, it should be less than $1 \, \text{W/cm}^2$ (Burnett, 1969). With this limitation, it is difficult to obtain a detectable signal or spectrum of the concentration of blood. However, nanostructures such as photonic crystals can be implanted within tissue, by which the scattered signal can be enhanced, thus providing a detectable spectrum (Florescu and Zhang, 2005; Gaponenko, 2002). Photonic crystals are low-loss periodic dielectric nanostructures, whose refractive index periodically changes in space, along one direction (1D), along two directions (2D), or along all three axes (3D) (Joannopoulos et al. 1995; Martinez et al. 2004). The refractive index modulation is in a scale that can be compared to the wavelength propagated in the device (Szymanski and Patela, 2005). These photonic crystals result in band gaps for photons, which are analogous to band gaps for electrons in semiconductors. Light with frequencies in the photonic band gap is forbidden to propagate inside the photonic crystal. Different topologies of the structure provide optical band gaps for different polarizations of light (Szymanski and Patela, 2005).

Spontaneous emission arises from the intricate interplay between a radiating system and its surrounding environment, and thus, the spontaneous emission in a photonic crystal can be enhanced, attenuated, or suppressed by changing the density of electromagnetic states at the transition frequency (Fan et al. 1997). It is possible to engineer the photonic density of states in a photonic crystal and this in turn enables enhancement of nonlinear Raman process without increasing the excitation intensity (Florescu and Zhang, 2005). This is possible because the dip in the photon density of states in the band gap coexists with enhanced density of states just outside it (Gaponenko, 2002). That is, the photon density of states is increased near the band edges (Enoch et al. 2002).

In hyperthermia, which is a noninvasive approach to cancer treatment, tissues are exposed to higher than normal temperatures to promote selective destruction of abnormal cells. Cancer cells are more susceptible to hyperthermic effects than healthy cells due to their higher metabolic rates (Li and Gu, 2009). Delivery of the thermal energy used in this approach can be achieved using several methods such as microwave radiation, radio-frequency pulses, and ultrasound. However, due to the diffusive nature of these energy forms, high fluences are required, which produce undesirable hyperthermic effects on surrounding tissues (Li and Gu, 2009). An alternative source of energy is NIR laser beams that can penetrate tissue with sufficient intensity and high spatial precision. However, this approach requires high levels of energy input to produce enough hyperthermic effects due to the low absorption of NIR light by tumors. Thus, to make this treatment method clinically safe and viable, the hyperthermic effects have to be intensified and highly localized, which makes it necessary to enhance the light absorption and energy conversion in the tumors (Li and Gu, 2009). For this purpose, localized hyperthermia with gold nanoparticles, which can quickly convert the absorbed energy into heat energy in the picosecond time domain, is being developed as an alternative to the conventional hyperthermia methods. Due to their good biocompatibility and size- and shape-dependent optical properties, gold nanoparticles have been extensively used in biomedical applications such as intracellular imaging and biosensing (Li and Gu, 2009).

In many biomedical and nanophotonics research studies, optical software simulation and computational modeling techniques are the only means to get a deeper understanding of the underlying scientific issues (Wilson et al. 2005).

11.2 Light Interaction with Tissue

In order to assist the development of optical treatment procedures in medical fields, it is necessary to understand the transmission behavior of light impulses incident on tissue (Yamada and Hasegawa, 1996). In this section, we introduce the optical properties of tissue and discuss what characteristics should a model used for light propagation through tissue have. The photon transport (PT) theory that is used to model light propagation through tissue is then introduced.

11.2.1 Optical Properties of Tissue

In minimally invasive medicine such as optical biopsy, functional imaging, and laser-induced thermotherapy, light in the range from 240 to 10,000 nm is used for diagnosis and therapy (Beuthan et al. 1996). Lasers, which are the most commonly used light source for biophotonics, are devices that produce highly coherent, highly directional, monochromatic, and intense beams of light (Prasad, 2003). The main challenge in optical diagnosis is to understand the changes in the optical properties of tissues with abnormalities (Beuthan et al. 1996). A change of physiological parameters will always change the amount and distribution of scattered light (Beuthan et al. 1996). The potential of diagnostic optical imaging has generated considerable interest in the optical properties of tissues and cells at NIR wavelengths where scattering is dominant over absorption (Dunn and Kortum, 1996).

Tissue is a complicated medium, which is treated as an absorbing and scattering (i.e., turbid) medium, and many of the optical–thermal events produced by laser radiation are interdependent (Welch and Gemert, 1995). However, methods have been proposed for calculating and measuring light propagation in tissue (Welch and Gemert, 1995). Both scattering and absorption provide important information about the physiological condition of tissue (Mourant et al. 1998). The epidermis of the skin absorbs and propagates light. The absorption property is mostly due to the natural pigment melanin (Baranoski and Krishnaswamy, 2004). When a biological cell is illuminated with laser light, the light is scattered in all directions to form a light-scattering pattern, which is dependent on the size and internal structure of the cell (Meyer, 1979). Therefore, appropriate measurements of this light-scattering pattern can provide morphological information about the cell (Meyer, 1979).

Light propagation in tissue depends on scattering and absorption properties of its components such as cells, cell organelles, and various fiber structures. The size, shape, density, and relative refractive index of these structures affect the propagation of light (Tuchin, 2007). Figure 11.1 from Tuchin (2007) shows major organelles and inclusions of the cell. This figure depicts the wide variety of structures within a cell that determine scattering of light in tissue. Scattering of light in tissue is due to scattering centers such as cells, nuclei, other organelles, and structures within organelles (Mourant et al. 1998). Multiple scattering and absorption in tissue result in laser beam broadening and decay as it travels through tissue (Tuchin, 2007). Cells and tissue structure elements can be as small as a few tenths of nanometers or as large as hundreds of micrometers (Tuchin, 2007).

An exact assessment of light propagation in tissue would require a model that characterizes the spatial and size distribution of tissue structures, their absorbing properties, and their refractive indices (Welch and Gemert, 1995). However, for real tissues, such as the skin, the task of creating a precise representation, either as a tissue phantom or as a computer simulation, is formidable if not totally impossible. Therefore, tissue is represented as an absorbing bulk material with scatterers randomly distributed over the volume (Welch and Gemert, 1995). Even though this approximation does not provide any information about the microscopic structure of tissue, it has provided accurate agreement with experimental measurements (Dunn, 1998; Boas et al. 1994). Two approaches are currently used for modeling tissues (Tuchin, 2007). The first approach is to model tissue as a medium with a continuous random spatial distribution of optical parameters, and the second approach is to consider tissue as a discrete ensemble of scatterers (Tuchin, 2007). The choice of the appropriate model of these two depends on the structure of the tissue under study and the kind of light-scattering characteristics that are to be obtained.

At a microscopic level, the tissue can be described as a medium with a spatially dependent refractive index. At this level, light interaction with tissue results in reflection, refraction, absorption, and diffraction processes (Beuthan et al. 1996). On a macroscopic scale, these

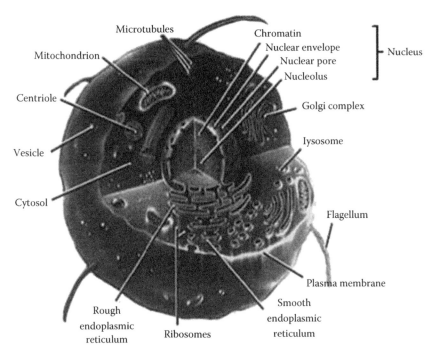

FIGURE 11.1
Major organelles and inclusions of the cell. (From Tuchin, V., *Tissue Optics: Light Scattering Methods and Instruments for Medical Diagnosis*, 2nd edn., SPIE Press, Washington, DC, 2007, Figure 1.1, p. 5. With kind permission from SPIE Press.)

processes are summarized as scattering (Beuthan et al. 1996). On the macroscopic level, mean optical values can be used instead of spatially dependent parameters to model light interaction with tissue. These mean values are the scattering coefficient, the absorption coefficient, and the phase function (Beuthan et al. 1996). Since the absorption and scattering of any tissue vary with wavelength, there are dramatic differences in the penetration depth of the radiation from the various lasers (Welch and Gemert, 1995). For example, light at either 193 nm or 2.96 μm is totally absorbed in the first micrometer of tissue owing to amino acid absorption in the UV band and water absorption in the IR band. In contrast, light from 600 nm to 1.1 μm can penetrate several millimeters in tissue. This is owing to the fact that within this red and NIR wavelength window, there is a lack of strongly absorbing tissue chromophores (Welch and Gemert, 1995). In general, in the UV and IR spectral regions, light does not penetrate deep into tissue because of high absorption and low scattering. Short-wave visible light penetrates as deep as 0.5–2.5 mm and light in wavelength range 600–1500 nm penetrates to a depth of 8–10 mm because scattering prevails over absorption (Tuchin, 1997).

Tissue optics involves two major tasks. The first task is finding the light energy per unit area per unit time that reaches a target chromophore at some position **r** (Welch and Gemert, 1995). The second task, which has so far been the most difficult one, is developing methods by which the absorbing and scattering properties of tissue can be measured. Such properties are called the optical properties of tissue and they are (i) the absorption coefficient; (ii) the scattering coefficient; (iii) the probability density function that scattering occurs from a certain direction (with unit vector Ω') into another direction (with unit vector Ω), sometimes also called the phase function of single particle scattering; and (iv) the index of refraction of the tissue (Welch and Gemert, 1995). The ultimate goal of the second task of tissue optics is to have methods available that can assess all optical properties noninvasively in living tissues (Welch and Gemert, 1995).

The reciprocal of the absorption coefficient is defined as the average distance a photon travels before being absorbed by the medium (Prahl et al. 1993). Similarly the reciprocal of the scattering coefficient is defined as the average distance a photon travels before being scattered by the medium (Prahl et al. 1993). Two dimensionless quantities can be used to characterize light propagation in a turbid medium. They are the albedo, α, and the optical thickness, τ, that are defined as (Prahl et al. 1993)

$$\alpha = \frac{\sigma_s}{\sigma_s + \sigma_a}, \qquad (11.1)$$

$$\tau = (\sigma_s + \sigma_a)d, \qquad (11.2)$$

where
   d is the physical thickness of the slab
   σ_s and σ_a are the scattering and absorption coefficients, respectively (Prahl et al. 1993)

Tissue scattering tends to be strongly forward-peaked with an anisotropy factor in the range 0.7–0.99 (Welch and Gemert, 1995). Thus, an accurate model for simulating light propagation in tissue should not have restrictions on the ratio of scattering to absorption and on the scattering anisotropy (Welch and Gemert, 1995).

For applications in which ultrashort pulses are used as the incident source, the medium can be treated as cold because the emission from the medium is negligible (Guo and Kumar, 2002). This is because the small amount of energy deposited per pulse is not sufficient to raise the temperature significantly, and hence, any emission can be neglected when compared to the intensity of the scattered incident pulse (Guo and Kumar, 2002). In addition, any emission from the medium will be at higher wavelengths due to the low temperature in the medium and therefore need not be included in the monochromatic analysis, which is conducted at the wavelength of the laser (Guo and Kumar, 2002). When the radiation propagates at the speed of light, the time is in the order of 1 ns or less. With such a time scale, the medium can be generally treated as cold because the heat-diffusion and heat-capacity effects are negligible (Guo and Kim, 2003).

11.2.2 Photon Transport Theory

In this section, we describe the PT theory, which is used to model light propagation through tissue. We discuss why the PT theory is used instead of Maxwell equations and introduce the photon transport equation (PTE).

11.2.2.1 Why Photon Transport Theory

The electromagnetic theory provides exact expressions for the absorption and scattering parameters of a uniform, nonscattering medium with ensembles of random scatterers. These scatterers consist of discrete scattering and absorbing "particles," possibly of different sizes, that are distributed randomly (Welch and Gemert, 1995). However, because of the inhomogeneity of biological tissue, analytic approaches using Maxwell equations do not lead to solvable equations for any case of practical interest in tissue (Cheong et al. 1990). Also, there has been little progress in developing even approximate solutions, which are applicable to optical propagation in biological media (Welch and Gemert, 1995). While the description of light propagation in tissue in terms of electromagnetic fields is intractable, the application of PT (radiative transfer) theory has proven to be a considerable success (Welch and Gemert, 1995), and most recent advances in describing the transfer of laser energy in tissue are based upon transport theory (Cheong et al. 1990). For wavelengths that are much longer than cell diameters, where there is little scattering from the cellular structures, the reflection, absorption, and transmission are described best using electromagnetic theory (Yoon et al. 1987). However, the electromagnetic spectrum of lasers lies in the IR to UV wavelength band, i.e., in the nanometer regime. Cell structures are also in nanometer scale. Thus, substantial multiple scattering in tissue occurs in this band due to the comparable size of cells with respect to the irradiation wavelength. For laser wavelengths used in nanophotonics medical applications, the PT theory provides a practical description for the optical propagation of light in tissue (Yoon et al. 1987).

In the PT formalism, light propagation is considered equivalent to the flow of discrete photons, which may be locally absorbed or scattered by the medium. The scattering is usually energy conserving (although effects such as fluorescence or Raman scattering, which involves wavelength and hence energy shift, can be incorporated). The photon–tissue interactions are described in terms of absorption and (elastic) scattering cross sections (or equivalent linear interaction coefficients) and the scattering angular distribution by phase functions (Welch and Gemert, 1995).

The PT theory is a heuristic model, which lacks the physical rigor of multiple-scattering electromagnetic theory. For example, it does not in itself include effects such as diffraction or interference, even though the absorption and scattering properties of the individual constituent particles may do so. It is fundamental to PT theory that there should be no correlation between the radiation fields. Only quantities such as power or intensity are considered, and the method ignores the behavior of the component wave amplitudes and phases (Welch and Gemert, 1995). However, PT has provided a self-consistent framework for nanophotonics studies of light propagation in tissues (Welch and Gemert, 1995). If the absorption and scattering properties of tissue are determined according to the conditions of the PT theory (i.e., only radiometric quantities, not wave amplitudes and phases, are measured) and, subsequently, these data are used as inputs to a PT model to calculate the spatial distribution of irradiance within a tissue volume for given incident irradiation and boundary conditions, then it is found in general that such distributions agree with experimental values (again measured ignoring light wave properties), although this has never been demonstrated rigorously (Welch and Gemert, 1995).

In past years, a number of investigators have reported values for the total attenuation coefficient, the effective attenuation coefficient, the effective penetration depth, the absorption and scattering coefficients, and the scattering anisotropy factor for a variety of tissues at a variety of light wavelengths. The majority of these results are based upon approximations to the PT theory (Cheong et al. 1990). However, there are some indications that the PT theory may break down in the case of highly structured tissues such as muscle, where the alignment of fibers may cause measurable wave interference effects and the scattering cannot be considered random (Welch and Gemert, 1995).

11.2.2.2 Basic Radiometric Quantities

Figure 11.2 shows the flow of radiative energy carried by a beam in the direction Ω through a transparent surface element dA. $\mathbf{n}$ is the surface normal such that $\cos\theta = \mathbf{n}\cdot\Omega$ (Thomas and Stamnes, 1999).

11.2.2.2.1 Radiance (I)

Radiance is defined as the radiant power per unit of solid angle about unit vector Ω and per unit area perpendicular to Ω (Welch and Gemert, 1995). That is, at a point on a surface and in a given direction, the radiant intensity of an element on the surface, divided by the area of the orthogonal projection of this element on a plane perpendicular to the given direction, is defined as the radiance. Thus, the radiance, I, can be expressed as

$$I(\mathbf{r},\Omega) = \frac{dP(\mathbf{r},\Omega)}{dA\cos\theta\,d\Omega}\ \left[\mathrm{W\ m^{-2}\ sr^{-1}}\right], \tag{11.3}$$

where $dP(\mathbf{r},\Omega)$ [W] is the power flowing through an infinitesimal area dA, located at $\mathbf{r}$, in the direction of the unit vector Ω (perpendicular to dA), within the infinitesimal solid angle $d\Omega$ (Welch and Gemert, 1995).

11.2.2.2.2 Net Flux Vector

Flux is a description of photon energy transfer per unit area. The net flux vector, $\mathbf{F(r)}$, is the vector sum of elemental flux vectors $I(\mathbf{r},\Omega)\Omega\,d\Omega$ (Welch and Gemert, 1995). That is,

$$\mathbf{F(r)} = \int_{4\pi} I(\mathbf{r},\Omega)\Omega\,d\Omega. \tag{11.4}$$

Hemispherical fluxes are defined as the energy flux through dA in either the forward direction $\mathbf{n}$ or the backward direction $-\mathbf{n}$. The hemispherical flux $\mathbf{F}_{n^+}(\mathbf{r})$ is an integral over the solid angle of 2π $(0 < \theta \leq \pi/2)$ and $\mathbf{F}_{n^-}(\mathbf{r})$ is an integral over the solid angle of 2π $(-\pi/2 < \theta \leq 0)$ (Welch and Gemert, 1995). Thus,

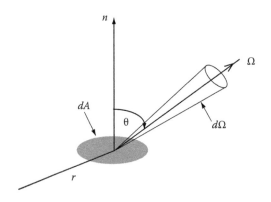

FIGURE 11.2
The flow of radiative energy carried by a beam in the direction Ω through surface element dA. (From Thomas, G.E. and Stamnes, K., *Radiative Transfer in the Atmosphere and Ocean*, Cambridge University Press, Cambridge, U.K., 1999, Figure 2.1, p. 36. With kind permission from Cambridge University Press.)

$$\mathbf{F}_{n^+}(\mathbf{r}) = \int\limits_{2\pi} I(\mathbf{r},\Omega)(\Omega \cdot \hat{\mathbf{n}})\,d\Omega, \tag{11.5}$$

and

$$\mathbf{F}_{n^-}(\mathbf{r}) = \int\limits_{2\pi} I(\mathbf{r},\Omega)(\Omega \cdot -\mathbf{n})d\Omega,$$

$$= -\int\limits_{2\pi} I(\mathbf{r},\Omega)(\Omega \cdot \mathbf{n})d\Omega. \tag{11.6}$$

From these forward and backward hemispherical fluxes, Equations 11.5 and 11.6, and the basic definition of the net flux vector, Equation 11.4, the net flux vector is related to the hemispherical fluxes as (Welch and Gemert, 1995)

$$\mathbf{F}(\mathbf{r})\hat{\mathbf{n}} = \mathbf{F}_{n^+}(\mathbf{r}) - \mathbf{F}_{n^-}(\mathbf{r}). \tag{11.7}$$

11.2.2.2.3 Irradiance ($\tilde{I}$)

Consider the flow of radiative energy across a surface element dA, located at a specific position, and having a unit normal $\mathbf{n}$ (as shown in Figure 11.2). The net rate of radiative energy flow, or power per unit area within a small spectral range, is called the spectral net flux or irradiance (Thomas and Stamnes, 1999) and can be expressed as

$$\tilde{I} = \frac{d^3 E}{dA\,dt\,dv} \left[\mathrm{W\,m^{-2}\,Hz^{-1}}\right]. \tag{11.8}$$

The irradiance (net flux) through a surface element dA depends upon the cumulative effect of all the angular beams crossing it in different directions. This quantity conveys little information about the directional dependence of the energy flow (Thomas and Stamnes, 1999).

However, an absorbing chromophore at location $\mathbf{r}$ inside the tissue can absorb photons irrespective of their direction of propagation, and therefore, the integral of the radiance over all directions (i.e., the irradiance) has more practical significance than the radiance itself (Welch and Gemert, 1995).

The radiance and irradiance can be related by (Thomas and Stamnes, 1999)

$$\tilde{I} = \int\limits_{2\pi} I \cos\theta\,d\Omega. \tag{11.9}$$

11.2.2.3 Photon Transport Equation

The propagation of visible or IR photons in a turbid medium, such as tissue, without change in energy (i.e., neglecting Raman scattering and fluorescence), is described by the PTE (Profio, 1989), which examines the change in radiance with distance in a particular direction Ω, at position $\mathbf{r} = (x, y, z)$ (Welch and Gemert, 1995).

When any wave phenomenon associated with light, such as diffraction and interference, is neglected, the computational modeling of light propagation in tissue is essentially equivalent to solving the full-time-dependent PTE (Welch and Gemert, 1995).

The transport equation relates the gradient of radiance, I, at position $\mathbf{r}$ in the direction Ω to losses owing to absorption and scattering and to a gain owing to light scattered from all other directions Ω' into the direction Ω (Welch and Gemert, 1995; Cheong et al. 1990; Edstrom, 2005; Sakami et al. 2002). For a medium with spatially constant refractive index, the PTE has the form

$$\frac{1}{v}\frac{\partial}{\partial t}I(\mathbf{r},\Omega,t)+\Omega\cdot\nabla_r I(\mathbf{r},\Omega,t)-\sigma_s\int_{4\pi} P(\Omega',\Omega)I(\mathbf{r},\Omega',t)d\Omega'$$

$$+\sigma_t I(\mathbf{r},\Omega,t)=F(\mathbf{r},\Omega,t), \tag{11.10}$$

where
  I [W/m^2 sr] is the radiance
  σ_t [1/m] is the extinction coefficient
  σ_s [1/m] is the scattering coefficient
  P [1/sr] is the phase (scattering) function
  F [W/m^3 sr] is the source of power generated at $\mathbf{r}$ in direction Ω

The speed of light in the medium is v.

The radiance in the transient PTE is dependent on the position, on local solid angle, as well as on time. The solid angle Ω can be written in terms of the azimuthal angle, ϕ, and the zenith angle, θ, as

$$d\Omega = \sin\theta\, d\theta\, d\phi. \tag{11.11}$$

The relationship given by Equation 11.11 is graphically illustrated in Figure 11.3. Thus, Equation 11.10 can be written in terms of θ and φ as (Menguc and Viskanta, 1985)

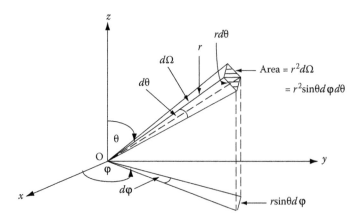

FIGURE 11.3
The relationship of the solid angle with the zenith and azimuthal angles.

$$\frac{1}{v}\frac{\partial}{\partial t}I(x,y,z,u,\phi,t)+\xi\frac{\partial}{\partial x}I(x,y,z,u,\phi,t)+\eta\frac{\partial}{\partial y}I(x,y,z,u,\phi,t)$$

$$+u\frac{\partial}{\partial z}I(x,y,z,u,\phi,t)-\frac{\sigma_s}{4\pi}\int_{0}^{2\pi}\int_{-1}^{1}P(u',\phi';u,\phi)I(x,y,z,u',\phi',t)du'\,d\phi'$$

$$+\sigma_t I(x,y,z,u,\phi,t)=F(x,y,z,u,\phi,t),\tag{11.12}$$

where
(x, y, z, θ, ϕ) are the standard coordinates
u, ξ, and η are direction cosines such that $u=\cos\theta$, $\xi=\sin\theta\cos\phi$, and $\eta=\sin\theta\sin\phi$
t is time

Equation 11.12 can be reduced to the 2D transient PTE, where the radiance is considered to be a function of (y, z, u, ϕ, t):

$$\frac{1}{v}\frac{\partial}{\partial t}I(y,z,u,\phi,t)+\eta\frac{\partial}{\partial y}I(y,z,u,\phi,t)+u\frac{\partial}{\partial z}I(y,z,u,\phi,t)+\sigma_t I(y,z,u,\phi,t)$$

$$-\frac{\sigma_s}{4\pi}\int_{0}^{2\pi}\int_{-1}^{1}P(u',\phi';u,\phi)I(y,z,u',\phi',t)du'\,d\phi'=F(y,z,u,\phi,t).\tag{11.13}$$

This can be reduced to the 1D transient PTE, where the radiance is considered to be a function of only (z, u, ϕ, t):

$$\frac{1}{v}\frac{\partial}{\partial t}I(z,u,\phi,t)+u\frac{\partial}{\partial z}I(z,u,\phi,t)+\sigma_t I(z,u,\phi,t)$$

$$-\frac{\sigma_s}{4\pi}\int_{0}^{2\pi}\int_{-1}^{1}P(u',\phi';u,\phi)I(z,u',\phi',t)du'\,d\phi'=F(z,u,\phi,t).\tag{11.14}$$

The source term, $F(\mathbf{r}, \Omega, t)$, can incorporate the irradiance but can also be used to represent fluorescence light generated within tissue or the internal light source during interstitial laser therapy (Welch and Gemert, 1995).

Several different variations of PTE for inhomogeneous media have been proposed in literature (Premaratne et al. 2005; Khan and Jiang, 2003; Dehghani et al. 2003; Bal, 2006; Shendeleva and Molly, 2007; Martí-López et al. 2006). However, most of these variations are not of a result of fundamental errors or differences but due to different assumptions about the medium or wave field properties. For example, references (Premaratne et al. 2005; Khan and Jiang, 2003; Bal, 2006) look at spatially slowly varying isotropic refractive index profiles in their work. Interestingly, the approach given in (Martí-López et al. 2006) is formulated to accommodate geometric optics approximations but ignores the wavefront curvature in their derivation. Wavefront curvature in the context of slowly varying refractive index approximation is considered in (Premaratne et al. 2005). Numerical considerations necessary to account for such slowly varying spatial refractive profiles are considered in Dehghani et al. (2003) and Shendeleva and Molly (2007).

The PTE for a medium with a spatially varying isotropic refractive index, in standard spherical coordinates, is (Premaratne et al. 2005)

$$\frac{n(\mathbf{r})}{c}\frac{\partial}{\partial t}I(\mathbf{r},\Omega,t)+\left(\frac{1}{R_1(s)}+\frac{1}{R_2(s)}\right)I(\mathbf{r},\Omega,t)+n^2(\mathbf{r})\frac{\partial}{\partial s}\left(\frac{I(\mathbf{r},\Omega,t)}{n^2(\mathbf{r})}\right)$$

$$+\sigma_t(\mathbf{r})I(\mathbf{r},\Omega,t)=\sigma_s(\mathbf{r})\int_{4\pi} P(\Omega,\Omega')I(\mathbf{r},\Omega',t)d\Omega'+F(\mathbf{r},\Omega,t), \qquad (11.15)$$

where
  $\mathbf{r}$ is the position vector of a point on a path of a ray
  s is the arc length along a ray
  $\Omega = d\mathbf{r}/ds$
  t is time
  $I(\mathbf{r}, \Omega, t)$ is the intensity
  $n(\mathbf{r})$ is the refractive index profile
  c is the speed of light in vacuum
  $R_1(s)$ and $R_2(s)$ are the principal radii of curvatures of the geometrical wavefronts
  σ_t is the attenuation coefficient with $\sigma_t = \sigma_a + \sigma_s$, σ_a is the absorption coefficient and σ_s is the scattering coefficient
  $P(\Omega, \Omega')$ is the phase function
  $F(\mathbf{r}, \Omega, t)$ represents sources inside the medium.

11.2.2.3.1 Phase Function of Scattering

The single-scattering phase function describes the amount of light scattered at an angle Θ from the incoming direction (Prahl et al. 1993). We assume that the scattering depends only on the angle Θ between unit vector directions Ω and Ω'. Figure 11.4 shows the incident and scattered directions and the scattering angle. Therefore, we assume that tissue is isotropic in terms of physical properties (such as refractive index and density) (Welch and Gemert, 1995). The phase function is normalized such that its integral over all directions is unity (Prahl et al. 1993).

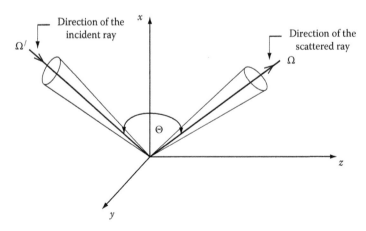

FIGURE 11.4
Illustration of the incident direction, scattered direction, and scattering angle.

Several phase functions, such as the Henyey–Greenstein, the modified Henyey–Greenstein, the Eddington, the delta-Eddington, the isotropic, the delta-isotropic, and a combination of the Rayleigh phase function and the Henyey–Greenstein phase function (Bevilacqua et al. 1999), have been postulated to represent single-scattering phase functions for tissue (Welch and Gemert, 1995; Cheong et al. 1990). Among these, the Henyey–Greenstein phase function can be used to model dermal and aortic tissues (Cheong et al. 1990) and this phase function has been used widely in modeling light propagation in tissue.

Isotropic scattering implies that the phase function is unity (Thomas and Stamnes, 1999); that is, $P(\Omega', \Omega) = 1$. Light scattering in tissue is not isotropic but strongly forward-directed (Welch and Gemert, 1995). A measure of the degree of anisotropy in scattering is the anisotropy factor g. Total forward scattering means $g = 1$ and isotropic scattering means $g = 0$, but for in vitro tissues at the visible and NIR wavelengths, g is found to be between about 0.7 and 0.99 (Welch and Gemert, 1995).

Mathematically, g is defined as the expectation value of the cosine of the scattering angle Θ (Welch and Gemert, 1995). That is,

$$g = \frac{\displaystyle\int_{4\pi} P(\Omega', \Omega)(\Omega' \cdot \Omega) d\Omega'}{\displaystyle\int_{4\pi} P(\Omega', \Omega) d\Omega'}. \tag{11.16}$$

The phase function is normalized such that (Welch and Gemert, 1995)

$$\int_{4\pi} P(\Omega', \Omega) d\Omega' = 1. \tag{11.17}$$

Therefore, Equation 11.16 reduces to

$$g = \int_{4\pi} P(\Omega', \Omega)(\Omega' \cdot \Omega) d\Omega'. \tag{11.18}$$

Calculations of light distribution based on the PTE require knowledge of the absorption and scattering coefficients and the phase function. Yet to arrive at these parameters, one must first have a solution of the PTE (Cheong et al. 1990). Typical optical properties are obtained by computational methods solving the PTE that express the optical properties in terms of readily measurable quantities (Prahl et al. 1993).

11.3 Computational Methods for Modeling Light Propagation through Tissue

Computational methods used to analyze the propagation of light in strongly scattering media are divided into two main types: statistical and deterministic (Yamada and Hasegawa, 1996). Deterministic methods are based on the PTE. Statistical methods require very long computation times in order to obtain statistically meaningful results. These methods are carried out by tracing the paths of photon bundles with simulation of the scattering and absorbing pattern of light. Therefore, it is almost impossible to obtain solutions within practical limits of computation time when the media are large in size and complex in configuration (Yamada and Hasegawa, 1996). On the other hand, deterministic methods are based on the PTE, which is in the form of an integrodifferential equation and is difficult to solve (Yamada and Hasegawa, 1996).

Numerical modeling of light propagation through tissue involves numerically solving the governing PTE. Many numerical techniques have been introduced by researchers in the field over the past decades. Most of these techniques developed in the last few decades have concentrated on solving the steady-state (i.e., time-independent) PTE, mainly focusing in the fields of astrophysical and atmospheric sciences. Only recently have researchers started developing models for the time-dependent PTE to be used in tissue optics. Each of these techniques has relative advantages and disadvantages.

11.3.1 Modeling Static Distributions

11.3.1.1 Monte Carlo Method

The Monte Carlo method is a stochastic (or statistical) method. It is sufficiently flexible to handle complex geometrical shapes, anisotropic scattering, and nonhomogeneous properties, but the results obtained by this method always have unavoidable random errors due to practical finite sampling. In contrast, deterministic methods do not suffer from this defect (Wu and Wu, 2000).

Monte Carlo simulations of photon propagation offer a flexible but rigorous approach to PT in tissues (Welch and Gemert, 1995). The principal idea of Monte Carlo simulations applied to absorption and scattering phenomena is to follow the optical path of a photon through the turbid medium. In this method, the rules of photon propagation are expressed as probability distributions for the incremental steps of photon movement between sites of photon–tissue interaction, for the angles of deflection in a photon's trajectory when a scattering event occurs, and for the probability of transmittance or reflectance at boundaries (Welch and Gemert, 1995). This method simulates the "random walk" of photons in a medium that contains absorption and scattering. It is based on a set of rules that govern the movement of a photon in tissue. The two key decisions are the mean free path for a scattering or absorption event and the scattering angle (Welch and Gemert, 1995).

The five principal steps of Monte Carlo simulations of absorption and scattering are (Niemz, 2004)

1. *Source photon generation*: Photons are generated at a surface of the considered medium. Their spatial and angular distributions can be fitted to a given light source (e.g., a Gaussian beam).

2. *Pathway generation*: After generating a photon, the distance to the first collision is determined. Absorbing and scattering particles in the turbid medium are assumed to be randomly distributed. Thus, the mean free path is $1/\tilde{n}\sigma_s$, where ϱ is the density of particles and σ_s is their scattering cross section. A random number ξ_1, such that $0 < \xi_1 < 1$, is generated by the computer, and the distance $L(\xi_1)$ to the next collision is calculated from

$$L(\xi_1) = -\frac{\ln(\xi_1)}{\varrho\sigma_s}. \tag{11.19}$$

Since

$$\int_0^1 \ln(\xi_1)\,d\xi_1 = -1 \tag{11.20}$$

the average value of $L(\xi_1)$ is indeed $1/\tilde{n}\sigma_s$. Hence, a scattering point has been obtained. The scattering angle is determined by a second random number, ξ_2, in accordance with a certain phase function. The corresponding azimuth angle ϕ is chosen as $\phi = 2\pi\xi_3$ where ξ_3 is a third random number between 0 and 1.

3. *Absorption*: To account for absorption, a weight is attributed to each photon. When entering a turbid medium, the weight of the photon is unity. Due to absorption (in a more accurate program also due to reflection), the weight is reduced by $\exp(-\alpha L(\xi_1))$, where α is the absorption coefficient. As an alternative to implementing a weight, a fourth random number ξ_4 between 0 and 1 can be drawn. Instead of assuming only scattering events in step 2, scattering takes place if $\xi_4 < a$, where a is the albedo. For $\xi_4 > a$, on the other hand, the photon is absorbed, which then is equivalent to step 4.

4. *Elimination*: This step only applies if a weight has been attributed to each photon. When this weight reaches a certain cutoff value, the photon is eliminated. A new photon is then launched, and the program proceeds with step 1.

5. *Detection*: After having repeated steps 1–4 for a sufficient number of photons, a map of pathways is calculated and stored in the computer. Thus, statistical statements can be made about the fraction of incident photons being absorbed by the medium as well as the spatial and angular distributions of photons having escaped from it.

In summary, the distance between two collisions is selected from a logarithmic distribution, using a random number generated by a computer. The absorption is accounted for by implementing a weight to each photon and permanently reducing this weight during propagation. If scattering is to occur, a new direction of propagation is chosen according to a given phase function and another random number is generated (Niemz, 2004). The whole procedure continues until the photon escapes from the considered volume or its weight reaches a given cutoff value (Niemz, 2004).

Even though the Monte Carlo method is rigorous, it is computationally intense. This method is basically statistical in nature and requires a computer to calculate the propagation of a large number of photons (Welch and Gemert, 1995). This is because the error bound of the Monte Carlo method is inversely proportional to the square root of the number of statistical samplings, and hence, it requires a large number of samples to reach the satisfactory accuracy (Sawetprawichkul et al. 2002).

Various variance reduction techniques are used to increase the efficiency of Monte Carlo simulations (Alerstam et al. 2008). The most fundamental of these is a technique in which absorption is modeled by reducing photon weights rather than by photon termination (Alerstam et al. 2008). Other techniques include photon splitting, electron history repetition, Russian roulette, and the use of quasi-random numbers (Kawrakow and Fippel, 2000). Variance reduction techniques are used to decrease the statistical fluctuations of Monte Carlo calculations without increasing the number of particle histories (Kawrakow and Fippel, 2000).

Graphics processing units (GPUs) can be used to dramatically decrease the computational time of Monte Carlo simulations of PT applications (Alerstam et al. 2008). The choice of the method of random number generation is a very important aspect of executing Monte Carlo simulations in parallel. This is due to the fact that the same random number generator with the same seed (e.g., when a timestamp is used as the seed) would most likely result in many threads performing exactly the same computations (Alerstam et al. 2008). Therefore, when Monte Carlo codes are executed in parallel, different threads should be properly and differently seeded (Alerstam et al. 2008).

11.3.1.2 Adding–Doubling Method

The adding–doubling method is a general numerical solution of the PTE (Prahl et al. 1993; van de Hulst, 1980). This method works naturally with layered media and yields reflection and transmission information readily. Reflectance is important for diagnostic applications using light. For measuring the optical properties of a sample, the only values needed are the total reflection and transmission of the sample (Welch and Gemert, 1995). The doubling method was introduced by van de Hulst to solve the PTE in a slab geometry (Welch and Gemert, 1995).

The adding–doubling method involves obtaining the reflection and the transmission matrices. Here, doubling refers to how one finds the reflection and transmission matrices of two layers with identical optical properties from those of the individual layers (Thomas and Stamnes, 1999). Thus, the doubling method assumes knowledge of the reflection and transmission properties of a single thin homogeneous layer. The reflection and transmission of a slab twice as thick are found by juxtaposing two identical slabs and summing the contributions from each slab (Welch and Gemert, 1995; Prahl et al. 1993). To start the doubling procedure, the initial layer is frequently taken to be thin enough that its reflection and transmission properties can be computed from single scattering (Thomas and Stamnes, 1999). The reflection and transmission from an arbitrarily thick slab are obtained by repeatedly doubling until the desired thickness is reached (Welch and Gemert, 1995; Thomas and Stamnes, 1999).

The adding method refers to the combination of two or more layers with different optical properties (Thomas and Stamnes, 1999). That is, the adding method extends the doubling method to dissimilar slabs, thus making it possible to simulate media with different layers and/or internal reflection at boundaries (Welch and Gemert, 1995). The solution technique for the adding–doubling method proceeds by first applying doubling to find the reflection and transmission matrices for each of the homogeneous layers, whereupon adding is subsequently used to find the solution for all the different layers combined (Thomas and Stamnes, 1999).

Figure 11.5 shows an illustration of the doubling concept. Basically the doubling concept starts with the notion that the emergent intensities $I^+(0)$ (the backscattered intensity at $z=0$) and $I^-(d)$ (the forward intensity at $z=d$) are determined by a reflection matrix, ρ, and a transmission matrix, τ, through the relations (Thomas and Stamnes, 1999)

$$I^-(d) = \tau I^-(0) + \rho I^+(d), \tag{11.21}$$

$$I^+(0) = \rho I^-(0) + \tau I^+(d), \tag{11.22}$$

for a homogeneous slab of thickness d.

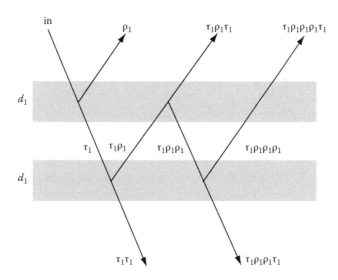

FIGURE 11.5
Illustration of the doubling concept. (From Thomas, G.E. and Stamnes, K., *Radiative Transfer in the Atmosphere and Ocean*, Cambridge University Press, Cambridge, U.K., 1999, Figure 8.12, p. 325. With kind permission from Cambridge University Press.)

Equations 11.21 and 11.22 can be written in matrix form as

$$\begin{pmatrix} \mathbf{I}^-(d) \\ \mathbf{I}^+(d) \end{pmatrix} = \begin{pmatrix} \mathbf{t} - \mathbf{r}\mathbf{t}^{-1}\mathbf{r} & \mathbf{r}\mathbf{t}^{-1} \\ -\mathbf{t}^{-1}\mathbf{r} & \mathbf{t}^{-1} \end{pmatrix} \begin{pmatrix} \mathbf{I}^-(0) \\ \mathbf{I}^+(0) \end{pmatrix}. \tag{11.23}$$

If $\boldsymbol{\tau}_1 = \boldsymbol{\tau}$ (d), $\boldsymbol{\rho}_1 = \boldsymbol{\rho}$ (d), $\boldsymbol{\tau}_2 = \boldsymbol{\tau}$ $(2d)$, and $\boldsymbol{\rho}_2 = \boldsymbol{\rho}$ $(2d)$, then comparing Equation 11.23 with the rearranged steady-state PTE in matrix form, it can be shown that (Thomas and Stamnes, 1999)

$$\begin{pmatrix} \mathbf{t}_2 - \mathbf{r}_2\mathbf{t}_2^{-1}\mathbf{r}_2 & \mathbf{r}_2\mathbf{t}_2^{-1} \\ -\mathbf{t}_2^{-1}\mathbf{r}_2 & \mathbf{t}_2^{-1} \end{pmatrix} = \begin{pmatrix} \mathbf{t}_1 - \mathbf{r}_1\mathbf{t}_1^{-1}\mathbf{r}_1 & \mathbf{r}_1\mathbf{t}_1^{-1} \\ -\mathbf{t}_1^{-1}\mathbf{r}_1 & \mathbf{t}_1^{-1} \end{pmatrix}^2. \tag{11.24}$$

Solutions for $\boldsymbol{\tau}_2$ and $\boldsymbol{\rho}_2$ are obtained by

$$\mathbf{t}_2 = \mathbf{t}_1(\mathbf{I} - \mathbf{r}_1^2)^{-1}\mathbf{t}_1, \tag{11.25}$$

$$\mathbf{r}_2 = \mathbf{r}_1 + \mathbf{t}_1\mathbf{r}_1(\mathbf{I} - \mathbf{r}_1^2)^{-1}\mathbf{t}_1, \tag{11.26}$$

where $\mathbf{I}$ is the identity matrix (Thomas and Stamnes, 1999). Equations 11.25 and 11.26 constitute the basic doubling rules from which the reflection and transmission matrices for a layer of thickness $2d$ are obtained from those of half the thickness, d (Thomas and Stamnes, (1999).

The adding–doubling method is sufficiently fast that iterated solutions are possible on microcomputers (Prahl et al. 1993). This method is also sufficiently flexible that anisotropic scattering and internal reflection at the boundaries may be included (Prahl et al. 1993). Internal reflection at the boundaries (caused by mismatched indices of refraction) can be included in the calculation by adding an additional layer for each mismatched boundary. The reflection and transmission of this layer are equal to the Fresnel reflection and transmission for unpolarized light incident on a plane boundary between two transparent media with the same indices of refraction (Prahl et al. 1993). For a medium characterized by any phase function, adding–doubling method can be used for any tissue optical thickness (Welch and Gemert, 1995).

This method assumes that the distribution of light is independent of time, samples have homogeneous optical properties, the sample geometry is an infinite plane-parallel slab of finite thickness, the tissue has a uniform index of refraction, internal reflection at boundaries is governed by Fresnel's law, the light is unpolarized, and the slab has no internal sources (Welch and Gemert, 1995; Prahl et al. 1993). The adding–doubling method has several advantages over most other techniques, because it permits asymmetric scattering, arbitrarily thick samples, Fresnel boundary conditions, and relatively fast computations (Baranoski and Krishnaswamy, 2004). In addition, in this method, only integrations over angle are required and physical interpretation of results can be made at each step. This method is equivalent for isotropic and anisotropic scattering and the results are obtained for all angles of incidence used in the integration (Welch and Gemert, 1995; Prahl et al. 1993). Two disadvantages of the adding–doubling method are that it is restricted to layered geometries with uniform irradiation and it is necessary that each layer has homogeneous optical properties (Welch and Gemert, 1995; Prahl et al. 1993). Since it is incapable of providing the radiance at arbitrary positions within the slab, its application to tissue optics is limited (Patterson et al. 1991).

11.3.1.3 Diffusion Approximation

The diffusion equation is derived as an approximate solution of the PTE (Welch and Gemert, 1995). It combines the scattering and the phase function in one parameter, called the reduced scattering coefficient (Baranoski and Krishnaswamy, 2004). The photon diffusion approximation is often employed to simplify the PTE into a partial differential equation with respect to space and time. However, the photon diffusion equation involves some errors due to the approximation used in its derivation (Yamada and Hasegawa, 1996). In this theory, light is principally described by particles with energy hv and velocity c (Welch and Gemert, 1995). These particles are scattered or absorbed by structures in turbid media and reflected at boundaries. These reflections can be determined by equations of Fresnel reflection (Welch and Gemert, 1995).

In diffusion theory, the radiance in the steady-state PTE is separated into unscattered, $I_c(\mathbf{r}, \Omega)$, and scattered, $I_d(\mathbf{r}, \Omega)$, components (Cheong et al. 1990):

$$I(\mathbf{r}, \Omega) = I_c(\mathbf{r}, \Omega) + I_d(\mathbf{r}, \Omega). \qquad (11.27)$$

The unscattered portion contains all the light that has not interacted with tissue and it satisfies the Beer's law (Cheong et al. 1990):

$$I(z) = I_0(1 - r_{sc})e^{-(\sigma_a + \sigma_s)z}, \qquad (11.28)$$

where
  $I(z)$ [W/m²] is the intensity (irradiance) of collimated light at position z in the tissue
  I_0 is the collimated irradiance
  r_{sc} is the Fresnel surface reflection of collimated light
  σ_a is the absorption coefficient
  σ_s is the scattering coefficient

The scattered portion contains all the light that has been scattered at least once and can be expressed exactly with an infinite sum of Legendre polynomials. However, the diffusion approximation truncates this sum to the first two terms (an isotropic and a slightly forward-directed term) (Cheong et al. 1990). This approximation simplifies the PTE to the diffusion equation given by

$$\left(\nabla^2 - \kappa^2\right)\Phi(\mathbf{r}) = -Q_0(\mathbf{r}),$$ (11.29)

where the constant κ is an approximation of the actual measured effective attenuation coefficient when absorption is dominated by scattering, the source term $Q_0(\mathbf{r})$ is generated by scattering of collimated normal irradiation, and the total scattered fluence rate $\Phi(\mathbf{r})$ is given by (Cheong et al. 1990):

$$\Phi(\mathbf{r}) = \int_{4\pi} I_d(\mathbf{r}, \mathbf{w}) \, d\mathbf{w}.$$ (11.30)

The diffusion model is approximately valid if the optical mean free path $(1/\sigma_t)$ is much smaller than the typical dimensions of the problem considered and if a photon is scattered many times before it is absorbed or leaves a medium (Welch and Gemert, 1995). Under these conditions, the density of photons at a given position is nearly uniform in all directions (Welch and Gemert, 1995). The accuracy of the diffusion equation is affected by the ratio of scattering and absorption, the scattering anisotropy, and the distance from light sources and boundaries (Cheong et al. 1990). When absorption is small compared to scattering, scattering is not very anisotropic, and the irradiance is not needed close to the source or a strong absorber or boundary, then diffusion theory may be used (Profio, 1989). Despite these assumptions, the fact that a 3D problem can be solved at all is a clear advantage of the diffusion theory (Welch and Gemert, 1995).

When the absorption coefficient is not significantly smaller than the scattering coefficient, the diffusion approximation provides a poor approximation of the PTE. It is worth noting that the human skin is characterized by the presence of pigments, such as melanin particles, which have a significant absorption cross section (Baranoski and Krishnaswamy, 2004). In addition, the diffusion approximation is not applicable when scattering is mostly forward-peaked, which is the usual case in tissue (Baranoski and Krishnaswamy, 2004). Diffusion theory can be derived from general principles using only macroscopic tissue properties and is therefore expected to hold with the restrictions involved in the approximation (Welch and Gemert, 1995). The time-dependent photon diffusion equation is believed to give inaccurate solutions at early times and in the vicinity of boundaries. In addition, there is controversy about how the diffusion coefficient should be specified (Yamada and Hasegawa, 1996).

11.3.1.4 Kubelka–Munk Theory

The Kubelka–Munk (K-M) theory describes the propagation of a uniform, diffuse irradiance through a 1D isotropic slab with no reflection at the boundaries (Cheong et al. 1990; Kubelka, 1948; Kubelka, 1954). This theory is a special case of the so-called many-flux theory, where the PTE is converted into a matrix differential equation by considering the radiance at many discrete angles (Niemz, 2004). The original K-M theory is considered to be a two-flux theory, which involves only two types of diffuse irradiance, forward irradiance and backward irradiance (Baranoski and Krishnaswamy, 2004). However, this theory has been recently extended to the many-flux theory, hence improving its applicability to tissue optics (Baranoski and Krishnaswamy, 2004).

The K-M expressions for reflection (R) and transmission (T) of diffuse irradiance on a slab of thickness t are (Cheong et al. 1990)

$$R = \frac{\sinh(S_{KM}yt)}{x\cosh(S_{KM}yt) + y\sinh(S_{KM}yt)}, \tag{11.31}$$

$$T = \frac{y}{x\cosh(S_{KM}yt) + y\sinh(S_{KM}yt)}, \tag{11.32}$$

where A_{KM} and S_{KM} are the K-M absorption and scattering coefficients, respectively, and have units of inverse length (m^{-1}). The parameters x and y are found using the equations given in the following (Cheong et al. 1990):

$$A_{KM} = (x-1)S_{KM}, \tag{11.33}$$

$$S_{KM} = \frac{1}{yt}\ln\left[\frac{1-R(x-y)}{T}\right], \tag{11.34}$$

$$x = \frac{1+R^2-T^2}{2R}, \tag{11.35}$$

$$y = +\sqrt{x^2-1}. \tag{11.36}$$

With these parameters, two differential equations can be formed (Niemz, 2004):

$$\frac{dJ_1}{dz} = -S_{KM}J_1 - A_{KM}J_1 + S_{KM}J_2, \tag{11.37}$$

$$\frac{dJ_2}{dz} = -S_{KM}J_2 - A_{KM}J_2 + S_{KM}J_1, \tag{11.38}$$

where
J_1 is the flux in the direction of the incident radiation
J_2 is the backscattered flux in the opposite direction
z denotes the mean direction of incident radiation

The general solutions to Equations 11.37 and 11.38 can be expressed by (Niemz, 2004)

$$J_1(z) = c_{11}e^{-\gamma z} + c_{12}e^{+\gamma z}, \tag{11.39}$$

$$J_2(z) = c_{21}e^{-\gamma z} + c_{22}e^{+\gamma z}, \tag{11.40}$$

where
$$\gamma = \sqrt{A_{KM}^2 + 2A_{KM}S_{KM}}.$$

One advantage of the K-M model is that the scattering and absorption coefficients may be directly expressed in terms of the measured reflection and transmission (Cheong et al. 1990). The simplicity of the K-M model has made it a popular method for measuring the optical properties of tissue. Unfortunately, the assumptions of isotropic scattering, matched boundaries, and diffuse irradiance are atypical of the interaction of laser light with tissue (Cheong et al. 1990).

K-M model is not a thorough model of PT (Baranoski and Krishnaswamy, 2004). Even though this model allows rapid determination of the optical parameters of tissue, its relative simplicity and the speed are achieved at the expense of the accuracy (Baranoski and Krishnaswamy, 2004). Despite attempts to extend the K-M model to collimated irradiance and anisotropic scattering, this method remains a poor approximation for laser light propagation in tissue (Cheong et al. 1990).

11.3.1.5 Discrete Ordinates Method

The discrete ordinates method (DOM) is a numerical technique in which the angular distribution and the spatial distribution are defined by a finite number of coordinates (cosines of the angle or dimensions of volume cells) rather than continuously (Profio, 1989). The essence of this technique is the conversion of the PTE to a system of linear algebraic equations suitable for numerical solution. The DOM can be carried out to any arbitrary order and accuracy (Modest, 2003). To do this, the radiance is represented only by its value at discrete values of the independent variables. A solution of the transport problem is found by solving the PTE for a set of discrete directions spanning the total solid angle range of 4π (Modest, 2003). In addition, the operations of differentiation and integration are replaced by their discrete counterparts, finite differences, and summation (or quadrature) (Patterson et al. 1991). The idea of discretizing the radiance was first proposed by Schuster (Schuster, 1905) who considered only the forward and backward fluxes. Chandrasekhar later generalized this scheme by using the Gaussian quadrature technique (Patterson et al. 1991).

The discrete ordinate approximation to the PTE is obtained by replacing the integrals by quadrature sums and thus transforming the integrodifferential equation into a system of coupled differential equations (Thomas and Stamnes, 1999) given by

$$\frac{1}{v}\frac{\partial}{\partial t}I(\mathbf{r},\mathbf{w}_i,t) + \mathbf{w}_i \cdot \nabla_\mathbf{r}I(\mathbf{r},\mathbf{w}_i,t) + \sigma_t I(\mathbf{r},\mathbf{w}_i,t) - \sigma_s \sum_{j=1}^{N} P(\mathbf{w}_j,\mathbf{w}_j)I(\mathbf{r},\mathbf{w}_j,t)$$

$$= F(\mathbf{r},\mathbf{w}_i,t), \tag{11.41}$$

where $i = 1, \ldots, N$.

For the discrete ordinate approximation, many quadrature rules, such as Gaussian, Lobatto, Chebyshev, or Fiveland (Mitra and Kumar, 1999), can be used. However, the use of the Gaussian quadrature is preferred because it ensures the correct normalization of the phase function, implying that the energy is conserved in the computation (Thomas and Stamnes, 1999). The main advantage of the discrete ordinates model over stochastic approaches using Monte Carlo methods is the speed, which is sustained using precomputation and compressing schemes (Baranoski and Krishnaswamy, 2004).

The two main drawbacks of the DOM are false scattering and the ray effect (Tan and Hsu, 2002; Modest, 2003). False scattering is due to spatial discretization errors.

When a single collimated beam is traced through an enclosure by the DOM, the beam will gradually widen as it moves farther away from its point of origin; this unphysical effect, even in the absence of real scattering, is called false scattering (Modest, 2003). False scattering can be reduced by using finer meshes (Modest, 2003). The ray effect is due to the errors in angular discretization and can be reduced by increasing the sizes of the meshes (Modest, 2003). Therefore, if a finer spatial mesh is used to reduce the false scattering, a finer angular quadrature scheme should be used to reduce the ray effect (Modest, 2003).

11.3.2 Modeling Transient Response

With the advent of short-pulse lasers and their rapid deployment in a variety of engineering applications such as ocean and atmosphere remote sensing, optical tomography, laser surgery, and combustion product characterization and combustion diagnostics, the traditional steady-state PT formulations cannot be used to analyze their interaction with participating media (Mitra et al. 1997).

Most of the methods developed to model the transient response deal with the 1D PTE. The 1D transient PTE for a medium without any internal sources is given by (Handapangoda et al. 2008a; Kim and Moscoso, 2002)

$$\frac{1}{v}\frac{\partial}{\partial t}I(z,u,\phi,t)+u\frac{\partial}{\partial z}I(z,u,\phi,t)-\frac{\sigma_s}{4\pi}\int_0^{2\pi}\int_{-1}^1 P(u',\phi';u,\phi)I(z,u',\phi',t)du'd\phi'$$

$$+\sigma_t I(z,u,\phi,t)=0, \tag{11.42}$$

where
  $I(z, u, \phi, t)$ is the light intensity (radiance)
  (z, θ, ϕ) are the standard spherical coordinates
  $u = \cos\theta$
  t denotes time
  σ_t and σ_s are attenuation and scattering coefficients, respectively

The speed of light in the medium is denoted by v and $P(u', \phi'; u, \phi)$ is the phase function. The following methods had been developed to solve the transient PTE.

11.3.2.1 Functional Expansion Methods

As in the DOM, the goal of this technique is to reduce the integrodifferential equation of PT to a set of coupled differential equations, which can be solved by standard techniques (Patterson et al. 1991). As opposed to the DOM, where a number of discrete directions of the radiance are considered, the angular, spatial, or temporal dependence of the radiance is here approximated by a finite series expansion of orthogonal functions (Patterson et al. 1991).

For example, Kim et al. (Kim and Moscoso, 2002) expanded the spatial dependence using Chebyshev polynomials. In their work, after representing the azimuthal dependence of the radiance in Fourier series, the spatial dependence is approximated by the Chebyshev spectral expansion as

$$I_n(z, u, t) \cong \sum_{k=0}^{N} a_k(u, t) T_k(z),$$

(11.43)

where
  $I_n(z, u, t)$ is the nth Fourier coefficient of the radiance
  $T_k(z)$ are the orthogonal Chebyshev polynomials (Kim and Moscoso, 2002)

Handapangoda et al. (Handapangoda et al. 2008a) approximated the temporal dependency by Laguerre polynomials as

$$I(z, u, t) = \sum_{k=0}^{N} B_k(z, u) L_k(t),$$

(11.44)

where
  $I(z, u, t)$ is the radiance
  $L_k(t)$ are the orthogonal Laguerre polynomials

11.3.2.2 P_1 Model

The P_1 model can be used when the scattered intensity is a linear function of u (Mitra and Kumar, 1999). In this model,

$$I(z, u, t) = u(z, t) + \frac{3}{4\pi} q(z, t) u,$$

(11.45)

and

$$u(z, t) = \frac{1}{4\pi} \int_{4\pi} I(z, u, t) \, d\Omega = \frac{1}{2} \int_{-1}^{1} I(z, u, t) \, du,$$

(11.46)

$$q(z, t) = \int_{4\pi} I(z, u, t) u \, d\Omega = 2\pi \int_{-1}^{1} I(z, u, t) u \, du,$$

(11.47)

where
  $u(z, t)$ is the average intensity over all angles
  $q(z, t)$ is the heat flux
  Ω is the solid angle

Equation 11.45 is then substituted in the transient PTE and two equations are obtained: the first by integrating the resulting equation with respect to Ω and the second by multiplying the resulting equation by u and then integrating with respect to Ω (Mitra and Kumar, 1999). Those two equations are then combined, which yields a hyperbolic equation, which indicates that the propagation speed of u along the z direction is $v/\sqrt{3}$ (Mitra and Kumar, 1999).

Mitra et al. (1997) used the P_1 approximation to analyze the 2D effects in a scattering–absorbing medium having a rectangular geometry. Steady-state studies have shown that the P_1 approximation for the intensity distribution is not as accurate as more sophisticated approximations and that the P_1 approximation fails to match the correct propagation speed (Mitra et al. 1997).

11.3.2.3 P_N Model

The general P_N method models the intensity by expanding it as a series of Legendre polynomials of u (Bayazitoglu and Higenyi, 1979):

$$I(z,u,t) = \sum_{m=0}^{N} I_m(z,t)p_m(u), \tag{11.48}$$

which is then substituted into the PTE. The PTE is subsequently multiplied by a Legendre polynomial P_k of order k less than or equal to N and integrated with respect to u (Mitra and Kumar, 1999). Use of the orthogonality property of the Legendre polynomials results in (Mitra and Kumar, 1999)

$$\frac{1}{v}\frac{\partial I_k}{\partial t} + \frac{k+1}{2k+3}\frac{\partial I_{k+1}}{\partial z} + \frac{k}{2k-1}\frac{\partial I_{k-1}}{\partial z} + \left(\sigma_t - \sigma_s\frac{a_k}{2k+1}\right)I_k = \frac{2k+1}{2}\int_{-1}^{1} SP_k\,du, \tag{11.49}$$

for intensity coefficient I_k where $0 \leq k \leq N$. In Equation 11.49, a_k are the Legendre coefficients of the phase function. Thus, $(N+1)$ coupled hyperbolic equations are obtained, one for each k (Mitra and Kumar, 1999).

11.3.2.4 Transformation to Steady-State Version

Handapangoda et al. (2008a) used the transformation

$$\tau = t - \frac{z}{vu}, \tag{11.50}$$

which maps the 1D transient PTE (Equation 11.42) to a moving reference frame with the pulse, which results in

$$u\frac{\partial}{\partial z}I(z,u,\phi,\tau) + \sigma_t I(z,u,\phi,\tau) - \frac{\sigma_s}{4\pi}\int_{0}^{2\pi}\int_{-1}^{1} P(u',\phi';u,\phi)I(z,u',\phi',\tau)\,du'\,d\phi' = 0, \tag{11.51}$$

subject to $I(z=0, u, \phi, \tau) = f(\tau)\delta(u-u_0)\delta(\phi-\phi_0)$. Use of the transformation (11.50) in the 1D transient PTE eliminates the time derivative term. Application of the DOM to the azimuthal angle, ϕ, and the cosine of the zenith angle, u, of Equation 11.51 results in (Handapangoda et al. 2008a)

$$u_i\frac{\partial}{\partial z}I(z,u_i,\phi_r,\tau) + \sigma_t I(z,u_i,\phi_r,\tau)$$

$$-\frac{\sigma_s}{4\pi}\sum_{j=1}^{L} w_j^\phi \sum_{k=1}^{K} w_k^u P(u_k,\phi_j;u_i,\phi_r)I(z,u_k,\phi_j,\tau) = 0, \tag{11.52}$$

where

$i = 1, \ldots, K$

$r = 1, \ldots, L$

φ_j is the jth quadrature point of φ

w_j^ϕ is the corresponding Gaussian weight

u_k is the kth quadrature point of u

w_k^u is the corresponding Gaussian weight

In order to remove the dependency on τ, $I(z, u_i, \phi_r, \tau)$ and $f(\tau)$ are expanded using Laguerre polynomials with respect to τ, such that $I(z, u_i, \phi_r, \tau) \approx \sum_{k=0}^{N} B_k(z, u_i, \phi_r) L_k(\tau)$ and $f(\tau) \approx \sum_{k=0}^{N} F_k L_k(\tau)$ where B_k and F_k are the coefficients corresponding to the Laguerre polynomial $L_k(\tau)$ (Handapangoda et al. 2008a).

Expanding the transformed 1D PTE, Equation 11.52, using a Laguerre basis and taking moments (i.e., multiplying by $L_n(\tau)e^{-\tau}$ and integrating over $[0, \infty)$), results in

$$u_i \frac{\partial}{\partial z} B_n(z, u_i, \phi_r) + \sigma_t B_n(z, u_i, \phi_r)$$

$$- \frac{\sigma_s}{4\pi} \sum_{j=1}^{L} w_j^\phi \sum_{k=1}^{K} w_k^u P(u_k, \phi_j; u_i, \phi_r) B_n(z, u_k, \phi_j) = 0, \qquad (11.53)$$

where $n = 1, \ldots, N$, $i = 1, \ldots, K$ and $r = 1, \ldots, L$.

This set of equations can be written in matrix form as (Handapangoda et al. 2008a)

$$\frac{d}{dz} \mathbf{A B}_n + \sigma_t \mathbf{B}_n - \frac{\sigma_s}{4\pi} \mathbf{P W B}_n = 0, \qquad (11.54)$$

where $\mathbf{B}_n = [B_n(z, u_i, \phi_r)]_{K \times L, 1}$, $\mathbf{P} = [P(u_k, \phi_j; u_i, \phi_r)]_{K \times L, K \times L}$, and $\mathbf{A}$ is a $(K \times L)$ by $(K \times L)$ diagonal matrix with diagonal elements u_1 to u_K repeating L times. The matrix $\mathbf{W}$ is also a $(K \times L)$ by $(K \times L)$ diagonal matrix with diagonal elements $w_r^\phi \times w_k^u$ with the pattern $w_1^\phi \times w_1^u, w_1^\phi \times w_2^u, \ldots, w_1^\phi \times w_K^u, w_2^\phi \times w_1^u, \ldots, w_L^\phi \times w_K^u$.

Rearranging, Equation 11.54 can be written as

$$\frac{d}{dz} \mathbf{B}_n = \mathbf{Y B}_n, \qquad (11.55)$$

where $\mathbf{Y} = \mathbf{A}^{-1} \left[\dfrac{\sigma_s}{4\pi} \mathbf{P W} - \sigma_t \mathbf{I} \right]$ and $\mathbf{I}$ is the identity matrix. Hence, the original transient PTE is reduced to a one-variable ordinary differential equation (Handapangoda et al. 2008a).

Equation 11.55 can be solved using the Runge–Kutta–Fehlberg (RKF) method (Chapra and Canale, 2002). There are a number of advantages of this technique over most of the other methods. Since the time dependence is expanded using a Laguerre basis, all the sampling points in the time domain are obtained in a single execution, as opposed to time marching techniques used in most of the other solution methods. This makes this method much faster when one requires the intensity profile at a particular point or a plane over a time interval. Also, since the RKF method is used to solve the final reduced equation, the intensity profile at several points or planes over the whole time spectrum can be obtained in one execution of the algorithm (Handapangoda et al. 2008a).

Handapangoda et al. (Handapangoda and Premaratne, 2008) discussed how this method can be adopted for modeling light propagation in inhomogeneous media with a spatially varying refractive index. This technique involves first tracing the ray paths defined by the refractive index profile of the medium by solving the eikonal equation using a Runge–Kutta integration algorithm. The PTE is solved only along these ray paths, minimizing the overall computational burden of the resulting algorithm.

An explicit solution to Equation 11.55 can be found using eigendecomposition as illustrated in (Handapangoda et al. 2011). The square matrix $\mathbf{Y}$ can be written as a product of three matrices composed of its eigenvalues and eigenvectors, as follows:

$$\mathbf{Y} = \mathbf{VDV}^{-1}, \tag{11.56}$$

where $\mathbf{V} = [\mathbf{v}_1 \mathbf{v}_2 \ldots \mathbf{v}_m]$. $\mathbf{v}_i$ is the ith eigenvector of $\mathbf{Y}$ and $\mathbf{D}$ is a diagonal matrix with its ith diagonal element equal to the ith eigenvalue, λ_i, of $\mathbf{Y}$. Using Equation 11.56 in Equation 11.55, we get

$$\frac{d}{dz} \mathbf{B}_n = \mathbf{YB}_n$$

$$= \mathbf{VDV}^{-1} \mathbf{B}_n$$

$$\mathbf{V}^{-1} \frac{d}{dz} \mathbf{B}_n = \mathbf{DV}^{-1} \mathbf{B}_n \tag{11.57}$$

$$\frac{d}{dz} \mathbf{X} = \mathbf{DX}$$

where

$$\mathbf{X} = \mathbf{V}^{-1} \mathbf{B}_n. \tag{11.58}$$

Since $\mathbf{D}$ in Equation 11.57 is a diagonal matrix, each row of Equation 11.57 can be solved independently resulting in

$$\mathbf{X} = \begin{pmatrix} x_1 \\ x_2 \\ \vdots \\ x_m \end{pmatrix} = \begin{pmatrix} x_1^0 e^{\lambda_1 z} \\ x_2^0 e^{\lambda_2 z} \\ \vdots \\ x_m^0 e^{\lambda_m z} \end{pmatrix}. \tag{11.59}$$

The column matrix $\mathbf{X}^0 = \mathbf{V}^{-1} \mathbf{B}_n^0$ is composed of the boundary values x_1^0 to x_m^0. Once $\mathbf{X}$ is evaluated using Equation 11.59, the solution to Equation 11.55 can be found using the transformation in Equation 11.58. That is, $\mathbf{B}_n = \mathbf{VX}$. Thus, the explicit solution to Equation 11.55 is given by

$$\mathbf{B}_n = x_1^0 e^{\lambda_1 z} \mathbf{v}_1 + x_2^0 e^{\lambda_2 z} \mathbf{v}_2 + \cdots + x_m^0 e^{\lambda_m z} \mathbf{v}_m.$$

Without loss of generality, here we have assumed that the eigenvalues of matrix **Y** are real and distinct, which is the case for most of the widely used phase functions in practice. However, if for a particular application the eigenvalues and eigenvectors contain complex values (in the form of complex conjugate pairs), a real solution to Equation 11.55 can be obtained as illustrated in (Edwards and Penney, 2004; Handapangoda et al. 2011).

11.3.2.5 Modeling Multidimensional Transient PTE

Many models have been developed to solve the 1D PTE, which assumes horizontally uniform plane-parallel media. However, in order to model 3D inhomogeneous media, the 3D PTE should be used (Evans, 1998). The modeling of the 3D PT is usually considered difficult, even for the steady-state case, because it involves solving integrodifferential equations of four or five variables (Guo and Kumar, 2002). However, many diverse methods for solving the multidimensional steady-state PT problem can be found in the research literature (Evans, 1998; Stephens, 1988), which discard the time-dependence complications of the transient problem considered here.

Most existing transient models have been developed for the 1D PTE, but some have been developed for the 2D PTE (Tan and Hsu, 2002). Only recently have researchers started to work on models for 3D transient PT in scattering media such as biological tissue.

Most recent developments on solving the 3D transient PTE rely on approximate methods such as diffusion approximation and spherical harmonic approximation, which are used to simplify the PTE (Tan and Hsu, 2002). These approximate methods fail to predict the correct propagation speed within the medium, and the solution accuracy is not satisfactory except for specific cases that consider optically thick media at asymptotically longer time scales (Tan and Hsu, 2002). Tan et al. (Tan and Hsu, 2002) proposed an integral equation technique to model the transient radiative transfer in 3D homogeneous as well as nonhomogeneous participating media. Guo et al. (Guo and Kumar, 2002) formulated a complete transient 3D DOM to solve the transient PTE in a nonhomogeneous turbid medium with a rectangular enclosure. Guo et al. (Guo and Kim, 2003) modeled heterogeneous biological tissues using the DOM, incorporating the Fresnel specularly reflecting boundary condition. They treated the laser intensity as having a diffuse part and a specular part. The reflectivity at the tissue–air interface was calculated using Snell's law and the Fresnel equations (Guo and Kim, 2003). Chai et al. (Chai et al. 2004) proposed a finite-volume method to calculate transient radiative transfer in a 3D enclosure, while Sawetprawichkul et al. (Sawetprawichkul et al. 2002) implemented the Monte Carlo method for modeling 3D transient PT in a parallel computer system.

Handapangoda et al. (Handapangoda and Premaratne, 2009) used the transformation

$$\tau = t - \frac{x}{3v\xi} - \frac{y}{3v\eta} - \frac{z}{3vu}, \tag{11.60}$$

which eliminates the partial derivative term with respect to time.

The radiance was then expanded using a Laguerre basis:

$$I(x, y, z, \mathbb{W}, \tau) = \sum_{k=0}^{N} B_k(x, y, z, \mathbb{W}) L_k(\tau), \tag{11.61}$$

where

$$\int_0^\infty L_n(\tau) L_m(\tau) e^{-\tau} d\tau = \delta_{mn}. \tag{11.62}$$

Then taking moments (i.e., multiplying by $L_n(\tau)e^{-\tau}$ and integrating over $[0,\infty)$), the time dependence can be removed, resulting in N uncoupled equations:

$$\xi \frac{\partial}{\partial x} B_n(x,y,z,\mathbf{w}) + \eta \frac{\partial}{\partial y} B_n(x,y,z,\mathbf{w}) + u \frac{\partial}{\partial z} B_n(x,y,z,\mathbf{w})$$

$$- \frac{\sigma_s(x,y,z)}{4\pi} \int_{4\pi} P(\mathbf{w}';\mathbf{w}) B_n(x,y,z,\mathbf{w}') d\mathbf{w}' + \sigma_t(x,y,z) B_n(x,y,z,\mathbf{w}) = 0, \tag{11.63}$$

where $n = 1, \dots, N$ and B_n is the nth Laguerre coefficient. Hence, the transient PTE is transformed to the steady-state version, which can be then solved using a finite-volume approach with the DOM as illustrated in (Handapangoda and Premaratne, 2009).

Since the time dependence is expanded using a Laguerre basis, all the sampling points in the time domain are obtained in a single execution of this algorithm, as opposed to the time marching techniques used in other solution methods. This makes it much faster when the intensity profile is required at a particular point or a plane over a time interval. In addition, the use of the Laguerre expansion to represent time dependency enables modeling the system with any arbitrary input pulse shape, using only a few Laguerre polynomials. Specifically, the Gaussian pulse shape used in many practical applications can be accurately represented using a few Laguerre polynomials, as opposed to the discrete sampling used in most other models. Also, this expansion implicitly imposes the causality of the system (Handapangoda and Premaratne, 2009).

11.4 Computational Methods for Modeling Light Propagation through Tissue with Nanostructure Implants

Theoretical results for obstacle scattering in biological tissues are limited (Kim, 2004). Kim (2004) studied light propagation in tissue containing an absorbing plate. He considered a perfectly absorbing plate of vanishing thickness placed inside a tissue specimen (Kim, 2004). Since biological tissues scatter light with a sharp forward peak, Kim used the Fokker–Planck equation instead of the PTE as the former is easier to solve than the latter (Kim, 2004). Although a validation of results from the Fokker–Planck equation with experimental data for biological tissues has not been carried out, Kim used this approach as it does not limit the analysis to weak absorption and optically thick media. Another reason for his using this approach is that it does not assume that the radiance is independent of the direction (Kim, 2004). In his analysis, Kim assumed a sharply forward-peaked phase function with $\Omega \cdot \Omega' \approx 1$ (Kim, 2004).

Feng et al. (Feng et al. 1995) presented an analytical perturbation analysis for studying the sensitivity of diffusive photon flux to the addition of a small spherical defect object in multiple-scattering media such as human tissues. They based their perturbation analysis on the diffusion theory for photon migration in tissue (Feng et al. 1995). In their study, they first analytically derived the photon migration path distributions and the shapes of the regions in which the photon migration paths are concentrated (the so-called banana regions). The sensitivity of detected photon flux densities to the inclusion of small spherical defects was then analyzed (Feng et al. 1995).

Handapangoda et al. (Handapangoda et al. 2008b) developed a technique for modeling light propagation through tissue with embedded implants, by mapping the PTE to Maxwell equations. In this section, we describe this technique and describe how light propagation through tissue with an implanted photonic crystal nanostructure can be modeled.

11.4.1 Two Sets of Governing Equations

Owing to its ability to accurately represent light propagation through tissue, wave propagation through biological tissue is modeled using the PTE (Premaratne et al. 2005; Handapangoda et al. 2008a). However, the interaction of electromagnetic energy through implanted structures can be best studied using Maxwell equations. Therefore, in order to model wave propagation through tissue with implanted foreign structures, a mapping of the PTE to Maxwell equations is required.

The PTE models light propagation using only the magnitude of the intensity (radiance), but Maxwell equations require both the magnitude and the phase of the electric and magnetic fields. Thus, for the mapping of these two sets of equations, a phase retrieval technique should be used in order to retrieve the phase information from the intensity profile.

11.4.2 Phase Retrieval

The transport-of-intensity equation (TIE) relates phase and intensity and hence can be used to construct the unknown phase from known intensity values.

The TIE is (Paganin 2006; Teague, 1983)

$$\nabla_{xy} \cdot \left(\tilde{I}(x,y,z) \nabla_{xy} \phi(x,y,z) \right) = -k \frac{\partial \tilde{I}(x,y,z)}{\partial z}. \tag{11.64}$$

It shows how the intensity (irradiance), $\tilde{I}$, and the phase, ϕ, are related, and this forms the basis of the phase construction.

To construct the phase, the TIE in Equation 11.64 is rewritten as

$$\tilde{I}(x,y,z) \nabla_{xy}^2 \phi(x,y,z) + \frac{\partial \tilde{I}(x,y,z)}{\partial x} \frac{\partial \phi(x,y,z)}{\partial x} + \frac{\partial \tilde{I}(x,y,z)}{\partial y} \frac{\partial \phi(x,y,z)}{\partial y} = -k \frac{\partial \tilde{I}(x,y,z)}{\partial z}. \tag{11.65}$$

Equation 11.65 can be solved for $\phi(x, y, z)$ numerically using a suitable technique such as the full multigrid algorithm (Gureyev et al. 1999; Allen and Oxley, 2001), a Green's function method (Teague, 1983), or a fast-Fourier-transform-based method (Gureyev and Nugent, 1996); Gureyev and Nugent, 1997). Of these techniques for solving the TIE, the full multigrid algorithm (Gureyev et al. 1999; Allen and Oxley, 2001; Press et al. 1992) solves the TIE exactly.

Equation 11.65 can be expressed as

$$\Gamma u = f, \tag{11.66}$$

where $\Gamma = \left(\tilde{I}(x,y,z) \nabla_{xy}^2 + \dfrac{\partial \tilde{I}(x,y,z)}{\partial x} \dfrac{\partial}{\partial x} + \dfrac{\partial \tilde{I}(x,y,z)}{\partial y} \dfrac{\partial}{\partial y} \right)$, $u = \phi(x, y, z)$, and $f = -k \dfrac{\partial \tilde{I}(x,y,z)}{\partial z}$.

In multigrid methods, the original equation is discretized on a uniform grid. Equation 11.65 can be discretized as follows:

$$\left(\frac{\tilde{I}_{i+1,j} - \tilde{I}_{i-1,j}}{2\Delta} \right)\left(\frac{\phi_{i+1,j} - \phi_{i-1,j}}{2\Delta} \right) + \left(\frac{\tilde{I}_{i,j+1} - \tilde{I}_{i,j-1}}{2\Delta} \right)\left(\frac{\phi_{i,j+1} - \phi_{i,j-1}}{2\Delta} \right)$$

$$+ \tilde{I}_{i,j}\left(\frac{\phi_{i-1,j} + \phi_{i,j-1} + \phi_{i+1,j} + \phi_{i,j+1} - 4\phi_{i,j}}{\Delta^2} \right) = f_{i,j}, \tag{11.67}$$

where $i = 1, \ldots, M$ and $j = 1, \ldots, M$ for $M \times M$ grid points. Also, $\tilde{I}_{i,j} = \tilde{I}(x_i, y_j, z)$, $\phi_{i,j} = \phi(x_i, y_j, z)$, $\Delta = x_{i+1} - x_i = y_{j+1} - y_j$, and $f_{i,j} = -k \dfrac{\partial \tilde{I}(x_i, y_j, z)}{\partial z}$. By solving the PTE on two closely separated planes $z = z$ and $z = z + \delta z$, two intensity profiles are obtained. Thus, the following approximations can be used in Equation 11.67:

$$\tilde{I}(x,y,z) \approx \frac{\tilde{I}(x,y,z+\delta z) + \tilde{I}(x,y,z)}{2}, \tag{11.68}$$

and

$$\frac{\partial \tilde{I}(x,y,z)}{\partial z} \approx \frac{\tilde{I}(x,y,z+\delta z) - \tilde{I}(x,y,z)}{\delta z}. \tag{11.69}$$

In Equations 11.68 and 11.69, $\tilde{I}$ represents the irradiance. However, the PTE solves for the radiance, I. $\tilde{I}$ and I are related by (Thomas and Stamnes, 1999; Ramamoorthi and Hanrahan, 2001)

$$\tilde{I} = \int_{2\pi} I \cos\theta \, d\Omega, \tag{11.70}$$

where
  θ is the zenith angle in a spherical coordinate system
  $d\Omega$ is an infinitesimal solid angle

This conversion of the intensity is required because in the PTE the ray model of optics is used, but in Maxwell equations, the wave model is used, and these two models deal with different definitions of intensity, radiance, and irradiance, respectively.

Since the intensity and its partial derivatives with respect to x, y, and z can be approximately calculated from the two intensity profiles, as shown in Equations 11.68 and 11.69, the only unknown in Equation 11.67 is $\phi_{i,j}$. Hence, the full multigrid algorithm can be used to solve Equation 11.67 for $\phi_{i,j}$, and thus, the phase can be retrieved on each grid point.

11.4.3 Modeling Light through Tissue with a Photonic Crystal Nanostructure Implant

Handapangoda (Handapangoda, 2009) described how light propagation through tissue with an implanted photonic crystal nanostructure can be modeled. In this section, we present this idea. This example focuses on obtaining the magnitude and the phase of the field from the intensity profile at $z = z_A$ and then converting these to the corresponding electric and magnetic fields, so that the field due to the photonic crystal nanostructure can be modeled. Then, at the exit of the photonic crystal nanostructure, the electric and magnetic fields can be converted back to the intensity profile so that the tissue layer beyond this plane can be modeled by solving the PTE.

Figure 11.6 shows a composite object composed of a layer of biological tissue and a layer of a photonic crystal nanostructure, and Figure 11.7 shows the end elevation of Figure 11.6.

A solution method described in Section 11.3.2 can be used to solve the PTE from $z = 0$ to $z = z_{A^-}$. Thus, the radiance profile at the plane just before the tissue–photonic crystal interface (i.e., at $z = z_{A^-}$) is obtained. However, in order to model the propagation of the laser pulse beyond this plane, Maxwell equations should be used. Maxwell equations require the phase of the field in addition to the magnitude. Thus, the phase information of the field at $z = z_{A^-}$ should be retrieved in order to model the light propagation through the photonic crystal nanostructure.

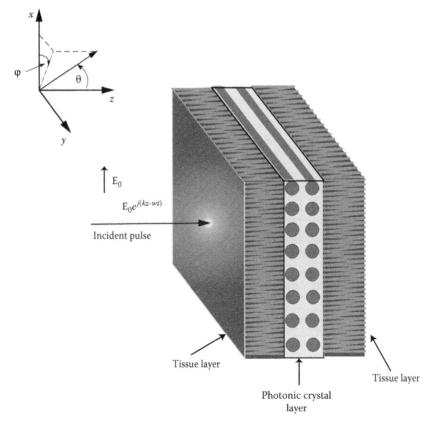

FIGURE 11.6
(See color insert.) Photonic crystal implanted in biological tissue.

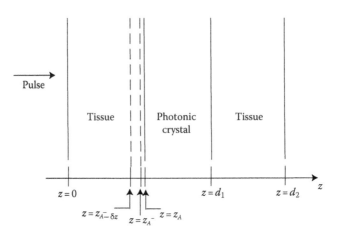

FIGURE 11.7
End elevation of the tissue–photonic crystal model.

In order to apply the phase retrieval technique, first, the radiance profile, I, obtained by solving the PTE should be converted to an irradiance profile, $\tilde{I}$, using the relationship given in Equation 11.70. Thus, the irradiance at $z = z_{A^-}$, I_{A^-}, and $z = z_{A^- - \delta z}$, $I_{A^- - \delta z}$, can be obtained by solving the PTE for radiance and integrating over the hemisphere. Then, the approximations

$$\tilde{I} \approx \frac{\tilde{I}_{A^-} + \tilde{I}_{A^- - \delta z}}{2},$$ (11.71)

and

$$\frac{\partial \tilde{I}}{\partial z} \approx \frac{\tilde{I}_{A^-} - \tilde{I}_{A^- - \delta z}}{\delta z}$$ (11.72)

are used in order to solve the TIE. The full multigrid algorithm (Gureyev et al. 1999; Allen and Oxley, 2001) is then used to solve the TIE, given by Equation 11.67, for the phase, $\phi(x, y, z)$. Thus, the phase at the tissue–photonic crystal interface is retrieved using the intensity values at two infinitesimally separated planes as described in Section 11.4.2.

Once the phase is retrieved, if the incident electric field is known, the field at the tissue–photonic crystal interface can be obtained. If the incident polarization vector is $\mathbf{E}_0$, as shown in Figure 11.6, the electric field at $z = z_A^+$ can be written as

$$\mathbf{E}_A = \sqrt{I_{A^-}}\, e^{j\phi_A} e^{j(kz - \omega t)} \mathbf{E}_0.$$ (11.73)

Then, the corresponding magnetic field at $z = z_A^+$ can be obtained from

$$\mathbf{H}_A = j\frac{1}{\omega}\nabla \times \mathbf{E}_A.$$ (11.74)

Thus, the incident electric and magnetic fields at the interface have been obtained.

Once the incident electric and magnetic fields at the interface have been obtained, in order to calculate the field distribution inside the photonic crystal nanostructure, the reflectance and transmittance at the tissue–photonic crystal interface should be known.

The photonic crystal nanostructure can be modeled using an effective refractive index, η_{eff}, which can be obtained using (Sakoda, 2005)

$$\nabla_k \omega = \frac{c}{\eta_{eff}} \hat{k}, \tag{11.75}$$

where
$$\nabla_k = \frac{\partial}{\partial k_x} \mathbf{u}_x + \frac{\partial}{\partial k_y} \mathbf{u}_y + \frac{\partial}{\partial k_z} \mathbf{u}_z$$

ω is the angular frequency
$\mathbf{k}$ is the wave vector
$\hat{k}$ is the unit vector parallel to $\mathbf{k}$
$\mathbf{u}_x$, $\mathbf{u}_y$, and $\mathbf{u}_z$ are unit vectors along x-, y-, and z-axes, respectively

The relationship between the angular frequency, ω, and the wave number, $\mathbf{k}$, is given by the dispersion relation of the photonic crystal (Sakoda, 2005). This relationship can be calculated and plotted in a band diagram. Figure 11.8 shows the band diagram of a square lattice composed of cylindrical dielectric rods, with a relative permittivity of 9 and a radius of 0.2 times the period of the lattice, in air, from reference (Lourtioz et al. 2008). It is evident from Equation 11.75 that the effective refractive index of the photonic crystal structure, corresponding to a particular wave number, can be found using the gradient of its band diagram.

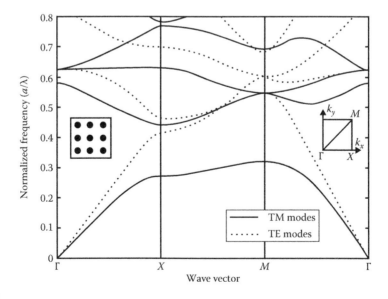

FIGURE 11.8
Band diagram of a square lattice composed of cylindrical dielectric rods, with a relative permittivity of 9 and a radius of 0.2 times the period of the lattice, in air. (From Lourtioz, J.M. et al., *Photonic Crystals: Towards Nanoscale Photonic Devices*, 2nd edn., Springer-Verlag, Berlin, Germany, 2008, Figure 1.9, p. 35. With kind permission from Springer Science+Business Media.)

Once the effective refractive index is calculated, the transmission coefficient can be approximated by that of a uniform medium (Sakoda, 2005). Let the amplitude of the incident electric vector be E and $E_\parallel$ and $E_\perp$ be the components parallel and perpendicular to the plane of incidence, respectively. Then, from Fresnel formulae and using Snell's law, the transmitted electric vector can be obtained by (Born and Wolf, 1999)

$$T_\parallel = \frac{2\eta_T \eta_{eff} \cos\theta_i}{\eta_{eff}^2 \cos\theta_i + \eta_T \sqrt{\eta_{eff}^2 - \eta_T^2 \sin^2\theta_i}} E_\parallel,$$

(11.76)

and

$$T_\perp = \frac{2\eta_T \cos\theta_i}{\eta_T \cos\theta_i + \sqrt{\eta_{eff}^2 - \eta_T^2 \sin^2\theta_i}} E_\perp,$$

(11.77)

where
η_{eff} is the effective refractive index of the photonic crystal
η_T is the refractive index of the tissue
θ_i is the angle of incidence of the wave

Hence, in order to find the transmission coefficients using the earlier equations, the angle of incidence, θ_i, of the wave should be found first. Figure 11.9 shows a plane wave of λ wavelength making an angle θ_i with the z-axis. In this figure, $PQ = \lambda$ and $\angle POQ = \theta_i$. If PO is taken to be α,

$$\alpha = \frac{\lambda}{\sin\theta_i},$$

(11.78)

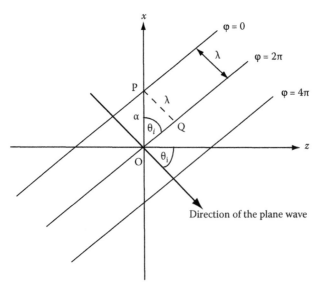

FIGURE 11.9
A plane wave in 2D geometry.

and

$$\frac{\partial \phi}{\partial x} = \frac{2\pi}{\alpha},$$

$$= \frac{2\pi}{\lambda} \sin \theta_i.$$

(11.79)

Therefore, for a 2D geometry,

$$\theta_i(x) = \sin^{-1}\left(\frac{\lambda}{2\pi} \frac{\partial \phi}{\partial x} \right),$$

$$\approx \frac{\lambda}{2\pi} \frac{\partial \phi}{\partial x} \quad \text{for small } \theta_i.$$

(11.80)

Similarly, for a 3D geometry,

$$\left| \theta_i(x, y) \right| = \sin^{-1}\left(\frac{\lambda}{2\pi} \left| \nabla_{xy} \phi \right| \right),$$

$$\approx \frac{\lambda}{2\pi} \left| \nabla_{xy} \phi \right| \quad \text{for small } \theta_i.$$

(11.81)

Hence, for a plane wave incident on the tissue–photonic crystal interface,

$$\left| \nabla_{xy} \phi_A \right| = \frac{2\pi}{\lambda} \sin \theta_i.$$

(11.82)

Thus, the angle of incidence can be obtained by

$$\left| \theta_i \right| = \sin^{-1}\left[\frac{\lambda}{2\pi} \left| \nabla_{xy} \phi_A \right| \right].$$

(11.83)

The phase profile at $z = z_A$, obtained using the phase retrieval technique discussed in Section 11.4.2, can be used to find the quantity $\left| \nabla_{xy} \phi_A \right|$ of Equation 11.83.

The field at each point on the tissue–photonic crystal interface can be resolved into parallel and perpendicular components, and Equations 11.76 and 11.77 can be used to find the transmitted field components. Then, the two transmitted components can be combined to find the resultant transmitted field at each point at this interface.

The next step is to calculate the field distribution inside the photonic crystal. Light propagation through photonic crystal structures is governed by Maxwell equations (Joannopoulos et al. 1995). The finite difference time domain technique (Taflove and Hagness, 2005) can be used to solve Maxwell equations, and hence, the field distribution at the exit of the photonic crystal layer can be obtained. Commercially available software for modeling photonic crystals may be used for this purpose.

Then, at the photonic crystal–tissue interface, the field will be again resolved to a transverse electric (TE) and transverse magnetic (TM) wave in order to find the proportion that is transmitted into the tissue layer. At this interface, the reflection coefficients are for the TE wave,

$$\Gamma'_{TE} = \frac{\eta_{eff}\cos\theta_i - \sqrt{\eta_T^2 - \eta_{eff}^2 \sin^2\theta_i}}{\eta_{eff}\cos\theta_i + \sqrt{\eta_T^2 - \eta_{eff}^2 \sin^2\theta_i}}, \tag{11.84}$$

and for the TM wave,

$$\Gamma'_{TM} = \frac{\eta_T^2\cos\theta_i - \eta_{eff}\sqrt{\eta_T^2 - \eta_{eff}^2 \sin^2\theta_i}}{\eta_T^2\cos\theta_i + \eta_{eff}\sqrt{\eta_T^2 - \eta_{eff}^2 \sin^2\theta_i}}. \tag{11.85}$$

Once the field transmitted into the tissue layer is obtained, the TE and TM components can be combined to obtain the resultant field. Then, the electric field, $(\mathbf{E}_{d_2})$, can be used to obtain the intensity (i.e., the irradiance) using the relationship

$$I = \frac{1}{2}v\varepsilon|\mathbf{E}|^2, \tag{11.86}$$

where v and ε are the propagation speed and the permittivity in the medium, respectively.

Once the irradiance profile on the plane $z = d_2$ is obtained, it should be converted back to a radiance profile so that the PTE can be used to model the light propagation beyond this plane. Figure 11.10 shows a strategy that can be used for mapping the irradiance profile to the radiance profile. In Figure 11.10, the axes (x, y, z) represent the global coordinate system used in solving the PTE; also shown is the ray-centered spherical coordinate system used to describe the irradiance-to-radiance mapping in the forward hemisphere. Based on

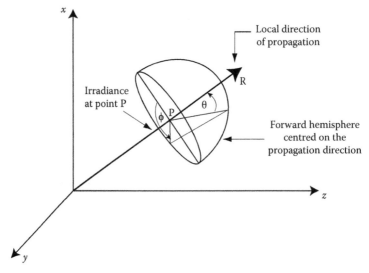

FIGURE 11.10
An illustration of the strategy used for mapping radiance to irradiance.

the work of Ramamoorthi et al. (Ramamoorthi and Hanrahan, 2001), this strategy uses a hemisphere positioned centrally at the ray propagation direction and uses the relationship between radiance and irradiance given by Equation 11.70 and spherical harmonic representation to achieve this task. These authors have shown that the irradiance can be represented as a simple convolution of the incident illumination (i.e., radiance and a clamped cosine transfer function) (Ramamoorthi and Hanrahan, 2001). Therefore, the radiance can be obtained by a deconvolution operation. They derived a simple closed-form formula for the irradiance in terms of spherical harmonic coefficients of the incident illumination (Ramamoorthi and Hanrahan, 2001).

Once the irradiance profile on the plane $z = d_2$ is converted back to a radiance profile at each point, in the forward hemisphere at the interface, the light propagation through the remaining layers of tissue can be modeled by solving the PTE using one of the techniques discussed in Section 11.3.2.

11.5 Conclusions

In this chapter, we discussed computational nanophotonics methods for modeling optical detection techniques in medicine. The importance and applications of computational modeling of light propagation through biological tissue and tissue with nanostructure implants were discussed. A concise introduction to light interaction with tissue was provided. The optical properties of tissue were introduced followed by a discussion about what characteristics should a model used for light propagation through tissue have. The PT theory, which is used to model light propagation through tissue, was then introduced.

Several statistical and deterministic computational methods used to model light propagation through tissue were then discussed together with their relative advantages and disadvantages. A computational method for modeling light propagation through tissue with nanostructure implants, by mapping the PTE and Maxwell equations using a phase retrieval technique, was then introduced. A detailed discussion about modeling light propagation through tissue with an implanted photonic crystal nanostructure was also presented.

References

Alerstam, E., T. Svensson, and S. A. Engels. 2008. Parallel computing with graphics processing units for high-speed Monte Carlo simulation of photon migration. *Journal of Biomedical Optics* **13**(6):060,504.

Allen, L. J. and M. P. Oxley. 2001. Phase retrieval from series of images obtained by defocus variation. *Optics Communications* **199**:65–75.

Bal, G. 2006. Radiative transfer equations with varying refractive index: A mathematical perspective. *Journal of the Optical Society of America A* **23**:1639–1644.

Baranoski, G. V. G. and A. Krishnaswamy. 2004. An introduction to light interaction with human skin. *RITA* **XI**(1):33–62.

Bayazitoglu, Y. and J. Higenyi. 1979. The higher order differential equations of radiative transfer: P3 approximation. *AIAA Journal* **17**:424–431.

Beuthan, J., O. Minet, J. Helfmann, M. Herrig, and G. Muller. 1996. The spatial variation of the refractive index in biological cells. *Physics in Medicine & Biology* **41**:369–382.

Bevilacqua, F., D. Piguet, P. Marquet, J. D. Gross, B. J. Tromberg, and C. Depeursinge. 1999. In vivo local determination of tissue optical properties: Applications to human brain. *Applied Optics* **38**:4939–4950.

Boas, D. A., M. A. O'Leary, B. Chance, and A. G. Yodh. 1994. Scattering of diffuse photon density waves by spherical inhomogeneities within turbid media: Analytic solution and applications. *Proceedings of the National Academy of Sciences, USA* **91**:4887–4891.

Born, M. and E. Wolf. 1999. *Principles of Optics*, 7th edn., Cambridge University Press, Cambridge, U.K.

Burnett, W. D. 1969. Evaluation of laser hazards to the eye and the skin. *American Industrial Hygiene Association Journal* **30**(6):582–587.

Chai, J. C., P. F. Hsu, and Y. C. Lam. 2004. Three-dimensional transient radiative transfer modeling using the finite-volume method. *Journal of Quantitative Spectroscopy & Radiative Transfer* **86**:299–313.

Chapra, S. C. and R. P. Canale. 2002. *Numerical Methods for Engineers*, 4th edn., McGraw-Hill, New York.

Cheong, W., S. A. Prahl, and A. J. Welch. 1990. A review of the optical properties of biological tissues. *IEEE Journal of Quantum Electronics* **26**(12):2166–2185.

Dehghani, H., B. Brooksby, K. Vishwanath, B. W. Pogue, and K. D. Paulsen. 2003. The effects of internal refractive index variation in near-infrared optical tomography: A finite element modeling approach. *Physics in Medicine and Biology* **48**:2713–2727.

Dunn, A. 1998. *Light Scattering Properties of Cells*, The University of Texas, Austin, TX.

Dunn, A. and R. R. Kortum. 1996. Three-dimensional computation of light scattering from cells. *IEEE Journal of Selected Topics in Quantum Electronics* **2**(4):898–905.

Edstrom, P. 2005. A fast and stable solution method for the radiative transfer problem. *SIAM Review* **47**(3):447–468.

Edwards, C. H. and D. E. Penney. 2004. *Differential Equations: Computing and Modeling*, 3rd edn., Prentice Hall, Upper Saddle River, NJ.

Enoch, S., B. Gralak, and G. Tayeb. 2002. Enhanced emission with angular confinement from photonic crystals. *Applied Physics Letters* **81**(9):1588–1590.

Evans, K. F. 1998. The spherical harmonics discrete ordinate method for three-dimensional atmospheric radiative transfer. *Journal of the Atmospheric Sciences* **55**:429–446.

Fan, S., P. R. Villeneuve, and J. D. Joannopoulos. 1997. High extraction efficiency of spontaneous emission from slabs of photonic crystals. *Physical Review Letters* **78**(17):3294–3297.

Feng, S., F. Zeng, and B. Chance. 1995. Photon migration in the presence of a single defect: A perturbation analysis. *Applied Optics* **34**(19):3826–3837.

Florescu, L. and X. Zhang. 2005. Semiclassical model of stimulated Raman scattering in photonic crystals. *Physical Review E* **72**:016,611.

Gaponenko, S. V. 2002. Effects of photon density of states on Raman scattering in mesoscopic structures. *Physical Review B* **65**:140,303(R).

Guo, Z. and K. Kim. 2003. Ultrafast-laser-radiation transfer in heterogeneous tissues with the discrete-ordinates method. *Applied Optics* **42**(16):2897–2905.

Guo, Z. and S. Kumar. 2002. Three-dimensional discrete ordinates method in transient radiative transfer. *Journal of Thermophysics and Heat Transfer* **16**(3):289–296.

Gureyev, T. E. and K. A. Nugent. 1996. Phase retrieval with the transport-of-intensity equation. II. Orthogonal series solution for non-uniform illumination. *Journal of Optical Society of America A* **13**:1670–1682.

Gureyev, T. E. and K. A. Nugent. 1997. Rapid quantitative phase imaging using the transport of intensity equation. *Optical Communications* **133**:339–346.

Gureyev, T. E., C. Raven, A. Snigireva, I. Snigireva, and S. W. Wilkins. 1999. Hard x-ray quantitative non-interferometric phase-contrast microscopy. *Journal of Physics D: Applied Physics* **32**:563–567.

Handapangoda, C. C. 2009. Modeling and numerical simulation of light propagation through biological tissue with implanted structures. PhD thesis, Monash University, Melbourne, Victoria, Australia.

Handapangoda, C. C., P. N. Pathirana, and M. Premaratne. 2011. Eigen decomposition solution to the one-dimensional time-dependent photon transport equation. *Optics Express* **19**:2922–2927.

Handapangoda, C. C. and M. Premaratne. 2008. An approximate numerical technique for characterizing optical pulse propagation in inhomogeneous biological tissue. *Journal of Biomedicine and Biotechnology* **2008**:784,354.

Handapangoda, C. C. and M. Premaratne. 2009. Implicitly causality enforced solution of multidimensional transient photon transport equation. *Optics Express* **17**(26):23,423–23,442.

Handapangoda, C. C., M. Premaratne, L. Yeo, and J. Friend. 2008a. Laguerre Runge-Kutta-Fehlberg method for simulating laser pulse propagation in biological tissue. *IEEE Journal of Selected Topics in Quantum Electronics* **14**:105–112.

Handapangoda, C. C., Malin Premaratne, David M. Paganin, and Priyantha R. D. S. Hendahewa. 2008b. Technique for handling wave propagation specific effects in biological tissue: Mapping of the photon transport equation to Maxwell's equations. *Optics Express* **16**(22):17,792–17,807

Joannopoulos, J. D., R. D. Meade, and J. N. Winn. 1995. *Photonic Crystals: Molding the Flow of Light*, Princeton University Press, Princeton, NJ.

Kawrakow, I. and M. Fippel. 2000. Investigation of variance reduction techniques for Monte Carlo photon dose calculation using XVMC. *Physics in Medicine and Biology* **45**:2163–2183.

Khan, T. and H. Jiang. 2003. A new diffusion approximation to the radiative transfer equation for scattering media with spatially varying refractive indices. *Journal of Optics A: Pure and Applied Optics* **5**:137–141.

Kim, A. D. 2004. Light propagation in biological tissues containing an absorbing plate. *Applied Optics* **43**(3):555–563.

Kim, A. D. and M. Moscoso. 2002. Chebyshev spectral methods for radiative transfer. *SIAM Journal on Scientific Computing* **23**:2074–2094.

Kubelka, P. 1948. New contributions to the optics of intensely light-scattering materials. Part I. *Journal of Optical Society of America* **38**:448–457.

Kubelka, P. 1954. New contributions to the optics of intensely light-scattering materials. Part II: Nonhomogeneous layers. *Journal of Optical Society of America* **44**:330–335.

Kumar, S., K. Mitra, and Y. Yamada. 1996. Hyperbolic damped-wave models for transient light-pulse propagation in scattering media. *Applied Optics* **35**:3372–3378.

Li, J. L., and M. Gu. 2009. Gold-nanoparticle-enhanced cancer photothermal therapy. *IEEE Journal of Selected Topics in Quantum Electronics* **16**(4):989–996.

Lourtioz, J. M., H. Benisty, V. Berger, and J. M. Gerard. 2008. *Photonic Crystals: Towards Nanoscale Photonic Devices*, 2nd edn. Springer Verlag, Berlin, Germany.

Martí-López, L., J. Bouza-Domínguez, R. A. Martínez-Celorio, and J. C. Hebden. 2006. An investigation of the ability of modified radiative transfer equations to accommodate laws of geometrical optics. *Optics Communications* **266**:44–49.

Martinez, A., H. Miguez, A. Griol, and J. Marti. 2004. Experimental and theoretical analysis of the self-focusing of light by a photonic crystal lens. *Physical Review B* **69**:165,119.

Menguc, M. P. and R. Viskanta. 1985. Radiative transfer in three-dimensional rectangular enclosures containing inhomogeneous, anisotropically scattering media. *Journal of Quantitative Spectroscopy and Radiative Transfer* **33**:533–549.

Meyer, R. A. 1979. Light scattering from biological cells: Dependence of backscatter radiation on membrane thickness and refractive index. *Applied Optics* **18**(5):585–588.

Mitra, K. and S. Kumar. 1999. Development and comparison of models for light-pulse transport through scattering-absorbing media. *Applied Optics* **38**(1):188–196.

Mitra, K., M. S. Lai, and S. Kumar. 1997. Transient radiation transport in participating media within a rectangular enclosure. *Journal of Thermophysics and Heat Transfer* **11**(3):409–414.

Modest, M. F. 2003. *Radiative Heat Transfer*, 2nd edn., Academic Press, San Diego, CA.

Mourant, J. R., J. P. Freyer, A. H. Hielscher, A. A. Eick, D. Shen, and T. M. Johnson. 1998. Mechanisms of light scattering from biological cells relevant to noninvasive optical-tissue diagnostics. *Applied Optics* 37(16):3586–3593.

Niemz, M. H. 2004. *Laser-Tissue Interactions: Fundamentals and Applications*, 3rd edn., Springer, Berlin, Germany.

Paganin, D. M. 2006. *Coherent X-Ray Optics*, Oxford University Press, New York.

Patterson, M. S., B. C. Wilson, and D. R. Wyman. 1991. The propagation of optical radiation in tissue 1. Models of radiation transport and their application. *Lasers in Medical Science* 6:155–168.

Prahl, S. A., J. C. van Gemert, and A. J. Welch. 1993. Determining the optical properties of turbid media by using the adding-doubling method. *Applied Optics* 32(4):559–568.

Prasad, P. N. 2003. *Introduction to Biophotonics*, John Wiley & Sons Inc., Hoboken, NJ.

Premaratne, M., E. Premaratne, and A. J. Lowery. 2005. The photon transport equation for turbid biological media with spatially varying isotropic refractive index. *Optics Express* 13(2):389–399.

Press, W. H., S. A. Teukolsky, W. T. Vetterling, and B. P. Flannery. 1992. *Numerical Recipes in C: The Art of Scientific Computing*, Cambridge University Press, Cambridge, U.K.

Profio, A. E. 1989. Light transport in tissue. *Applied Optics* 28(12):2216–2222.

Ramamoorthi, R. and P. Hanrahan. 2001. On the relationship between radiance and irradiance: Determining the illumination from images of a convex Lambertian object. *Journal of Optical Society of America A* 18:2448–2459.

Sakami, M., K. Mitra, and P. F. Hsu. 2002. Analysis of light pulse transport through two-dimensional scattering and absorbing media. *Journal of Quantitative Spectroscopy and Radiative Transfer* 73:169–179.

Sakoda, K. 2005. *Optical Properties of Photonic Crystals*, Springer, Berlin, Germany.

Sawetprawichkul, A., P. F. Hsu, and K. Mitra. 2002. Parallel computing of three-dimensional Monte Carlo simulation of transient radiative transfer in participating media. 1–10. *Eighth AIAA/ASME Joint Thermophysics and Heat Transfer Conference*, American Institute of Aeronautics and Astronautics St. Louis, MO.

Schuster, A. 1905. Radiation through a foggy atmosphere. *Astrophysical Journal* 21:1–22.

Shendeleva, M. L. and J. A. Molly. 2007. Scaling property of the diffusion equation for light in a turbid medium with varying refractive index. *Journal of the Optical Society of America A* 24:2902–2910.

Stephens, G. L. 1988. Radiative transfer through arbitrarily shaped optical media. Part I: A general method of solution. *Journal of the Atmospheric Sciences* 45(12):1818–1836.

Szymanski, D., and S. Patela. 2005. Modeling of photonic crystals. *International Students and Young Scientists Workshop: Photonics and Microsystems, IEEE*, Dresden, Germany, pp. 79–82.

Taflove, A. and S. C. Hagness. 2005. *Computational Electrodynamics: The Finite-Difference Time-Domain method*, 3rd edn., Artech House, Inc., Boston, MA.

Tan, Z. M. and P. F. Hsu. 2002. Transient radiative transfer in three-dimensional homogeneous and non-homogeneous participating media. *Journal of Quantitative Spectroscopy & Radiative Transfer* 73:181–194.

Teague, M. R. 1983. Deterministic phase retrieval: A Green's function solution. *Journal of the Optical Society of America* 73:1434–1441.

Thomas, G. E. and K. Stamnes. 1999. *Radiative Transfer in the Atmosphere and Ocean*, Cambridge University Press, Cambridge, U.K.

Tuchin, V. 2007. *Tissue Optics: Light Scattering Methods and Instruments for Medical Diagnosis*, 2nd edn., SPIE Press, Washington, DC.

Tuchin, V. 1997. Light scattering study of tissues. *Physics-Uspekhi* 40(5):495–515.

van de Hulst, H. C. 1980. *Multiple Light Scattering*, Vol. 1, Academic Press, New York.

Waynant, R. W. and V. M. Chenault. 1998. Overview of non-invasive fluid glucose measurement using optical techniques to maintain glucose control in diabetes mellitus. *IEEE LEOS Newsletter* 12(2):3–6.

Welch, A. J. and M. J. C. Van Gemert. 1995. *Optical-Thermal Response of Laser-Irradiated Tissue*, Plenum Press, New York.

Wilson, B. C., V. V. Tuchin, and S. Tanev. 2005. *Advances in Biophotonics*, IOS Press, Amsterdam, the Netherlands.

Wu, C. Y. and S. H. Wu. 2000. Integral equation formulation for transient radiative transfer in an anisotropically scattering medium. *International Journal of Heat and Mass Transfer* **43**:2009–2020.

Yamada, Y. and Y. Hasegawa. 1996. Time-dependent FEM analysis of photon migration in biological tissues. *JSME International Journal Series B* **39**(4):754–761.

Yoon, G., A. J. Welch, M. Motamedi, and M. C. J. van Gemert. 1987. Development and application of three-dimensional light distribution model for laser irradiated tissue. *IEEE Journal of Quantum Electronics* **23**:1721–1733.

12

Defense Applications for Nanophotonics

Suman Shrestha, Mohit Kumar, Aditi Deshpande, and George C. Giakos

CONTENTS

12.1 Introduction

Nanophotonics is a study of characterizing, building, and manipulating optically active nanostructures and nanoparticles for producing new capabilities in instrumentation for the nanoscale, chemical and biomedical sensing, information and communication technologies, enhanced solar cells and lighting, disease treatment, environmental remediation, and many other applications. Photonics at the nanoscale, or nanophotonics, is defined as *"the science and engineering of light-matter interactions that take place on wavelength and sub-wavelengths scale where the physical, chemical or structural nature of natural or artificial nano-structures matter controls the interactions"* [1]. This field deals with a number of important and interesting topics in optics and material structures at nanometer scale and their applications and uses in general. Nanophotonics aims at controlling the optical energy and its conversion on the nanometer-scale range by combining different properties of metals, semiconductors, polymers, and dielectric materials to create new states of light and matter. These new, combined materials formed as a result of this combination are called metamaterials. Some of the targets of the scientist in this field are pointed out [2]:

- Controlled quantum coupling at the nanoscale
- Understanding ultrafast processes at ultrasmall length scales
- New routes to functional nanophotonic materials
- Efficient energy transport in plasmonic nanostructures

The evolution of optical and electromagnetic computational techniques, such as boundary element methods (BEM) and finite-difference frequency domain (FDFD), contributed significantly to the development and enhancement of already existing fabrication techniques; accessibility to high-resolution patterning and pattern transfer processes have produced photonic crystal, microdisk, and ring resonator with exceptional performance [1].

Over the last few years, there has been an enormous rise in the number of diverse photonics applications, which has resulted in significant advances in computational design, fabrication techniques, and development of new optical and structural characterization methods.

Nanophotonics has significant applications in the field of lasers, imaging, flat-panel displays, defense and military systems, medicine, optical microscopy, harvesting solar energy, optoelectronic chips, telecommunications, etc.

Many applications are on the way of being researched and commercialized like the resonant cavity quantum well lasers and microcavity-based single-photon sources, culminating tremendous research opportunities in the area of communications, biochemical sensing, and quantum cryptography. The various processes and techniques in nanophotonics that have been classified into three branches are shown in Figure 12.1.

Nanostructures have distinctive, expedient, and tunable optical properties that crop up due to their nanoscale size from the fact that they are smaller than the wavelength for which they are usually designed, and these properties are essential in order to determine optical response. The sub-wavelength optical confinement and low optical loss of nano-photonic devices boost the interaction between light and matter within these structures. Localization of light and applications to metallic mirrors, photonic crystals, optical waveguides, microresonators, and plasmonics are gaining incredible practical interest.

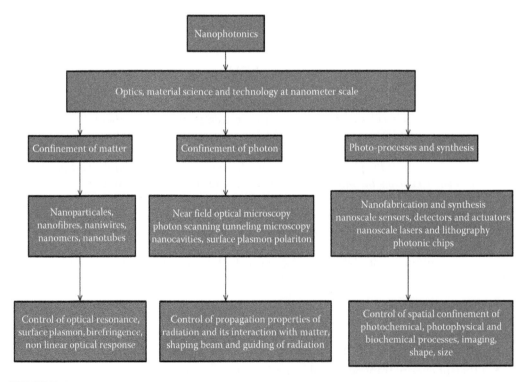

FIGURE 12.1
A block diagram representing various major processes and techniques available in nanophotonics. (From Ghoshal, S.K. et al., Nanophotonics for 21st century, *Optoelectronics-Devices and Applications*, P. Predeep (ed.), ISBN: 978-953-307-576-1, 2011, Intech, Available from: http://www.intechopen.com/books/optoelectronics-devices-and-applications/nanophotonics-for-21st-century)

The areas of nanophotonics can be exemplified based on their physical phenomenon and modulation of index of refraction in the nanoscale material as photonic crystals, metamaterials, plasmonics, and confined semiconductor structures [2].

The main objective of nanophotonics research is to manage optical energy and its conversion on nanometer scale by combining the properties of metal, organic, semiconductor, organometallic, polymers, and dielectric materials to generate new states of light and matter called metamaterials. Nanophotonics has also been benefited by the access to optical nanoscale characterization tools like scanning near-field optical microscopy (SNOM) as these tools have become user-friendly and also from structural characterization tools such as atomic force microscopy (AFM), nano-secondary ion mass spectrometry (nano-SIMS), scanning electron microscopes (SEMs), and transmission electron microscopes (TEMs), as these tools had major impact on the ability to correlate size, atomic structure, and spatial arrangements of nanostructures (Figure 12.2) [1].

The advancement in simulation and fabrication of nanophotonic structures has resulted in formation of ultrahigh-quality optical structures that allow researchers to examine different optical states, confine and slow the velocity of light, and create efficient light-emitting sources and strongly coupled light–matter interactions ensuing in new quantum mechanical states (Figure 12.3) [1].

The vital characteristics for building nanophotonic devices are the following: nanoscale quantum dots (QDs) should be formed in high density, QDs are supposed to be in isolated condition when not in photoexcited state, nanorods or nanowires ought to have large quantum yield and aspect ratio, highly excited level energy must be controlled by material mixture ratio and dimension, and the light element excitation should be at stable room temperature when high-excited level do not degenerate (Figure 12.4) [2].

Understanding the fundamentals of nanophotonics is important because to fabricate a test structure, it is necessary to focus on a window of parameters and to evaluate this new knowledge from both novel science and latent applications [2].

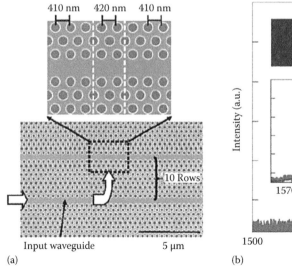
(a)

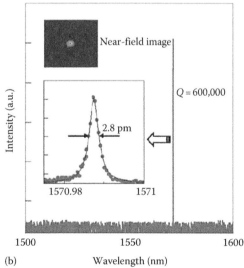
(b)

FIGURE 12.2
(a) SEM of fabricated photonic crystal structure. (b) Spectrum of cavity with Q=600,000. (From Song, B.-S.S. et al., *Nat. Mater.*, 4, 207, 2005.)

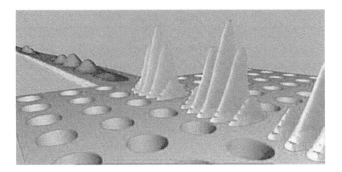

FIGURE 12.3
Illustration of pulse compression and increase in pulse intensity after entering into slow-light regime. (From Krauss, T., *J. Phys. D: Appl. Phys.*, 40(9), 2666, 2007.)

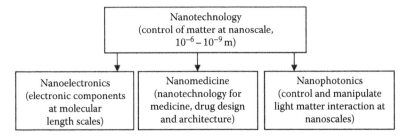

FIGURE 12.4
Block diagram showing the three major branches of nanotechnology. (From Ghoshal, S.K. et al., 2011, Nanophotonics for 21st century, *Optoelectronics-Devices and Applications*, P. Predeep (Ed.), ISBN: 978-953-307-576-1, Intech, Available from: http://www.intechopen.com/books/optoelectronics-devices-and-applications/nanophotonics-for-21st-century)

12.2 Nanomaterials and Surface Plasmonics

Nanoscale confinement of matter is used to direct the band gap and the optical resonance; in addition to this, it also manages the excitation dynamics, and moreover, it is also used to produce periodic optical domains that lead to photon localization [4]. Nanoscale confinement of domains produces QDs, quantum wells, and quantum wires in inorganic semiconductors that are eminent and extensively deliberated (Figure 12.5) [5,6]

Surface-enhanced Raman spectroscopy (SERS) being one of the applications of metallic nanostructures in 1970s, the field of plasmonics has grown eventually in the late 1990s and 2000s and has gained importance as popular components of metamaterials. This field was enhanced by the development of surface plasmon resonance (SPR)-based sensor in 1991 [7]. A diverse set of plasmonics applications has emerged in the last decade, which includes development of high-performance near-field optical microscopy (NSOM) and biosensing methods (Figure 12.6) [1].

More recently, many new technologies have emerged, which include thermally assisted magnetic recording [8], thermal cancer treatment [9], catalysis and nanostructure growth [10], solar cells [11,12], and computer chips [13,14], which efficiently use plasmonics.

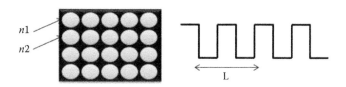

FIGURE 12.5
Basic structure of a photonic crystal.

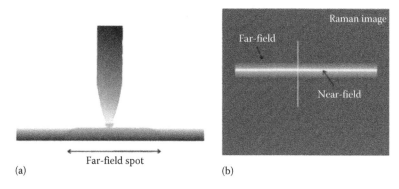

FIGURE 12.6
(a) Schematic on tip of sample of a 1D nanostructure, (b) optical field enhancement Raman spectroscopy. (From Roy, D. et al., *J. Appl. Phys.*, 105, 013530, 2009.)

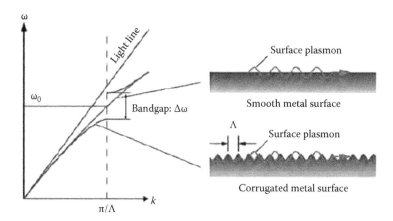

FIGURE 12.7
SPR on flat and conjugated metal surfaces. (From Ghoshal, S.K. et al., 2011, Nanophotonics for 21st century, *Optoelectronics-Devices and Applications*, P. Predeep (Ed.), ISBN: 978-953-307-576-1, Intech, Available from: http://www.intechopen.com/books/optoelectronics-devices-and-applications/nanophotonics-for-21st-century)

High-dielectric constant materials can be effectively used as antennas, waveguides, and resonators (Figure 12.7) [15,16]

Materials that exhibit strong optical resonances are predominantly motivating because they can exhibit high positive, negative, and near-zero magnitudes of dielectric constant [1]. After apprehending the fact that long-distance information transport on chips with plasmonic waveguides would suffer strongly from heating effects [17], it has been recognized that modulators and detectors that meet the power, speed, and material requirements

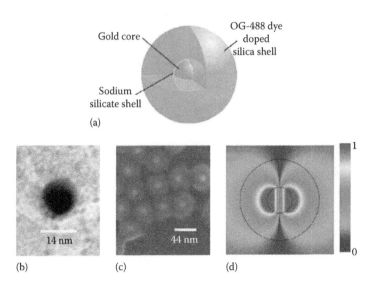

FIGURE 12.8

Nonlinear optical effects from surface plasmons. (a) Diagram of the hybrid nanoparticle architecture (not to scale), indicating dye molecules throughout the silica shell, (b) transmission electron microscope image of Au core, (c) scanning electron microscope image of Au/silica/dye core-shell nanoparticles and, (d) spaser mode (in false color), with $\lambda = 525$nm and $Q = 14.8$; the inner and the outer circles represent the 14-nm core and the 44-nm shell, respectively. The field strength color scheme is shown on the right. (From Noginov, M.A. et al., *Nature*, 460, 1110, 2009.)

essential to incorporate plasmonics with complementary metal-oxide-semiconductor (CMOS) technology can be attained (Figure 12.8) [1].

Plasmonic sources that have potential of proficiently coupling quantum emitters to single, well-defined optical node find numerous applications in the field of quantum plasmonics and power-efficient chip-scale optical sources [18,19], and the recent prophecy and awareness of coherent nanometallic light sources constitutes an extremely significant development [20–23].

12.3 Design Techniques and Nanotechnologies

There have been many emerging technologies for nanophotonics in the past few years. Composite photonic structures based on silica and polymer artificial opals obtain 3D omnidirectional photonic band gap for optical frequencies. Various infiltration techniques such as chemical bath deposition (CBD), electrodeposition, sol-gel method, atomic layer deposition, and chemical vapor deposition (CVD) are used to insert different substances into pores of initial templates resulting in composite photonic crystal that exhibits all desirable properties (Figure 12.9) [24,25].

Semiconductor nanostructures with anisotropic optical and electronic properties present vital material class for bottom-up design of nanodevices and optical functionalization of hybrid optically active nanostructures [26]. Polymer photonic devices aim to manipulate light at micro- and nanoscale to create highly resourceful cost-efficient systems such as nanoimprint lithography using low-cost polymer materials that hold application in areas of communications, environmental lighting, sensing, and optofluidics [27,28]. Optical integration, functional 1D confined hybrid organic–inorganic nanotechnology promises high speed and greater device versatility [7].

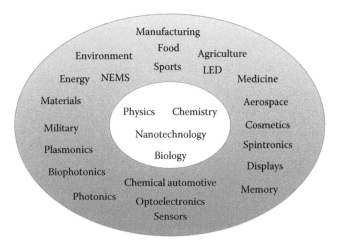

FIGURE 12.9
Different branches of nanotechnology in the field of science and engineering. (From Ghoshal, S.K. et al., 2011, Nanophotonics for 21st century, *Optoelectronics-Devices and Applications*, P. Predeep (Ed.), ISBN: 978-953-307-576-1, Intech, Available from: http://www.intechopen.com/books/optoelectronics-devices-and-applications/nanophotonics-for-21st-century)

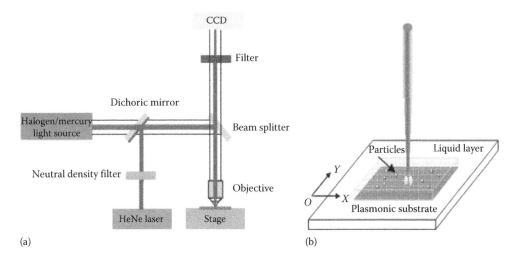

FIGURE 12.10
Experimental setup on plasmonic substrate. (From Ghoshal, S.K. et al., 2011, Nanophotonics for 21st century, *Optoelectronics-Devices and Applications*, P. Predeep (Ed.), ISBN: 978-953-307-576-1, Intech, Available from: http://www.intechopen.com/books/optoelectronics-devices-and-applications/nanophotonics-for-21st-century)

The experimental design setup for trapping and concentration experiments on plasmonic substrate is as shown in Figure 12.10.

The length, flexibility, and strength of these structures facilitate their manipulation and attachment to surfaces including optical linking of nanoribbon and other nanostructured elements to form networks. The bottom-up assembly of 1D confined nanostructure on processable CMOS-compatible substrates with nanowire light sources and detectors could constitute as a considerable step toward building nanostructured electronic and photonic hybrid circuitry [29].

12.4 Defense Applications

The general class of the impending applications of nanophotonics includes sensing, command, control, communications, computing, countermeasures, and power. Sensing holds large applications among all these classes and it subcategorizes into imaging sensors that include night vision as well as long and mid-wave infrared (IR) sensors, chemical sensors, biosensing and advanced spectroscopy for sensing, and materials characterization. Mission-specific areas of nanophotonics include health monitoring, tagging, tracking, locating or eavesdropping, and situational awareness [30].

Command, control, and communications applications can be facilitated for a variety of potential needs such as platform avionics, remote actuation of devices, and emissive displays for monitoring, and the field of secure communications includes quantum key distribution. Computing applications are anticipated in robust computation, computing systems, microprocessors, data storage, and optical signal processing. Countermeasure applications include sensor protection, bioremediation, and decontamination. Power applications include micropower, solar cells, and energy harvesting [30].

The two new companies, Konarka and Nanosolar, offer nanostructured PV materials. Konarka focuses on expansion and encroachment of nano-enabled polymer photovoltaic materials that are lightweight, flexible, and more adaptable than traditional solar materials, whereas Nanosolar integrates nanostructured material components as basis of large-area printable photovoltaic structures (Figure 12.11) [30].

As nanophotonics senses a broad range of chemicals like explosives, organisms like viruses, as well as physical effects on materials such as temperature, strain, and vibrations, remote sensing uses a device to gather information about the environment from a distance (Figure 12.12) [30].

Photonic nanostructures can be integrated into diverse media such as fibers, nanocomposites, and fabrics (Figure 12.13) [30].

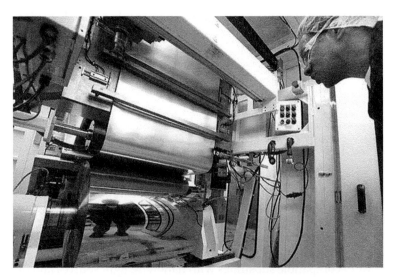

FIGURE 12.11
Nanostructured roll-to-roll process solar materials. (Image courtesy of Nanosolar, Inc.)

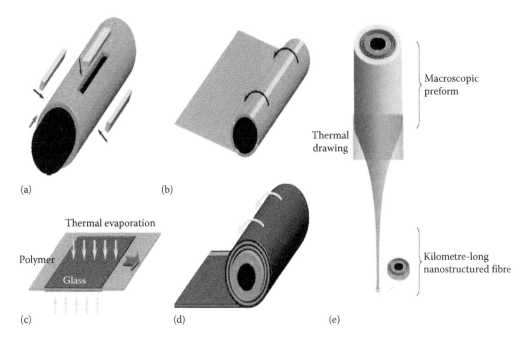

FIGURE 12.12
Nanostructured integrated fiber devices. (From Abouraddy, A.F. et al., *Nat. Mater.*, 6, 336, 2007.)

Emissive displays have two nanophotonic components, which include use of nanoparticles as emitters and use of photonic band-gap structures. Periodic dielectric structures operate as large-area vertically emitting lasers that do not necessitate an external cavity, and it can be used to control bandwidth and directionality of light emitted by a variety of sources.

Near-field optics allows focusing electromagnetic radiation far beyond the diffraction limit that can be oppressed for ultrahigh-density data storage applications. Be it mapping of computation bits or communications to quantum mechanical states, the use of phase of state holds applications in securing communications or quantum cryptography. The IR spectral region holds a vital role in military systems. Night vision refers to high-gain systems that amplify and detect near-IR ambient radiation [30].

Raman spectrum of each molecule is a fingerprint that exceptionally recognizes the molecule. Surface plasmons have been used to boost the surface sensitivity (Figure 12.14) [30].

The large enhancement factor for SERS facilitates the detection of molecules with enormously weak Raman cross sections.

12.5 Nanophotonics in Space Defense

The implementation of nanophotonics in space research will strongly depend on the development of commercial space activities and the realization of demanding scientific missions in the future.

Defense applications have previously involved radars or more specifically advanced synthetic aperture radars (SAR) [31]. Hooper et al. used an airborne imaging system consisting of multiple cameras to take images remotely in several spectral bands [32].

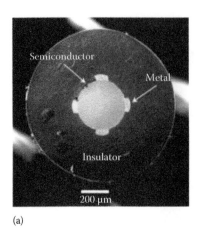

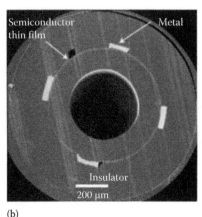

(a) (b)

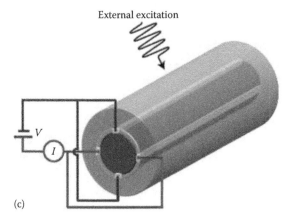

(c)

FIGURE 12.13
(See color insert.) Insulated fiber devices. (a) SEM micrograph of a cross-section (the semiconductor is As$_{40}$Se$_{50}$Te$_{10}$Sn$_5$, the insulator polymer is PES, and the metal is Sn). (Reprinted from Abouraddy, A.F. et al., *Nat. Mater.*, 6, 336, 2007.) (b) SEM micrograph of a thin-film fiber device (the semiconductor is As$_2$Se$_3$, the insulator polymer is PES, and the metal is Sn). (c) Electrical connection of the four metal electrodes at the periphery of the fiber to an external electrical circuit. (From Abouraddy, A.F. et al., *Nat. Mater.*, 6, 336, 2007.)

Also, Wang et al. used polarimetric properties to estimate refractive indices of interrogated objects [33]. Giakos used a multispectral imaging technique for image enhancement [34,35]. The polarimetric Bidirectional Reflectance Distribution Function (pBRDF) principles aimed at the development of efficient polarimetric space surveillance LADAR sensor architectures for enhanced object detection and characterization are presented in [36,37].

Image formation through detection of the polarization states of light offers distinct advantages for a wide range of detection and classification problems and has been explored by a number of authors, given the intrinsic potential of optical polarimetry to offer high-contrast, high-specificity images under low-light conditions [33–39].

Blending polarimetry with nanotechnology would lead to the development of enhanced resolution, functionality, and high-discrimination potential imaging sensors. Interestingly enough, innovative and efficient photon detection trends, based on nanotechnology design principles for space-based surveillance and situational awareness missions, have been introduced by the Cardimona group at air force research laboratory (AFRL) [47].

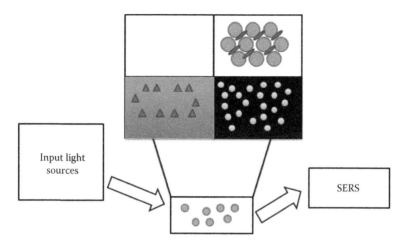

FIGURE 12.14
SERS.

Nanostructured optoelectronics (e.g., QD lasers and photonic crystals) offers wide and promising applications in the field of space surveillance research in the field of optical satellite communications or sensor technology (IR sensor, high-resolution charge coupled device (CCD) cameras, etc.). For the data transmission, extremely frequency-stable solid-state lasers are used, which are pumped with diode lasers [34].

QD lasers offer a new degree of freedom in selecting the working wavelength of photonic elements. They almost cover the entire spectral region ranging from the ultraviolet to far IR region with a small number of substrate materials. Other advantages that the QD lasers offer are the small energy consumption through low threshold current densities, a high modulation range for high-speed applications, as well as improved temperature stability. Due to their radiation hardness and the low energy consumption, QD lasers in principle are relevant for space applications and hence proving one of the applications of nanophotonics in space surveillance. However, appropriate measures have to be taken by the space industry for the specification, system integration, and space qualification to apply the potential application of QDs in space technology applications [43,48].

Harsh and extreme ambient conditions in space as well as intense high-energy cosmic radiation, extreme temperatures, high vacuum, and extreme mechanical and thermal conditions during launch and reentry pose significant technological challenges on the widespread diffusion of nanotechnology by limiting the inherent advantages of certain nanomaterials and nanocomponents. However, there are nanocomponents operating on inherently robust physical principles such as quantum-dot-based detectors and sensors while offering at the same time unique miniaturization capabilities [46]. For instance, there exist a variety of nanocomponents that can be operated on magnetoelectronic memories, QD lasers, QD refrigeration, quantum interference, frequency agile detection, and photonic crystals offering distinct advantages for space applications [48]. As a result, the extreme and harsh operational requirements in spaceflight do not represent a general barrier for nanotechnology products but should be regarded differently for each respective component [43]. Although miniaturization of space systems can be perceived as a limiting factor of inherent functionalities of the optical imagery systems and space instrumentation, in reality, it can offer substantial advantages such as cost reduction, energy saving, and reduced form factor.

Several polarimetric systems have been developed that are capable for the characterization of objects. All the systems developed have their own level of automation that can perform the measurements quickly and efficiently. Liu et al. [40] developed a system utilizing liquid crystal optical devices, PC control, and CCD images. This system could extract details not seen by non-polarization-sensitive imaging. A polarimetric system that was mostly focused on automated calibration techniques is reported in [41], while a polarimetric system utilizing manual insertion of components is described in [42], while the design principles of an automated system liquid crystal polarimetric system of a multidomain LIDAR system for space surveillance have been presented in [43–45].

The interaction of light with space materials can yield numerous potential applications in defense and security. Different electronics have been developed that are very much useful for space research. Space-based electronics systems have been an important research aspect in the field of nanophotonics in the last few years. Applications of nanophotonics in space technology appear to be feasible in a short- to medium-term time horizon, which could lead to major improvements in the area of lightweight and strong space structures, improved systems and components of energy production and storage, data processing and transmission, sensor technology, as well as life support systems.

As the size of the system becomes smaller, the design constraints regarding optical instrumentation and radiation shielding to protect against the cosmic radiation become more difficult and challenging. Therefore, miniaturized radiation resilient electronics and high-strength materials, as well as materials with programmable intrinsic sensing and compensating properties, are needed to meet the demands of new space systems designs. All the technologies for the fabrication of nanophotonics electronics require the following [48] (Figure 12.15):

1. Higher precision tools to control size, shape, and placement
2. Larger reproducibility
3. Improved control over material's purity
4. New ways to integrate and interface with electronics and conventional optics

Another application of nanophotonics in space research is the use of photonic crystals, which offer potential application in optical data communication. It exhibits a periodic refractive index with a photonic band gap for certain frequency in the visible and IR wavelength range. The lattice constant of photonic crystals lies in the range of half the wavelength of the light in the medium. For visible light, this means that for the production of photonic crystals, a precision within the range of 10 nm is necessary. Nowadays, 2D forms of these crystals can be manufactured with high precision. However, great efforts are being applied for the development of 3D forms with the utilization of lithography and self-organization procedures, in which nanoscale colloids (e.g., from polymers or silicates) arrange spontaneously to a cubic lattice. This development would open up new possibilities in optical data communication (light could be guided and branched to arbitrary directions) and offer in principle the potential for the realization of purely optical circuits (optical computing). Such photonic transistors are however at present still very far from realization [35]. In the long run, photonic crystals will find applications in optical satellite communications.

12.5.1 Computational Nanophotonics with MATLAB®

In the following example, a Mueller matrix polarimeter is considered; a light source (from a laser) interrogates an object, consisting of nanoparticles or a group of molecules, with a

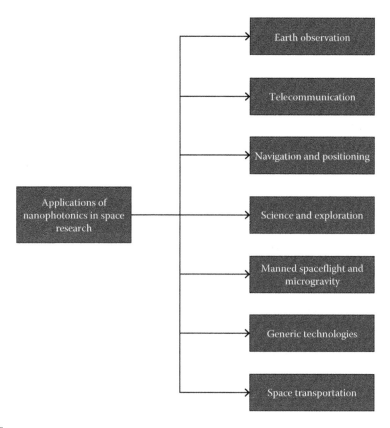

FIGURE 12.15
Applications of nanophotonics in space research. (From Giakos, G., Future directions: Opportunities and challenges, in *Computational Nanotechnology: Modeling and Applications with MATLAB®*, S.M. Musa (ed.), Taylor & Francis, Boca Raton, FL, 2012, Chapter 13.).

series of generator states g_i, and analyzes the light that has interacted with the object with a set of analyzer states a_i recording a set of flux measurements arranged in a vector P. The Mueller matrix of the object will be calculated [38,39,49–52].

As well known from [49], the output Stokes vectors S' are related to the input S via the Mueller matrix of the sample:

$$\begin{bmatrix} S'_0 \\ S'_1 \\ S'_2 \\ S'_3 \end{bmatrix} = \begin{bmatrix} m_{11} & m_{12} & m_{13} & m_{14} \\ m_{21} & m_{22} & m_{23} & m_{24} \\ m_{31} & m_{32} & m_{33} & m_{34} \\ m_{41} & m_{42} & m_{43} & m_{44} \end{bmatrix} \cdot \begin{bmatrix} S_0 \\ S_1 \\ S_2 \\ S_3 \end{bmatrix} \tag{12.1}$$

A data reduction technique [39,49] is applied to calculate the Mueller matrix of the object. For this, we must first know the configurations of the polarization state generator and the polarization state analyzer for each of the q intensity measurements. These configurations lead to a q × 16 polarimetric measurement matrix, **W.** Let G_q denote the Stokes vector for the generator states and A_q denote the Stokes vector of the analyzer states where "q" represents the states of the experiment. Also let P_q denote the measured intensity of each of the states. Let M denote the Mueller matrix of the object. Then,

$$P_q = A_q^T.M.S_q = \begin{bmatrix} A_{q,1} & A_{q,2} & A_{q,3} & A_{q,4} \end{bmatrix} \begin{bmatrix} m_{11} & m_{12} & m_{13} & m_{14} \\ m_{21} & m_{22} & m_{23} & m_{24} \\ m_{31} & m_{32} & m_{33} & m_{34} \\ m_{41} & m_{42} & m_{43} & m_{44} \end{bmatrix} \begin{bmatrix} G_{q,1} \\ G_{q,2} \\ G_{q,3} \\ G_{q,4} \end{bmatrix}$$

$$= \sum_{j=1}^{4} \sum_{k=1}^{4} A_{q,j} m_{j,k} G_{q,k} \tag{12.2}$$

The analyzer and generator matrix can be ordered as in Equation 12.1 where it can be expressed as a 16×1 measurement vector for the qth state [9]:

$$W_q = \begin{bmatrix} A_{q,1}G_{q,1} & A_{q,1}G_{q,2} & A_{q,1}G_{q,3} & \cdots & A_{q,4}G_{q,4} \end{bmatrix}^T \tag{12.3}$$

Then, the Mueller matrix can be into a 16×1 vector thus giving 16 intensities for each state represented by P_q. Thus,

$$P_q = \begin{bmatrix} A_{q,1}G_{q,1} \\ A_{q,1}G_{q,2} \\ A_{q,1}G_{q,3} \\ . \\ . \\ A_{q,4}G_{q,4} \end{bmatrix} \begin{bmatrix} m_{11} \\ m_{12} \\ m_{13} \\ . \\ . \\ m_{44} \end{bmatrix} \tag{12.4}$$

since there are 16 measurements to be taken that makes $q = 16$. Thus, W_q turns to be a 16×16 matrix, which when multiplied by 16×1, Mueller vector gives 16×1 intensity vector, each element of a vector representing an intensity for each state. Thus, the Mueller matrix of the object can be determined as

$$M = W^{-1} \cdot P \tag{12.5}$$

If W contains 16 linearly independent rows, all the elements of the Mueller matrix can be determined using Equation 12.5. But if $q > 16$ and M is overdetermined, then the optimal polarimetric data reduction equation uses the pseudoinverse W_p^{-1} of W and is given by Equation 12.6:

$$M = [W^T W]^{-1} W^T P = W_p^{-1} P \tag{12.6}$$

The advantages of the polarimetric data reduction technique are as follows:

1. This method can be easily understood and implemented.
2. It is independent of the specific type of light source or detector used. If the Stokes vector associated with the generator and analyzer arm are determined, the effect of the nonideal elements is corrected.
3. It can be used for overdetermined measurement states, that is, $q > 16$, using Equation 12.6.

Mueller matrix of an element consists of various optical properties of an element. In order to know about these properties, Mueller matrix can be decomposed into three other matrices that can be used to further characterize the materials. These three matrices are the depolarization, diattenuation, and retardance matrix. This algorithm of decomposing the Mueller matrix into these three matrices is called Lu–Chipman decomposition [49–51] and is given by Equation 12.7:

$$M = M_\Delta \cdot M_R \cdot M_D \tag{12.7}$$

where
M_Δ is the depolarization matrix
M_R is the retardance matrix
M_D is the diattenuation matrix

For the algorithm of the decomposition of Mueller matrix to the three matrices, refer to [39–41]. The code following the algorithm has been shown in the following section.

MATLAB code:

```
%Define elements of a an analyzer (A, 16×4) and a generator vector(G,4×16)
where each row of%the analyzer and column of the generator matrix
represent a measurement state(qi).A minimum%of 16 states (i=1 to 16) are
required to determine the 16 elements of the target Mueller
matrix.%For q=16:
```

```
fprintf('Input Generator matrix');
Generator=[S1; S2; S3; S4; S5; S6;;S7; S8; S9; S10; S11; S12; S13; S14;
S15; S16]
fprintf('Input Analyser matrix');
Analyser=[A1; A2; A3; A4; A5; A6; A7; A8; A9; A10; A11; A12; A13; A14;
A15; A16]'
```

```
%Calculation of W matrix (Q×16 matrix, polarimetric measurement matrix)
```

```
W=zeros(16,16);
for q=1:16
   k=1;
   for i=1:4
      for j=1:4
      W(q,k)=Analyser(i,q)*Generator(q,j);
      k=k+1;
      end;
   end;
end;
```

```
fprintf(`W matrix');
%- - - - - - - - - - - - - - - - - - - - - - - - - - - - - - - - -
%Calculation of Mueller matrix
M=zeros(4,4);
   tempM=inv(transpose(W)*W)*transpose(W)*P(:,s);
   tempM=tempM';
   t=1;
   for k=1:4
   for j=1:4
      M(k,j)=tempM(t);
   t=t+1;
   end;
   end;
   MaxM=max(M); %Returns the vector containing maximum element
   MaxMM=max(MaxM); %Returns largest element in the matrix
   format short;
   digits(2);
   M=M./MaxMM; %Normalized Mueller matrix of the target
   fprintf(`The Mueller matrix of the target:');
%Putting the matrix elements in column form
   M_column=[M(1,1); M(1,2); M(1,3); M(1,4); M(2,1); M(2,2);…
      M(2,3); M(2,4); M(3,1); M(3,2); M(3,3); M(3,4);…
      M(4,1); M(4,2); M(4,3); M(4,4)];
%M is the final Mueller matrix of the object.
   After the Mueller matrix has been computed, it can be decomposed to
   three other matrix as discussed before.
%M will be taken of the form [1 D'
%                             P m] for decomposition done below
%- - - - - - - - - - - - - - - - - - - - - - - - - - - - - - - - -
%Decomposition of Mueller matrix into Diattenuation, Depolarization and
%Retardance.
%(1) Diattenuation
%Diattenuation matrix is of the form [1 D'
%                                     D mD]
%mD=aI3+b(vector(D).vector(D)')
%D=mag(vector(D)),a=sqrt(1-D^2) and b=(1 - sqrt(1-D^2))/D^2
%D=M(1,2)^2+M(1,3)^2+M(1,4)^2
M_temp=M./M(1,1); %Normalizing the Mueller matrix for
                   %Diattenuation
D_vect=[M_temp(1,2) M_temp(1,3) M_temp(1,4)]';
mag_D_vect=sqrt(D_vect(1)^2+D_vect(2)^2+D_vect(3)^2);
a=sqrt(1 - mag_D_vect^2);
b=(1 - a)/mag_D_vect^2;
I3=[1 0 0;0 1 0;0 0 1];
mD=a*I3+b*(D_vect * D_vect');
%Diattenuation matrix
MD=[1 D_vect';D_vect mD];
%Putting the matrix elements in column form
MD_column=[MD(1,1); MD(1,2); MD(1,3); MD(1,4);…
           MD(2,1); MD(2,2);MD(2,3); MD(2,4);…
           MD(3,1); MD(3,2); MD(3,3); MD(3,4);…
           MD(4,1); MD(4,2); MD(4,3); MD(4,4)];
%MD is the decomposed diattenuation matrix
%- - - - - - - - - - - - - - - - - - - - - - - - - - - - - - - - -
```

```
%(2) Depolarization
% Depolarization matrix is of the form[1    0'
%                                       Pdelta mdelta]
%mdelta = +,-inv((mprime*mprime'+k2I))+(k1*mprime*mprime'+k3I)
%depending on the determinant of mprime.
%k1,k2,k3 are values determined with eigen values of mprime*mprime'.
%mprime is the mprime(2:4,2:4) of matrix Mprime.
%Mprime = M*inverse(MD)
%Pdelta = (vector(P) - m*vector(D))/1 - D^2.
Mprime = M_temp*inv(MD);
mprime = Mprime(2:4,2:4);
mdelta_temp = mprime*mprime';
%Eigen values
```

12.6 Conclusion

A comprehensive study aiming at identifying potential nanophotonics applications as applied to defense was presented. The overall consensus is that nanophotonics is a prominent technology to the defense arena aimed at developing highly integrating solutions and miniaturized devices operating at reduced power with excellent electrical and optical signal figures of merit.

References

1. Roco, Mihail C., Mirkin, Chad A., Hersam, Mark C. 2011. *Nanotechnology Research Directions for Societal Needs in 2020*. Science Policy Reports, 1: 417–444, doi: 10.1007/978-94-007-1168-6-10.
2. Ghoshal, S.K., Sahar, M.R., Rohani, M.S. et al. 2011. Nanophotonics for 21st century. *Optoelectronics-Devices and Applications*, P. Predeep (ed.), ISBN: 978-953-307-576-1, Intech, doi: 10.5772/1036 Available from: http://www.intechopen.com/books/optoelectronics-devices-and-applications/nanophotonics-for-21st-century
3. Krauss, T. 2007. Slow light in photonic crystal waveguides. *J. Phys. D: Appl. Phys.*, 40(9): 2666–2670, doi: 10.1088/0022-3727/40/9/S07.
4. Shen, Y., Friend, C.S., Jiang, Y. et al. 2000. Nanophotonics: Interactions, materials and applications. *J. Phys. Chem. B*, 104: 7577–7587, ISSN 1089–5647.
5. Weisbuch, C., Bensity, H., and Houdre, R.J. 2000. Acoustic-phonon-assisted localization of free excitons due to interface roughness in quantum wells. *J. Lumin.*, 85: 271.
6. Lal, M., Kumar, N.D., Joshi, M.P. et al. 1998. Polymerization in a Reverse Micelle Nanoreactor: Preparation of Processable Poly(p-phenylenevinylene) with Controlled Conjugation Length. *Chem. Mater.*, 10: 1065.
7. Brongersma, M.L. and Kik, P.G. 2007. *Surface Plasmon Nanophotonics*. Dordrecht, the Netherlands: Springer, ISBN 978-1-4020-4349-9.
8. Challener, W.P., Peng, C., Itagi, A.V. et al. 2009. Heat-assisted magnetic recording by a near-field transducer with efficient optical energy transfer. *Nat. Photon.*, 3: 220–224, doi: 10.1038/nphoton.2009 2.

9. Hirsch, L.S., Stafford, R.J., Bankson, J.A. et al. 2003. Nanoshell-mediated near-infrared thermal therapy of tumors under magnetic resonance guidance. *Proc. Natl. Acad. Sci. USA*, 100(23): 3549–13554, doi: 10.1073/pnas. 2232479100.

10. Cao, L.B., Barsic, D.N., and Guichard, A.R. 2007. Plasmon-assisted local temperature control to pattern individual semiconductor nanowires and carbon nanotubes. *Nano Lett.*, 7(11): 3523–3527.

11. Pala, R.W., White, J., Barnard, E. et al. 2009. Design of photonic thin-film solar cells with broadband absorption enhancements. *Adv. Mater.*, 21: 3504–3509, doi: 10.1002/cdma. 200900331.

12. Atwater, H.P. and Polman, A. 2009. Plasmonics for improved photovoltaic devices. *Nat. Mater.*, 9(3): 205–213.

13. Cai, W.W., White, J.S., and Brongersma, M.L. 2009. Compact, high-speed and power-efficient electro optic plasmonic modulators. *Nano Lett.*, 9(12): 4403–4411.

14. Tang, L.S.E., Kocabas, S., Latif, A.K. et al. 2008. Nanometer-scale germanium photo detector enhanced by a near-infrared dipole antenna. *Nat. Photon.*, 2: 226–229, doi: 10.1038/nphoton. 2008 30.

15. Cao, L.W., White, J.S., Park, J.-S. et al. 2009. Engineering light absorption in semiconductor nanowires devices. *Nat. Mater.*, 8: 643–647, doi: 10.1038/nmat2477.

16. Schuller, J.A., Taubner, T., and Brongersma, M.L. 2009. Optical antenna thermal emitters. *Nat. Photon.*, 3: 658–661, doi: 10.1038/nphoton.2009 188.

17. Zia, R., Schuller, J.A., Chandran, A. et al. 2006. Plasmonics: The next chip-scale technology. *Mater. Today*, 9: 20–27.

18. Akimov, A.M., Mukherjee, A., Yu, C.L. et al. 2007. Generation of single optical plasmonics in metallic nanowired coupled to quantum dots. *Nature*, 450: 402–406.

19. Hryciw, A.J., Jun, Y.C., and Brongersma, M.L. 2010. Electrifying plasmonics on silicon. *Nat. Mater.*, 9: 3–4, doi: 10.1038/nmat 2598.

20. Bergsman, D.S. and Stockman, M.I. 2003. Surface plasmon amplification by stimulated emission of radiation. *Phys. Rev. Lett.*, 90: 027402.

21. Hill, M.T., Oei, Y.-S., Smalbrugge, B. et al. 2007. Lasing in metallic-coated nanocavities. *Nat. Photon.*, 1: 589–594, doi: 10.1038/nphoton. 2007 171.

22. Noginov, M.Z., Zhu, G., Belgrave, A.M. et al. 2009. Demonstration of a spaser-based nanolaser. *Nature*, 460: 1110–1112, doi: 10.1038/nature 08318.

23. Oulton, R.F., Sorger, V.J., Zentgraf, T. et al. 2009. Plasmon lasers at deep subwavelength scale. *Nature*, 461: 629–632, doi: 10.1038/nature 08364.

24. PhOREMOST Network of Excellence. 2008. *Emerging Nanophotonics*. Cork, Ireland: PhOREMOST.

25. Pevtsov, A.B., Kurdyukov, D.A., Golubev, V.G. et al. 2007. Ultrafast stop band kinetics in a three-dimensional opal- VO_2 photonic crystal controlled by a photoinduced semiconductor-metal phase transition. *Phys. Rev. B*, 75: 153101.

26. Le Thomas, N., Allione, M., Fedatik, Y. et al. 2006. Multiline spectra of single CdSe/ZnS core-shell nanorods. *Appl. Phys. Lett.*, 89: 263115.

27. Reboud, V., Kehagias, N.C., Sotomayor, M. et al. 2007. *Appl. Phys. Lett.*, 90: 011114. Selected in: *Virtual J. Nanoscale Sci. Technol.*, 15(3). http://apl.aip.org/resource/1/applab/v90/i1/p011114_s1?isAuthorized=no

28. Kehagias, N., Zelsmann, M., Pfeiffer, K. et al. 2007. Reverse-contact UV nanoimprint lithography for multilayered structure fabrication. *Nanotechnology*, 18: 175303.

29. Dwyer, C.O., Lavayen, V., Mirabal, N. et al. 2007. Surfactant-mediated variation of band-edge emission in CdS nanocomposites. *Photon. Nanostruct: Fundam. Appl.*, 5: 45.

30. National Research Council of the National Academics [NRC]. 2008. *Nanophotonics: Accessibility and Applicability*. Washington, DC: National Academies Press.

31. October 2000. Space-borne imaging, *IEEE Aerosp. Electron. Syst. Mag.*, 15(10): 118–124.

32. Hooper, B.A., Baxter, B., Piotrowski, C. et al. 2009. An airborne imaging multispectral polarimeter (AROSS-MSP), *OCEANS 2009, MTS/IEEE Biloxi—Marine Technology for Our Future: Global and Local Challenges*, Bremen, Germany, pp. 1–10, October 26–29, 2009.

33. Qingsong, W., Creusere, C.D., Thilak, V. et al. 2009. Active polarimetric imaging for estimation of scene geometry, *13th Digital Signal Processing Workshop and 5th IEEE Signal Processing Education Workshop, 2009, DSP/SPE 2009 IEEE*, Marco Island, FL, pp. 659–663, January 4–7, 2009.

34. Giakos, G.C. 2004. Multispectral, multifusion, laser polarimetric imaging principles, imaging systems and techniques, 2004 (IST). *IEEE International Workshop on Imaging Systems and Techniques*, Stresa, Italy, pp. 54–59, May 14, 2004.

35. Giakos, G.C. 2006. Multifusion multispectral lightwave polarimetric detection principles and systems, instrumentation and measurement, *IEEE Trans. Instrum. Meas.*, 55(6): 1904–1912, December 2006.

36. Giakos, G.C., Picard, R.H., Dao, P.D. et al. 2009. Object detection and characterization by monostatic ladar Bidirectional Reflectance Distribution Function (BRDF) using polarimetric discriminants. *SPIE Eur. Remote Sens.*, 7482, September 2009.

37. Giakos, G.C., Picard, R.H., and Dao, P.D. 2008. Super resolution multispectral imaging polarimetric space surveillance ladar sensor design architectures. *SPIE Eur. Remote Sens.*, 7107, September 15–18, 2008.

38. National Technical Information Services, Department of Defense, Air Force Research Laboratory, Munitions Directorate, (2003, February), Polarization Signature Research, Eglin AFB FL, D. Goldstein and Collet, Polarized Light, CRC Press, 2003.

39. Goldstein, D. 2005. Monostatic mueller matrix laser reflectometer, *Proc. SPIE*, 5888: 58770.

40. Liu, G.L., Li, Y., and Cameron, B.D. 2002. Polarization-based optical imaging and processing techniques with application to cancer diagnostics. *Proc. SPIE*, 4617: 208–220.

41. Baba, J.S., Chung, J., Delaughter, A.H. et al. 2002. Development and calibration of an automated mueller matrix polarization imaging system. *J. Biomed. Opt*, 7(3): 341–349, July 2002.

42. Bueno, J.M. 2000. Polarimetry using liquid crystal variable retarders: Theory and calibration. *J. Opt. A: Pure Appl. Opt*, 2: 216–222.

43. Giakos G. C., 2009–2013. Detection of LADAR targets using multidomain mueller matrix polarimetric bidirectional reflectance distribution function (BRDF) and fractals for space surveillance, DOD/Airforce Research Laboratory (AFRL) contract F 8718–09–C–0040.

44. Giakos, G.C., Picard, R.H., Dao, P.D. et al. 2009. Object detection and characterization by monostatic ladar Bidirectional Reflectance Distribution Function (BRDF) using polarimetric discriminants, SPIE Proceedings. *Electro-Opt. Remote Sens. Photon. Technol. Appl.*, 7482, September 2009.

45. Petermann, J. 2012. Design of a fully automated polarimetric imaging system for remote characterization of space materials. Master's thesis, The University of Akron, Akron, OH, May 2012.

46. Reynolds, J.G. and Hart, B.R. 2004. Nanomaterials and their application to defense and homeland security. *JOM J. Miner. Met. Mater. Soc.*, 56(1): 36–39, January 2004.

47. Alsing, P.M., Cardimona, D.A., Huang, D.H. et al. 2007. Advanced space-based detector research at the air force research laboratory. *Infrared Phys. Technol.*, 50: 89–94.

48. Giakos, G. 2012. Future directions: Opportunities and challenges, in *Computational Nanotechnology: Modeling and Applications with MATLAB®*, S.M. Musa (ed.). Boca Raton, FL: Taylor & Francis, Chapter 13.

49. Chipman, R.A. 1994. Polarimetry. *Handbook of Optics*, 2nd edn., Vol. 2. New York: McGraw Hill.

50. Lu, S.-Y. and Chipman, R.A. 1996. Interpretation of Mueller matrices based on polar decomposition. *J. Opt. Soc. Am. A*, 13(5): 1106–1113, May 1996.

51. Ghosh, N., Wood, M.F.G., and Vitkin, I.A. 2008. Mueller matrix decomposition for extraction of individual polarization parameters from complex turbid media exhibiting multiple scattering, optical activity, and linear birefringence. *J. Biomed. Opt.*, 13(4): 044036.

52. Giakos, G.C., Marotta, S., Narayan, C. et al. 2011. Polarimetric phenomenology of photons with lung cancer tissue. *Meas. Sci. Technol.*, 22(11): 114018.

53. Song, B.-S.S., Noda, T.A., and Akahane, Y. 2005. Ultra-high-Q photonic double-heterostructure nanocavity. *Nat. Mater.*, 4: 207–210, doi: 10.1038/nmat 1320.

54. Roy, D., Wang, J., and Williams, C. 2009. Novel methodology for estimating the enhancement factor for tip-enhanced Raman spectroscopy. *J. Appl. Phys.*, 105: 013530, http://dx.doi.org/10.1063/1.3056155

55. Noginov, M.A., Zhu, G., Belgrave, A.M. et al. 2009. Demonstration of a Spaser-based nanolaser. *Nature*, 460: 1110–1112, doi: 10.1038/nature 08318.

56. Abouraddy, A.F., Shapira, O., Bayindir, M. et al. 2007. Towards multimaterial multifunctional fibres that see, hear, sense and communicate. *Nat. Mater.*, 6: 336.

13

Future Trends in Nanophotonics: Medical Diagnostics and Treatment, Nanodevices, and Photovoltaic Cells

Chaya Narayan, Aditi Deshpande, Suman Shrestha, Tannaz Farrahi, Jennifer Syms, Chris Mela, Yinan Li, Anandi Mahadevan, Lin Zhang, Ryan Koglin, and George C. Giakos

CONTENTS

13.1 Introduction

The interaction of light with matter integrates major technologies such as photonics, nanotechnology, optical electronics, microfluidics, and electrochemistry, leading to the shaping of new research frontiers such as nanodiagnostics, polymer nanophotonics, space optics, biophotonics, and biomolecular nanophotonics [1–48]. Nanoscale systems and nanostructured materials cover a wide spectrum of materials from inorganic and organic amorphous or crystalline nanoparticles (NPs) over nanocolloids suspensions up to nanostructured carbon compounds such as fullerenes and carbon nanotubes, such as

- Colloidal and epitaxial NPs
- Colloidal quantum dots
- Quantum dots, nanocrystals, and metamaterials
- Biomedical markers
- Metallic nanorods and nanoshells
- Carbon nanotubes
- Nanoarrays
- Nanowires
- Nano electromechanical systems (NEMS) and bio nano electromechanical systems (BIONEMS) devices

The search for efficient medical imaging techniques capable of providing physiological information, at the molecular level, is an important area of research. Advanced metabolic and functional imaging techniques, operating on multiple physical principles, using high-resolution, high-selectivity nanoimaging techniques, making use of quantum dots, NPs, biomarkers, nanostructures, probes encapsulated by biologically localized embedding (PEBBLE) nanosensors, imaging microarray chips, and nanoclinics for optical diagnostics and targeted therapy, can play an important role in the diagnosis and treatment of cancer, as well as provide efficient drug-delivery imaging solutions for disease treatment with increased sensitivity and specificity. Emphasis is on the design of the following:

- New in vivo imaging and therapeutic techniques
- Nanoarray technology for DNA and protein
- Upconverting nanophores for photodynamic treatment
- Lab-on-a-chip systems for biomedical and scientific in situ detection
- Molecular motors that move fluids in chip-sized laboratories
- Nanoinstrumentation imagery

NP technologies may lead to potential commercial applications that would revolutionize the environmental and energy technology. Future directions include the following:

- LEDs
- Solar cells

- Molecular motors that move fluids in chip-sized laboratories (lab-on-a-chip)
- Raman, LIBS, and LIDARs

Special emphasis will be paid on future directions and challenges regarding the discovery and development of new nanomaterials, metamaterials, and photonic nanocrystals for

- Space-based electronic systems such as communication satellites and interplanetary space probes (semiconductors and optoelectronics for data communication)
- Space optical instrumentation and systems for missile defense and air force applications
- Nanoelectronics (data processing and communication systems with minimized energy consumption, highly integrated nanodevices for miniaturized space systems, etc.)
- Imaging sensors and focal planar arrays (FPAs)

Future trends and applications of nanotechnology have emphasis on

- Photonic and nanophotonic detectors.
- Remote sensing systems and standoff detection with potential applications in the area of homeland security are anticipated.

The impact of nanophotonics on several technological areas is highlighted in Figure 13.1.

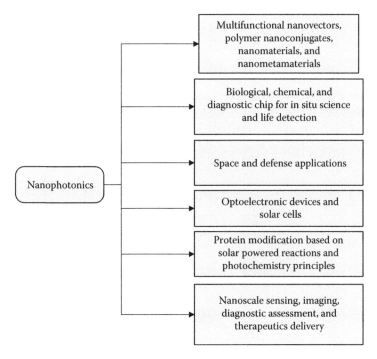

FIGURE 13.1
Impact of the nanophotonics on several technical areas.

13.2 Metallic Nanoparticles Classical Theory

Metallic NPs have been found to exhibit unique tunable optical properties compared to the properties of their bulk counterparts. Plasmons occurring within these NPs give them the ability to localize incident radiation to regions smaller than their wavelength and to enhance the intensity of the radiation. Plasmons in particles ranging in size greater than about 10 nm are successfully modeled as collective oscillations of a free electron gas excited by an incident electromagnetic field. The electromagnetic field induces a polarization charge along the surface of the metallic NPs, and a restoring force causes the plasmonic oscillation. The electric field induced within the particle under the influence of an applied electric field is given by

$$E_i = E_0 \frac{3\varepsilon_m}{\varepsilon + 2\varepsilon_m} \tag{13.1}$$

where
  ε is the dielectric function of the NP
  ε_m is the dielectric function of the surrounding medium

The classical interpretation of the excitation of plasmons by polarization of metallic NPs is shown in Figure 13.2.

According to the quasistatic approximation for small particles, the localized surface plasmon resonance (LSPR) peak arises when the particle's polarizability is maximized. The static polarizability of a sphere is given by [2]

$$\alpha = 4\pi \varepsilon_0 R^3 \frac{\varepsilon - \varepsilon_m}{\varepsilon + 2\varepsilon_m} \tag{13.2}$$

This comes from the solution of Laplace's equation. From this equation, it can be concluded that the polarizability is maximized when $|\varepsilon + 2\varepsilon_m|$ is minimized. This shows the strong relation between the dielectric function of the particle and its plasmonic properties.

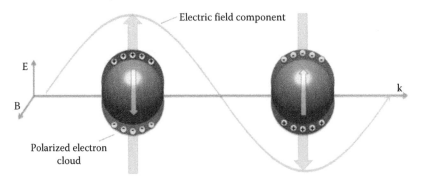

FIGURE 13.2
Classical interpretation of the excitation of plasmons by polarization of metallic NPs. Note that at the resonance frequency, the plasmons are oscillating with a $\pi/2$ phase difference. (From Trugler, A., Optical properties of metallic nanoparticles, PhD thesis, Institut fur Physik, Berlin, Germany, 2011.)

The Drude–Sommerfeld model is used to find the dielectric response of a free electron gas against a positive ion core background and describes the frequency-dependent permittivity of the particle. The Drude dielectric function is given by [1]

$$\varepsilon_d(\omega) = \varepsilon_\infty - \frac{\omega_p^2}{\omega^2 + i\lambda\omega} \tag{13.3}$$

where

   ω is the angular frequency of the incident light
   ω_p is the bulk plasma frequency of the incident light
   λ is the factors in the scattering frequency of the electron travelling through the metal
   ε_∞ accounts for the ionic background of the metal at high frequencies

If this model is simplified to $\varepsilon_d = 1 - (\omega_p^2/\omega^2)$, it may be seen that for $\omega > \omega_p$, ε_d is positive. Thus, the refractive index (RI) is a real number and the electromagnetic wave is transmitted into the metal. When $\omega > \omega_p$, ε_d is negative resulting in an imaginary value of the RI, indicating that the electromagnetic wave is reflected. It is interesting to note that the value of ω_p corresponding to most metals is in the ultraviolet region, which explains their shimmery quality. It is also worth noting that at higher energies, valence electrons from a lower band (d-band) may transit into the conduction band. The contribution of these lower band electrons may be accounted for by replacing ε_∞ with a frequency-dependent term ε_{IB} [3].

13.2.1 Metallic Nanoparticles Quantum Theory

Plasmons occurring in particles ranging in size from about 2 to 10 nm are associated with quantum mechanical effects and described by electron transitions between occupied and unoccupied energy levels. These energy levels become discretized, only allowing certain electron transitions. Therefore, a more accurate model for these NPs is to envision the electrons constrained within an infinite potential barrier around the particle. According to quantum mechanics, the discrete set of energy levels in which an electron may exist becomes increasingly separated from one another as the particle size decreases. This explains the increase in the plasmon peak frequency exhibited in particles of decreasing size below about 10 nm. The allowed transition energies of conduction electrons from occupied states, i, to unoccupied states, j, is characterized by the transition frequency, ω_{if}, given by [3]

$$\omega_{if} = \frac{E_f - E_i}{h} \tag{13.4}$$

Therefore, the analytic quantum permittivity calculation for NPs exhibiting quantum LSPR shifts with decreasing particle size is given by [3]

$$\varepsilon(\omega) = \varepsilon_{IB} + \omega_p^2 \sum_i \sum_f \frac{S_{if}}{\omega_{if}^2 - \omega^2 - i\lambda\omega} \tag{13.5}$$

where S_{if} is an oscillator strength term. This analysis may be used to model different metals and different NP shapes by adjusting the material parameters and the quantum wave functions [3], as shown in Figure 13.3.

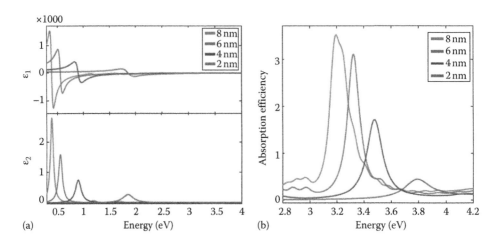

FIGURE 13.3
(a) Plot of the analytic quantum theory of particle permittivity with real (ε_1) and imaginary (ε_2) dielectric functions of silver NPs with various diameters, and (b) plot of the analytic quantum theory of absorption spectra of the particles according to the Mie theory. Note that as particle size decreases, the plasmon peak blueshifts and exhibits reduced absorption efficiencies. (From Scholl, J.A. et al., *Nature*, 483, 421, 2012.)

13.3 SPR Principles

Due to the significant contribution of the dielectric parameter of the material surrounding metal NPs and the boundary conditions for the propagation of a surface plasmon polariton at a dielectric–metal interface, a surface plasmon biosensor has become an intriguing application to the use of metallic NPs.

Understanding the physical and optical concepts of surface plasmon resonance (SPR) begins with the phenomenon of total internal reflection (TIR) and the occurrence of an evanescent field wave. This is depicted in Figure 13.3. TIR occurs at the boundary between two materials with a different index of refraction. When light travelling through a medium with a high index of refraction, such as a prism, comes into contact with a medium of lower index of refraction, such as a dielectric solution, TIR occurs at the boundary at an incident angle greater than a critical angle. The critical angle θ_c is given by [4]

$$\theta_c = \sin^{-1}\left(\frac{n_2}{n_1}\right) \tag{13.6}$$

where n_2 and n_1 are the RIs of the solution and prism, respectively. The incident light beam leaks an electromagnetic field wave called the evanescent wave into the lower RI medium. The intensity $I(z)$, or amplitude, of the evanescent field wave decays exponentially as a function of the distance z from the interface given by [4]

$$I(z) = I_0 * e^{(-z/d)} \tag{13.7}$$

where
I_0 is the initial intensity
d is the depth given by

$$d = \frac{\lambda_0}{4\pi \left[n_1^2 \sin^2(\theta) - n_2^2 \right]^{(-1/2)}} \tag{13.8}$$

for angles of incidence $\theta > \theta_c$ and light wavelength in vacuum λ_0.

If the TIR interface is coated with an NP conductor film, such as gold, the p-polarized component of the evanescent field wave may interact with the free electrons of the gold surface in contact with the lower RI medium. The energy and momentum of the incident light photons are converted into surface plasmons. The resonance creates an enhanced evanescent field wave.

The conversion of light photons into plasmons may happen only when the wave vector of the incident light parallel to the conductor surface is equal to the wave vector of the surface plasmons. A drop in the intensity of the reflected light occurs under this condition as depicted in Figure 13.4.

The wave vector of the incident light, k_x, is given by [5]

$$k_x = \left(\frac{2\pi}{\lambda_0} \right) n_1 \sin(\theta) \tag{13.9}$$

and the wave vector of the plasmon wave, k_{sp}, is given by [5]

$$k_{sp} = \left(\frac{2\pi}{\lambda_0} \right) \left(\frac{n_{gold}^2 n_2^2}{(n_{gold}^2 + n_2)} \right)^{(1/2)} \tag{13.10}$$

where
  λ_0 is the wavelength of the incident light in a vacuum
  n_1, n_2, and n_{gold} are the RIs of the prism, solution, and gold film, respectively

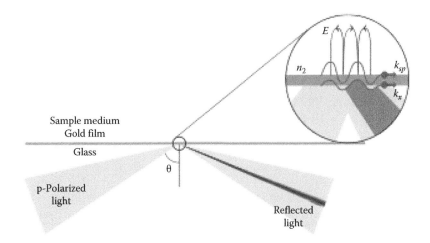

FIGURE 13.4
SPR excited by p-polarized light (after undergoing total internal reflection) at a glass/metal film interface, the surface plasmon enhancing the evanescent field wave, E. (From Biacore, Surface plasmon resonance technology note 1 [online], Available FTP: protein.iastate.edu Directory: seminars/BIACore File: Surface Plasmon Resonance, 2001.)

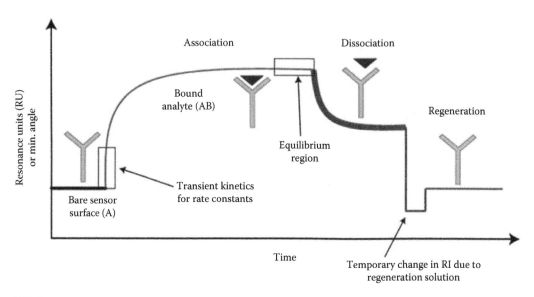

FIGURE 13.5
Sample SPR sensorgram showing binding, dissociation, and regeneration of an analyte to the biosensor surface. (From Linman, M.J. et al., *Analyst*, 135, 2759, 2010.)

According to Equation 13.10, the wave vector of the plasmon wave is dependent upon the RI of the solution. An increase in the RI of the solution penetrated by the plasmon-enhanced evanescent field increases the wave vector of the plasmon wave. The wave vector of the incident light may be changed to equate the wave vector of the plasmon wave by varying the angle of incidence, θ, or the wavelength of the incident light. This idea is depicted in Figure 13.5, which shows the angle variation associated with the RI change of a sensor when binding, dissociation, and regeneration occur.

13.3.1 SPR Biological Applications

SPR is used in studying biological interactions in real time. This includes protein–carbohydrate, DNA–protein, and protein–lipid interactions. Protein–carbohydrate interactions are important when studying cellular adhesion, recognition, and development events, viral and bacterial infections, and cancer and metastasis [6–10]. Aptamer-based SPR sensors recognizing DNA–protein interactions are important because of the high selectivity of aptamers with their molecules, increased stability compared to antibodies, and low cost. Aptamers are used to study corresponding protein interactions and DNA recombination, replication, and transcription [6,7,11–13]. Finally, protein–lipid interactions are important when studying cellular structure, function, communication, adhesion, and signal transduction cascades. The analysis of the lipid bilayer that forms the cellular membrane is a more recent area of study due to preparation difficulties in the past and an absence of adequate membrane model systems to be used on the sensor, which have recently become available [6,7,14,15].

The ligands are fixed to the metal–solution interface and a biomolecule solution flows over the substrate. Binding events change the RI of the solution within the range of the enhanced evanescent field wave. The angle of incidence that satisfies the conditions of resonance changes as molecular binding events occur. For specific examples of the utility of an SPR system, I would like to refer the reader to Refs. [6] and [7]. These are very well-written articles by Linman and Cheng, who are very knowledgeable not only on the subject of SPR but also on the many biological applications that are discussed in detail.

13.3.2 SPR Limitations

The SPR technique is discussed and has been applied by a wide variety of research groups. The future for this technology is showing rapid improvements due to its intrinsic capabilities and general excitement of the research community for its usefulness in studying biological behavior. This includes advancements in sensitivity, surface chemistry of the metal substrate, and studies utilizing SPR in combination with other techniques.

Although the potential of SPR imaging over fluorescence assays is appealing especially due to its label-free quality, an obstacle that must be overcome is its lack of sensitivity. Enhancing the signal may be done through chemical and biological means. One biological method proposed is by enzymatic amplification [6]. Chemical methods of improving signal quality are through the use of new substrate materials. These new substrates involve coating the gold surface with materials that offer greater stability and increased control over the interactions taking place. Glass-coated gold substrates have been shown to be a useful surface for the attachment of lipids and do not reduce the sensitivity of the process. Exploration of the use of stimuli-responsive polymers coating the gold substrate is an exciting topic. This is because of the ability to control conformational and chemical changes of the polymers by outside means such as modulating the temperature, pH, and electric fields [7].

The use of negative-index materials, or metamaterials, has attracted considerable attention. A material's property may be characterized by the electric permittivity ε and the magnetic permeability μ. Permittivity is a measure of a material's ability to polarize in response to an applied electric field. Permeability is a measure of a material's ability to magnetize in response to an applied magnetic field. The permittivity and permeability are used to define the RI of a material, according to [16]:

$$n = \sqrt{\varepsilon\mu} \qquad (13.11)$$

Most materials in nature have both a positive permittivity and permeability. A few have one positive and one negative permittivity or permeability. Man-made materials may be designed using different architectures and substrate materials in order to achieve a metamaterial with both a negative permittivity and permeability simultaneously at a given frequency. Negative-index materials exhibit new phenomena not found in naturally occurring materials including the ability to recover, enhance, and convert evanescent waves into propagating waves. Metamaterials capable of supporting surface plasmons have been discussed [17–19]. If used in an SPR system, these materials could cause the propagation of surface plasmons in the normal direction. This would increase the depth of penetration of the evanescent waves and therefore the sensitivity of the technique [20]. Reference [21] describes a study of a biosensor based on the use of a plasmonic nanorod metamaterial. Reference [22] discusses the use of metamaterials for plasmon resonance sensors at microwave frequencies. Figure 13.6 summarizes these and other ideas for optional materials used to build the SPR apparatus.

Ideas of SPR in combination with other analytical techniques hold many advantages for a fully functional system with advanced capabilities over the use of SPR alone. Mass spectrometry (MS) is a technique that uses ionization to charge molecules and then measures

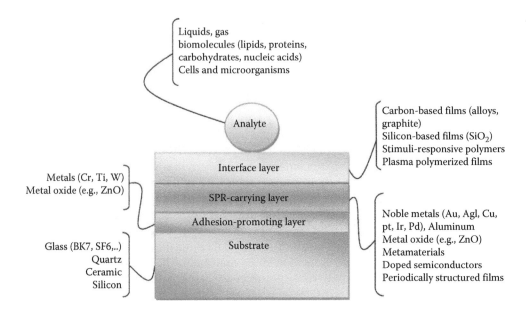

FIGURE 13.6
Possible materials to be used as the different layers of the SPR apparatus. (From Abbas, A., *Biosens. Bioelectron.*, 26, 1815, 2011.)

their mass to charge ratio in order to determine the chemical structure and composition of the molecules. The realization of SPR with MS allows for a unique approach to screening and identification confirmation of the molecules after SPR has been observed [23]. High-performance liquid chromatography (HPLC) is used to dissolve a mixture in fluid and then carry it through another material that separates the constituents. This separation is achieved because of the differing travel rates of the mixture components through the material. Upon separation, the measurements are used in identification and quantification of the proportions in which the components exist. HPLC combined with SPR has been used in the analytical detection of samples of protein [24]. Another integrative approach involves the excitation of a fluorescent signal by surface plasmons; although this technique allows for greater sensitivity, it does require labeling of the biomolecules with fluorescent molecules such as quantum dots discussed in the next section. Microscopy methods used to image the surface of a material in conjunction with SPR have been shown useful in measuring the positions of the molecules on the material substrate [20]. Investigation of other imaging and identification techniques that create a fully integrated platform offers a neat perspective for evolving trends in the functionality of SPR.

13.4 Semiconducting Nanoparticles

13.4.1 Quantum Dot Principles

The optically tunable fluorescent properties of semiconducting NPs, also known as quantum dots, are similar to their metallic counterparts discussed in analytical detail previously. Quantum dots are semiconducting fluorescent crystals that are made between

1 and 10 nm in diameter, which fluoresce when stimulated by an external energy source. These crystals confine the motion of electrons in the three spatial dimensions. Confinement occurs because the diameter of the dot is about equal to the distance between electrons in the conduction band and holes in the valence band. The dot radius is less than the excitation Bohr radius, which is the approximate size of a hydrogen atom in its ground state. The Bohr radius, *r*, is given in the following equation:

$$r = \frac{h^2}{mke^2}$$ (13.12)

where
  e is the charge of an electron
  m is the mass of an electron
  h is Planck's constant
  k is Coulomb's constant

This introduces the effect of quantum confinement in which electron and hole movement is constrained. The energy levels an electron can assume are discrete. The electron in the confined state exhibits standing wave behavior. As the confining measurement decreases, the wavelength absorbed or emitted by the dot moves toward the blue end of the spectrum. This is because in a bulk semiconductor material, an electron–hole pair created upon energy (photon) absorption can move around within the material, although, due to coulomb attraction, the electron will orbit the hole at a certain radius. The creation of an electron–hole pair is depicted in Figure 13.7. In the quantum dot, due to the confinement, the energy states of the electron–hole pair shift to higher levels and the dot is made smaller. This effect is noticed when the smaller the dot is made, the higher the energy of the photon it will absorb or emit. This is similar to a particle-in-a-box model in which the smaller the box, the higher the energy levels within that box according to $1/L^2$ where L is the length of the box.

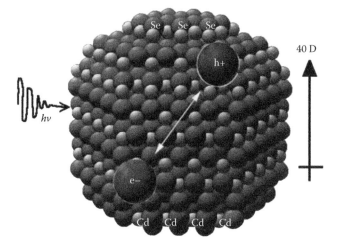

FIGURE 13.7
(See color insert.) An electron–hole pair is created upon photon absorption. The electron orbits the hole at a certain radius dependent upon the material, and when recombination occurs, the quantum dot emits energy viewed as fluorescence. (From Rosenthal, S.J. et al., *Chem. Biol.*, 18, 10, 2011.)

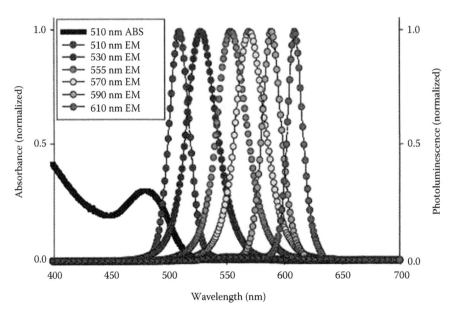

FIGURE 13.8

The narrow emission spectra of six different sizes of a CdSe quantum dot coated with ZnS. Note that the emission wavelength increases as the size of the dot increases. (From Salman Ogli, A., *Cancer Nano*, 2, 1, 2011.)

Atomic arrangement in quantum dots is similar to that of a bulk material but with a much higher surface-to-volume ratio. Typical quantum dots consist of a semiconducting core, usually made of a toxic material, and an outer shell coating of a higher band-gap material. The coating is used to decrease toxicity and to minimize an effect called blinking. Blinking occurs due to imperfections of the crystal causing a loss of energy. In biological applications, the dot will be coated with an additional layer in order to prepare it for use in a solution and for the attachment of molecules.

The nature of quantum dots provide many advantages for their use due to tunability, narrow emission bandwidth, long fluorescent lifetimes, resistance to photo-bleaching, high surface-to-volume ratio, and wide absorption spectrum. Figure 13.8 shows the narrow emission spectra of a CdSe quantum dot coated with ZnS of six different sizes.

13.4.2 Quantum Dot Biological Applications

Quantum dots have many imaging applications. They are currently being used in place of fluorophores for imaging cells and tissues because of their many useful properties. An additional advantage of the dots is the ability to stimulate their fluorescence without the use of a light source using self-illuminating bioluminescence resonance energy transfer (BRET) quantum dot NPs [27,28]. BRET is a naturally occurring phenomenon in which a light-emitting protein transfers energy non-radiatively to a fluorescent protein or quantum dot in close proximity. This excitation causes the quantum dot to fluoresce, eliminating the need for an external light source [29]. Another advantage is the development of quantum dots with fluorescence in the near-infrared (NIR) range. Tissue penetration is the highest in this range with low interference from tissue auto-fluorescence when used for in vivo fluorescence imaging [30]. They are also implicated in cell labeling, which allows tracking of individual cell types [31,32]. Tracking is accomplished by tagging a specific type of

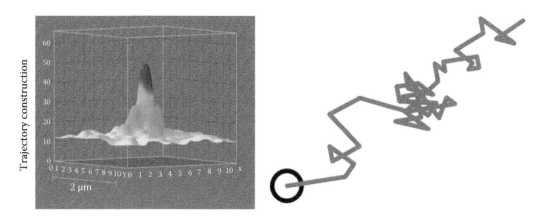

FIGURE 13.9
Example of the determination of the trajectory of a molecule tagged by a quantum dot. Intensity values were used to create a Gaussian distribution and then used to determine the trajectory shown on the right. (From Rosenthal, S.J. et al., *Chem. Biol.*, 18, 10, 2011.)

molecule with a quantum dot. Their movement is monitored using an optical fluorescent microscope system. The position trajectory, estimated over time by the intensity of the signal, is shown in Figure 13.9.

Quantum dots have become especially important in studying the motion of motor proteins using optical microscopy. Their resistance to photo-bleaching is important to this research in order to follow the motion for long periods of time [33]. Along with the ability to track single molecules, different emission spectral colors of the quantum dots can code multiple molecules simultaneously, such as in DNA analysis. DNA is composed of four base nucleotides, adenine, thymine, guanine, and cytosine. Four different color quantum dots may be attached to each of the nucleotides and stimulated under one source frequency to produce narrow emission bands [34]. Quantum dots have also been successfully used in determining the precise location of protein–DNA binding sites [35]. Another specific application of quantum dots worth noting is functioning as a probe in studying neurotransmitters released in the synaptic cleft between two neurons or a neuron and a cell in order to visualize, understand, and identify methods of neurotransmitter release [25].

13.4.3 Quantum Dot Limitations

The most widely discussed limitation of quantum dots concerns their toxicology. The materials of which they are composed have raised concerns of their effect on biological specimen, especially when being studied in their natural environment. The toxicology of each quantum dot is characterized by such properties as its size, materials, concentration, stability, etc. [31]. Reference [36] gives a thorough overview of the toxicology of quantum dots and the many different contributing factors.

Although current technology allows for quantum dots to be made with a diameter less than 14 nm, their mass becomes larger after additional coatings for biocompatibility and attachment of small molecules. This causes an inability of use in small locations with a cluster of other molecules [37]. Also, when imaging dynamic processes or use in tracking, their size may hinder the ability to follow the motion to provide an accurate quantification of the trajectory of the molecules [25].

Applications of quantum dots in intracellular events are challenging. Novel ways of administering the dots into the cytoplasm or organelles must be developed. Injection with a microneedle can be very time consuming because each cell must be injected individually. An electric pulse may be used to create temporary permeability of the cell membrane for the dots to diffuse into [37]. Endocytosis is the process by which cells engulf small particles in the extracellular matrix. This is a very promising method of delivery of quantum dots, but some controlled manipulation is needed in order to release the dots from the carrying vesicle upon entry [33,37,38]. Receptors along the cell membrane allow active or passive transport of small molecules into the cell but with some selectivity. This is a possible method of delivery of the quantum dots into the cell [37,38].

The effect of blinking may be viewed as a useful property of the quantum dots or a serious limitation to their applicability in studying dynamic processes. Blinking is caused by variations in the photoluminescence intensity of a single nanocrystal viewed as changing between on and off states [39]. This is a major disadvantage when studying dynamic processes due to sometimes long-lived intermittent off states that disrupt the imaging sequence. It is also a property that may be exploited for differentiation purposes when multiple quantum dots aggregate in clusters [25,33]. Suppression of the blinking phenomenon has been reported in CdSe/ZnS quantum dots in the presence of TiO_2 NPs [39].

13.5 Nanoshells

13.5.1 Nanoshell Principles

This last type of NP mentioned in this chapter has emerged as another method of tuning the LSPR peak. It consists of a dielectric core surrounded by a thin shell of metal, typically gold or silver, with a negative real part of its dielectric function within the visible spectrum. The outer and inner surfaces of the shell exhibit two SPRs, the surface and cavity resonances, respectively. These plasmons interact with each other with increasing strength as the shell thickness is decreased. Thus, the strength of the interaction between the surface and cavity resonances shifts relative to the strength of the individual resonances. At a certain increase in thickness of the shell, the cavity resonance will no longer be stimulated by incident radiation and the particle can be modeled approximately as a solid sphere. Decreasing below this thickness, the position of the plasmon peak, occurring as a result of the cavity and surface resonance interaction, may be tuned by altering the ratio of the core radius to the shell radius, or the shell aspect ratio. It may also be changed by altering the dielectric constant of the core, shell, or surrounding medium, as realized in the following equation for the resonance condition where the dielectric constant of the surrounding medium is unity [40]:

$$\varepsilon_1 = -2\varepsilon_2 \left[\frac{\varepsilon_2(1-f)+(2+f)}{\varepsilon_2(2f+1)+2(1-f)} \right] \tag{13.13}$$

where
   ε_1 and ε_2 are the dielectric functions of the inner core and outer shell, respectively
   f is the fraction of the total particle volume occupied by the core given by

$$f = \left(\frac{r_1}{r_2} \right)^3 \tag{13.14}$$

where r_1 and r_2 are the inner and outer shell radii, respectively.

Such nanoshells may be designed with plasmon peak tunability from the visible to the infrared spectrum. This is very important for in vivo biological applications due to the transparency of tissue to incident radiation from 700 to 900 nm. Within this region, light may penetrate tissue from a few millimeters to centimeters depending on the different types of tissues [41]. The theoretical calculations of the change in thickness of a gold shell with an SiO_2 core, where the peak resonance red shifts from visible to infrared by increasing the shell aspect ratio, are depicted in Figure 13.10.

Along with the ability to tune the plasmon resonance peak, nanoshells exhibit the ability to tune their absorption to scattering ratio by adjusting the outer radius of the nanoshell and keeping the shell thickness constant [41], or also for certain shell aspect ratio configurations [43]. Scattering dominates in larger nanoshells, and absorption dominates in smaller nanoshells. This allows for interesting dual scatter–based imaging and photothermal therapy applications.

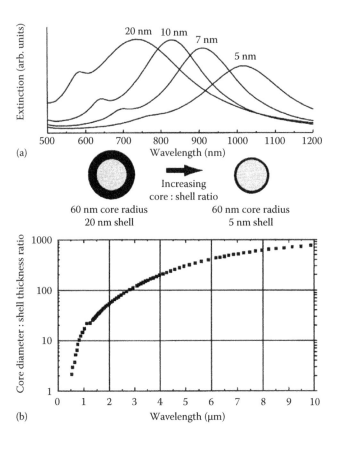

FIGURE 13.10
(a) The theoretically calculated resonances of a silica core with gold-shell metal nanoshells, (b) calculation of the optical resonance wavelength versus shell aspect ratio for silica core with gold-shell metal nanoshells. (From Oldenburg, S.J., *Chem. Phys. Lett.*, 288, 243, 1998.)

13.5.2 Key Nanoshell Biological Applications

13.5.2.1 Imaging Contrast Agents

As mentioned in the section on the principles of nanoshells, they may be fabricated to preferentially scatter and may be used as optical contrast agents for imaging applications. They may also be engineered to target specific cell types, such as cancer, by the conjugation of antibodies or other small molecules to the outer shell. Nanoshells overcome the limitations of previous contrast agents due to their tunability, enabling the use of multiple laser wavelengths within the NIR spectrum. They also offer large resonant scattering cross-sectional capabilities, minimized vulnerability to chemical or thermal denaturation, and increased biocompatibility. The following is a description of the work of two groups, indicating the utility of nanoshells in imaging applications. Gao et al. have configured nanoshells with peak scattering resonance at 800 nm. They used the two-photon excitation microscopy imaging modality for successful in vivo imaging of blood vessels in mice [44]. Zhang et al. demonstrate the use of multiple nanoshell configurations with similar excitation and emission wavelengths but differing emission lifetimes. The emission signals of the nanoshells were able to be distinguished from one another on cell images. This allows for detection of multiple target areas within a cell bound to nanoshells of various emission lifetimes [45].

13.5.2.2 Photothermal Therapy

The imaging applications use nanoshells designed to preferentially scatter, but photothermal therapy applications take advantage of their strong absorption and localization of heat. The concept of photothermal ablation is to generate a lethal dose of heat to a confined tissue volume without damaging the surrounding tissue not set as a target. Nanoshells with an absorption peak in the NIR spectrum, corresponding to minimal absorption by tissue, may be conjugated with biomolecular agents allowing accumulation in a target site. Upon NIR radiation exposure, the nanoshells efficiently convert the energy into localized heat, causing thermal destruction of tissue in the immediate area. Liu et al. show that gold nanoshells functionalized with a peptide preferentially targeting hepatocarcinoma cells offer selective apoptosis of the cells upon irradiation with NIR light [46]. Gobin et al. take an interesting approach to the NIR absorbance properties of nanoshells. This group explores their ability to assist in NIR laser-tissue welding. The study reveals the success of a nanoshell solder to weld muscle tissue incisions made on rats [47]. Another interesting approach to photothermal targeting used to induce cell apoptosis is the use of nanoshells as small-molecule capsules. Zhang et al. describe the release of small-molecule fluorophores from nanoshells with a silica core and silver shell upon laser irradiation [48]. Further research in this area may demonstrate nanoshells used as drug capsules with specific biomolecular targets to localize the release and control the delivery by irradiation.

13.6 Solar Cells

With an increase in the global demand for power due to a growing population and greater industrialization comes an increased need for alternative power sources. Crude oil, natural gas, and coal are all in limited supply. Their utilization also leads to the production of airborne hydrocarbons that are considered to be a major pollutant and a greenhouse gas [51,52]. Nuclear power is both more efficient and cleaner than the burning of fossil fuels; however, despite improved security measures, there is still a stigma surrounding nuclear reactors as being unsafe. These reactors also produce a waste that can be contained and partially recycled, but not eliminated [53]. Clean energy sources provide electricity without a by-product, outside of those required to create the energy sources themselves. They are fueled by natural and renewable sources such as the sun, the wind, or flowing water. While wind turbines can be loud and potentially disruptive to avian migration patterns and hydroelectric dams can disrupt the local marine ecology, solar panels are quiet and unobtrusive to nature [54–56]. Their greatest defaults are their monetary costs and lack of efficiency [57,58]. This chapter deals with current advances in solar cell technology made to address those concerns [59].

Solar cells operate due to the photovoltaic effect in a semiconducting medium [51]. By this effect, when an incident photon of energy greater than that of the semiconducting material's bandgap is absorbed by the material, an electron is excited and promoted from its valence band to the conduction band of the medium (Figure 13.11). This creates an electron/hole pair. Solar cells are commonly created with a PN junction where p-type silicon is doped near one surface with an n-type material. For some thin-film cells, a PIN junction is created to drive the current flow. In these designs, the electrons migrate to the n-type material near one surface while the holes aggregate in the p-type on the other, thus creating an electric field to drive the current flow (Figure 13.12) [51]. Thin films provide a cost reduction over thicker films due to their use of much less material, although their absorption efficiencies are less [1,9].

13.6.1 Materials

Silicon is the most commonly used semiconductor medium today. This is due in part to its relatively low defect count, the fact that it is a mature and stable technology, its abundance and nontoxicity, and that its band-gap energy allows for a relatively high efficiency by the Shockley–Queisser limit [51,60]. This theoretical limit dictates the maximum conversion efficiency an unmodified single-layer semiconducting material can have based on its bandgap without the use of a light concentrator. The maximum efficiency is set to 33.7% at about a 1.4 mV gap energy at room temperature, 300 K [11]. Silicon has a 1.1 mV gap energy, giving it a max efficiency of close to that mark at 31%. This limit is for pure

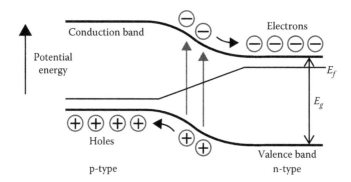

FIGURE 13.11
PN-junction energy diagram. (From Marshall, E., *Nanophotonics—Revolutionising Power Generation*, H.H. Wills Physics Laboratory, Bristol, U.K., 2011.)

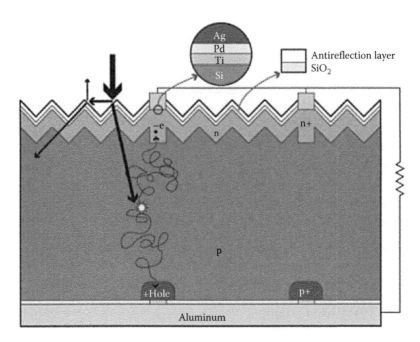

FIGURE 13.12
Typical commercial silicon solar cell. (From Marshall, E., *Nanophotonics—Revolutionising Power Generation*, H.H. Wills Physics Laboratory, Bristol, U.K., 2011.)

crystalline silicon (c-Si); amorphous silicon (a-Si) has a lower max efficiency at about 25% with a bandgap of 1.7 mV [61].

Unlike c-Si, a-Si is prepared by a process of gaseous deposition [51,62]. This process results in a structure composed of a silicon/hydrogen alloy that creates a glassy, amorphous bulk material embedded with microcrystals. As such, it lacks both the regular crystal lattice structure and purity of c-Si. This makes it a more efficient absorber than c-Si, due to a shorter absorption length [51,62]. The overall photon to electron conversion efficiency is less than that of c-Si, however, at around 13% for multijunction pin cells, compared to 25% for crystalline under the global AM 1.5 spectrum [64]. Also, a-Si does not absorb red to infrared wavelengths as well [62]. a-Si also undergoes some degradation in performance over the first several hundred hours of use, where the absorption efficiency will drop from its initial level to some steady-state value. This is due to the Staebler–Wronski effect, which is believed to be due to some level of dangling bonds between the silicon and hydrogen atoms in the a-Si structure [63]. a-Si solar devices make up for their lack of efficiency, in part, by being able to be made into thin films that can be flat, curved, or even flexible, as well as being less expensive and easier to produce than c-Si [51].

Other commonly used semiconducting media for solar cells are cadmium telluride, CdTe, and gallium arsenide, GaAs [59,61]. While not as mature technologies as silicon, these devices benefit from being able to utilize a direct band transition at the optimal band-gap energy of 1.4 mV. This gives these cells the maximum potential for absorption efficiency. CdTe cells are formed into thin films, like a-Si, and like a-Si they combine lower costs with lower efficiencies, compared to c-Si [51]. Gallium arsenide can be formed into thin films as well or be used as a multijunction device for higher efficiencies [59]. Such devices have garnered much attention as they have demonstrated conversion efficiencies of over 20% for thin films and over 30% for multijunction under the AM 1.5 [53].

Dyed nanocrystal–oxide solar cells have been introduced as an inexpensive, flexible, and easy to manufacture alternative to conventional solid-state cells [51,54]. These devices are constructed and are created from a metal-oxide thin film composed of metal-oxide NPs, commonly ZnO or TiO_2 [54,55]. The film is soaked in an organic dye, typically a heteroleptic polypyridyl ruthenium complex, which will bond with the metal-oxide molecules and act to absorb incident photons [54]. The dye composition can be varied to absorb different portions of the optical spectrum. The thin film is sealed between two glass plates that act as the electrodes for the device. On either side of the film, within cavities in the glass, is a liquid electrolyte. Light absorbed by the dye/metal-oxide molecule is converted into an electron/hole pair, or an exciton. The electron is passed from the molecular valence, or the highest occupied molecular state, to the conduction band, or lowest unoccupied molecular state, of the metal-oxide molecules where it is passed through the electrodes to the metal contacts. The holes left in the dye are filled from electrons in the ions of the electrolytic liquid [54]. This device is limited by the range of broadband absorption of the dye. Conversion efficiencies have been reported up to about 11% [53].

Organic or polymer cells have also won recent attention. Such cells are of interest because they are highly configurable, as they can be made from a wide range of polymers and can utilize different junction types [51]. The manufacturing procedures for such a cell are also variable and convenient, as these materials can be made into flexible rolls or even sprayed or printed onto a substrate [51]. This is a newer technology, however, and as such the current conversion efficiencies are low, at typically 3%–5% although a 10% cell was observed under the AM 1.5 spectrum [51,59,63,66].

The previously discussed efficiencies for solar cells based on the Shockley–Queisser limit are all theoretical and only apply to unmodified semiconducting cells [51]. Often times, even modified cells will not attain the efficiencies predicted by this limit. The typical efficiencies for commercial a-Si and CdTe thin-film cells are around 8%–10%, while efficiencies for very pure commercial c-Si cells can be as high as 18%–20% [59]. The conversion efficiencies listed are for modified laboratory cells (see references for further details). Some of the defects that limit the efficiency of these cells will now be discussed along with treatments for each deficiency (Figure 13.13).

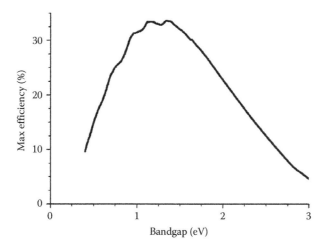

FIGURE 13.13
The Shockley–Queisser limit curve. (From Trugler, A., Optical properties of metallic nanoparticles, PhD thesis, Institut fur Physik, Berlin, Germany, 2011.)

13.6.2 Limiting Factors in Solar Cell Efficiency

Factors that can limit solar cell efficiency are many. These include defects or impurities in the semiconducting medium [51], a minority carrier diffusion length that is less than the cell thickness [56], absorption lengths that are longer than cell thickness resulting in a large amount of photons escaping the cell without being absorbed [67], and surface reflection resulting in photons being deflected away from the absorbing material [67–69].

Impurities in the atomic structure of a semiconducting material such as c-Si can limit cell efficiency by absorbing electrons, shortening the minority carrier lifetime [2]. The energy of the quenched atom will be lost to heat rather than emitted into the electric current. Such defects come in the form of non-silicon atoms that diffuse into the medium, displacing a silicon [70]. This can also lead to crystal lattice structural defects such as line shifts. Such defects, along with other structural imperfections such as microfractures, can also reduce device efficiency by inserting trapping levels that can reduce the number of electrons in the conduction band [70]. This makes the use of thinner absorptive materials a more attractive option, the idea being that the thinner material will have less defects.

Having a minority carrier diffusion length that is less than the thickness of the cell will result in an increased amount of electrons and holes recombining prior to reaching the contacts [58]. The length should be long enough for an electron to travel from the far edge of the p-type material of the cell to the n-doped section for collection [52]. Likewise, the holes should be able to travel from the edge of the n-type region to the contacts at the end of the p-type region. Lesser carrier lifetimes will result in a lessened collection efficiency. This demonstrates another advantage of thin-film materials; their carrier lifetimes don't have to be as long.

One of the biggest issues in solar cell operation deals with light trapping. Absorption lengths that approach or exceed the cell "optical thickness" will result in more photons passing through the cell without interacting and therefore without transferring their energy [58]. This is an example of the benefit of a thicker cell and one of the major problems of a thinner cell that must be dealt with. A compromise is often made between the absorptive abilities of a thick cell and the lower costs of a thin one. Reflection of photons off the surface of the cell is another issue preventing photons from transferring their energy to the cell. This occurs due the difference in RIs seen by the light going from the air to the material at the cell's surface, which is often an oxide, for example, SiO_2 [57,71].

The following sections will look into various techniques that are being developed to deal with the aforementioned deficiencies in solar cell operation.

13.6.3 Plasmonics and Solar Cell Technology

One way to improve solar cell efficiency by addressing the issue of inefficient light trapping in thin materials is through plasmonics [57,58]. Plasmons are created when the incident light excites the electrons in a metal or low dielectric constant material into the conduction band. Through the successive excitation and relaxation of these electrons, a collective oscillation of conduction band electrons will occur. Either surface or bulk plasmons can be created. Bulk plasmons occur as guided waves within the material. Surface plasmons occur along the interface of the high-index material with a dielectric. Solar cells often use NP plasmons to boost performance. For an NP, the depth of penetration of the light is about the same as the particle depth in at least one dimension. Therefore, bulk and surface excitations become one. Then the optical excitation of the electrons up to the conduction band results in an electric field that can span the entire bulk of the structure.

This field will create a charged surface on opposing sides of the particle. This creates a local electric field that will act to restore the electrons back to their initial positions [57,58]. The photonic excitation and electric field interactions create charge oscillations within the particle with characteristics defined by the electron density of the material and NP shape and dimensions [72]. When the incident light acts on the NP at its resonant frequency, the electromagnetic field around the particle can be greatly enhanced. This enhanced field will extend further into the semiconducting material of the solar cell, improving absorption by increasing the local density of optical states [57].

Metal NPs can also improve solar cell efficiency by scattering light into the absorptive material. The light incident on particles at an interface between high and low dielectric media will preferentially scatter in the direction of the high dielectric, or into the absorber [58]. If there is a back reflector on the rear of the cell, the light may be trapped within the cell for several passes before being transmitted, greatly increasing the cell's effective optical thickness.

It has been found that the size and shape of the NPs used will have a certain effect on the spectrum and intensity of light seen by the absorptive layer [51,58,72]. Smaller particles will couple more incident light energy into the semiconductive material, though too small a particle will incur ohmic losses [58]. It has also been determined that taller NPs, in the perpendicular direction to the incident plane, will trap light better due to a higher light scattering coefficient at the top surface of the cell [72]. This provides for better light coupling efficiency between the particles and the absorptive material. Shape also makes a difference with coupling of scattered energy as columnar and hemispherical particles have been found to scatter more effectively than spherical particles [58]. Also, more faces on the NP, that is, spherical, relate to a higher peak intensity, a narrower band, and a blueshift of the incident spectrum. Conversely, a particle shape with less faces, that is, square, will result in a lower peak intensity, a broader bandwidth, and a red shift of the incident light [31].

The spacing between NPs will also make a difference in the effect the particles have on cell efficiency. Typically, it is desirable for the particles' local plasmon fields to interact constructively with each other in order to increase the absorption cross section of the cell. Spacing between particles can be tuned based on particle size and material. Particles spaced too closely can result in destructive interference between their fields, resulting in more absorption [58].

13.6.4 Anti-Reflection

The traditional way of dealing with reflection of incident light on the cell involves applying an anti-reflection coating (ARC) applied to the surface of the cell [68]. Typically, this coating is a thin film of a quarter-wavelength dielectric. Also, the surface under the ARC is usually roughening or textured [surface text, graded]. This causes some of the light that is reflected away from the incident plane of the cell to be scattered forward into the cell instead, or for light passing through the bottom of the cell to be scattered back into the cell. The downside of this method is an increase in minority carrier recombination near the surface that can decrease efficiency [68,69]. Efforts have been made to decrease the minority recombination while improving the forward scattering of photons into the cell by adjusting the profile of textured surfaces. Such profiles experimented include irregular, irregular periodic, symmetric and asymmetric triangular, and Fourier series [69]. The Fourier series models were found to be the most effective in this study, though a pyramidal scheme is most common in industry.

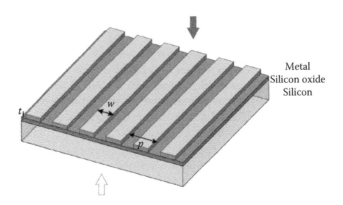

FIGURE 13.14

Thin-film Si solar cell with a 1D metallic nanograting. Light can be incident from the top (top arrow) or the bottom (bottom arrow). (From Birnie, D., Thin film amorphous silicon solar cells, Rep. 14:150:491, Rutgers, Newark, NJ, 2005.)

Another way of reducing the effects of reflection involves the use of a metallic grid layer on the top or bottom surfaces of the cell, as shown in Figure 13.14 [58,52,72]. When on the top of the cell, a grid or array of metallic nanostructures can both reflect light into the cell, rather than away from, and create a resonant electromagnetic field that will increase the absorption cross section of the semiconducting material. A top surface grid, which is on the cell's incident plane, can also block some light from getting through to the absorptive material. For this reason, top grids with minimal widths are generally used [62]. The pattern of the grid for a particular NP shape also makes a difference. The placement of the particles should be made so that the enhanced resonant electromagnetic waves resulting from the particle/light interactions interfere constructively with each other. In order to prevent the blocking of light by the surface grid is to place the grid on the bottom of the cell [62]. This grid can benefit more from wider particles that will reflect unabsorbed light back into the cell. The grid will also function to enhance the cells absorbance, utilizing plasmons like the surface grid.

Since reflection at the cell surface results from a sharp change in the index of refraction at the air/dielectric boundary, the amount of reflection can be mitigated by lessening that difference [69]. To do this, a graded refractive-index structure (GRIS) must be created (Figure 13.15). With such a structure, the light will first contact a dielectric layer with an index of refraction closer to that of air. To avoid reflection at the interface of this layer and the next, the light must then come into contact with another dielectric layer with an index of refraction slightly above that of the prior layer. This grating of RIs, layer by layer, will continue until the light can be passed into the absorptive material with minimal reflection [69].

13.6.5 Thermal Losses

Another limitation in solar cell efficiency is the release of energy in excess of the bandgap to heat [52–54]. The photon-accepting material can absorb light with energy greater than or equal to its bandgap. If the energy is greater than, however, the remaining energy will be lost. One way to help mitigate such losses is to use a capping material on the cell capable of downconverting higher photon energies into lower energies [72,74]. This process would occur prior to photon interaction with the absorptive material and so could be used with

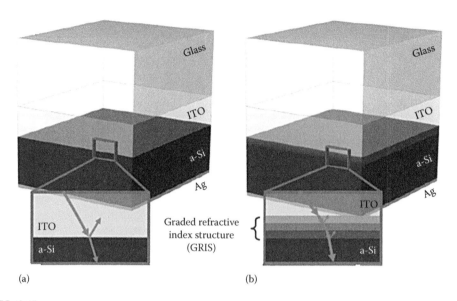

FIGURE 13.15
(a) Traditional ARC over an a-Si cell. (b) GRIS with an ARC over the top as it further decreases reflection. (From Jang, S.J., *Opt. Express*, 19(S2), A108, 2011.)

various material types. Rare earth ion-doped materials have been looked into for such an application. The energy of the incident photon is "cut" or given up to a host material, sensitizer, which transfers that energy to a rare earth ion, activator [72,74]. A method has been developed where the ion is both the sensitizer and the activator. Upon relaxation, the activator will then release the stored energy as two or more lower-energy photons [54]. It has also been determined that combining a downconverter cap with an upconverter at the rear of the cell can further boost performance [72]. The upconverter would combine the energy of two transmitted low-energy photons and reemit a higher-energy photon capable of being absorbed, as shown in Figure 13.16.

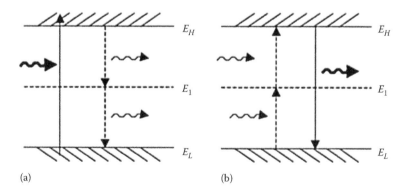

FIGURE 13.16
Simple energy diagrams of (a) downconversion of one high-energy photon into two lower-energy photons, and (b) the upconversion of two low-energy photons into one higher-energy photon. (From del Cañizo, C. et al., Photon conversion potential to improve Si solar cell performance, in *20th European Photovoltaic Solar Energy Conference*, June 6–10, 2005, Barcelona, Spain, pp. 1309–1312.)

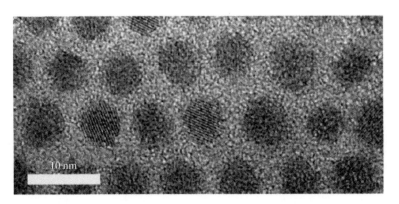

FIGURE 13.17
Spherical PbTe nanocrystals. (From Maynard, J.A. et al., *Biotechnology*, 11, 1542, 2009.)

Thermophotovoltaic cells have also been looked into for utilizing both the optical energy of light directly absorbed into the cell material and the additional thermal energy from high-energy photons. Some methods attempt to utilize the excess thermal energy for water or home heating purposes, when applicable [59]. Others attempt to directly harness the energy for electron generation [59,73]. One such method concentrates the light onto a selective absorber in order to generate large amounts of heat at that location. The absorber is thermally coupled to a selective emitter that radiates electromagnetic energy. A method described in a paper by Bermel et al. utilizes metamaterials to concentrate the light across a prescribed range of incident angles and frequencies [73]. This method improves efficiency by decreasing infrared losses and boosting absorbance through resonance waveguiding.

Another way of dealing with high-energy photons involves the use of nanocrystals. These small semiconductors with a diameter much less than the wavelength of light, like quantum dots, are embedded into a host material, as shown in Figure 13.17. Previously utilized configurations include indium arsenide dots in a GaAs matrix and PbTe dots in an unreported matrix [74,75]. They have the interesting property of being able to emit two or more electrons for a single photon absorbed. The number of electrons produced depends on the energy of the photon as compared to the bandgap. Ideally, a photon with energy X times the bandgap will produce X electrons in the conduction band. This process usually does not work in bulk semiconducting material, where the excess energy is given up as heat, although sometimes an additional electron can be freed in a process called impact ionization [54].

13.6.6 Metamaterials and Photovoltaic Applications

Metamaterials are an emerging field of study with potential photovoltaic applications. These artificially engineered materials are designed to have unnatural anisotropic optical properties [76–78]. One such property is a negative RI that allows the materials to focus or diverge incident light onto or around an object through waveguides [76]. This could benefit solar cells by directing the optical energy into the cell, limiting scattering and reflection at the surface. A common disadvantage to metamaterials is that their designs tend to be narrowband or wavelength specific, meaning that the material will only guide incident light having a wavelength within a narrow band of some predetermined tunable central

(a)

(b)

(c)

FIGURE 13.18
Two patterns for varying the metamaterial central wavelength over the cell surface. (a) Circular holes in the metamaterial, (b) square holes, and (c) the general structure of the metamaterial cell with an indium tin oxide anti-reflecting cap, Ag-etched mask over an a-Si absorber, and Ag back reflector. (From Cui, Y. et al., *Nano Lett.*, 12, 1443, 2012.)

wavelength [77,78]. This wavelength is determined as the resonant frequency between the incident light and the metamaterial. This resonance suppressed reflection at that wavelength and guides the light through the material as a bulk plasmon [78]. Relative broadband waveguides have been reported, although the accepted wavelengths for these are still not all encompassing for most solar absorptive materials; for example, one reported design will accept most of the infrared spectrum [76]. Another design has been created to utilize the entire visible spectrum by varying the metamaterial construction over the surface of the solar cell, like a grating. By this method, one end of the cell will absorb blue light and the other red, as shown in Figure 13.18 [78].

Radial PN junctions have been designed to decouple light absorption from minority carrier collection [52,79,80]. For optimal efficiency in solar cells, a thick material for high photon absorption and a long minority carrier diffusion length for optimal electron collection is desired [keys]. Radial junctions create an option in which the minority lifetime does not have to be very long. The direction of minority carrier diffusion is perpendicular to the direction of light absorption (Figure 13.19). Such devices have been created using nanowires with a c-Si core surrounded by an a-Si shell. The inclusion of both types of silicon provides for a broader range of photon energy absorbance than either material would have alone. The distance from the center of each wire to the a-Si/c-Si junction can be tuned to be very short, while the length of each wire in the direction of the incident light is longer [52,80]. Simulations have determined that 30 nm long nanowires provide for optimal

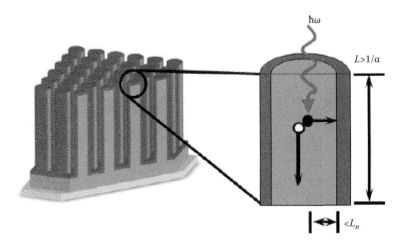

FIGURE 13.19
Nanowire array. A-Si shell in black and c-Si core in white. (From Kayes, B.M., Radial pn junction, wire array solar cells, PhD dissertation, *Appl. Phys.*, California Institute of Technology, Pasadena, CA, August 2008.)

FIGURE 13.20
ZnO nano-tree array. (a) SEM picture, (b) authors depiction of length-grown (LG) and branch-grown (BG) nano-trees. (From Ko, S.H. et al., *Nano Lett.*, 11, 666, 2011.)

absorbance in this scheme [30]. Absorption is also improved by the nanowire array through scattering of incident light between the nanowires, suppressing reflection, and by leaky-mode resonances, or the interaction between the incident and reemitted light between the sub-wavelength structures [80], as shown in Figure 13.19. A similar study was conducted utilizing zinc oxide "trees" in a dye-based cell, as shown in Figure 13.20. The trees acted in the role of the nanowires but also allowed for increased spatial coverage, scattering, as well as electron current flow between the trees by means of their branches [65].

Overall, the low efficiency in solar cell conversion of absorbed photons into electrical energy can be largely attributed to several issues. These are a minority carrier diffusion length that is less than the cell thickness, an absorption length that is greater than the cell thickness, surface reflection of incident light, and thermal losses. The use of thermophotovoltaic cells, ARCs, plasmonics, metamaterials, and different cell structures, like columnar arrays, can help mitigate losses associated with these deficiencies. New, multijunction solar cell materials may also lead to higher efficiency performance.

13.7 Computational Examples Using MATLAB®

Referring to the equations of Chapter 3, the critical angle with two given indices of refraction is given by

$$\theta_c = \sin^{-1}\left(\frac{n_2}{n_1}\right) \tag{13.15}$$

where n_1 and n_2 are the index of refraction of two different media.

Keeping n_1 fixed and varying n_2 for different media, the critical angle, θ_c, can be obtained. n_2 is varied from 1 to 1.45 in steps of 0.05 for 10 different media. The plot of the critical angle for different values of n_2 keeping n_1 fixed is as shown in Figure 13.21.

The plot clearly shows that the values for critical angle increases as the index of refraction for the second medium (n_2) is increased.

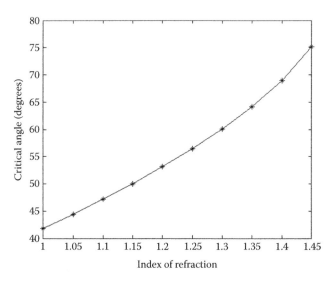

FIGURE 13.21
Plot of critical angles for different values of index of refraction (n_2) with n_1 fixed at 1.0.

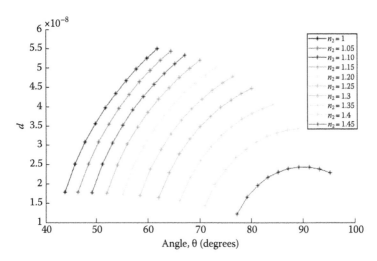

FIGURE 13.22
Plot of d versus θ for different values of n_2.

After the calculation of different values of critical angle for different values of n_2, let us consider an equation:

$$d = \frac{\lambda_0}{4\pi\left[n_1^2\sin^2\theta - n_2^2\right]^{-1/2}} \qquad (13.16)$$

where $\theta > \theta_c$ and λ_0 is the intrinsic wavelength in air.

Next, assuming $\lambda_0 = 800$ nm, the value of d versus θ is plotted for different values of n_2. The value of θ is achieved by increasing θ in small increments with respect to θ_c as

$$\theta = \theta_c + 2°,$$

$$\theta = \theta_c + 4°, \qquad (13.17)$$

$$\theta = \theta_c + 6°, \text{ and so on.}$$

For the plot, a total of 10 increments are made, that is, until $\theta = \theta_c + 10°$. This plot is made for 10 different values of n_2 as obtained from equation (1). This plot is shown in Figure 13.22.

The plot in Figure 13.22 shows that the value of d increases as the value of θ increases. Also, the values of d are higher for lower values of n_2.

MATLAB® code

```
%%- - - - - - - - - - - - - - - - - - - - - - - - - - - - - - - - - - - -
%%Critical Angle Analysis

%%- - - - - - - - - - - - - - - - - - - - - - - - - - - - - - - - - - - -
%%Critical Angle = inv(sin(n2/n1))--Equation (1)
%%- - - - - - - - - - - - - - - - - - - - - - - - - - - - - - - - - - - -
clear all;
```

```
close all;
clc;
n1 = 1.5;
n2 = 1:0.05:1.45;
for i = 1:10
   theta_c_rad(i) = asin(n2(i)/n1);
   theta_c_deg(i) = theta_c_rad(i)*(180/pi);
end;
plot(n2,theta_c_deg,'k-*');
xlabel('Index of Refraction');
ylabel('Critical Angle (Degrees)');
title('Plot of critical angle for different values of index of
refraction');

%%- - - - - - - - - - - - - - - - - - - - - - - - - - - - - - - - - -
%%Value of 'd'——Equation (2)
%%- - - - - - - - - - - - - - - - - - - - - - - - - - - - - - - - - -

lambda_0 = 800e-9;
for i = 1:1:10
   figure(2);
   hold on;
   t = 1;
   for j = 1:1:10
      inc = j*2;
      theta(t,j) = theta_c_deg(i) + inc;
      d(t,j) = lambda_0./(4*pi*power(power(n1,2).*power(sind(theta(t,j)),2)…
         - power(n2(i),2),-0.5));
   end;
   plot(theta,d,'-*');
   xlabel('Angle,theta (Degrees)');
   ylabel('d');
   title('Plot of d vs theta for different values of index of
   refraction');
   legend('n2 = 1','n2 = 1.05','n2 = 1.10','n2 = 1.15','n2 = 1.20',…
      'n2 = 1.25','n2 = 1.3','n2 = 1.35','n2 = 1.4','n2 = 1.45');
end;

%%- - - - - - - - - - - - - - - - - - - - - - - - - - - - - - - - - -
%%      END
%%- - - - - - - - - - - - - - - - - - - - - - - - - - - - - - - - - -
```

13.8 Medical Diagnostics and Theranostics

The field of nanophotonics opens the horizon to new and efficient treatment areas by combining diagnostics with therapeutics for medical purposes (theranostics). It includes patient prescreening and therapy monitoring as a potentially powerful tool for medical diagnostics, treatment, and drug delivery. The objective of this section is to explore a few methods, systems, and applications in the area of bionanophotonics and highlight the significance of computer modeling and simulations of some complex nanoscale systems. We will first look at an overview of the use of nanophotonics in highly targeted early cancer diagnosis, its therapy, post-therapy monitoring, and drug-delivery therapeutics. We will also look at the utilization of sensors based on nanophotonics for the simultaneous detection and discrimination of specific gene-associated cancerous stages from normal conditions, invention of light-guided nanodevices, and light-activated therapy with the ability to monitor real-time drug activity. A review of the achievements in this field will help us understand cancer better and provide opportunities to develop clinically critical research tools.

Nanophotonics is the science of light–matter interaction at the nanometer scale (<1 µm to ≥1 nm). It has become one of the fastest-growing fields in nanotechnology spanning multiple areas such as microelectronics, magnetic recording, environmental remediation, homeland security, biomedicine, and life sciences, ranging from the development of novel nanobiotechnologies and nanobiomaterials to minimally invasive optical diagnostics and therapeutics [81,82]. By integrating multifunctional nanovectors (NVs) and nanoconjugates (NCs) with imaging, we get reconfigurable and scalable photonic devices and imaging systems that have been used in the early diagnosis, accurate staging, treatment, and minimally invasive monitoring of cancer and better facilitation of localized surgical interventions. As a result, we have been able to see a dramatic improvement in the current poor survival rate of patients with a variety of tumors [83–86]. Several diagnostic and therapeutic modalities rely on using nanostructures with enhanced functionality contributing synergistically to the development of nanophotonics in clinical areas such as

1. Targeting cancer cells with NPs
2. Gold NPs and photodynamic therapy in cancer detection
3. Quantum dots for imaging and treatment of cancer
4. Molecular and functional imaging applications
5. Nanomagnetic particles and cancer detection
6. Optically tunable NP contrast agents for early cancer detection
7. NIR fluorescence imaging of cancer with NP-based probes
8. Polymer-conjugated NPs
9. Polymer nanostructures and surface plasmonics

For instance, certain cancer therapies utilize nanoscale metallic plasmonic elements especially gold-coated nanoshells for targeted detection and destruction of tumor tissue. In particular, gold-coated nanoshells act as optical contrast agents to discriminate cancerous cells from normal cells. The mixture of antibodies with gold-coated nanoshells is applied to map the expression of relevant biomarkers for molecular imaging under confocal reflectance microscopy. After injection of the cocktail containing gold-coated nanoshells with a special biological coating into the bloodstream (Figure 13.23), the nanoshells disperse inside the body, seek out and bind to tumor cells via antibodies stuck to their surface, and are microsurgically removed from vessels for further genetic analysis or noninvasively eradicated directly in blood vessels by laser irradiation through the skin, which is safe for normal blood cells [87].

Each gold nanoshell is about 10,000 times smaller than a white blood cell. Each tumor cell is covered by about 20 nanoshells, which are functionally enabled by antibodies [82]. Once each tumor cell is roughly covered, the nanoshells would be illuminated by brief exposure to harmless NIR light through the tissue. Gold nanoshells elicit an optical contrast to discriminate between cancerous and normal cells [83], and the delivery of a more

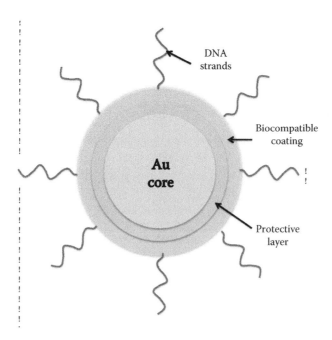

FIGURE 13.23
Structure of a gold-coated nanoshell used in cancer detection and treatment.

intense NIR dose will be selectively directed toward bound tumor cells. Free-floating electrons on the outer surface of the gold shells concentrate the intensified NIR energy to heat and destroy the tumor cells selectively [87]. Nanophotonics-based achievements in oncology successfully unite three discrete stages including diagnostics, therapy, and therapy guidance, all in one single theranostic process. These exciting findings and explorations in nanophotonics dramatically improve early cancer diagnosis and treatment and prevent metastasis.

NPs containing optical probes, light-activated therapeutic agents, and specific carrier groups can identify cancerous cells or tissues in order to provide efficient targeted drug delivery, offering at the same time unique opportunities for real-time monitoring of drug efficacy (Figure 13.24). Another novel application of nanophotonics is nanocages, a class of plasmonic NPs for drug-delivery vehicles [49,90]. Nanocages are promising targeted drug-release carriers. Drugs delivered via NP or molecular complexes target specific sites within the human body. In response to the NIR radiation, the conformation of complex biomolecules changes, thereafter triggering the opening of "cages" to release their contents. Nanocages also provide a novel approach to treat cancer due to their ability to effectively convert light into heat for photothermal destruction of cancerous cells and tissues [90].

NP-based strategies can be used for biosensing using plasmonic nanosensors such as metal NPs functionalized with nucleic acid strands for colorimetric assays and biobar codes for protein detection or intense labels for immunoassays [91]. Some NP systems can also be used for sensing by exploring a typical fluorescent resonance energy transfer (FRET) system or can be surrounded with Raman reporters in order to provide in vivo detection and tumor targeting. In fact, NPs symbolize an important class of materials with unique features suitable for biomedical imaging applications such as increased sensitivity in detection and high quantum yields for fluorescence. Alternatively, NPs can survey

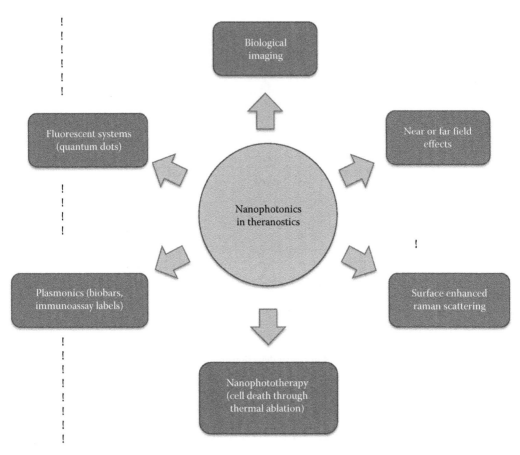

FIGURE 13.24
Nanophotonics used in theranostics.

near/far field enhancing qualities that hold promise for a bounty of novel applications in optics and photonics. Engineered NPs can also act as phototherapeutic agents that can be attached to specific targets for selective damage to cancer cells [91].

13.8.1 Optical Multifunctional Nanovectors and Nanoconjugates

With the emergence of nanotechnology, the development of novel multifunctional materials at the nanoscale range to aid in imaging enhancement and targeted theranostics payload delivery such as NVs or NCs provides new and powerful tools for imaging, diagnosis, and therapy. Nanoimaging based in optical multifunctional NVs and NCs has become the next generation of cancer imaging with great potential for early cancer detection [92].

13.8.1.1 Nanovectors

An NV is a multifunctional organic or inorganic nano-entity, like NPs, nanowires, and nanotubes that could potentially improve both the delivery of therapeutic drugs and the localized killing of targeted cells. A typical NV consists of an NP core coated with a targeting agent specific to the target cells and a biomolecule with designated functionality. NVs can successfully overcome a series of systemic, tissue, and cellular barriers, such as bypassing the recognition of the reticuloendothelial system (RES); extravasating at the leaky tumor vasculatures; penetrating and homogeneously distributing in solid tumor tissues, introduced by the target cancer cells; penetrating cellular and subcellular membranes; and then releasing their drug cargo in the cytoplasm of the target cancer cells in a sustained manner with minimal side effects to healthy tissue [93]. As carriers for therapeutic and imaging contrast agents, NVs possess projected advantages of drug localization at disease lesions and the ability to circumvent the biological barriers encountered between the point of administration and the projected target [94].

An NV can historically be classified into three main generations [95]. First generation of stage 1 mesoporous silicon particles (S1MPs) and second generation of stage 2 nanoparticles (S2NPs) both are single-entity particles. Each generation is designed to negotiate one or more biological barriers. Third-generation NVs decouple multiple tasks together using separate nanocomponents, where the S2NPs are loaded into the S1MPs and acting in a synergistic fashion so that the biodegradable and biocompatible S1MPs are able to host, protect, and deliver S2NPs on intravenous injection. This third generation of NVs aims at sequentially overcoming the biological barriers en route to the target delivery site by separating and assigning tasks to the coordinated logic-embedded vectors constituting it.

S1MPs were made in the clinics more than 15 years ago. They were rationally designed and fabricated in a nonspherical geometry to enable superior blood visualization and increase cell surface adhesion. They could efficiently transport NPs that were loaded into their porous structure and protect them during transport from the administration site to the disease lesion. Each task was based on the permeability of tumor-associated neovasculature by a mechanism known as enhanced permeation and retention (EPR) [96]. S1MPs are fabricated by semiconductor processing and electrochemical etching. These two techniques ensure the exquisite control and precise reproducibility of S1MP physical characteristics such as size, shape, and porosity [97]. Particles with 20–50 nm pores were formed by selective electrochemical etch of a silicon-rich silicon nitride (SiN)-masked array of 2 μm cylindrical trenches in the silicon in a mixture of hydrofluoric acid (49% HF) and ethanol (3:7 v/v). The etch was performed initially by applying a current density of 100 mA cm^{-2} for 40 s, followed by a high-porosity layer applying a current density of 380 mA cm^{-2} for 6 s. After removing the SiN layer by HF, particles were released by ultrasound in isopropyl alcohol (IPA) for 1 minute and preserved in a sealed centrifuge tube at 4°C in controlled humidity [98,99].

The surface of S1MPs can be chemically modified with negatively/positively charged groups, PEG, and other polymers, covalently conjugated and electrostatically attached to a vast spectrum of dyes, fluorophores, fluorescent tags, radioactive molecules, and other imaging contrast agents. They also easily attach with biologically active targeting moieties such as antibodies, peptides, aptamers, and phage based on electrostatic interactions. All of these multifunctional nanoassemblies allow the incorporation of imaging components such as NIR dye, single-photon emission computed tomography (SPECT) agent, and positron emission tomography (PET) agent onto the S1MP surface. NIR is currently being used in clinic for its minimal intrinsic absorption from tissue and molecules in this spectral region.

Hemispherical or disk-shaped nanoporous silicon S1MPs are engineered to exhibit an enhanced ability to distinguish within blood vessels and adhere to disease-associated endothelium. Once positioned at the disease site, the S1MP can release the drug-/

siRNA-loaded S2NP to achieve the desired therapeutic effect, prior to complete biodegradation of the carrier particle. They can also release an imaging agent or external energy-activated S2NP such as gold NPs or nanoshells. Another possible mechanism of action is cell-based delivery of the MSVs into the disease loci followed by triggered release of the S1MP/S2NP from the cells.

13.8.1.2 Nanocojugates

An NC is a chemically modified NP made of therapeutic biomolecules and polymer [82,85,94]. They serve as a type of drug-delivery vehicle, which means they can carry a drug through the body to the place where it is needed to treat the disease. Similar to attractive drug-delivery vehicles—NVs—NCs can successfully overcome all those systemic, tissue, and cellular barriers. However, many NVs aggregate in solution or in a lyophilized state. Aggregation can destroy the utility of these compositions by making them unable to accomplish one or more of the previously mentioned objectives. Accordingly, NCs have been developed, which can be prepared, stored, and systemically administered without aggregation. NCs are more capable of circumventing many translational barriers and are eventually able to provide clinically useful drug formulations efficiently.

Both NVs and NCs have provided promising results in cancer research. By combining the power of innovation in nanomaterials, cancer biology, and photonics, the invention of optical multifunctional NVs and NCs comprising optical contrast agents such as optically active metal nanospheres, fluorescent dye, and an NP comprising a magnetic or superparamagnetic iron oxide core covered by a metal or metal-oxide NP shell when conjugated to a chemotherapeutic drug greatly improve early cancer detection by enhancing the optical contrast between cancerous and healthy tissue. Meanwhile, the advanced resolution of nanophotonics in sensing and imaging technical aspects significantly helps probe, monitor, and understand changes in tissues/microenvironments that are crucial to preventive strategies for cancer treatments. Integration of several single molecular imaging modalities together according to specific purposes provides opportunities for novel multimodal molecular imaging platform [82,88,100–102]. With particular focus on molecular and functional contrast agents for nanophotonics, the researchers are able to develop safer, more efficient, and specific platforms for clinical settings in order to facilitate early diagnosis and treatment. An overview of imaging and therapeutic modalities that use theranostic nanoshells is shown in Figure 13.25.

13.8.2 Applications of SPR in Medical Diagnostics

SPR is a phenomenon that results from the illumination of a metallic surface such as gold, by visible or NIR radiation from a monochromatic light source via a hemispherical prism, which exits to a detector typically a photodiode array at an angle related to the RI. The resultant oscillation of free electrons generates surface plasmons or electromagnetic waves that resonate and absorb light. The specific wavelength/angle at which this occurs is a function of the RI in the proximity of the gold surface and relates to the mass on the chip surface. A change in mass, effected by the immobilization of a ligand and, subsequently, further interactions that take place when analytes are passed over the modified sensor surface, causes a shift in the resonance to a longer wavelength. This introduces an RI change [103,111,114,116]. Figure 13.26 shows the working principle of the SPR method.

When the wave vector closely matches that of the surface plasmon at the metal–sample interface, reflected light is significantly attenuated. Electronic surface plasmons obey the

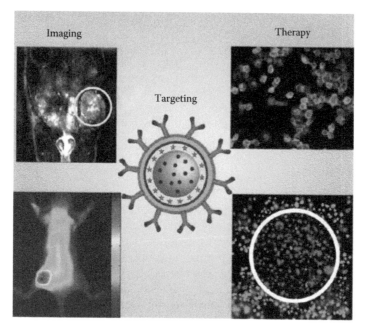

FIGURE 13.25
Theranostic nanoshells: from probe design to imaging and treatment of cancer. (From Bardhan, R. et al., *Adv. Func. Mater.*, 19, 3901, 2009.)

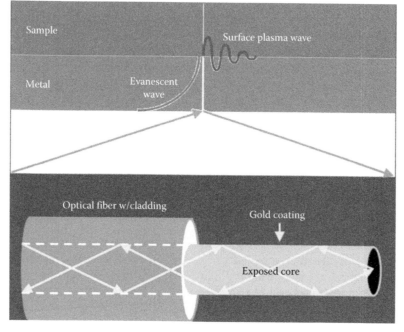

FIGURE 13.26
Concept of SPR. (From http://www.mdpi.com/1424–8220/9/3/1339/pdf)

following dispersion relations [114] where ε_m and ε_s are the dielectric constants of the metal and the dielectric medium, respectively:

$$k_{spw} = k_0 \sqrt{\frac{\varepsilon_s \varepsilon_m}{\varepsilon_s + \varepsilon_m}} \tag{13.18}$$

$$k_x = k_0 n_{sub} \sin(\theta_{inc}) \tag{13.19}$$

$$k_0 = \frac{2\pi}{\lambda} \tag{13.20}$$

As an example of an immunologic application of SPR, consider the case of the bacteria *Escherichia coli*. It is an enteric pathogen that poses a serious health risk and must be monitored closely in the food supply. Conventional methods rely on culturing, which takes 1–3 days, a high degree of expertise, and specialized facilities. Newly designed SPR systems are able to identify the *E. coli* bacteria with the same accuracy as the traditional methods. GE Healthcare has introduced a set of SPR-based sensors to identify the *E. coli* with a high degree of accuracy. The Biacore-based SPR analytical platform is one of them, produced by GE, which has demonstrated the ability of the researcher to capture, sandwich, or subtract-inhibition assays.

Hearty and colleagues [14] produced a murine monoclonal antibody, which was shown to be specific for the surface-located *Listeria monocytogenes* internalin A (InA) protein in native and recombinantly expressed forms. When this antibody was immobilized on a CM5 surface through EDC/NHS coupling, a limit of detection of 1×10^7 CFU/mL was observed in testing of *L. monocytogenes*. Cross-reactivity studies clearly demonstrated the specificity of this monoclonal antibody, with minimal binding to *E. coli*, *Bacillus cereus*, and *Listeria innocua* (the latter selected due to the non-expression of the InA protein) observed. This further illustrates the importance of this antibody as a species-specific reagent [112–114].

Many laboratory-grade SPR detection systems for immunosensing and DNA hybridization detection exist, most notably the Biacore from GE Healthcare. Recently, efforts have focused on reducing the size and complexity of SPR sensors using integrated microfluidic systems. The Sensata Spreeta SPR sensor is a commercially available, low-cost SPR sensor that can be adapted to a variety of applications. The handheld device is shown in Figure 13.27 [103,105,107,108,117]. The device provides an impressive LOD of 80 pM when tested with IgG. Waswa et al. demonstrated the use of the Spreeta for the immunological detection of *E. coli* O157:H7 in milk, apple juice, and ground beef extract [103,109]. They showed an LOD of 102 colony-forming units per mL and demonstrated that the sensor was nonresponsive to other organisms (*E. coli* K12 and Shigella). The Spreeta sensor was able to reliably quantify *E. coli* in 30 min.

Another briefcase-style sensor was developed by the same group that was capable of simultaneously detecting small molecules, proteins, viruses, bacteria, and spores. Recently, Feltis et al. demonstrated a low-cost handheld SPR-based immunosensor for the toxin ricin [106,117]. They demonstrated an LOD of 200 ng/mL. Although this is substantially higher than the LOD obtained with the Biacore sensor (10 ng/mL), it is quite sufficient for this particular application (200 ng/mL is 2500 times lower than the minimum lethal dose), and the result is available in 10 min.

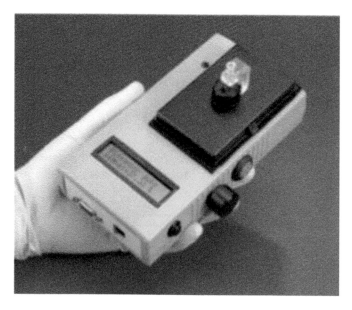

FIGURE 13.27
The Sensata Spreeta SPR sensor used for *E. coli* detection. (From http://users.path.ox.ac.uk/~vdmerwe/internal/spr.pdf)

13.8.3 Advantages, Disadvantages, and Limitations

Some of the advantages of the optical systems are that they possess high sensitivity and can be used to detect small molecules using advanced techniques of amplification. They are very small in size and can be used as portable instruments.

The advantages of SPR systems are as mentioned in the following:

- High specificity
- Ease of miniaturization
- Real-time monitoring
- Adaptability for remote sensing
- Good for small analytes, high-throughput assays, and evaluation of macromolecules [103]

Some of the disadvantages of the SPR systems are as follows:

- Limited to choice of metal that results in SPR.
- Any mismatch between the wavelength of the light either too short or too long, and the gold will fail to exhibit the plasmon oscillation.
- The matching wavelength is also dependent on the permittivity of the material on the opposite side of the gold film
- Sample preparation—attaching probe to metal surface can prove difficult.
- RI is temperature dependent.

13.9 Computational Calculations Using MATLAB®

A Mueller matrix can be decomposed as a sequence of three matrix factors, namely, the depolarization, retardance, and diattenuation [118,119]. This concept can be expanded by expressing them in terms of the aspect angle of an object, θ, so that

$$M(\theta) = M_{depol}(\theta)\, M_{ret}(\theta)\, M_{diat}(\theta) \qquad (13.21)$$

The diattenuation, d, is dependent on the first-row vector of the muller matrix (MM). This vector describes differential attenuation for both linear and circular polarization states, and the diattenuation can be expressed in terms of

$$d(\theta) = \frac{1}{m_{00}(\theta)} \times \sqrt{m_{01}^2(\theta) + m_{02}^2(\theta) + m_{03}^2(\theta)} \qquad (13.22)$$

where m_{nm} are the first-row Mueller matrix elements, a function of the aspect angle of the sample. A computational algorithm of the diattenuation, using MATLAB, is offered in the following text:

Diattenuation matrix has the following form (assuming matrix of diattenuation (MD) is not singular):

```
% MD= Tu* [ 1 D';
%          D mD ]

% Diattenuation (scalar)
 Diat_s = sqrt(M(1,2)^2+M(1,3)^2+M(1,4)^2)/M(1,1);
% Diattenuation vector is the first row of the Mueller matrix
 Diat_v= (1/M(1,1))*[M(1,2) M(1,3) M(1,4)]';
% Magnitude of Diattenuation Vector
 magDiat = sqrt(Diat_v(1).^2+Diat_v(2).^2+Diat_v(3).^2);
% Diattenuation Unit Vector
 Diatunit = Diat_v/magDiat;
% 3x3 Identity Matrix
 I = [1 0 0; 0 1 0; 0 0 1];
% submatrix mD
 mDiat = sqrt(1-Diat_s.^2)*I+(1-sqrt(1-Diat_s.^2))*Diatunit*Diatunit';

%disp('DIATTENUATION MATRIX:')
% Creates the FINAL DIATTENUATION MATRIX
 MD = M(1,1)*[1 Diat_v'; Diat_v mDiat];
% Re-write in a 16x1 column vector to be written to the output xls:
 MD_V = [MD(1,1); MD(1,2); MD(1,3); MD(1,4);…
        MD(2,1); MD(2,2); MD(2,3); MD(2,4);…
        MD(3,1); MD(3,2); MD(3,3); MD(3,4);…
        MD(4,1); MD(4,2); MD(4,3); MD(4,4)];
```

13.10 Conclusion

The design, engineering and operation of several nanophotonic components, devices, systems, and techniques were presented, relating the technical challenges to computational modeling. Key paradigms of nanophotonics were introduced and their design operational principles within different technological domains were presented, analyzed, and discussed. Novel methodologies, systems, techniques, and applications in the area of bionanophotonics for medical diagnostics, treatment, and drug delivery aimed at highlighting the significance of computer modeling and simulations of complex nanoscale systems were presented. Bionanophotonics is expected to open tremendous opportunities toward the design of novel and efficient diagnostic/therapeutic techniques and devices within the biological, biochemical, and medical device field, tending to become a potentially powerful tool for medical diagnostics, treatment, and drug delivery.

References

1. A. Trugler, Optical properties of metallic nanoparticles, PhD thesis, Institut fur Physik, Berlin, Germany, 2011.
2. S.K. Ghosh and T. Pal, Interparticle coupling effect on the surface plasmon resonance of gold nanoparticles: From theory to applications, *Chem. Rev.*, 107, 4797–4862, 2007.
3. J.A. Scholl, A.L. Koh, and J.A. Dionne, Quantum plasmon resonances of individual metallic nanoparticles, *Nature*, 483, 421–427, 2012.
4. D. Axelrod, Total internal reflection fluorescence, *Ann. Rev. Biophys. Bioeng.*, 13, 247–68, 1984.
5. Biacore, Surface plasmon resonance technology note 1 [online]. Available FTP: Protein.iastate.edu Directory: Seminars/BIACore File: Surface Plasmon Resonance, 2001, accessed on 14, October, 2012.
6. M.J. Linman and Q.J. Cheng, Surface plasmon resonance: New biointerface designs and high-throughput affinity screening, *Opt. Guided-Wave Chem. Biosens. I*, 7, 133–153, 2010.
7. M.J. Linman, A. Abbas, and Q. Cheng, Interface design and multiplexed analysis with surface plasmon resonance (SPR) spectroscopy and SPR imaging, *Analyst*, 135, 2759–2767, 2010.
8. G. Safina, Application of surface plasmon resonance for the detection of carbohydrates, glycoconjugates, and measurement of the carbohydrate-specific interactions: A comparison with conventional analytical techniques. A critical review, *Anal. Chim. Acta.*, 712, 9–29, 2012.
9. M.J. Linman, J.D. Taylor, H. Yu, X. Chen, and Q. Chen, Surface plasmon resonance study of protein-carbohydrate interactions using biotinylated sialosides, *Anal. Chem.*, 80, 4007–4013, 2008.
10. K.M. Bolles, F. Cheng, J. Burk-Rafel, M. Dubey, and D.M. Ratner, Imaging analysis of carbohydrate-modified surfaces using ToF-SIMS and SPRi, *Materials*, 3, 3948–3964, 2010.
11. J. Pollet, F. Delporta, K. Janssena, K. Jansb, G. Maesc, H. Pfeiffer, M. Weversd, and J. Lammertyna, Fiber optic SPR biosensing of DNA hybridization and DNA-protein interactions, *Biosens. Bioelectron.*, 25, 864–869, 2009.
12. S. Song, L. Wang, J. Li, J. Zhao, and C. Fan, Aptamer-based biosensors, *Trends Anal. Chem.*, 27, 108–117, 2008.
13. A. Sassolas, L.J. Blum, and B.D. Leca-Bouvier, Optical detection systems using immobilized aptamers, *Biosens. Bioelectron.*, 26, 3725–3736, 2011.
14. J.A. Maynard, N.C. Lindquist, J.N. Sutherland, A. Lesuffleur, A.E. Warrington, M. Rodriguez, and S. Oh, Next generation SPR technology of membrane-bound proteins for ligand screening and biomarker discovery, *Biotechnology*, 11, 1542–1558, 2009.
15. M.P. Besenicar and G. Anderluh, Preparation of lipid membrane surfaces for molecular interaction studies by surface plasmon resonance biosensors, *Methods Mol. Biol.*, 627, 191–200, 2010.
16. B. Anne, T. Sergei, B. Philippe, S. Toralf, K. Volodymyr, and B. Iris, *Nanostructured Metamaterials*, Publications Office of the European Union, Luxembourg, 2010.
17. M.A. Noginov, G. Zhu, M. Mayy, B.A. Ritzo, N. Noginova, and V.A. Podolskiy, Stimulated emission of surface plasmon polaritons, *Phys. Rev. Lett.*, 101, 226806, 2008.

18. M. Zeller, M. Cuevas, and R.A. Depine, Surface plasmon polaritons in attenuated total reflection systems with metamaterials: Homogeneous problem, *J. Opt. Soc. Am.*, 28, 2042–2047, 2011.

19. W.Q. Hu, E.J. Liang, P. Ding, G.W. Cai, and Q.Z. Xue, Surface plasmon resonance and field enhancement in #-shaped gold wires metamaterial, *Opt. Express*, 17, 21843–21849, 2009.

20. A. Abbas, M.J. Linman, Q. Cheng, New trends in instrumental design for surface plasmon resonance-based biosensors, *Biosens. Bioelectron.*, 26, 1815–1824, 2011.

21. A.V. Kabashin, P. Evans, S. Pastkovsky, W. Hendren, G.A. Wurtz, R. Atkinson, R. Pollard, V.A. Podolskiy, and A.V. Zayats, Plasmonic nanorod metamaterials for biosensing, *Nat. Mater.*, 8, 867–871, 2009.

22. A. Ishimaru, S. Jaruwatanadilok, and Y. Kuga, Generalized surface plasmon resonance sensors using metamaterials and negative index materials, *Prog. Electromagnet. Res.*, 51, 139–152, 2005.

23. A. Madeira, E. Ohman, A. Nilsson, B. Sjogren, P. Andre, and P. Svenningsson, Coupling surface plasmon resonance to mass spectrometry to discover novel protein-protein interactions, *Nat. Protoc.*, 4, 1023–1037, 2009.

24. C. Jungar, M. Strandh, St. Ohlson, and C. Mandenius, A liquid chromatography-SPR immunosensing system, *Biojournal*, 1, 19–21, 2001.

25. S.J. Rosenthal, J.C. Chang, O. Kovtun, J.R. McBride, and I.D. Tomlinson, Biocompatible quantum dots for biological applications, *Chem. Biol.*, 18, 10–24, 2011.

26. A. Salman Ogli, Nanobio applications of quantum dots in cancer: Imaging, sensing, and targeting, *Cancer Nano*, 2, 1–19, 2011.

27. M.K. So, C. Xu, A.M. Loening, S.S. Gambhir, and J. Rao, Self-illuminating quantum dot conjugates for in vivo imaging, *Nat. Biotechnol.*, 24, 339–343, 2006.

28. G.A. Quinones, S.C. Miller, S. Bhattacharyya, D. Sobek, and J.P. Stephan, Ultrasensitive detection of cellular protein interactions using bioluminescence resonance energy transfer quantum dot-based nanoprobes, *J. Cell. Biochem.*, 113, 2397–2405, 2012.

29. Y. Xing, M.K. So, A.L. Koh, R. Sinclair, and J. Rao, Improved QD-BRET conjugates for detection and imaging, *Biochem. Biophys. Res. Commun.*, 372, 388–394, 2008.

30. E.H. Sargent, Infrared quantum dots, *Adv. Mater.*, 17, 515–522, 2005.

31. S. Jin, Y. Hu, Z. Gu, L. Liu, and H. Wu, Application of quantum dots in biological imaging, *J. Nanomater.*, 1–13, 2011.

32. J.C. Chang and S.J. Rosenthal, Real-time quantum dot tracking of single proteins, *Methods Mol. Biol.*, 726, 51–62, 2011.

33. A.M. Smith, M. M. Wen, and S. Nie, Imaging dynamic cellular events with quantum dots: The bright future, *Biochemistry*, 32, 1–11, 2010.

34. S. Abramowitz, Quantum dots: Their use in biomedical research and clinical diagnostics, *Handbook of Biosensors and Biochips*, John Willey & sons, Ltd., pp. 1–5, 2007.

35. U. Ebenstein, N. Gassman, S. Kim, J. Antelman, Y. Kim, S. Ho, R. Samuel, X. Michalet, and S. Weiss, Lighting up individual DNA binding proteins with quantum dots, *Nano Lett.*, 9, 1598–1603, 2009.

36. R. Hardman, A toxicologic review of quantum dots: Toxicity depends on physicochemical and environmental factors, *Environ. Health Perspect.*, 114, 165–172, 2005.

37. J. Weng and J. Ren, Luminescent quantum dots: A very attractive and promising tool in biomedicine, *Curr. Med. Chem.*, 13, 897–909, 2006.

38. W. Wei and Z. Jun-Jie, Optical applications of quantum dots in biological system, *Sci. China Chem.*, 54, 1177–1184, 2011.

39. M. Hamada, S. Nakanishi, T. Itoh, M. Ishikawa, and V. Biju, Blinking suppression in CdSe/ZnS single quantum dots by TiO_2 nanoparticles, *ACS Nano*, 4, 4445–4454, 2010.

40. N. Harris, M.J. Ford, P. Mulvanev, and M.B. Cortie, Tunable infrared absorption by metal nanoparticles: The case for gold rods and shells, *Chem. Mater. Sci.*, 41, 5–14, 2008.

41. T.A. Erickson and J.W. Tunnell, Gold nanoshells in biomedical applications, *Nanomater. Life Sci.*, 3, 1–40, 2009.

42. S.J. Oldenburg, Nanoengineering of optical resonances, *Chem. Phys. Lett.*, 288, 243–247, 1998.

43. C. Loo, A. Lin, L. Hirsch, M.H. Lee, J. Barton, N. Halas, J. West, and R. Drezek, Nanoshell-enabled photonics-based imaging and therapy of cancer, *Technol. Cancer Res. Treat.*, 3, 33–40, 2004.

44. L. Gao, T.J. Vadakkan, and V. Nammalvar, Nanoshells for in vivo imaging using two-photon excitation microscopy, *Nanotechnology*, 22, 1–9, 2011.

45. J. Zhang, Y. Fu, and J.R. Lakowicz, Fluorescent metal nanoshells: Lifetime-tunable molecular probes in fluorescent cell imaging, *J. Phys. Chem.*, 115, 7255–7260, 2011.

46. S. Liu, Z. Liang, F. Gao, S. Luo, G. Lu, In vitro photothermal study of gold nanoshells functionalized with small targeting peptides to liver cancer cells, *J. Mater. Sci. Mater. Med.*, 21, 665–674, 2010.

47. A.M. Govin, D.P. O'Neal, D.M. Watkins, N.J. Halas, R.A. Drezek, and J.L. West, Near infrared laser-tissue welding using nanoshells as an exogenous absorber, *Lasers Surg. Med.*, 37, 123–129, 2005.

48. J. Zhang, Y. Fu, F. Jiang, and J.R. Lakowicz, Metal nanoshell-capsule for light driven release of a small molecule, *J. Phys. Chem.*, 114, 7653–7659, 2010.

49. Y. Liu, H. Miyoshi, and M. Nakamura, Nanomedicine for drug delivery and imaging: A promising avenue for cancer therapy and diagnosis using targeted functional nanoparticles, *Int. J. Cancer*, 120(12), 2527–2537, 2007.

50. P.N. Prasad, *Introduction to Biophotonics*, John Wiley & Sons, Hoboken, NJ, 2003.

51. E. Marshall, *Nanophotonics—Revolutionising Power Generation*, H.H. Wills Physics Laboratory, Bristol, U.K., March 2011.

52. B.M. Kayes, Radial pn junction, wire array solar cells, PhD dissertation, *Appl. Phys.*, California Institute of Technology, Pasadena, CA, August 2008.

53. A.S. Paschoa, Environmental effects of nuclear power generation, in *Interactions: Energy/Environment, in Encyclopedia of Life Support Systems*, J. Goldemberg, Ed., Oxford, U.K.: Eolss Publishers, 2004.

54. D. Castaldi, E. Chastain, M. Windram, and L. Ziatyk, A study of hydroelectric power: From a global perspective to a local application, Rep. CAUSE 2003, Penn State University, University Park, PA, 2003.

55. The European Wind Association, Wind power and the environment—Benefits and challenges, *Wind Directions*, 25(4), 26–40, July/August 2006.

56. Energy Center of Wisconsin, Wind power and the environment, Rep. 433-3, Madison, WI, 2000.

57. M.A. Green and S. Pillai, Harnessing plasmonics for solar cells, *Nat. Photonics*, 6, 130–132, March 2012.

58. H.A. Atwater and A. Polman, Plasmonics for improved photovoltaic devices, *Nat. Mater.*, 9, 205–213, March 2010.

59. B. Paridaa, S. Iniyanb, and R. Goicc, A review of solar photovoltaic technologies, *Renewable Sustainable Energy Rev.*, 15, 1625–1636, 2011.

60. M.L. Brongersma, Plasmonic and high index nanostructures for efficient solar energy conversion, *FiO/Ls Technical Digest*, 2011, pp. 1–2.

61. B. Liao and W. Hsu, An investigation of Shockley-Queisser limit of single p-n junction solar cells, Rep. 2.997, MIT, Cambridge, MA, 2011.

62. D. Birnie, Thin film amorphous silicon solar cells, Rep. 14:150:491, Rutgers, Newark, NJ, 2005.

63. M.A. Green et al., Solar cell efficiency tables (version 39), *Prog. Photovolt. Res. Appl.*, 20, 12–20, 2012.

64. Q. Zhang et al., Light scattering with oxide nanocrystallite aggregates for dye-sensitized solar cell application, *J. Nanophoton.*, 4, 041540:1–23, 2010.

65. S.H. Ko et al., Nanoforest of hydrothermally grown hierarchical ZnO nanowires for high efficiency dye-sensitized solar cell, *Nano Lett.*, 11, 666–671, 2011.

66. J. Yang et al., Plasmonic polymer tandem solar cell, *Am. Chem. Soc.*, 5(8), 6210–6217, 2011.

67. S. Xiao, E. Stassen, and N.A. Mortensen, Ultrathin silicon solar cells with enhanced photocurrents assisted by plasmonic nanostructures, *J. Nanophoton.*, 6, 061503:1–7, 2012.

68. X. Sheng, S.G. Johnson, J. Michel, and L.C. Kimerling, Optimization-based design of surface textures for thin-film Si solar cells, *Opt. Express*, 19(S4), A841–A850, July 2011.
69. S.J. Jang, Antireflective property of thin film a-Si solar cell structures with graded refractive index structure, *Opt. Express*, 19(S2), A108–A117, March 2011.
70. S. Riepe et al., Research on efficiency limiting defects and defect engineering in silicon solar cells—Results of the German research cluster Solar Focus, *Phys. Status Solidi C*, 8(3), 733–738, 2011.
71. S. Mokkapati et al., Resonant nano-antennas for light trapping in plasmonic solar cells, *J. Phys. D: Appl. Phys.*, 44, 185101:1–9, March 2011.
72. C. del Cañizo, I. Tobías, Y. Kugimiya, and A. Luque, Photon conversion potential to improve Si solar cell performance, in *20th European Photovoltaic Solar Energy Conference*, June 6–10, 2005, Barcelona, Spain, pp. 1309–1312.
73. P. Bermel et al., Tailoring photonic metamaterial resonances for thermal radiation, *Nanoscale Res. Lett.*, 6(549), 1–5, 2011.
74. M. Miritello, R.L. Savio, P. Cardile, and F. Priolo, Enhanced down conversion of photons emitted by photoexcited Er_xY_2-xSi_2O_7 films grown on silicon, *Phys. Rev. B*, 81, 041411:1–4, January 2010.
75. J. Wua et al., Surface plasmon enhanced intermediate band based quantum dots solar cell, *Solar Ener. Mater. Solar Cells*, 102, 44–49, 2012.
76. Y. Cui et al., Ultrabroadband light absorption by a sawtooth anisotropic metamaterial slab, *Nano Lett.*, 12, 1443–1447, February 2012.
77. Y. Liu et al., Study of energy absorption on solar cell using metamaterials, *Sol. Energy*, 86, 1586–1599, March 2012.
78. Y. Wang, Metamaterial-plasmonic absorber structure for high efficiency amorphous silicon solar cells, *Nano Lett.*, 12, 440–445, December 2011.
79. B.C.P. Sturmberg et al., Modal analysis of enhanced absorption in silicon nanowire arrays, *Opt. Express*, 19(S5), A1067–A1081, September 2011.
80. W.Q. Xie, W.F. Liu, J.I. Oh, and W.Z. Shen, Optical absorption in c-Si/a-Si:H core/shell nanowire arrays for photovoltaic applications, *Appl. Phys. Lett.*, 99, 03310:1–3, 2011.
81. M. Ohtsu, K. Kobayashi, T. Kawazoe, T. Yatsui, and M. Naruse, *Principles of Nanophotonics, Series in Optics and Optoelectronics*, Taylor & Francis, CRC Press, Boca Raton, FL, 2008.
82. Y. Shen and P.N. Prasad, Nanophotonics: A new multidisciplinary frontier, *Appl. Phys. B*, 74(7–8), 641–645, 2002.
83. K. Riehemann et al., Nanomedicine—Challenge and perspectives, *Angew. Chem. Int. Ed.*, 48, 872–897, 2009.
84. J.R. Heath and M.E. Davis, Nanotechnology and cancer. *Annu. Rev. Med.*, 59, 251–653, 2008.
85. D. Peer et al., Nanocarriers as an emerging platform for cancer therapy. *Nat. Nanotech.*, 2, 751–60, 2007.
86. M. Ferrari., Cancer nanotechnology: Opportunities and challenges. *Nat. Rev. Cancer*, 5, 161–171, 2005.
87. K. Kelleher, Engineers Light Up Cancer Research. Emerging medicine: Scientists design gold "nanoshells" that seek and destroy tumors. PopSci. Posted 6 November 2003. Retrieved July 10, 2012 from http://www.popsci.com/scitech/article/2003-11/engineers-light-cancer-research
88. R. Bardhan, W. Chen, C. Perez-Torres, M. Bartels, R.M. Huschka, L. Zhao, E. Morosan, R. Pautler, A. Joshi, and N.J. Halas, Nanoshells with targeted, simultaneous enhancement of magnetic and optical imaging and photothermal therapeutic response, *Adv. Func. Mater.*, 19, 3901–3909, 2009.
89. W. Chen, R. Bardhan, M. Bartels, C. Perez-Torres, R.G. Pautler, N.J. Halas, and A. Joshi, A molecularly targeted theranostic probe for ovarian cancer, *Mol. Cancer Ther.*, 9, 1028, 2010.
90. C.M. Cobley, L. Au, J. Chen, and Y. Xia, Targeting gold nanocages to cancer cells for photothermal destruction and drug delivery, *Exp. Opin. Drug Deliv.*, 7, 577–587, 2010.
91. J. Conde, J. Rosa, J.C. Lima, and P.V. Baptista, Nanophotonics for molecular diagnostics and therapy applications, *Int. J. Photoener.*, 2012.

92. D.J. Lewis, C. Bruce, S. Bohic, P. Cloetens, S.P. Hammond, D. Arbon, S. Blair-Reid, Z. Pikramenou, and B. Kysela, Intracellular synchrotron nanoimaging and DNA damage/genotoxicity screening of novel lanthanide-coated nanovectors, *Nanomed. (Lond)*, 5(10), 1547–1557, 2010.

93. B. Godin, E. Tasciotti, X. Liu, R.E. Serda, and M. Ferrari, Multistage nanovectors: From concept to novel imaging contrast agents and therapeutics, *Acc. Chem. Res.*, 44, 979–989, 2011

94. R.K. Jain and T. Stylianopoulos, Delivering nanomedicine to solid tumors, *Nat. Rev. Clin. Oncol.*, 7, 653–664, 2010.

95. B. Godin, W.H. Driessen, B. Proneth, S.Y. Lee, S. Srinivasan, R. Rumbaut, W. Arap, R. Pasqualini, M. Ferrari, and P. Decuzzi, An integrated approach for the rational design of nanovectors for biomedical imaging and therapy, *Adv. Genet.*, 69, 31–64, 2010.

96. H. Maeda and Y. Matsumura, EPR effect based drug design and clinical outlook for enhanced cancer chemotherapy, *Adv. Drug Deliv. Rev.*, 63, 129–130, 2010.

97. C. Chiappini, E. Tasciotti, J.R. Fakhoury, D. Fine, L. Pullan, Y.C. Wang, L. Fu, X. Liu, and M. Ferrari, Tailored porous silicon microparticles: Fabrication and properties, *ChemPhysChem*, 11, 1029–1035, 2010.

98. E. Tasciotti et al., Mesoporous silicon particles as a multistage delivery system for imaging and therapeutic applications, *Nat. Nanotechnol.*, 3, 151–157, 2008. doi:10.1038/nnano. 2008.34.

99. E. Tasciotti, B. Godin, J.O. Martinez, C. Chiappini, R. Bhavane, X. Liu, and M. Ferrari, Near-infrared imaging method for the in vivo assessment of the biodistribution of nanoporous silicon particles, *Mol. Imaging*, 10, 56–68, 2011.

100. V.P. Torchilin, Recent advances with liposomes as pharmaceutical carriers, *Nat. Rev. Drug Discov.*, 4, 145–160, 2005.

101. P. Debbage, Targeted drugs and nanomedicine: Present and future, *Curr. Pharm. Design*, 15, 153–172, 2009.

102. F.B. Myers and L.P. Lee, Innovations in optical microfluidic technologies for point-of-care diagnostics, October 2008, http://biopoems.berkeley.edu/fbmyers/documents/loc08_review.pdf, accessed on 28, October 2012.

103. Antonella Badia, Surface Plasmon Resonance (SPR) Spectroscopy Theory, Instrumentation & Applications, January 26, 2007, http://csacs.mcgill.ca/francais/docs/CHEM634/SPR_Badia.pdf

104. P. Anton Vander Merwe, Surface Plasmon Resonance, http://users.path.ox.ac.uk/~vdmerwe/internal/spr.pdf

105. M.Z. Zheng, J.L. Richard, and J. Binder, A review of rapid methods for the analysis of mycotoxins, *Mycopathologia*, 161, 261–273, 2006. doi:10.1007/s11046-006-0215-6.

106. H.M. Hiep et al., A localized surface plasmon resonance based immunosensor for the detection of casein in milk, *Sci. Technol. Adv. Mater.*, 8, 331, 2007.

107. T.M. Chinowsky, J.G. Quinn, D.U. Bartholomew, R. Kaiser, and J.L. Elkind, Performance of the Spreeta 2000 integrated surface plasmon resonance affinity sensor, *Sens. Actuat. B: Chem.*, 91, 266–274, 2003.

108. J. Waswa, J. Irudayaraj, and C. DebRoy, Principles of Bacterial Detection: Biosensors, Recognition Receptors, and Microsystems, *LWT—Food Sci. Technol.*, 40, 187–192, 2007.

109. T.M. Chinowsky, M.S. Grow, K.S. Johnston, K. Nelson, T. Edwards, E. Fu, and P. Yager, Methods in Bioengineering: Biomicrofabrication and Biomicrofluidics, *Biosens. Bioelectron.*, 22, 2208–2215, 2007.

110. E. Fu, T. Chinowsky, K. Nelson, K. Johnston, T. Edwards, K. Helton, J. Grow, J. Miller, and P. Yager, N. Ann, SPR imaging-based salivary diagnostics system for the detection of small molecule analytes, *Acad. Sci*, 1098, 335–344, 2007.

111. T.M. Chinowsky, S.D. Soelberg, P. Baker, N.R. Swanson, P. Kauffman, A. Mactutis, M.S. Grow, R. Atmar, S.S. Yee, and C.E. Furlong, Portable 24-analyte surface plasmon resonance instruments for rapid, versatile biodetection, *Biosens. Bioelectron.*, 22, 2268–2275, 2007.

112. S. Hearty, P. Leonard, J. Quinn, R. O'Kennedy, Production, characterisation and potential application of novel monoclonal antibody for rapid identification of virulent *Listeria monocytogenes*, *J. Microbiol. Methods*, 66, 294–312, 2006.

113. B. Byrne, E. Stack, N. Gilmartin, and R. O'Kennedy, Antibody-based sensors: Principles, problems and potential for detection of pathogens and associated toxins, *Sensors*, 9, 4407–4445, 2009. doi:10.3390/s90604407.

114. Karl Booksh, Densie Wilson, Surface Plasmon Resonance Portable Biochemical Systems, http://www.ee.washington.edu/research/denise/www/Lab/research_present/spr1.ppt

115. Vesna Hodnik, Toxin Detection by Surface Plasmon Resonance, 26 Feburary, 2009, Sensors 2009, 9, 1339–1354; doi:10.3390/s9031339, http://www.mdpi.com/1424-8220/9/3/1339/pdf

116. Daniel R. Sommers, Design and Verification of a Surface Plasmon Resonance Biosensor, Master Thesis, Sep 2003, http://citeseerx.ist.psu.edu/viewdoc/download?doi=10.1.1.114.4355&rep=rep1&type=pdf

117. B. Feltis, B. Sexton, F. Glenn, M. Best, M. Wilkins, and T. Davis, A hand-held surface plasmon resonance biosensor for the detection of ricin and other biological agents, *Biosens. Bioelectron.*, 23, 1131–1136, 2008.

118. R.A. Chipman, *Polarimetry Handbook of Optics*, 2nd edn., vol. 2, chapter 22, McGraw-Hill, New York, February 2010.

119. S.-Y. Lu and R.A. Chipman, Interpretation of Mueller matrices based on polar decomposition, *J. Opt. Soc. Am. A*, 13, 1106–1113, 1996.

Appendix A: Material and Physical Constants

A.1 Common Material Constants

TABLE A.1

Approximate Conductivity at 20°C

Material	Conductivity (S/m)
1. Conductors	
Silver	6.3×10^7
Copper (standard annealed)	5.8×10^7
Gold	4.5×10^7
Aluminum	3.5×10^7
Tungsten	1.8×10^7
Zinc	1.7×10^7
Brass	1.1×10^7
Iron (pure)	10^7
Lead	5×10^7
Mercury	10^6
Carbon	3×10^7
Water (sea)	4.8
2. Semiconductors	
Germanium (pure)	2.2
Silicon (pure)	4.4×10^{-4}
3. Insulators	
Water (distilled)	10^{-4}
Earth (dry)	10^{-5}
Bakelite	10^{-10}
Paper	10^{-11}
Glass	10^{-12}
Porcelain	10^{-12}
Mica	10^{-15}
Paraffin	10^{-15}
Rubber (hard)	10^{-15}
Quartz (fused)	10^{-17}
Wax	10^{-17}

TABLE A.2

Approximate Dielectric Constant and Dielectric Strength

Material	Dielectric Constant (or Relative Permittivity) (Dimensionless)	Strength, E (V/m)
Barium titanate	1200	7.5×10^6
Water (sea)	80	—
Water (distilled)	8.1	—
Nylon	8	—
Paper	7	12×10^6
Glass	5–10	35×10^6
Mica	6	70×10^6
Porcelain	6	—
Bakelite	5	20×10^6
Quartz (fused)	5	30×10^6
Rubber (hard)	3.1	25×10^6
Wood	2.5–8.0	—
Polystyrene	2.55	—
Polypropylene	2.25	—
Paraffin	2.2	30×10^6
Petroleum oil	2.1	12×10^6
Air (1 atm)	1	3×10^6

TABLE A.3

Relative Permeability

Material	Relative Permeability, μ_r
1. Diamagnetic	
Bismuth	0.999833
Mercury	0.999968
Silver	0.9999736
Lead	0.9999831
Copper	0.9999906
Water	0.9999912
Hydrogen (STP)	≈ 1.0
2. Paramagnetic	
Oxygen (STP)	0.999998
Air	1.00000037
Aluminum	1.000021
Tungsten	1.00008
Platinum	1.0003
Manganese	1.001
3. Ferromagnetic	
Cobalt	250
Nickel	600
Soft iron	5000
Silicon iron	7000

TABLE A.4

Approximate Conductivity for
Biological Tissue

Material	Conductivity (S/m)	Frequency
Blood	0.7	0 (DC)
Bone	0.01	0 (DC)
Brain	0.1	10^2–10^6 Hz
Breast fat	0.2–1	0.4–5 GHz
Breast tumor	0.7–3	0.4–5 GHz
Fat	0.1–0.3	0.4–5 GHz
	0.03	10^2–10^6 Hz
Muscle	0.4	10^2–10^6 Hz
Skin	0.001	1 kHz
	0.1	1 MHz

TABLE A.5

Approximate Dielectric Constant for Biological
Tissue

Material	Dielectric Constant (Relative Permittivity)	Frequency
Blood	10^5	1 kHz
Bone	3000–10,000	0 (DC)
Brain	10^7	100 Hz
	10^3	1 MHz
Breast fat	5–50	0.4–5 GHz
Breast tumor	47–67	0.4–5 GHz
Fat	5	0.4–5 GHz
	10^6	100 Hz
	10	1 MHz
Muscle	10^6	1 kHz
	10^3	1 MHz
Skin	10^6	1 kHz
	10^3	1 MHz

A.2 Physical Constants

Quantity	Best Experimental Value	Approximate Value for Problem Work
Avogadro's number (/kg mol)	6.0228×10^{26}	6×10^{26}
Boltzmann constant (J/k)	1.38047×10^{-23}	1.38×10^{-23}
Electron charge (C)	-1.6022×10^{-19}	-1.6×10^{-19}
Electron mass (kg)	9.1066×10^{-31}	9.1×10^{-31}
Permittivity of free space (F/m)	8.854×10^{-12}	$\dfrac{10^{-9}}{36\pi}$
Permeability of free space (H/m)	$4\pi \times 10^{-7}$	12.6×10^{-7}
Intrinsic impedance of free space (Ω)	376.6	120π
Speed of light in free space or vacuum (m/s)	2.9979×10^{8}	3×10^{8}
Proton mass (kg)	1.67248×10^{-27}	1.67×10^{-27}
Neutron mass (kg)	1.6749×10^{-27}	1.67×10^{-27}
Planck's constant (J s)	6.6261×10^{-34}	6.62×10^{-34}
Acceleration due to gravity (m/s²)	9.8066	9.8
Universal constant of gravitation (m²/kg s²)	6.658×10^{-11}	6.66×10^{-11}
Electron volt (J)	1.6030×10^{-19}	1.6×10^{-19}
Gas constant (J/mol K)	8.3145	8.3

Appendix B: Photon Equations, Index of Refraction, Electromagnetic Spectrum, and Wavelength of Commercial Laser

B.1 Photon Energy, Frequency, Wavelength

Photon energy (J)	Planck's constant × frequency
Photon energy (eV)	$\dfrac{\text{Planck's constant} \times \text{frequency}}{\text{Electron charge}}$
Photon energy (cm^{-1})	$\dfrac{\text{Frequency}}{\text{Speed of light in vacuum}}$
Photon frequency (Hz)	$\dfrac{1\,(\text{cycle})}{\text{Period (s)}}$
Photon wavelength (µm)	$\dfrac{\text{Speed of light in free space}}{\text{Frequency}}$

B.2 Index of Refraction for Common Substances

Substance	Index of Refraction
Air	1.000293
Diamond	2.24
Ethyl alcohol	1.36
Fluorite	1.43
Fused quartz	1.46
Crown glass	1.52
Flint glass	1.66
Glycerin	1.47
Ice	1.31
Polystyrene	1.49
Rock salt	1.54
Water	1.33

B.3 Electromagnetic Spectrum

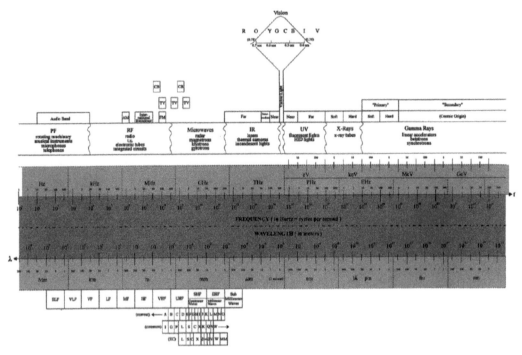

FIGURE B.1
Simplified chart of the electromagnetic spectrum. (From Whitaker, J.C., *The Electronics Handbook*, CRC Press, Boca Raton, FL, 1996.)

TABLE B.1

Approximate Common Optical
Wavelength Ranges of Light

Color	Wavelength
Ultraviolet region	10–380 nm
Visible region	380–750 nm
Violet	380–450 nm
Blue	450–495 nm
Green	495–570 nm
Yellow	570–590 nm
Orange	590–620 nm
Red	620–750 nm
Infrared	750 nm–1 mm

B.4 Wavelengths of Commercially Available Lasers

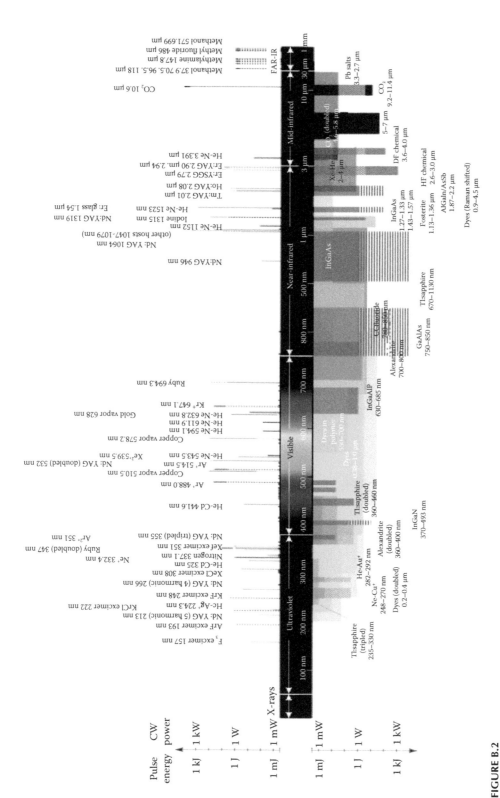

FIGURE B.2

(See color insert.) Wavelengths of commercially available lasers. (From Weber, M.J., *Handbook of Laser Wavelengths*, CRC Press, Boca Raton, FL, 1999, ISBN 0-8493-3508-6.)

References

1. Weber, M.J., *Handbook of Laser Wavelengths*, CRC Press, Boca Raton, FL, 1999, ISBN 0-8493-3508-6.
2. Whitaker, J.C., *The Electronics Handbook*, CRC Press, Boca Raton, FL, 1996.

Appendix C: Symbols and Formulas

C.1 Greek Alphabet

Uppercase	Lowercase	Name
A	α	Alpha
B	β	Beta
Γ	γ	Gamma
Δ	δ	Delta
E	ε	Epsilon
Z	ζ	Zeta
H	η	Eta
Θ	θ, ϑ	Theta
I	ι	Iota
K	κ	Kappa
Λ	λ	Lambda
M	μ	Mu
N	ν	Nu
Ξ	ξ	Xi
O	o	Omicron
Π	π	Pi
P	ρ	Rho
Σ	σ	Sigma
T	τ	Tau
Υ	υ	Upsilon
Φ	ϕ, φ	Phi
X	χ	Chi
Ψ	ψ	Psi
Ω	ω	Omega

C.2 International System of Units (SI) Prefixes

Power	Prefix	Symbol	Power	Prefix	Symbol
10^{-35}	Stringo	—	10^{0}	—	—
10^{-24}	Yocto	y	10^{1}	Deca	da
10^{-21}	Zepto	z	10^{2}	Hecto	h
10^{-18}	Atto	A	10^{3}	Kilo	k
10^{-15}	Femto	f	10^{6}	Mega	M
10^{-12}	Pico	p	10^{9}	Giga	G
10^{-9}	Nano	n	10^{12}	Tera	T
10^{-6}	Micro	μ	10^{15}	Peta	P
10^{-3}	Milli	m	10^{18}	Exa	E
10^{-2}	Centi	c	10^{21}	Zetta	Z
10^{-1}	Deci	d	10^{24}	Yotta	Y

C.3 Trigonometric Identities

$$\cot\theta = \frac{1}{\tan\theta}, \quad \sec\theta = \frac{1}{\cos\theta}, \quad \operatorname{cosec}\theta = \frac{1}{\sin\theta}$$

$$\tan\theta = \frac{\sin\theta}{\cos\theta}, \quad \cot\theta = \frac{\cos\theta}{\sin\theta}$$

$$\sin^2\theta + \cos^2\theta = 1, \tan^2\theta + 1 = \sec^2\theta, \cot^2\theta + 1 = \csc^2\theta$$

$$\sin(-\theta) = -\sin\theta, = \cos(-\theta) = \cos\theta, \tan(-\theta) = -\tan\theta$$

$$\csc(-\theta) = -\csc\theta, \sec(-\theta) = \sec\theta, \cot(-\theta) = -\cot\theta$$

$$\cos(\theta_1 \pm \theta_2) = \cos\theta_1 \cos\theta_2 \pm \sin\theta_1 \sin\theta_2$$

$$\sin(\theta_1 \pm \theta_2) = \sin\theta_1 \cos\theta_2 \pm \cos\theta_1 \sin\theta_2$$

$$\tan(\theta_1 \pm \theta_2) = \frac{\tan\theta_1 \pm \tan\theta_2}{1 \mp \tan\theta_1 \tan\theta_2}$$

$$\cos\theta_1 \cos\theta_2 = \frac{1}{2}\left[\cos(\theta_1 + \theta_2) + \cos(\theta_1 - \theta_2)\right]$$

$$\sin\theta_1 \sin\theta_2 = \frac{1}{2}\left[\cos(\theta_1 - \theta_2) + \cos(\theta_1 + \theta_2)\right]$$

$$\sin\theta_1 \cos\theta_2 = \frac{1}{2}\left[\sin(\theta_1 + \theta_2) + \sin(\theta_1 - \theta_2)\right]$$

$$\cos\theta_1 \sin\theta_2 = \frac{1}{2}\left[\sin(\theta_1 + \theta_2) + \sin(\theta_1 - \theta_2)\right]$$

$$\sin\theta_1 + \sin\theta_2 = 2\sin\left(\frac{\theta_1 + \theta_2}{2}\right)\cos\left(\frac{\theta_1 - \theta_2}{2}\right)$$

$$\sin\theta_1 - \sin\theta_2 = 2\cos\left(\frac{\theta_1 + \theta_2}{2}\right)\sin\left(\frac{\theta_1 - \theta_2}{2}\right)$$

$$\cos\theta_1 + \cos\theta_2 = 2\cos\left(\frac{\theta_1 + \theta_2}{2}\right)\cos\left(\frac{\theta_1 - \theta_2}{2}\right)$$

$$\cos\theta_1 - \cos\theta_2 = 2\sin\left(\frac{\theta_1 + \theta_2}{2}\right)\sin\left(\frac{\theta_1 - \theta_2}{2}\right)$$

$$a\sin\theta - b\cos\theta = \sqrt{a^2 + b^2}\,\cos(\theta + \phi), \quad \text{where} \quad \phi = \tan^{-1}\left(\frac{b}{a}\right)$$

$$a\sin\theta - b\cos\theta = \sqrt{a^2 + b^2}\,\sin(\theta + \phi), \quad \text{where} \quad \phi = \tan^{-1}\left(\frac{b}{a}\right)$$

$$\cos(90° - \theta) = \sin\theta, \ \sin(90° - \theta) = \cos\theta, \ \tan(90° - \theta) = \cot\theta$$

$$\cot(90° - \theta) = \tan\theta, \ \sec(90° - \theta) = \operatorname{cosec}\theta, \ \operatorname{cosec}(90° - \theta) = \sec\theta$$

$$\cos(\theta \pm 90°) = \mp\sin\theta, \ \sin(\theta \pm 90°) = \pm\sin\theta, \ \tan(\theta \pm 90°) = -\cot\theta$$

$$\cos(\theta \pm 180°) = -\cos\theta, \ \sin(\theta \pm 180°) = -\sin\theta, \ \tan(\theta \pm 180°) = \tan\theta$$

$$\cos 2\theta = \cos^2\theta - \sin^2\theta, \ \cos 2\theta = 1 - 2\sin^2\theta, \ \cos 2\theta = 2\cos^2\theta - 1$$

$$\sin 2\theta = 2\sin\theta\cos\theta, \quad \tan 2\theta = \frac{2\tan\theta}{1 - \tan^2\theta}$$

$$\cos 3\theta = 4\cos^3\theta - 3\sin\theta$$

$$\sin 3\theta = 3\sin\theta - 4\sin^3\theta$$

$$\sin\frac{\theta}{2} = \pm\sqrt{\frac{1-\cos\theta}{2}}, \quad \cos\frac{\theta}{2}\pm\sqrt{\frac{1+\cos\theta}{2}},$$

$$\sin\theta = \frac{e^{j\theta} - e^{-j\theta}}{2j}, \quad \cos\theta = \frac{e^{j\theta} + e^{-j\theta}}{2}\left(j = \sqrt{-1}\right), \quad \tan\theta = \frac{e^{j\theta} + e^{-j\theta}}{j(e^{j\theta} + e^{-j\theta})}$$

$$e^{\pm j\theta} = \cos\theta \pm j\sin\theta \text{ (Euler's identity)}$$

$$1 \text{ rad} = 57.296°$$

$$\pi = 3.1416$$

C.4 Hyperbolic Functions

$$\cosh x = \frac{e^x + e^{-x}}{2}, \quad \sinh x = \frac{e^x - e^{-x}}{2}, \quad \tanh x = \frac{\sinh x}{\cosh x}$$

$$\cosh x = \frac{1}{\tanh x}, \quad \operatorname{sech} x = \frac{1}{\cosh x}, \quad \operatorname{cosech} x = \frac{1}{\sinh x}$$

$$\sin jx = j\sinh x, \cos jx = \cosh x$$

$$\sinh jx = j\sin x, \cosh jx = \cos x$$

$$\sin(x \pm jy) = \sin x \cosh y \pm j\cos x \sinh y$$

$$\cos(x \pm jy) = \cos x \cosh y \pm j\sin x \sinh y$$

$$\sinh(x \pm y) = \sinh x \cosh y \pm \cosh x \sinh y$$

$$\cosh(x \pm y) = \cosh x \cosh y \pm \sinh x \sinh y$$

$$\sinh(x \pm jy) = \sinh x \cos y \pm j\cosh x \sin y$$

$$\cosh(x \pm jy) = \cosh x \cos y \pm j\sinh x \sin y$$

$$\tanh(x \pm jy) = \frac{\sinh 2x}{\cosh 2x + \cos 2y} \pm j\frac{\sin 2y}{\cosh 2x + \cos 2y}$$

$$\cosh^2 x - \sinh^2 x = 1$$

$$\text{sech}^2 x + \tanh^2 x = 1$$

C.5 Complex Variables

A complex number can be written as

$$z = x = jy = r\angle\theta = re^{j\theta} = r(\cos\theta + j\sin\theta),$$

where,

$x = \text{Re } z = r \cos\theta$
$y = \text{Im } z = r \sin\theta$
$r = |z| = \sqrt{x^2 + y^2}, \quad \theta = \tan^{-1}\left(\frac{y}{x}\right)$

$j = \sqrt{-1}, \quad \dfrac{1}{j} = -j, \quad j^2 = -1$

The complex conjugate of $z = z^* = x - jy = r \angle - \theta = re^{-j\theta} = r(\cos\theta - j\sin\theta)$
$(e^{j\theta})^n = e^{jn\theta} = \cos n\theta + j\sin n\theta$ (de Moivre's theorem).
If $z_1 = x_1 + jy_1$ and $z_2 = x_2 + jy_2$, then only if $x_1 = x_2$ and $y_1 = y_2$,

$$z_1 \pm z_2 = (x_1 + x_2) \pm j(y_1 + y_2)$$

$$z_1 z_2 = (x_1 x_2 - y_1 y_2) + j(x_1 y_2 + x_2 y_1) = r_1 r_2 e^{j(\theta_1 + \theta_2)} = r_1 r_2 \angle\theta_1 + \theta_2$$

$$\frac{z_1}{z_2} = \frac{(x_1 + jy_1)}{(x_2 + jy_2)} \cdot \frac{(x_2 + jy_2)}{(x_2 + jy_2)} = \frac{x_1 x_2 + y_1 y_2}{x_2^2 + y_2^2} + j\frac{x_2 y_1 - x_1 y_2}{x_2^2 + y_2^2} = \frac{r_1}{r_2} e^{j(\theta_1 + \theta_2)} = \frac{r_1}{r_2}\angle\theta_1 - \theta_2$$

$\ln(re^{j\theta}) = \ln r + \ln e^{j\theta} = \ln r + j\theta + j2m\pi$ $(m = \text{integer})$

$$\sqrt{z} = \sqrt{x + jy} = \sqrt{r}\left(e^{j\theta/2}\right) = \sqrt{r}\angle\theta/2$$

$$z^n = (x + jy)^n = r^n e^{jn\theta} = r^n\angle n\theta \quad (n = \text{integer})$$

$$z^{1/n} = (x + jy)^{1/n} = r^{1/n} e^{j\theta/n} = r^{1/n}\angle\theta/2 + 2\pi m/n, \quad (m = 0, 1, 2, \ldots, n-1)$$

C.6 Table of Derivatives

$y =$	$\dfrac{dy}{dx} =$
c(constant)	0
cx^n(n any constant)	cnx^{n-1}
e^{ax}	ae^{ax}
$a^x (a > 0)$	$a^x \ln a$
$\ln x (x > 0)$	$\dfrac{1}{x}$
$\dfrac{c}{x^a}$	$\dfrac{-ca}{x^{a+1}}$
$\log_a x$	$\dfrac{\log_a e}{x}$
$\sin ax$	$a \cos ax$
$\cos ax$	$-a \sin ax$
$\tan ax$	$-a \sec^2 ax = \dfrac{a}{\cos^2 ax}$
$\cot ax$	$-a \operatorname{cosec}^2 ax = \dfrac{-a}{\sin^2 ax}$
$\sec ax$	$\dfrac{a \sin ax}{\cos^2 ax}$
$\operatorname{cosec} ax$	$\dfrac{-a \cos ax}{\sin^2 ax}$
$\arcsin ax = \sin^{-1} ax$	$\dfrac{a}{\sqrt{1 - a^2 x^2}}$
$\arccos ax = \cos^{-1} ax$	$\dfrac{-a}{\sqrt{1 - a^2 x^2}}$
$\arctan ax = \tan^{-1} ax$	$\dfrac{a}{1 + a^2 x^2}$
$\operatorname{arccot} ax = \cot^{-1} ax$	$\dfrac{-a}{1 + a^2 x^2}$
$\sinh ax$	$a \cosh ax$
$\cosh ax$	$a \sinh ax$
$\tanh ax$	$\dfrac{a}{\cosh^2 ax}$
$\sinh^{-1} ax$	$\dfrac{a}{\sqrt{1 - a^2 x^2}}$
$\cosh^{-1} ax$	$\dfrac{a}{\sqrt{a^2 x^2 - 1}}$
$\tanh^{-1} ax$	$\dfrac{a}{1 - a^2 x^2}$
$u(x) + v(x)$	$\dfrac{du}{dx} + \dfrac{dv}{dx}$
$u(x)\,v(x)$	$u\dfrac{dv}{dx} + v\dfrac{du}{dx}$
$\dfrac{u(x)}{v(x)}$	$\dfrac{1}{v^2}\left(v\dfrac{du}{dx} - u\dfrac{dv}{dx} \right)$

$$\frac{1}{\upsilon(x)} \qquad \frac{-1}{\upsilon^2}\frac{d\upsilon}{dx}$$

$$y(\upsilon(x)) \qquad \frac{dy}{d\upsilon}\frac{d\upsilon}{dx}$$

$$y(\upsilon(u(x))) \qquad \frac{dy}{d\upsilon}\frac{d\upsilon}{du}\frac{du}{dx}$$

C.7 Table of Integrals

$$\int a\,dx = ax + c\,(c\text{ is an arbitrary constant})$$

$$\int x\,dy = xy - \int y\,dx$$

$$\int x^n dx = \frac{x^{n+1}}{n+1} + c, \quad (n \neq -1)$$

$$\int \frac{1}{x}dx = \ln|x| + c$$

$$\int e^{ax}dx = \frac{e^{ax}}{a} + c$$

$$\int a^x dx = \frac{a^x}{\ln a} + c \quad \text{for} \quad (a > 0)$$

$$\int \ln x\,dx = x\ln x - x + c \quad \text{for} \quad (x > 0)$$

$$\int \sin ax\,dx = \frac{-\cos ax}{a} + c$$

$$\int \cos ax\,dx = \frac{\sin ax}{a} + c$$

$$\int \tan ax\,dx = \frac{-\ln|\cos ax|}{a} + c$$

$$\int \cot ax\,dx = \frac{-\ln|\cos ax|}{a} + c$$

$$\int \sec ax\,dx = \frac{-\ln\left(\dfrac{1-\sin ax}{1+\sin ax}\right)}{2a} + c$$

$$\int \operatorname{cosec} ax \, dx = \frac{-\ln\left(1 - \cos ax / 1 + \cos ax\right)}{2a} + c$$

$$\int \frac{1}{x^2 + a^2} \, dx = \frac{\tan^{-1}(x/a)}{a} + c$$

$$\int \frac{1}{x^2 - a^2} \, dx = \frac{\ln(x - a/x + a)}{2a} + c \quad \text{or} \quad \frac{\tanh^{-1}(x/a)}{a} + c$$

$$\int \frac{1}{a^2 - x^2} \, dx = \frac{\ln(x + a/x - a)}{2a} + c$$

$$\int \frac{1}{\sqrt{a^2 - x^2}} \, dx = \sin^{-1}(x/a) + c$$

$$\int \frac{1}{\sqrt{a^2 - x^2}} \, dx = \frac{\sinh^{-1}(x/a)}{a} + c \quad \text{or} \quad \ln\left(x + \sqrt{x^2 + a^2}\right) + c$$

$$\int \frac{1}{\sqrt{x^2 - a^2}} \, dx = \ln\left(x + \sqrt{x^2 + a^2}\right) + c$$

$$\int \frac{1}{x\sqrt{x^2 - a^2}} \, dx = \frac{\sec^{-1}(x/a)}{a} + c$$

$$\int x e^{ax} \, dx = \frac{(ax - 1) e^{ax}}{a^2} + c$$

$$\int x \cos ax \, dx = \frac{\cos ax + ax \sin ax}{a^2} + c$$

$$\int x \sin ax \, dx = \frac{\sin ax + ax \cos ax}{a^2} + c$$

$$\int x \ln x \, dx = \frac{x^2}{2} \ln x - \frac{x^2}{4} + c$$

$$\int x e^{ax} \, dx = \frac{e^{ax}(ax - 1)}{a^2} + c$$

$$\int e^{ax} \cos bx \, dx = \frac{e^{ax}(a \cos bx + b \sin bx)}{a^2 + b^2} + c$$

$$\int e^{ax} \sin bx \, dx = \frac{e^{ax}\left(-b\cos bx + a\sin bx\right)}{a^2+b^2} + c$$

$$\int \sin^2 x \, dx = \frac{x}{2} - \frac{\sin 2x}{4} + c$$

$$\int \cos^2 x \, dx = \frac{x}{2} - \frac{\sin 2x}{4} + c$$

$$\int \tan^2 x \, dx = \tan x - x + c$$

$$\int \cot^2 x \, dx = \cot x - x + c$$

$$\int \sec^2 x \, dx = \tan x + c$$

$$\int \operatorname{cosec}^2 x \, dx = -\cot x + c$$

$$\int \sec x \tan x \, dx = \sec x + c$$

$$\int \operatorname{cosec} x \cot x \, dx = -\operatorname{cosec} x + c$$

C.8 Table of Probability Distributions

1. Discrete Distribution	Probability $P(X=x)$	Expectation (Mean) μ	Variance σ^2
Binomial $B(n,p)$	$\binom{n}{r} p^r (1-p)^{n-r} = \frac{n! p^r q^{n-r}}{r!(n-r)!},$ $r = 0,1,\dots,n$	np	$np(1-p)$
Geometric $G(p)$	$(1-p)^{r-1} p$	$\dfrac{1}{p}$	$\dfrac{1-p}{p^2}$
Poisson $p(\lambda)$	$\dfrac{\lambda^n e^{-\lambda}}{n!}$	λ	λ
Pascal (negative binomial) $NB(r,p)$	$\binom{x \quad -1}{r \quad -1} p^r (1-p)^{x-r},$ $x = r, r+1, \dots$	$\dfrac{r}{p}$	$\dfrac{r(1-p)}{p^2}$
Hypergeometric $H(N,n,p)$	$\dfrac{\binom{Np}{r}\binom{N-Np}{n-r}}{\binom{N}{n}}$	np	$np(1-p)\dfrac{N-n}{N-1}$

2. Continuous Distribution	Density $f(x)$	Expectation (Mean) μ	Variance σ^2
Exponential $E(\lambda)$	$\begin{cases} \lambda e^{-\lambda x}, & x \geq 0 \\ 0 & x < 0 \end{cases}$	$\dfrac{1}{\lambda}$	$\dfrac{1}{\lambda^2}$
Uniform $U(a,b)$	$\begin{cases} \dfrac{1}{b-a}, & a < x < b \\ 0, & \text{elsewhere} \end{cases}$	$\dfrac{a+b}{2}$	$\dfrac{(b-a)^2}{12}$
Standardized normal $N(0,1)$	$\varphi(x) = \dfrac{e^{-x^2/2}}{\sqrt{2\pi}}$	0	1
General normal	$\dfrac{1}{\sigma}\varphi\left(\dfrac{x-\mu}{\sigma}\right)$	μ	σ^2
Gamma $\Gamma(n,\lambda)$	$\dfrac{\lambda^n}{\Gamma(n)}x^{n-1}e^{-\lambda x}$	$\dfrac{n}{\lambda}$	$\dfrac{n}{\lambda^2}$
Beta $\beta(p,q)$	$a_{p,q}x^{p-1}(1-x)^{q-1}, 0 \leq x \geq 1$ $$a_{p,q} = \dfrac{\Gamma(p+q)}{\Gamma(p)\Gamma(q)}, \quad p > 0, \quad q > 0$$	$\dfrac{p}{p+q}$	$\dfrac{pq}{(p+q)^2(p+q+1)}$
Weibull $W(\lambda, \beta)$	$\lambda^{\beta}\beta x^{\beta-1}e^{-(\lambda x)^{\beta}}, x \geq 0$ $$F(x) = 1 - e^{-(\lambda x)^{\beta}}$$	$\dfrac{1}{\lambda}\Gamma\left(1+\dfrac{1}{\beta}\right)$	$\dfrac{1}{\lambda^2}(A-B)$ $$A = \Gamma\left(1+\dfrac{2}{\beta}\right)$$ $$B = \Gamma^2\left(1+\dfrac{1}{\beta}\right)$$
Rayleigh $R(\sigma)$	$\dfrac{x}{\sigma^2}e^{-x^2/2\sigma^2}, \quad x \geq 0$	$\sigma\sqrt{\dfrac{\pi}{2}}$	$2\sigma^2\left(1-\dfrac{\pi}{4}\right)$

C.9 Summations (Series)

1. Finite element of terms

$$\sum_{n=0}^{N} a^n = \frac{1-a^{N+1}}{1-a}; \quad \sum_{n=0}^{N} na^n = a\left(\frac{1-(N+1)a^N + Na^{N+1}}{(1-a)^2}\right)$$

$$\sum_{n=0}^{N} n = \frac{N(N+1)}{2}; \quad \sum_{n=0}^{N} n^2 = \frac{N(N+1)(2N+1)}{6}$$

$$\sum_{n=0}^{N} n(n+1) = \frac{N(N+1)(N+2)}{3};$$

$$(a+b)^N = \sum_{n=0}^{N} NC_n a^{N-n} b^n, \quad \text{where} \quad NC_n = NC_{N-n} = \frac{NP_n}{n!} = \frac{N!}{(N-n)!n!}$$

2. Infinite element of terms

$$\sum_{n=0}^{\infty} x^n = \frac{1}{1-x}, (|x| < 1); \quad \sum_{n=0}^{\infty} nx^n = \frac{1}{(1-x)^2}, \quad (|x| < 1)$$

$$\sum_{n=0}^{\infty} n^k x^n = \lim_{a \to 0}(-1)^k \frac{\partial^k}{\partial a^k}\left(\frac{x}{x-e^{-a}}\right), \quad (|x| < 1); \quad \sum_{n=0}^{\infty} \frac{(-1)^n}{2n+1} = 1 - \frac{1}{3} + \frac{1}{5} - \frac{1}{7} + \ldots = \frac{1}{4}\pi$$

$$\sum_{n=0}^{\infty} \frac{1}{n^2} = 1 + \frac{1}{2^2} + \frac{1}{3^2} + \frac{1}{4^2} + \ldots = \frac{1}{6}\pi^2$$

$$e^x = \sum_{n=0}^{\infty} \frac{x^n}{n!} = 1 + \frac{1}{1!}x + \frac{1}{2!}x^2 + \frac{1}{3!}x^3 + \ldots$$

$$a^x = \sum_{n=0}^{\infty} \frac{(\ln a)^n x^n}{n!} = 1 + \frac{(\ln a)x}{1!} + \frac{(\ln a)^2 x^2}{2!} + \frac{(\ln a)^3 x^3}{3!} + \ldots$$

$$\ln(1 \pm x) = \sum_{n=1}^{\infty} \frac{(\pm 1)^n x^x}{n} = \pm x - \frac{x^2}{2} \pm \frac{x^3}{3} - \ldots, \quad (|x| < 1)$$

$$\sin x = \sum_{n=0}^{\infty} \frac{(-1)^n x^{2n+1}}{(2n+1)!} = x - \frac{x^3}{3!} + \frac{x^5}{5!} - \frac{x^7}{7!} + \ldots$$

$$\cos x = \sum_{n=0}^{\infty} \frac{(-1)^n x^{2n}}{(2n)!} = 1 - \frac{x^2}{2!} + \frac{x^4}{4!} - \frac{x^6}{6!} + \ldots$$

$$\tan x = x + \frac{x^3}{3} + \frac{2x^5}{15} + \ldots, \quad (|x| < 1)$$

$$\tan^{-1} x = \sum_{n=0}^{\infty} \frac{(-1)^n x^{2n+1}}{(2n+1)} = x - \frac{x^3}{3} + \frac{x^5}{5} - \frac{x^7}{7} + \ldots, \quad (|x| < 1)$$

C.10 Logarithmic Identities

$$\log_e a = \ln a \text{ (natural logarithm)}$$

$$\log_{10} a = \log a \text{ (common logarithm)}$$

$$\log ab = \log a + \log b$$

$$\log \frac{a}{b} = \log a - \log b$$

$$\log a^n = n \log a$$

C.11 Exponential Identities

$$e^x = 1 + x + \frac{x^2}{2!} + \frac{x^3}{3!} + \frac{x^4}{4!} + ..., \quad \text{where} \quad e \simeq 2.7182$$

$$e^x e^y = e^{x+y}$$

$$(e^x)^n = e^{nx}$$

$$\ln e^x = x$$

C.12 Approximations for Small Quantities

If $|a| \ll 1$, then

$$\ln(1 + a) \simeq a$$

$$e^a \simeq 1 + a$$

$$\sin a \simeq a$$

$$\cos a \simeq 1$$

$$\tan a \simeq a$$

$$(1 \pm a)^n \simeq 1 \pm na$$

C.13 Matrix Notation and Operations

1. Matrices

A *matrix* is a rectangular array of elements arranged in rows and columns. The array is commonly enclosed in brackets. Let a matrix A (expressed in boldface as **A** or in bracket as $[A]$) have m rows and n columns; then the matrix can be expressed by

$$\mathbf{A} = [A] = \begin{bmatrix} a_{11} & a_{12} & \cdot & \cdot & \cdot & a_{1j} & \cdot & \cdot & \cdot & a_{1n} \\ a_{21} & a_{22} & \cdot & \cdot & \cdot & a_{2j} & \cdot & \cdot & \cdot & a_{2n} \\ \cdot & \cdot & \cdot & \cdot & \cdot & \cdot & \cdot & \cdot & \cdot & \cdot \\ \cdot & \cdot & \cdot & \cdot & \cdot & \cdot & \cdot & \cdot & \cdot & \cdot \\ \cdot & \cdot & \cdot & \cdot & \cdot & \cdot & \cdot & \cdot & \cdot & \cdot \\ a_{i1} & a_{i2} & \cdot & \cdot & \cdot & a_{ij} & \cdot & \cdot & \cdot & a_{in} \\ \cdot & \cdot & \cdot & \cdot & \cdot & \cdot & \cdot & \cdot & \cdot & \cdot \\ \cdot & \cdot & \cdot & \cdot & \cdot & \cdot & \cdot & \cdot & \cdot & \cdot \\ \cdot & \cdot & \cdot & \cdot & \cdot & \cdot & \cdot & \cdot & \cdot & \cdot \\ a_{m1} & a_{m2} & \cdot & \cdot & \cdot & a_{mj} & \cdot & \cdot & \cdot & a_{mn} \end{bmatrix}$$

where the element a_{ij} has two subscripts, the first refers to the row position of the element in the array and the second refers to the column position. A matrix with m rows and n columns, $[A]$, is defined as a matrix of order or size $m \times n$ (m by n), or an $m \times n$ matrix. A vector is a matrix that consists of only one row or one column.

Location of an element in a matrix:

$$\text{Let } A = \begin{bmatrix} a_{11} & a_{12} & a_{13} & a_{14} \\ a_{21} & a_{22} & a_{23} & a_{24} \\ a_{31} & a_{32} & a_{33} & a_{34} \\ a_{41} & a_{42} & a_{43} & a_{44} \end{bmatrix} \text{ is matrix with size } 4 \times 4$$

where,

a_{11} is the element a at row 1 and column 1.
a_{12} is the element a at row 1 and column 2.
a_{32} is the element a at row 3 and column 2.

2. Special common types of matrices
   a. If $m \neq n$, then the matrix $[A]$ is called *rectangular matrix.*
   b. If $m = n$, then the matrix $[A]$ is called *square matrix of order n.*
   c. If $m = 1$ and $n > 1$, then the matrix $[A]$ is called *row matrix or row vector.*
   d. If $m > 1$ and $n = 1$, then the matrix $[A]$ is called *column matrix or column vector.*
   e. If $m = 1$ and $n = 1$, then the matrix $[A]$ is called a scalar.

f. A *real matrix* is a matrix whose elements are all real.

g. A *complex matrix* is a matrix whose elements may be complex.

h. A *null matrix* is a matrix whose elements are all zero.

i. An *identity* (or *unit*) *matrix*, [*I*] or **I**, is a square matrix whose elements are equal to zero except those located on its *main diagonal* elements, which are unity (or one). *Main diagonal* elements have equal row and column subscripts. The main diagonal runs from the upper left corner to the lower right corner. If the elements of an identity matrix are denoted as e_{ij}, then

$$e_{ij} = \begin{cases} 1 & i = j \\ 0 & i \neq j \end{cases}$$

j. A *diagonal matrix* is a square matrix which has zero elements everywhere except on its main diagonal. That is, for diagonal matrix, $a_{ij}=0$ when $i \neq j$ and not all a_{ii} are zero.

k. A *symmetric matrix* is a square matrix whose elements satisfy the condition $a_{ij}=a_{ji}$ for $i \neq j$.

l. An *antisymmetric* (or *skew symmetric*) *matrix* is a square matrix whose elements $a_{ij}=-a_{ji}$ for $i \neq j$, and $a_{ii}=0$.

m. A *triangular matrix* is a square matrix whose all elements on one side of the diagonal are zero. There are two types of triangular matrices; first, an upper triangular **U** whose elements below the diagonal are zero, and second, a lower triangular **L**, whose elements above the diagonal are all zero.

n. A *partitioned* (or *block*) *matrix* is a matrix which is divided by horizontal and vertical lines into smaller matrices called submatrices or blocks.

3. Matrix operations

a. Transpose of a matrix

The *transpose* of a matrix $\mathbf{A}=[a_{ij}]$ is denoted as $\mathbf{A}^T=[a_{ji}]$ and is obtained by interchanging the rows and columns in matrix **A**. Thus, if a matrix **A** is of order $m \times n$, then $\mathbf{A}^T$ will be of order $n \times m$.

b. Addition and subtraction

Addition and subtraction can only be performed for matrices of the same size. The addition is accomplished by adding corresponding elements of each matrix. For addition, $\mathbf{C}=\mathbf{A}+\mathbf{B}$ implies that $c_{ij}=a_{ij}+b_{ij}$.

Now, the subtraction is accomplished by subtracting corresponding elements of each matrix. For subtraction, $\mathbf{C}=\mathbf{A}-\mathbf{B}$ implies that $c_{ij}=a_{ij}-b_{ij}$ where c_{ij}, a_{ij}, and b_{ij} are typical elements of the **C**, **A**, and **B** matrices, respectively.

If matrices **A** and **B** are both of the same size $m \times n$, the resulting matrix **C** is also of size $m \times n$.

Matrix addition and subtraction are associative:

$$\mathbf{A}+\mathbf{B}+\mathbf{C}=(\mathbf{A}+\mathbf{B})+\mathbf{C}=\mathbf{A}+(\mathbf{B}+\mathbf{C})$$

$$\mathbf{A}+\mathbf{B}-\mathbf{C}=(\mathbf{A}+\mathbf{B})-\mathbf{C}=\mathbf{A}+(\mathbf{B}-\mathbf{C})$$

Matrix addition and subtraction are commutative:

$$\mathbf{A}+\mathbf{B}=\mathbf{B}+\mathbf{A}$$

$$\mathbf{A}-\mathbf{B}=-\mathbf{B}+\mathbf{A}$$

c. Multiplication by scalar

A matrix is multiplied by a scalar by multiplying each element of the matrix by the scalar. The multiplication of a matrix $\mathbf{A}$ by a scalar c is defined as

$$c\mathbf{A}=[ca_{ij}]$$

The scalar multiplication is commutative.

d. Matrix multiplication

The product of two matrices is $\mathbf{C}=\mathbf{AB}$ if and only if the number of columns in $\mathbf{A}$ is equal to the number of rows in $\mathbf{B}$. The product of matrix $\mathbf{A}$ of size $m \times n$ and matrix $\mathbf{B}$ of size $n \times r$ results in matrix $\mathbf{C}$ of size $m \times r$. Then, $c_{ij} = \sum_{k=1}^{n} a_{ik}b_{kj}$.

That is, the (ij)th component of matrix $\mathbf{C}$ is obtained by taking the dot product

$$c_{ij} =(i\text{th row of } \mathbf{A})\cdot (j\text{th column of } \mathbf{B})$$

Matrix multiplication is associative:

$$\mathbf{ABC}=(\mathbf{AB})\mathbf{C}=\mathbf{A}(\mathbf{BC})$$

Matrix multiplication is distributive:

$$\mathbf{A}(\mathbf{B}+\mathbf{C})=\mathbf{AB}+\mathbf{AC}$$

Matrix multiplication is not commutative:

$$\mathbf{AB}\neq\mathbf{BA}$$

e. Transpose of matrix multiplication

The transpose of matrix multiplication is usually denoted $(\mathbf{AB})^T$ and is defined as

$$(\mathbf{AB})^T =\mathbf{B}^T\mathbf{A}^T$$

f. Inverse of square matrix

The inverse of a matrix $\mathbf{A}$ is denoted by $\mathbf{A}^{-1}$. The inverse matrix satisfies

$$\mathbf{AA}^{-1}=\mathbf{A}^{-1}\mathbf{A}=\mathbf{I}$$

A matrix that possesses an inverse is called *nonsingular matrix (or invertible matrix)*. A matrix without an inverse is called a *singular matrix*.

g. Differentiation of a matrix

The differentiation of a matrix is differentiation of every element of the matrix separately. To emphasize, if the elements of the matrix **A** are a function of t, then

$$\frac{d\mathbf{A}}{dt} = \left[\frac{da_{ij}}{dt}\right]$$

h. Integration of a matrix

The integration of a matrix is integration of every element of the matrix separately. To emphasize, if the elements of the matrix **A** are a function of t, then

$$\int \mathbf{A}dt = \left[\int a_{ij}dt\right]$$

i. Equality of matrices

Two matrices are equal if they have the same size and their corresponding elements are equal.

4. Determinant of a matrix

The determinant of a square matrix **A** is a scalar number denoted by $|\mathbf{A}|$ or det **A**. The value of a second-order determinant is calculated from

$$\det\begin{bmatrix} a_{11} & a_{12} \\ a_{21} & a_{22} \end{bmatrix} = \begin{vmatrix} a_{11} & a_{12} \\ a_{21} & a_{22} \end{vmatrix} = a_{11}a_{22} - a_{12}a_{21}$$

By using the sign rule of each term, the determinant is determined by the first

row in the diagram $\begin{vmatrix} + & - & + \\ - & + & - \\ + & - & + \end{vmatrix}$.

The value of a third-order determinant is calculated as

$$\det\begin{bmatrix} a_{11} & a_{12} & a_{13} \\ a_{21} & a_{22} & a_{23} \\ a_{31} & a_{32} & a_{33} \end{bmatrix} = \begin{vmatrix} a_{11} & a_{12} & a_{13} \\ a_{21} & a_{22} & a_{23} \\ a_{31} & a_{32} & a_{33} \end{vmatrix} =$$

$$a_{11}\begin{vmatrix} a_{22} & a_{23} \\ a_{32} & a_{33} \end{vmatrix} - a_{12}\begin{vmatrix} a_{21} & a_{23} \\ a_{31} & a_{33} \end{vmatrix} + a_{13}\begin{vmatrix} a_{21} & a_{22} \\ a_{31} & a_{32} \end{vmatrix}$$

C.14 Vectors

1. Vector derivative

   a. Cartesian coordinates

Coordinates	(x,y,z)
Vector	$A = A_x\,a_x + A_y\,a_y + A_z\,a_z$
Gradient	$\nabla A = \dfrac{\partial A}{\partial x}\,a_x + \dfrac{\partial A}{\partial y}\,a_y + \dfrac{\partial A}{\partial z}\,a_z$
Divergence	$\nabla \cdot A = \dfrac{\partial A_x}{\partial x} + \dfrac{\partial A_y}{\partial y} + \dfrac{\partial A_z}{\partial z}$
Curl	$\nabla \times A = \begin{vmatrix} a_x & a_y & a_z \\ \dfrac{\partial}{\partial x} & \dfrac{\partial}{\partial y} & \dfrac{\partial}{\partial z} \\ A_x & A_y & A_z \end{vmatrix}$
	$= \left(\dfrac{\partial A_z}{\partial y} - \dfrac{\partial A_y}{\partial z} \right) a_x + \left(\dfrac{\partial A_x}{\partial z} - \dfrac{\partial A_z}{\partial x} \right) a_y + \left(\dfrac{\partial A_y}{\partial x} - \dfrac{\partial A_x}{\partial y} \right) a_z$
Laplacian	$\nabla^2 A = \dfrac{\partial^2 A}{\partial x^2} + \dfrac{\partial^2 A}{\partial y^2} + \dfrac{\partial^2 A}{\partial z^2}$

   b. Cylindrical coordinates

Coordinates	(ρ,ϕ,z)
Vector	$A = A_\rho\,a_\rho + A_\phi\,a_\phi + A_z\,a_z$
Gradient	$\nabla A = \dfrac{\partial A}{\partial \rho}\,a_\rho + \dfrac{1}{\rho}\dfrac{\partial A}{\partial \phi}\,a_\phi + \dfrac{\partial A}{\partial z}\,a_z$
Divergence	$\nabla \cdot A = \dfrac{1}{\rho}\dfrac{\partial}{\partial \rho}(\rho A_\rho) + \dfrac{\partial A_\phi}{\partial \phi} + \dfrac{\partial A_z}{\partial z}$
Curl	$\nabla \times A = \dfrac{1}{\rho}\begin{vmatrix} a_\rho & \rho a_\phi & a_z \\ \dfrac{\partial}{\partial \rho} & \dfrac{\partial}{\partial \phi} & \dfrac{\partial}{\partial z} \\ A_\rho & \rho A_\phi & A_z \end{vmatrix}$
	$= \left(\dfrac{1}{\rho}\dfrac{\partial A_z}{\partial \phi} - \dfrac{\partial A_\phi}{\partial z} \right) a_\rho + \left(\dfrac{\partial A_\rho}{\partial z} - \dfrac{\partial A_z}{\partial \rho} \right) a_\phi + \dfrac{1}{\rho}\left(\dfrac{\partial}{\partial x}(\rho A_\phi) - \dfrac{\partial A_\rho}{\partial \rho} \right) a_z$
Laplacian	$\nabla^2 A = \dfrac{1}{\rho}\dfrac{\partial}{\partial \rho}\left(\rho \dfrac{\partial A}{\partial \rho} \right) + \dfrac{1}{\rho^2}\dfrac{\partial^2 A}{\partial \phi^2} + \dfrac{\partial^2 A}{\partial z^2}$

C. Spherical coordinates

Coordinates	(r,θ,ϕ)
Vector	$A = A_r\, a_r + A_\theta\, a_\theta + A_\phi\, a_\phi$
Gradient	$\nabla A = \dfrac{\partial A}{\partial r}\, a_r + \dfrac{1}{r}\dfrac{\partial A}{\partial \theta}\, a_\theta + \dfrac{1}{r\sin\theta}\dfrac{\partial A}{\partial \phi}\, a_\phi$
Divergence	$\nabla \cdot A = \dfrac{1}{r^2}\dfrac{\partial}{\partial r}(r^2 A_r) + \dfrac{1}{r\sin\theta}\dfrac{\partial}{\partial \theta}(A_\theta \sin\theta) + \dfrac{1}{r\sin\theta}\dfrac{\partial A_\phi}{\partial \phi}$

Curl

$$\nabla \times A = \frac{1}{r^2 \sin\theta}\begin{vmatrix} a_r & r a_\theta & (r\sin\theta)a_\phi \\ \dfrac{\partial}{\partial r} & \dfrac{\partial}{\partial \theta} & \dfrac{\partial}{\partial \phi} \\ A_r & r A_\theta & (r\sin\theta)A_\phi \end{vmatrix}$$

$$= \frac{1}{r\sin\theta}\left(\frac{\partial}{\partial \theta}(A_\phi \sin\theta) - \frac{\partial A_\theta}{\partial \phi}\right)a_r + \frac{1}{r}\left(\frac{1}{\sin\theta}\frac{\partial A_r}{\partial \phi} - \frac{\partial}{\partial r}(rA_\phi)\right)a_\theta +$$

$$\frac{1}{r}\left(\frac{\partial}{\partial r}(rA_\theta) - \frac{\partial A_r}{\partial \theta}\right)a_\phi$$

Laplacian

$$\nabla^2 A = \frac{1}{r^2}\frac{\partial}{\partial r}\left(r^2\frac{\partial A}{\partial r}\right) + \frac{1}{r^2 \sin\theta}\frac{\partial}{\partial \theta}\left(\sin\theta\frac{\partial A}{\partial \theta}\right) + \frac{1}{r^2 \sin\theta}\frac{\partial^2 A}{\partial \phi^2}$$

2. Vector identity
a. Triple products

$$A(B \times C) = B\,(C \times A) = C \cdot (A \times B)$$

$$A(B \times C) = B(A \cdot C) - C(A \cdot B)$$

b. Product rules

$$\nabla(fg) = f(\nabla g) + g(\nabla f)$$

$$\nabla(A \cdot B) = A \times (\nabla \times B) + B \times (\nabla \times A) + (A \cdot \nabla)B + (B \times \nabla)A$$

$$\nabla \cdot (fA) = f(\nabla \cdot A) + A \cdot (\nabla f)$$

$$\nabla(A \times B) = B \cdot (\nabla \times A) - A \cdot (\nabla \times B)$$

$$\nabla \times (f\mathbf{A}) = f(\nabla \times \mathbf{A}) - \mathbf{A} \times (\nabla f) = \nabla \times (f\mathbf{A}) = f(\nabla \times \mathbf{A}) + (\nabla f) \times \mathbf{A}$$

$$\nabla \times (\mathbf{A} \times \mathbf{B}) = (\mathbf{B} \cdot \nabla)\mathbf{A} - (\mathbf{A} \cdot \nabla)\,\mathbf{B} + \mathbf{A}(\nabla \cdot \mathbf{B}) - (\nabla \cdot \mathbf{A})$$

c. Second derivative

$$\nabla \cdot (\Delta \times \mathbf{A}) = 0$$

$$\nabla \times (\nabla f) = 0$$

$$\nabla \cdot (\nabla f) = \nabla^2 f$$

$$\nabla \times (\nabla \times \mathbf{A}) = \nabla(\nabla \cdot \mathbf{A}) - \nabla^2 \mathbf{A}$$

d. Addition, division, and power rules

$$\nabla(f + g) = \nabla f + \nabla g$$

$$\nabla \cdot (\mathbf{A} + \mathbf{B}) = \nabla \cdot \mathbf{A} + \nabla \cdot \mathbf{B}$$

$$\nabla \times (\mathbf{A} \times \mathbf{B}) = \nabla \times \mathbf{A} + \nabla \times \mathbf{B}$$

$$\nabla\left(\frac{f}{g}\right) = \frac{g(\nabla f) - f(\nabla g)}{g^2}$$

$$\nabla f^n = nf^{n-1}\nabla f (n = \text{integer})$$

3. Fundamental theorems

a. Gradient theorem

$$\int_a^b (\nabla f) \cdot dl = f(b) - f(a)$$

b. Divergence theorem

$$\int_{volume} (\nabla \cdot \mathbf{A})dv = \oint_{surface} \mathbf{A} \cdot ds$$

c. Curl (Stokes) theorem

$$\int\limits_{surface} (\nabla \times \mathbf{A}) \cdot ds = \oint\limits_{line} \mathbf{A} \cdot dl$$

d. $$\oint\limits_{line} f dl = - \oint\limits_{surface} \nabla f \times ds$$

e. $$\oint\limits_{surface} f ds = - \oint\limits_{volume} \nabla f dv$$

f. $$\oint\limits_{surface} \mathbf{A} \times ds = - \int\limits_{volume} \nabla \times \mathbf{A} dv$$

Index

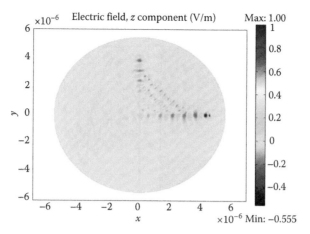

FIGURE 1.7
The 2D surface plot for the z component of the electric field.

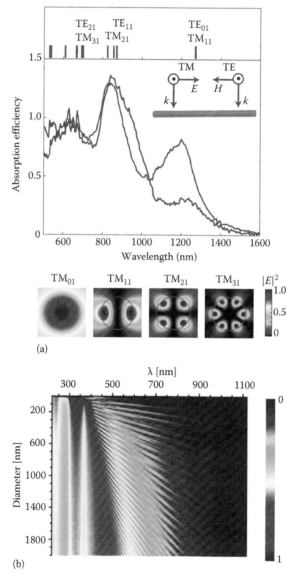

(a)

(b)

FIGURE 3.7
(a) Measured absorption spectrum of a Ge nanowire with a diameter of 220 nm and the electric field intensity profiles for several resonant modes. (Adapted from Cao, L. et al., *Nat. Mater.*, 8, 643, 2009.) (b) Calculated absorption spectrum of a Si nanowire as a function of nanowire size. (Adapted from Bronstrup, G. et al., *ACS Nano*, 4, 7113, 2010.)

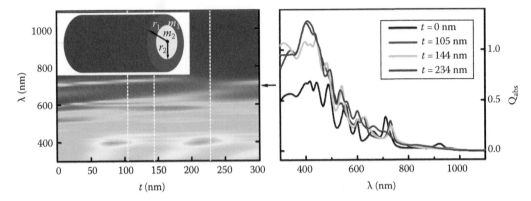

FIGURE 3.15
The absorption spectrum of a Si/SiO$_2$ core/shell nanowire (inset) calculated using Lorenz–Mie theory. The diameter of Si core is 300 nm and the thickness of SiO$_2$ shell is t. (Adapted from Liu, W. and Sun, F., *Adv. Mater. Res.*, 391–392, 264, 2012.)

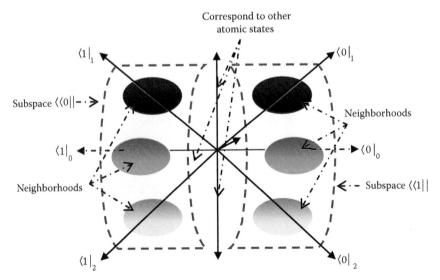

(a)　　　　　　　　　　　　　　　　(b)

FIGURE 3.29
Schematics of plasmonic cavities. (a) A CdS nanowire on silver substrate. (Adapted from Oulton, R.F. et al., *Nature*, 461, 629, 2009.) (b) A gold/dye-doped silica core/shell nanosphere. (Adapted from Noginov, M.A. et al., *Nature*, 460, 1110, 2009.)

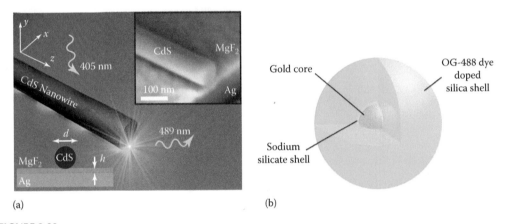

FIGURE 6.15
Neighborhoods in the subspaces S_0 and S_1 of the Hilbert space.

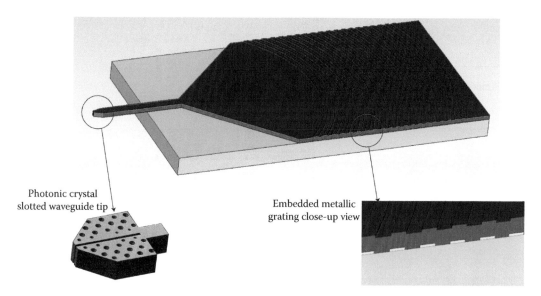

FIGURE 7.1
Schematic of the proposed NSOM tip composed of PhC tip and embedded metallic grating coupler as a whole system.

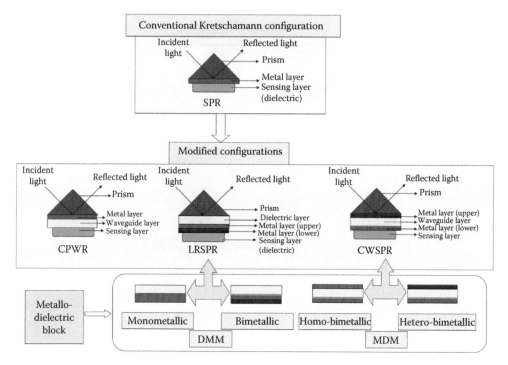

FIGURE 9.2
Classification of different nanoplasmonic structures.

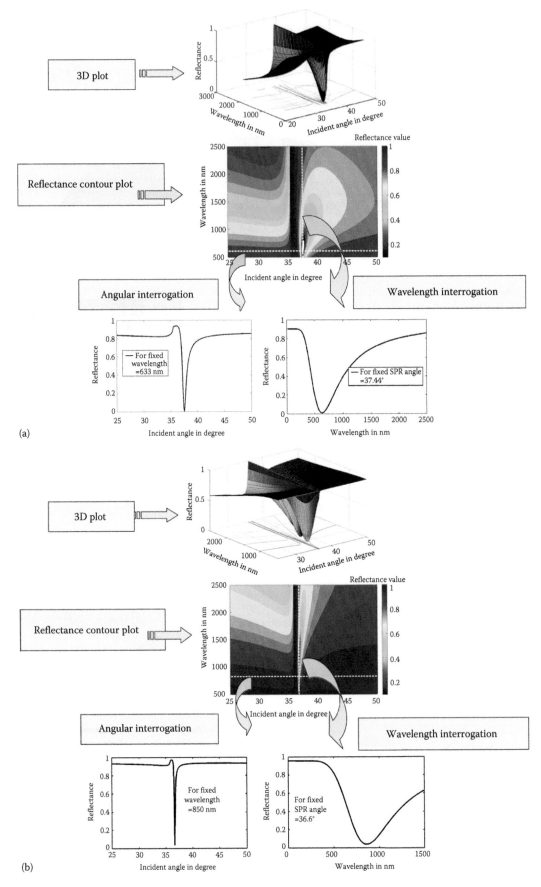

FIGURE 9.4
3D resonance curve with default contour plot and reflectance contour plot with specified reflectance value for conventional SPR structure with angular and wavelength interrogation in 2D representation for (a) 633 and (b) 850 nm.

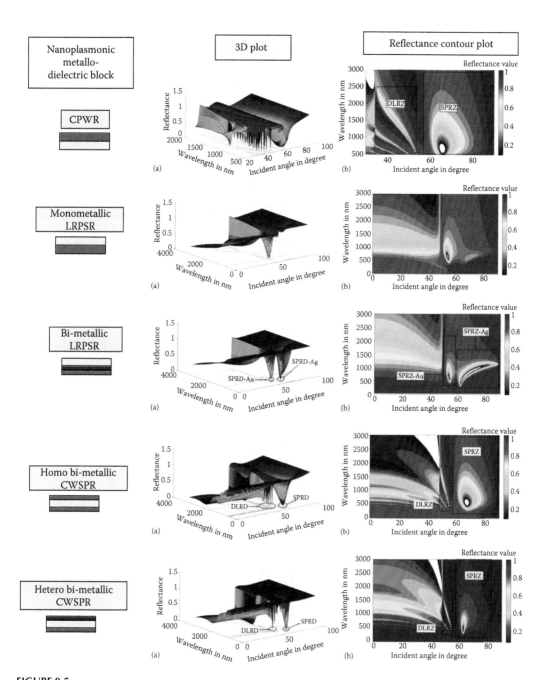

FIGURE 9.5

(a) 3D plot and (b) reflectance contour plot for different modified nanoplasmonic structures with their specific metallo-dielectric block.

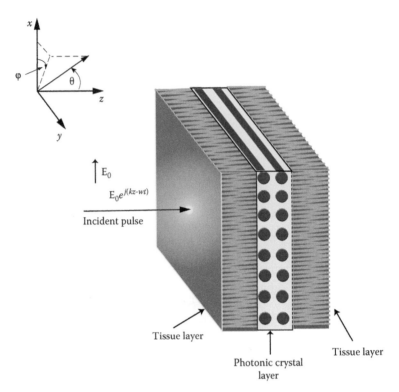

FIGURE 11.6
Photonic crystal implanted in biological tissue.

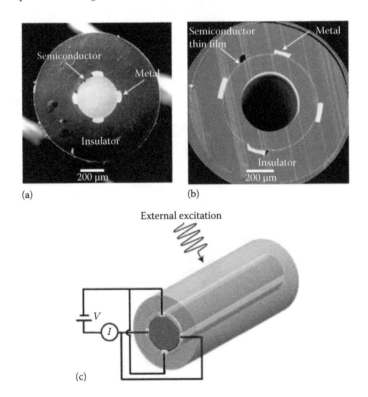

FIGURE 12.13
Insulated fiber devices. (a) SEM micrograph of a cross-section (the semiconductor is $As_{40}Se_{50}Te_{10}Sn_5$, the insulator polymer is PES, and the metal is Sn). (Reprinted from Abouraddy, A.F. et al., *Nat. Mater.*, 6, 336, 2007.). (b) SEM micrograph of a thin-film fiber device (the semiconductor is As_2Se_3, the insulator polymer is PES, and the metal is Sn). (c) Electrical connection of the four metal electrodes at the periphery of the fiber to an external electrical circuit. (From Abouraddy, A.F. et al., *Nat. Mater.*, 6, 336, 2007.)

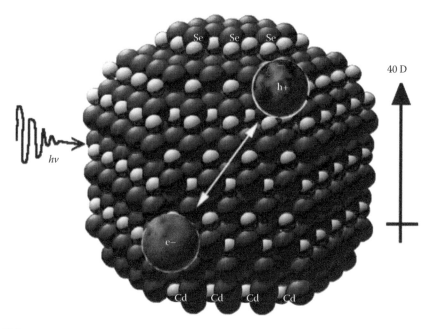

FIGURE 13.7
An electron–hole pair is created upon photon absorption. The electron orbits the hole at a certain radius dependent upon the material, and when recombination occurs, the quantum dot emits energy viewed as fluorescence. (From Rosenthal, S.J. et al., *Chem. Biol.*, 18, 10, 2011.)

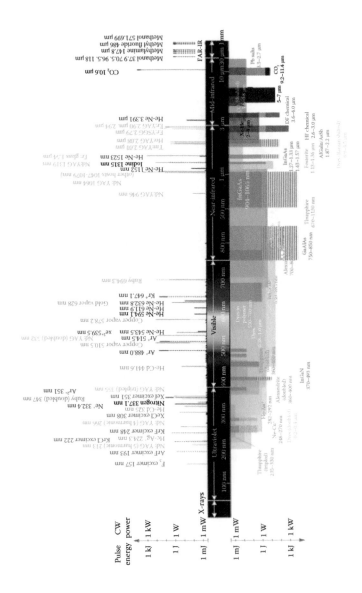

FIGURE B.2
Wavelengths of commercially available lasers. (From Weber, M.J., *Handbook of Laser Wavelengths*, CRC Press, Boca Raton, FL, 1999, ISBN 0-8493-3508-6.)

Printed and bound by CPI Group (UK) Ltd, Croydon, CR0 4YY

18/10/2024

01776253-0014